Materials Horizons: From Nature to Nanomaterials

Series Editor

Vijay Kumar Thakur, School of Aerospace, Transport and Manufacturing, Cranfield University, Cranfield, UK

Materials are an indispensable part of human civilization since the inception of life on earth. With the passage of time, innumerable new materials have been explored as well as developed and the search for new innovative materials continues briskly. Keeping in mind the immense perspectives of various classes of materials, this series aims at providing a comprehensive collection of works across the breadth of materials research at cutting-edge interface of materials science with physics, chemistry, biology and engineering.

This series covers a galaxy of materials ranging from natural materials to nanomaterials. Some of the topics include but not limited to: biological materials, biomimetic materials, ceramics, composites, coatings, functional materials, glasses, inorganic materials, inorganic-organic hybrids, metals, membranes, magnetic materials, manufacturing of materials, nanomaterials, organic materials and pigments to name a few. The series provides most timely and comprehensive information on advanced synthesis, processing, characterization, manufacturing and applications in a broad range of interdisciplinary fields in science, engineering and technology.

This series accepts both authored and edited works, including textbooks, monographs, reference works, and professional books. The books in this series will provide a deep insight into the state-of-art of Materials Horizons and serve students, academic, government and industrial scientists involved in all aspects of materials research.

More information about this series at https://link.springer.com/bookseries/16122

Bibhu Prasad Swain
Editor

Nanostructured Biomaterials

Basic Structures and Applications

Springer

Editor
Bibhu Prasad Swain
Department of Physics
National Institute of Technology Manipur
Imphal, India

ISSN 2524-5384 ISSN 2524-5392 (electronic)
Materials Horizons: From Nature to Nanomaterials
ISBN 978-981-16-8398-5 ISBN 978-981-16-8399-2 (eBook)
https://doi.org/10.1007/978-981-16-8399-2

This Springer imprint is published by the registered company Springer Nature Singapore Pte Ltd.
The registered company address is: 152 Beach Road, #21-01/04 Gateway East, Singapore 189721, Singapore

Preface

The discovery of COVID-19 allows the innovation of several vaccines which permits hundreds of millions of people to be vaccinated within a short period. There are no direct analogies to describe COVID-19 vaccines. It is expected that in the upcoming days the number of severe COVID-19 cases will have plummeted and patients with high risks will no longer die. Moreover, after the vaccination of the younger generation, the COVID-19 cases may retreat as well. The development in science and technology has given us the paramount opportunities to compose a book that identifies the materials and molecules utilized in a broad range of severe life-threatening diseases. The purpose behind the representation of underlines the advancement of materials properties, natural reaction, and plant materials embedded with graphene nanocomposite. This also expanded the immense utilization of materials such as Azole, Thiazole, Imidazole, Pyrazole, Tetrazole, Benzimidazole, Oxazole, Isoxazoles, and 1,2,4-oxadiazoles, 1,2,3-Triazole, derivatives in medicinal chemistry and drug discovery research.

The physicochemical properties of surfaces influence virus attachment and persistence on surfaces. The current practices and applications of antiviral and virucidal materials and coatings in consumer products, personal protective equipment, healthcare, and public settings. Chapter 1 deals with a brief history, pathophysiology, the transmission of the SARS-CoV-2 virus, and recent advances on transition metal complexes and nanocomposites as the potent antiviral agents from COVID-19 perspectives. Nanocarriers use to circumvent the problems associated with conventional antitumor drug delivery systems, including their nonspecificity, severe side effects, burst release, and damage the normal cells. Nanocarriers improve the bioavailability and therapeutic efficiency of antitumor drugs while providing preferential accumulation at the target site. Hence, Chap. 2 deals with nanocarriers as drug delivery vectors. Cocrystallization of a drug substance with a conformer is a promising and emerging approach to improve the performance of pharmaceuticals, such as solubility, dissolution profile, pharmacokinetics, and stability. Chapter 3 deals with Cocrystals and their induced activity of drugs. The design, synthesis, and antimicrobial activity of azole derivatives are investigated and have become one of the highly active highlights in recent years. In particular, a large number of azole-based antibacterial and antifungal agents are penetratingly studied as clinic candidates and the great potential and development value of azole compounds. Chapter 4 deals with the evolution of azole derivatives in medicinal chemistry. Several artificial paths and varied Physico-chemical factors of such thiazoles made especial consideration of medicinal chemists to yield combinatorial library and carry out thorough efforts in the search of thiazoles. Chapter 5 deals with an overview of the biological activities of Thiazole derivatives. The imidazole derivatives possess an extensive spectrum of biological activities such as antibacterial, anticancer, antitubercular, antifungal, analgesic, and anti-HIV activities. Chapter 6 deals with an overview of the biological activities of Imidazole derivatives. Pyrazoles are reported to possess a wide range of biological activities such as an anti-microbial, anti-fungal, anti-tubercular, anti-inflammatory, anticonvulsant, anticancer, anti-viral, Angiotensin-Converting Enzyme (ACE) inhibitory, neuroprotective, cholecystokinin-1 receptor antagonist, and Estrogen Receptor (ER) ligand activity, etc. Chapter 7 deals with an overview of the biological activities of Pyrazole derivatives. Tetrazoles are important ligands for many useful transformations and precursors for a variety of nitrogen-containing heterocycles. The toxic properties of a drug can decrease through the introduction of a tetrazole ring into the molecule. The tetrazole moiety is also generally accepted to exhibit stronger resistance to in vivo metabolization than the carboxylate group, which confers to the corresponding drug's longer bioavailability lifetimes in blood. Chapter 8 deals with an overview of the biological evaluation of tetrazole derivatives. The benzimidazole and its derivatives play a very important role as a therapeutic agent e.g. antiulcer and anthelmintic drugs. Apart from this, the benzimidazole derivatives exhibit pharmacological activities such as antimicrobial, antiviral, anticancer, anti-inflammatory, analgesic, etc. Hence, Chap. 9 deals with an overview of the biological activity of benzimidazole derivatives. 1,2,4-oxadiazole heterocycle has been widely explored bringing a vast number of compounds exhibiting diverse biological activities such as anticancer, anti-inflammatory, anticonvulsant, antiviral, antibacterial, antifungal, antidepressant, antiangiogenic, analgesic, anti-insomnia, anti-oedema, antiparasitic, and anti-Alzheimer. Chapter 10 deals with an overview of the biological activities of Oxazole, Isoxazoles, and 1,2,4-oxadiazoles Derivatives. 1,2,3-triazoles have found broad applications in drug discovery, organic synthesis, polymer chemistry, supramolecular chemistry, bioconjugation, chemical biology, fluorescent imaging, and materials science. Therefore, the development of a facile and straightforward methodology for the synthesis of 1,2,3-triazoles is of noteworthy interest. Chapter 11 deals with an overview of the biological activities of 1,2,3-Triazole derivatives. graphene and its derivatives, Graphene oxide, reduced graphene oxide, and graphene quantum dots are possible nanostructured materials for coronaviruses prevention. The antiviral mechanisms of graphene materials can be related to events such as the inactivation of virus and/or the host cell receptor, electrostatic trapping, and the physicochemical destruction of viral species which are enhanced by functionalization and/or decoration of carbons with species that enhances graphene-virus interactions. Hence, Chap. 12 deals with graphene-derived nanomaterials and their application in COVID-19 related prevention, treatment, and diagnosis.

Imphal, India Bibhu Prasad Swain

Contents

About the Editor

Dr. Bibhu Prasad Swain is currently an associate professor and HOD at the Department of physics, National Institute of Technology Manipur, Langol, Imphal. He obtained his B.Sc. Physics (Hons.) from Utkal University, Bhubaneswar, and M.Sc. (Physics), M.Tech (Materials Science) and Ph.D. from the National Institute of Technology Rourkela, Barkatullah University, and Indian Institute of Technology, Bombay, respectively. His major areas of research interests include high bandgap semiconductors thin films for device applications, biocompatibility coating for artificial heart valve coating and stent applications, silicon and carbon-based alloys of nanostructured materials and titanium nitride-based mechanical hard coating applications. He has published more than 100 papers in reputed international journals. Dr. Swain received the Japan Society of Promotion of the Science (JSPS) Fellow at the National Institute of Advanced Industrial Science and Technology (AIST) Japan; National Research Foundation (NRF), Fellowship, at University of Cape Town; and Brain Korea 21 Fellowship at Seoul National University. He also served in various administrative posts such as Dean academic and IIC president in NIT Manipur. Currently, he is a reviewer of more than 25 international journals and editorial board member of Nanoscience and Nanotechnology Asia.

Chapter 1
Brief History, Pathophysiology, Transmission of SARS-CoV-2 Virus, and Recent Advances on Transition Metal Complexes and Nanocomposites as the Potent Antiviral Agents from COVID-19 Perspectives

Dulal Musib, Maynak Pal, Uday Sankar Allam, and Mithun Roy

1 Introduction

Coronavirus (nCoves) is belonging to the family of *Coronaviridae* viruses which is generally a positive-sensed, single-stranded RNA virus [1]. The SARS-CoV-2 is the latest modified form of coronavirus, which broke out in late 2019. The outbreak attracted global attention and lead to the greatest global health crisis on 30th January 2020 affecting more than 220 countries with 174 individual infections and 3.7 morbidities [2, 3]. On 11th February 2020, WHO designated the disease as novel coronavirus disease or COVID-19.

Further, it was declared as a pandemic by WHO on 11th March 2020. More than 63 million individuals are affected globally, and more than 1.4 million people are deceased until 30th November 2020. The numbers are still increasing exponentially with days, as shown in Fig. 1.1 [4, 5]. The death rate of SAR-CoV-2 is 3% worldwide, and the death rate varies from different countries is approximately 3–5% [5–7]. Social distancing was proposed by WHO as the most effective strategy to break the chain of SARS-CoV-2 transmission, and about 200 countries and several territories came across with the lockdown and curfews [8–11]. The outbreak risked the global economy under threat with GDP drop, unemployment, and socio-economic imbalance [12–17].

D. Musib · M. Pal · M. Roy (✉)
Department of Chemistry, National Institute of Technology Manipur, Imphal West, Langol, Manipur 795004, India

U. S. Allam
Department of Biotechnology, Vikrama Simhapuri University, Kakutur, Nellore, Andhra Pradesh 524 320, India

B. P. Swain (ed.), *Nanostructured Biomaterials*, Materials Horizons: From Nature to Nanomaterials, https://doi.org/10.1007/978-981-16-8399-2_1

(a)

(b)

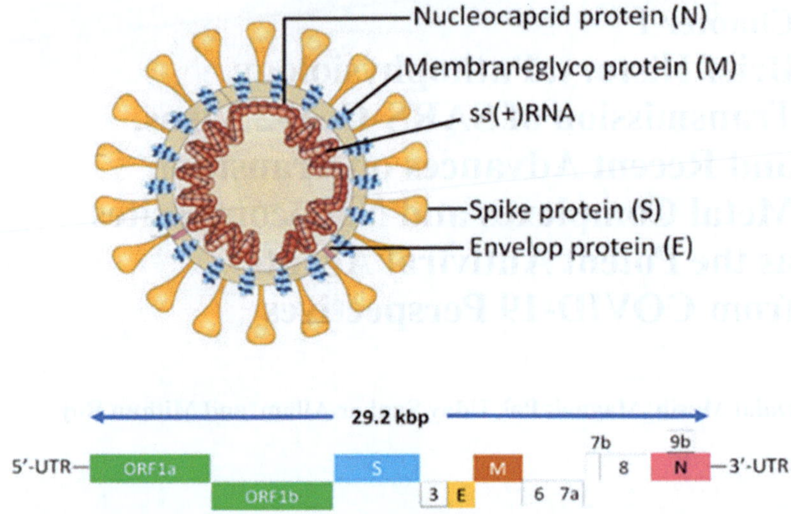

Fig. 1.1 **a** The structure of SARS-CoV-2, **b** the genome organization of SARS-CoV-2

The virus is spread through close contact or touching a contaminated surface [18]. The common symptoms are high fever (88%), shortness of breath, dry cough (68%), fatigue, loss of smell, muscle and joint pain, and acute respiratory distress [19]. The symptoms can be visible within 3–14 days. Self-isolation of the infected individual, maintaining the social distancing, and wearing a face mask are the commonly recommended protocols to prevent the spreading of SARS-CoV-2 [20, 21]. According to the latest WHO guideline the use of antiviral/antimalarial drugs including lopinavir/ritonavir/remdesivir/ribavirin or chloroquine/hydroxychloroquine as the repurposing strategy of plasma therapy is the recommended strategy for the treatment of COVID-19 [22]. Several vaccines against the SARS-CoV-2 virus are rolling over the different parts of the world to prevent the SARS-CoV-2 virus infection [23]. However, vaccines are reported to be less effective against the several mutated strains of the SARS-CoV-2 virus [24]. Nevertheless, potent therapeutic solutions for COVID-19 are still an elusive goal.

Although carbon-based small molecules or biologically derived compounds dominate the pharmaceutical domain, metal-based drugs have emerged as the alternative therapeutic option due to their unique and fascinating properties like

(i) Wide range of coordination number and geometry depending upon the nature of central metal atom or the ligands,
(ii) A broad spectrum of accessible oxidation states,
(iii) Tunable redox properties,
(iv) Unpaired electron spins,
(v) Remarkable luminescent properties,
(vi) Tunable thermodynamic and kinetic properties,

(vii) Unique chemical reactions in their photo-activated states, etc. endow the transition metal complexes with a wide range of bioactivities in biological processes.

For instance, metal ions bearing several accessible oxidation states facilitate to exhibit diverse action in a biological mechanism. The ligand tunability helps to modulate the thermodynamic and kinetic properties of the transition metal complexes according to the desired pharmacological properties. The luminescence properties exhibited by several metal complexes have diagnostic potential. Paramagnetic and radioactive metal complexes have been applied for magnetic resonance imaging (MRI) contrasting agents, positron emission tomography (PET), and single-photon emission computed tomography (SPECT) in clinics. Therefore, these properties provide metal complexes with an adaptable platform for drug design [25, 26]. The metal nanocomposites also have emerged as smart tools for medicinal applications relating to the excellent drug delivery platform to therapeutic potentials. There is a remarkable advancement of potent application of transition metal complexes and the metal nanocomposites for antiviral applications which are reviewed in this chapter. One year of extensive research, since the outbreak of the SARS-CoV-2 virus, to understand the structural and molecular biology of SARS-CoV-2, prompted us to explore or understand the possible therapeutic role of transition metal complexes or the metal nanocomposites in pandemic COVID-19. The present chapter illustrates the history and source of the COVID-19 pandemic, pathophysiology transmission of SARS-CoV-2 virus, repurposing of present FDA-approved antiviral drugs in COVID-19 treatment, recent advances on transition metal or nanocomposite-based antiviral agents, and the perspective on the potential and viable application of inorganic compounds in the treatment of COVID-19.

2 Brief History and Origin of SARS-CoV-2 (Novel Coronavirus)

See Table 1.1.

2.1 *HCoV-229E, HCoV-OC43, HCoV-NL63 and HCoV-HKU1 (Viral Strain)*

The first coronavirus was isolated in the name of alpha-HCoV-229E in 1965 at the University of Chicago. Consequently, another type of coronavirus named beta-HCoV-OC43 was discovered during organ culture in 1967 [30]. Both alpha-HCoV-229E and beta-HCoV-OC43 viruses can transmit to humans via respiratory droplets. The symptoms were typical cold and cough. Approximately 20% of people had sneezing, headache, sore throat, and fever. These two viruses (HCoV-OC43 and

Table 1.1 A brief history of coronaviruses

Disease	Virus/stain	Generic-lineage	Year of discovery	Cellular receptor in host	Hosts	Intermediate host	Respiratory symptom
Severe Acute Respiratory Syndrome	HCoV-229E	Alpha (α)	1966	CD13 (Aminopeptidase N)	Bats/humans	Camelids	Mild
	HCoV-OC43	Beta (β)-A	1967	9-*O*-SA (Acetylated sialic acid)	Rodents/humans	Cattle	Mild
	SARS-CoV	Beta (β)-B	2003	ACE2	Bats/humans	Masked palm civets	Severe acute
	HCoV-NL63	Alpha (α)	2004	ACE2	Bats/humans	–	Mild
	HcoV-HKU1	Beta (β)-A	2005	9-*O*-SA (Acetylated sialic acid)	Rodents/humans	–	Mild
Middle East Respiratory Syndrome	MERS-CoV	Beta (β)-C	2012	DPP4	Bats/humans	Camels	Severe acute
COVID-19	SARS-CoV-2	Beta (β)-B	2019	ACE2	Bats/humans	Pangolin?	Severe acute

Based on the structural differences, governed by the protein sequences, corona viruses (CoVs) are categorized into four genera (α-CoV, β-CoV, γ-CoV, and δ-CoV), among which the beta-CoV (HCoVs) can further be subdivided into four lineages (A, B, C, and D) [27, 28]. The genetic origin of all coronaviruses was rodents, bats, and other domestic animals. A total of seven types of coronavirus species are known to infect humans as shown in Table 1.1 [29].

HCoV-229E) spread globally, and the rate of infection was higher in cold climates [31]. The incubation period of these viruses in an infected individual was one week, and they left the human body after approximately two weeks of illness.

At the end of 2004, a new coronavirus HCoV-NL63 was first detected in a seven-month-old baby in the Netherlands. Palm civets and bats were the origins of the virus [32]. The virus was also dominant during cold climates. Like other coronaviruses (HCoV-229E, HCoV-OC43), the common symptoms were cold, and cough and this virus infected the lower respiratory tract. As a result, cough, fever, rhinitis, sore throat, and pneumonia were common symptoms in humans. The HCoV-NL63 virus spread globally via person to person through direct contact. The HCoV-NL63 was not a life-threatening virus. Generally, pain relievers and medication for fever were referred for treatment [33].

In 2005, human coronavirus HKU1 was discovered in Hong Kong [34]. The HCoV-HKU1 virus originated from mice and was transmitted to a 71 years old man initially. The HCoV-HKU1 was found similar to HCoV-HKU1, HCoV-OC43 and HCoV-229E, HCoV-NL63 viruses. It affected the upper raspatory system. The cold was the typical symptom and, in few cases, pneumonia and bronchiolitis were observed [35].

2.2 *SARS-CoV*

The World Health Organisation (WHO) reported a new coronavirus case named Severe Acute Respiratory Syndrome Coronavirus (beta-SARS-CoV) which was first broken out at Guangdong province in China in April 2003 [36]. The virus was discovered from the biopsy of the lung of the patient. The natural source of SARS-CoV was bats, and the intermediate host was palm civets. The SARS-CoV belonged to the lineage B (Sarbecovirus) of the β-CoVs family. The SARS-CoV was a positive single-stranded RNA virus. The virus was outbroken across 29 countries with 8,096 reported cases and 774 (9.6%) deaths. The primary symptoms of this viral infection were fever, myalgia, headache, dyspnea, cough, and respiratory distress [37]. The virus took 4–7 days to incubate, and it left 10–12 days of illness. The clinically approved drug for SARS-CoV patients was Ribavirin (dose: 1000 μg/mL), oseltamivir, foscarnet, intravenous immunoglobulin [38].

2.3 *MERS-CoV*

A new coronavirus, known as Middle East Respiratory Syndrome (MERS) Coronavirus (MERS-CoV), was first discovered in Saudi Arabia. The first infected patient was a 60 years old person, and later in 2015, 186 confirmed cases were reported in South Korea [39]. The MERS-CoV belonged to the lineage C (Sarbecovirus) of β-CoVs family. Similar to the SARS-CoV, bats were the natural source, and dromedary camels were intermediate hosts of the MERS-CoV. The death rate of MERS-CoV was significantly high (34%), and the typical symptoms were myalgia, fever, cough, sore throat, diarrhoea and vomiting, chills, and dyspnea. The MERS-CoV spread out by contact with dromedary camels [40]. On February 14, 2020, laboratories confirmed over 2500 cases, infected by MERS-CoV, bearing a high rate of fatality of 34.4%. The clinically approved drugs for MERS-CoV-infected patients were the combination of ribavirin and interferon, mycophenolic acid (MPA), and IFN-β [41].

2.4 SARS-CoV-2

In the month of November 2019, the SARS-CoV-2 virus was initially found in Wuhan, Hubei Province of China. The SARS-CoV-2 virus was characterized as a descendent to the lineage B (Sarbecovirus) of the β-CoVs family. The SARS-CoV and SARS-CoV-2 both are similar in their 82% nucleotide sequence and 50% similar sequence with MERS-CoV [42]. The SARS-CoV-2 is more transmissible but less pathogenic as compared to MERS-CoV and SARS-CoV viruses. The SARS-CoV-2 also results in severe respiratory dysfunction and the common symptoms are, for example, fever, dry cough, headache, myalgia, and dyspnea, and few patients have Diarrhea [43]. The important concern is that SARS-CoV-2-infected patients have some time to become asymptomatic. According to previous literature reports, the average incubation period of this virus was supposed to be 3–6 days [44]. From the latest WHO guideline antiviral drugs, including Remdesivir, ribavirin, and lopinavir/ritonavir, may be used for the treatment of COVID-19 [45–47].

3 The Genomic Structure of SARS-CoV-2

Recently the genome sequence data of SARS-CoV-2 has been published by the NCBI SARS-CoV-2 database and NGDC Genome Warehouse (Fig. 1.1) [48]. From the genome sequence data, it is evident that SARS-CoV-2 belongs to lineage B of genus Beta-coronavirus. In the nucleotide sequence, SARS-CoV-2 is approximately ~79% similar to SARS-CoV, ~96% similar to bat CoV-RaTG13, and 88% similar to bat SL-CoVZC45. SARS-CoV virus (Sarbecovirus) and MERS-CoV virus (Merbecovirus) are of different subgenus groups [49]. The virus particle possesses 60–100 nm diameter and seems to be round or oval. The SARS-CoV-2 belongs to the coronavirus family with a large positive-sense and single-stranded RNA genome of around 29.2 kb. The genome is typically constituted of a 5′-methylguanosine cap at the 5′-end, a 3′poly-A tail at the 3′-end. This order of their genes is highly conserved. Genome analysis revealed that the genome of SARS-CoVs contains several numbers (6–11) of open reading frames (ORFs). The primary ORF (ORF1a/b) gradually translates two polyproteins (pp1a and pp1b) encoding 16 non-structural proteins (NSP), and the remaining ORFs encode accessory and structural proteins. The remaining viral genomes contain four other essential structural proteins, including spike (S) glycoprotein, envelope (E) protein, membrane (M) protein, and nucleocapsid (N) protein, along with several other accessory proteins, which generally hinders the host-innate response. Although the SARS-CoV-2 genes are mutated differently depending on the provinces, there is some genomic and phylogenetic similarity to SARS-CoV, specifically in the S-glycoprotein gene and receptor-binding domain (RBD), that became very important for human-to-human viral infection [50, 51].

3.1 Functions of Non-structural and Structural Proteins

3.1.1 Functions of Non-structural Proteins (NSPs)

The purpose of genomic RNA is to be translated into polyprotein pp1a/pp1ab, encoding 16 non-structural proteins (NSPs) into the replication-transcription complex (RTC). These non-structural proteins (NSPs) play a vital role in viral transmission. The nps-1 participates in mRNA degradation and IFN signal inhibition [52]. The nps-2 takes part in the transition of the signalling pathway related to the survival of host cells. The nps-3 is responsible for the breakage of C-terminus of the replicase polyprotein and nps-4 gathers (virally induced) double-membrane vesicles required for viral replication. The nps-6 induces autophagosomes primarily from the endoplasmic reticulum. The nps-7 & nps-8 from a hexadecameric hybrid and take part in viral replication with the help of nps-9. The nps-10 takes a crucial role in viral transcription. The nps-11 became converted in nps-12 which helps in replication and transcription of the viral genome. The nps-13 helps in the unwinding of RNA and DNA duplex, nps-14 possesses exoribonuclease activity and N7-guanine methyl transfer activity. The nps-16 has a vital role in viral mRNA cap methylation. A total of 16 naps plays vital roles at several biochemical cascades in the pathogenesis of the SARS-CoV-2 virus in the host cell [53].

3.1.2 Functions of Structural Proteins

The β-SARS-CoV-2 virus contains four essential structural glycoproteins named, spike (S) protein, membrane (M) glycoprotein, envelope (E) glycoprotein, nucleocapsid (N) protein, and several accessory proteins encoded by ORF3a, ORF6, ORF7a, ORF7b, and ORF8 genes [54].

Coronavirus enters into the human cells via virus surface spike protein. Spike protein contains receptor-binding domain (RBD), and its subunit mediates fusion between the viral and host cell membranes. The spike (S) glycoprotein is found in the outer membrane of the virus, and the average molecular weight is of each subunit of S-protein is 150 kDa [55]. The cleavage of spike glycoprotein could occur by the host cell furan-like protease to form 2 subunits (S1 and S2). The S1 subunit is present on the upper side of spike protein, and it contains 14–685 amino acid residues. The negotiation with the host receptor results in the attachment between spike protein S1 of the virion and the host cell membrane [56]. Attachment of human ACE2 and CLEC4M/DC-SIGNR receptors occurs, and the virus is incorporated into the endosomes of the host cell through the endocytosis, causing conformational modifications of the S glycoprotein. The S1 subunit determines the host-virus range along with cellular-tropism while another part, S2 subunit is mainly responsible for virus fusion to host cells. The S2 subunit is present at the terminal side of the spike protein, and it contains 686–1273 amino acid residues. The fusion of the virion's membrane occurs with the help of the S2 subunit, where S2 subunit behaves like a class I viral fusion protein. From the latest reported model of SARS-CoV-2 proteins, it may be concluded that it contains at least three conformational states: (i) pre-fusion native state, (ii) pre-hairpin intermediate state, and (iii) post-fusion hairpin state [57]. At first, spike protein produces homotrimers in the virus surface and binds enveloped viruses with host cells by expressed angiotensin-converting enzyme 2 (ACE2). Previous studies revealed that the S-protein of SARS-CoV-2 binds human ACE2 with 10–20 times higher affinity than SARS-CoV [58].

Another essential structural protein is nucleocapsid (N) protein, which is found in the endoplasmic reticulum-Golgi complex, and it is structurally bound to the viral RNA [59]. The N-protein binds with RNA and is further involved in the viral replication cycle as well as in the host cell's response to viral infections. The viral RNA converts into a helical ribonucleocapsid (RNP). The helical ribonucleocapsid (RNP) plays a crucial role during virion gathering by their interactions with the membrane protein (M) and viral RNA, which leads to increase in the activity of subgenomic viral RNA transcription and replication [60].

The shape of the virus envelope is determined by M protein. This M protein has the purpose of binding with all supplementary structural proteins [61]. Binding of the M protein leads to stabilization of the N proteins and eventually promotes the fulfilment of viral accumulation via stabilization of N-protein-RNA complex [62].

Finally, the last one is the E protein, which implies an important role in the production and cultivation of the virus. The E protein contains the smallest number of amino acid residues in the SARS-CoV structure [63]. It takes the primary role of the virus morphogenesis and assembly. It plays as viroporin and it self-accumulates in host membranes. This self-assembly in the host membrane causes the formation of pentameric protein-lipid pores which are responsible for ion transport [64]. It also initiates apoptosis. The host NLRP3 inflammation is activated by this protein resulting in IL-1 beta overproduction [65].

4 Pathogenesis and Life Cycle of SARS-CoV-2

Based on several experimental proofs, the disease is classified into three different clinical stages. The transmission of SARS-CoV-2 has occurred via direct contact, respiratory droplet, and airborne routes [66–69].

Stage-II

In The 2nd stage, the SARS-CoV-2 virus spreads into the human lower respiratory tract, and in the epithelial cells of the lungs, the replication of the virus occurs [70]. The incubation period of SARS-CoV-2 virus in host is 2–14 days. In the incubation period, the virus affects the lungs and migrates to different organs, e.g. liver, lung, heart, and kidney [71]. Then it starts to disrupt the cell membrane, and systemic viral sepsis occurs within 7–8 days. According to literature, at this stage, 75–80% of infected patients have observed moderate symptoms, and the infection is restricted to the conducting airways [72]. In this stage, the patient can be recovered at home with systematic therapy and typical viral medication [73].

Stage-III

About 20–25% of severely infected patients moved to the 3rd stage. After attacking the lower respiratory tract, the virus then generates the overactivation of T cells [74]. This stage is the most vital because the immune function is destroyed, and the damage of multiple organs occurs. The open binding domain accelerates the binding of S-protein to the ACE2 receptor. This open binding causes structural conformational changes. This structural conformational change results in the termination of the membrane and the establishment of the viral genome into the host cell [75]. Then, Acute Respiratory Distress Syndrome (ARDS) occurs by preventing enough oxygen from getting into the lungs and from circulating oxygen. The whole process is causing most respiratory disorders and resulting in acute lung injury [76, 77].

5 Cell Cycle

Bats are the natural source of SARS-CoV 2, and pangolin is the intermediate host, then the virus is transferred to the human body [78]. The SARS-CoV-2 coronavirus passes through the mucous membranes, and then it enters into the human body through its receptors angiotensin-converting enzyme 2 (ACE2) which are accessible to different organs such as lungs, heart, kidneys, etc. In this way, the viral entry is facilitated into target cells (Fig. 1.2) [79]. The probable steps of the SARS-CoV-2 life cycle are (i) The SARS-CoV-2 infiltrates into the host cell by the attachment of the Spike (S) glycoprotein with the ACE2 receptor, of the host cells (such as in type II pneumocytes in the lungs). Such attachment process transpires in the binding region of Spike protein of SARS-CoV-2 receptors which contain 331 to 524 residues, and this S-protein can be attached firmly to human ACE2 and bat ACE2. (ii) The envelope of SAR-CoV-2 is removed, and the genomic RNA is released into the cytoplasm, which contains the ORF1a and ORF1b RNAs. (iii) The ORF1a/1b are eventually translated to pp1a and pp1ab proteins, respectively. (iv) The polyprotein of pp1a and ppa1b is hydrolyzed by a protease to generate 16 non-structural proteins (nps). Few nps form replication and transcription complex (RTC) (RNA-dependent RNA polymerase by using the (+) strand genomic RNA ((+) gRNA) as a moult. (v) The genome

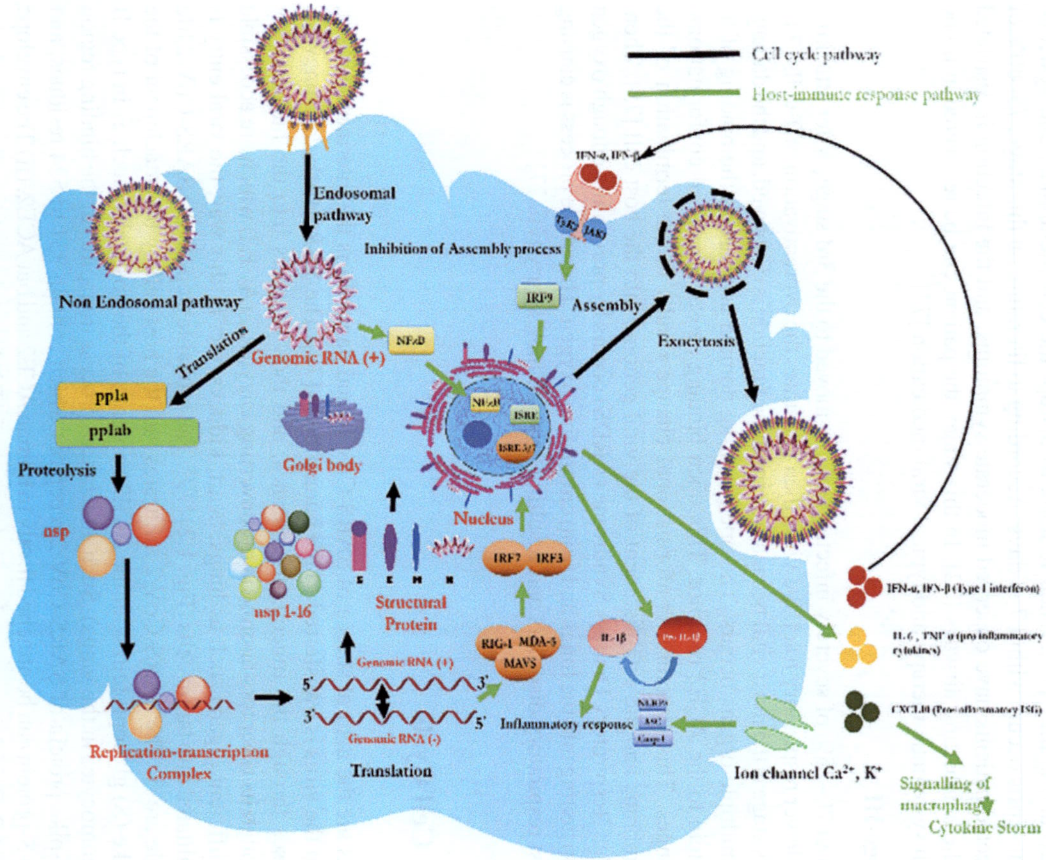

Fig. 1.2 Life cycle of SARS-CoV-2 in human alveolar epithelial cells (host) and host-innate responses (cytokine storm)

of the new virus particle is converted from the (+) strand genomic RNA generated through the replication process. (vi) The structural protein (S: spike protein, E: envelope protein, M: membrane protein, and N: nucleocapsid protein) construct a viral particle generated from subgenomic RNA which are produced through the transcription process. (vii) Endoplasmic reticulum allows spike enveloped and M protein to enter into itself and the nucleocapsid protein is joined with the genomic RNA to form nucleoprotein complex. (viii) These combine to form the virus particle in the ER-Golgi complex chamber, and finally, the virus is discharged to the extracellular area through the Golgi chamber and the vesicle (Fig. 1.2) [80, 81].

6 Repurposed FDA-Approved Antiviral Drugs for the Treatment of COVID-19

Currently, the finding of effective treatment for COVID-19 viral diseases is one of the difficult challenges in the twenty-first century across the whole world because there is no specific drug that cures SARS-CoV-2 viral infection [82]. According to the literature survey, more than 200 clinical trials are ongoing worldwide for the treatment of COVID-19. Here we highlight all the WHO-recommended antiviral drugs that are repurposed for the treatment of COVID-19 as shown in Scheme 1.1, Table 1.2 [83].

6.1 Single Drug

Ribavirin: Ribavirin is the most specific antiviral drug which is used for hepatitis C, RSV infection, and few viral fevers [84]. Ribavirin is a prodrug that is absorbed into nucleoside analogues and prevents viral mRNA capping and viral RNA synthesis. In 2003, ribavirin was clinically used against the SARS-CoV and in 2013 MERS-CoV. Ribavirin was effectively used for the treatment of SAR-CoV-2 viral infection along with or without the use of steroids recently in China. The SARS-CoV-2 replication process in vitro could be inhibited by the collaborated effect of Ribavirin and IFN-β. Determination of appropriate dose for ribavirin to the SARS-CoV-2 infected patient is the critical part of its use [85].

Remdesivir: Remdisivir has shown its activity against SARS-CoV-2 in human cells as well as in mice infected with the virus [86]. Remdesivir mainly targets the main viral proteins participating in producing new viruses. The United States first used remdesivir for the treatment of COVID-19. In vitro study results showed that remdesivir is more active against SARS-CoV-2 virus with the EC_{50}, 0.77 μM. It was used for the development of blockage against infection by coronaviruses and even ebola

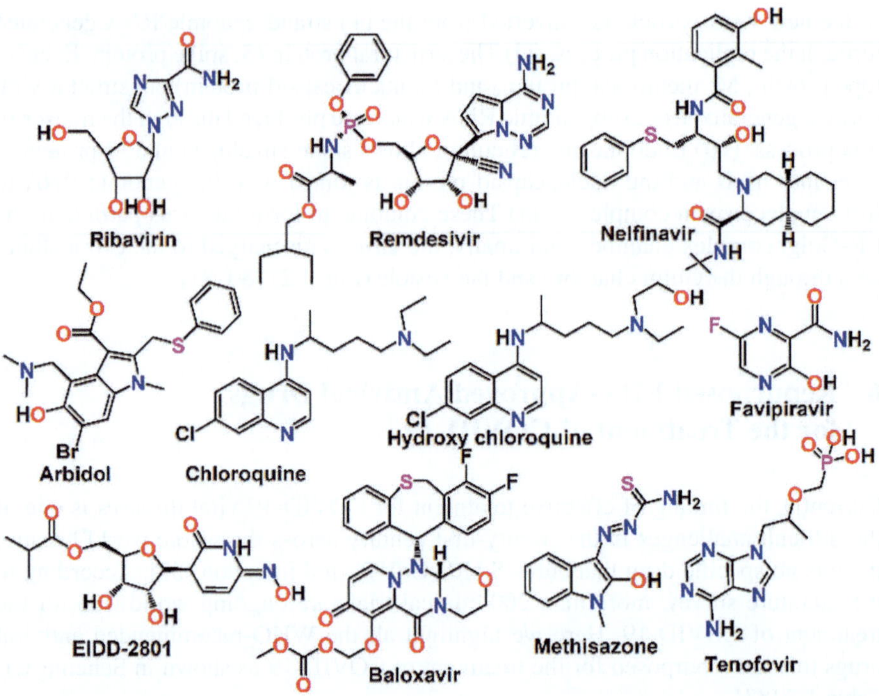

Scheme 1.1 Repurposed FDA-approved organic drugs currently used for the treatment of COVID-19

intravenously. The WHO recommended the use of Remdisivir against SARS-CoV-2 no severe side effects were reported with remdesivir [87].

Nelfinavir: Nelfinavir is one of the highly effective drugs of HIV protease. It also has robust antiviral activity against SARS-CoV-2 viral infection [88].

Arbidol: Arbidol prevents viral influenza viruses. Arbidol and its derivative, arbidolmesylate have inhibited the SARS-CoV-2 infection at μM concentration [89].

Chloroquine: Chloroquine has excellent biochemical properties, and it is used during anti-malaria and rheumatoid arthritis decreases [90]. In January 2020, chloroquine has shown potent inhibition of the SARS-CoV-2 virus by interfering with the ACE2 binding domain of spike protein, and it also has controlled the viral replication. Nevertheless, the disadvantage of the use of chloroquine is that it has various systemic side effects, and it also blocked the immune response [91].

Hydroxy chloroquine: Hydroxychloroquine is an antimalarial drug, and it is also used for the treatment of lupus and rheumatoid arthritis [92]. Hydroxychloroquine was recommended for clinical trials by Chinese Clinical society against COVID-19 infected patients. Hydroxychloroquine has remarkably reduced viral load in nasal swabs. However, the major drawback of hydroxychloroquine intake is a heart attack,

Table 1.2 Selected drugs repurposed for COVID-19 treatment

Drug (AHFS class)	Possible target	Dosage	Duration of treatment	$EC_{50}/\mu M$	Refs.
Ribavirin (antiviral)	RNA dependent RNA polymerase	500 mg/3 times per day	Upto 10 days	109.50	[84]
Remdesivir (antiviral)	RNA dependent RNA polymerase	200 mg for 1st day and 100 mg once daily upto 10 days	5–10 days	0.77	[85]
Nelfinavir (anti-HIV)	Protease inhibitor	750 mg (three 250-mg tablets) three times daily	5–10 days	1.13	[86]
Arbidol (antiviral)	S-protein and inhibitor of ACE2	200 mg three times daily	7–10 days	4.11	[88]
Chloroquine (anti-malaria)	Endosome and inhibitor of ACE2	500 mg weekly	Upto 4 weeks	1.13	[89]
Hydroxy chloroquine (anti-malaria and rheumatoid arthritis)	Inhibitor of ACE2	100–200 mg 2–3 times/week	Upto 4 weeks	0.72	[91]
IFN-α	Inhibitor of ACE2	5 million U each time, two times/day	Upto 10 days	5 IU/ml	[92]
EIDD-2801 (antiviral)	RNA dependent RNA polymerase		10–12 days	–	[94]
Baloxavir (antiviral)	Inhibitor of ACE2	First, fourth and seventh day: 80 mg	7 days	–	[96]
Favipiravir	Inhibits viral RNA-dependent RNA polymerase	The first day 1600 mg twice other days (up to seven days) 600 mg twice daily	7–10 days	61.88	[97]
Marboran/Methisazone (smallpox virus inhibitor)	Protease inhibitor	–	–	–	[98]
Tenofovir (anti-HIV)	Reverse transcriptase inhibitor	–	–	–	[99]
Lopinavir/ritonavir (HIV infection)	Coronavirus main protease 3CLpro	400 and 100 mg, two times per day	Upto 10 days	17.1	[101]
Nitazoxanide/Azithromycin (antiparasitic)	RNA synthesis inhibitor	600 mg twice daily for 5 days	Upto 5 days	–	[103]

(continued)

Table 1.2 (continued)

Drug (AHFS class)	Possible target	Dosage	Duration of treatment	$EC_{50}/\mu M$	Refs.
Hydroxy chloroquine/azithromycin (macrolides)	Inhibitor of ACE2	First day: 500 mg, Day 2–5: 250 mg in conjunction with 10-day regimen of hydroxychloroquine	Upto 10 days	–	[104]
Lopinavir/ritonavir/ribavirin (HIV infection/antiviral)	Protease inhibitor	400 mg/100 mg/400 mg every 12 h	Upto 14 days	0.72	[106]

Table 1.3 Metal-based antiviral agents potentially important as the repurposed medicine in COVID-19

Complex	Class of virus	Target protein	EC_{50}	Refs.
Vanadium-based antiviral agents				
1	HIV virus	HIV-1 RT (reverse transcriptase) (RT) enzyme	5 μM (EC_{97})	[11]
2	HIV virus	HIV-1 RT (reverse transcriptase) (RT) enzyme	5 μM (EC_{97})	[111]
3	HIV virus	HIV-1 RT (reverse transcriptase) (RT) enzyme	5 μM (EC_{97})	[112]
4	HIV virus	HIV-1 RT (reverse transcriptase) (RT) enzyme	–	[113]
5	HIV virus	HIV-1 RT (reverse transcriptase) (RT) enzyme	–	[113]
Manganese-based antiviral agents				
6	*A. aegypti larvae*	–	0.011 g/L (LC_{50})	[114]
7	*A. aegypti larvae*	–	0.011 g/L (LC_{50})	[114]
8	*A. aegypti larvae*	–	0.011 g/L (LC_{50})	[114]
9	Human immunodeficiency virus (HIV)	gp120 HIV protein	>100 μM	[115]
10	Human immunodeficiency virus (HIV)	gp120 HIV protein	70 μM	[115]
11	Human immunodeficiency virus (HIV)	gp120 HIV protein	>125 μM	[115]
Iron-based antiviral agents				
12	HIV (human immunodeficiency virus)	gp120 HIV protein	40 μM	[115]
13	HIV (human immunodeficiency virus)	gp120 HIV protein	>20 μM	[115]
14	HIV (human immunodeficiency virus)	gp120 HIV protein	60 μM	[115]
15	Hepatitis C virus (HCV)	HCV envelope glycoproteins E1 and E2	1 μM	[116]

(continued)

Table 1.3 (continued)

Complex	Class of virus	Target protein	EC_{50}	Refs.
16	SARS-CoV-1 virus	–	4.9 μM	[117]
17	SARS-CoV-1 virus	–	1.9 μM	[117]
18	SARS-CoV-1 virus	–	3.6 μM	[117]
19	Zika virus, Influenza A virus, HIV virus and Enterovirus 71 (EV71)	–	–	[118]
20	Hepatitis C virus (HCV)	HCV genotype 1a and 1b replicons	7 pM	[119]
21	Hepatitis C virus (HCV)	HCV genotype 1a and 1b replicons	5 pM	[119]
Cobalt-based antiviral agents				
22	Herpes simplex virus type 1 (HSV-1)	–	–	[120]
23	Human Immunodeficiency Virus (HIV)	–	–	[122]
24	Dengue virus and hepatitis C virus)	NS2B/NS3 protease and NS5B polymerase (2WCX	−11.21 kcal/mol and −10.88 kcal/mol (binding energy)	[123]
25	HIV (human immunodeficiency virus)	HIV-PR (HIV protease)	6 μM	[124]
26	HIV (human immunodeficiency virus)	HIV-PR (HIV protease)	6 μM	[124]
27	HIV (human immunodeficiency virus)	HIV-PR (HIV protease)	3 μM	[124]
28	HIV (human immunodeficiency virus)	HIV-PR (HIV protease)	0.25 μM	[124]
29	HIV (human immunodeficiency virus)	HIV-PR (HIV protease)	110 nM	[125]
30	HIV (human Immunodeficiency Virus)	HIV-PR (HIV protease)	70 nM	[125]
31	HIV (human immunodeficiency virus)	HIV-PR (HIV protease)	50 nM	[126]

(continued)

Table 1.3 (continued)

Complex	Class of virus	Target protein	EC_{50}	Refs.
Nickel-based antiviral agents				
32	Herpes simplex virus type 1 (HSV-1)	–	5 μM	[127]
33	Herpes simplex virus type I and II (HSV)	–	1.8 and 4.3 μM	[128]
34	Human immunodeficiency virus (HIV)	–	2.86 μM	[129]
35	Dengue virus and hepatitis C virus	NS2B/NS3 protease and NS5B polymerase (2WCX	−10.45 kcal/mol and −10.29 kcal/mol (Binding energy)	[123]
36	HIV-1 and HIV-2	gp120 HIV protein	60, 19 μg/mL	[130]
37	HIV-1 and HIV-2	gp120 HIV protein	75, 70 μg/mL	[130]
38	HIV-1 and HIV-2	gp120 HIV protein	1 and 85 μg/mL	[130]
39	Human immunodeficiency virus (HIV)	CXCR4 co-receptor	–	[131]
Copper-based antiviral agents				
40	Human immunodeficiency virus (HIV-1)	–	CC50 is >10 μg/mL	[136]
41	Vesicular stomatitis virus	–	4 μg/mL	[137]
42	Human immunodeficiency virus (HIV-1)	–	5 μM	[138]
43	Human immunodeficiency virus (HIV-1)	–	5 μM	[138]
44	Influenza M2	WT (Wild Type) and the amantadine-resistant M2S31N protein	3.7 μM	[139]
45	Influenza M2	WT (Wild Type) and the amantadine-resistant M2S31N protein	1.1 μM	[139]
46	Dengue virus and hepatitis C virus)	NS2B/NS3 protease and NS5B polymerase (2WCX)	−11.19 kcal/mol and −11.11 kcal/mol (Binding energy)	[123]
47	Gammaherpesvirus 68 (MHV-68)	–	–	[129]
48	Human immunodeficiency virus (HIV-1)	–	8.36 μM	[139]

(continued)

Table 1.3 (continued)

Complex	Class of virus	Target protein	EC_{50}	Refs.
49	Herpes simplex virus (HSV-1; HSV-2)	–	–	[128]
Zinc-based antiviral agents				
50	Dengue virus and Japanese encephalitis virus	–	0.5 μM	[140]
51	Human rhinovirus, coxsackievirus, and mengovirus	–	–	[141]
52	Picornavirus	–	–	[142]
53	Human immunodeficiency virus (HIV)	–	25 μM	[143]
54	SARS-CoV-1	SARS-CoV 3CL protease	–	[144]
55	SARS-CoV-1	SARS-CoV 3CL protease	–	[144]
Platinum and palladium-based antiviral agents				
56	Cytomegalovirus	–	2.9 μM	[145]
57	Cytomegalovirus	–	5 μM	[145]
58	Type 1 poliovirus	–	32.5 ± 3.7 IU/mL	[146]
59	Type 1 poliovirus	–	–	[146]
60	HSV-1 and HSV-II	–	0.01 μM	[147]
61	HSV-1 and HSV-II	–	1 μM	[147]
62	Herpes simplex virus (HSV)	–	0.01 μM	[148]
63	Herpes simplex virus (HSV)	–	10 μM	[148]
64	Dengue virus type 2	–	–	[149]
Ruthenium-based antiviral agents				
65	Herpes simplex and polio viruses	–	–	[150]
66	Herpes simplex and polio viruses	–	–	[150]
67	HSV-1	–	0.8 μM	[151]
68	HSV-1	–	30 μM	[151]
69	HSV-1	–	>100 μM	[151]

(continued)

Table 1.3 (continued)

Complex	Class of virus	Target protein	EC_{50}	Refs.
70	Human immunodeficiency virus (HIV)	Nucleoside reverse transcriptase inhibitors	41.9 μg/mL	[152]
71	Cytomegalovirus, herpes simplex virus-1&2, varicella zoster virus, and vaccinia virus	–	11 μM (cytomegalovirus)	[153]
72	Cytomegalovirus, herpes simplex virus-1&2, varicella zoster virus, and vaccinia virus	–	>20 μM (cytomegalovirus)	[153]
Gold-based antiviral agents				
73	HIV-1	HIV-1 RT enzyme	100 μM (ED_{52})	[156]
74	HIV-1	HIV-1 RT enzyme	–	[156]
75	HIV-1	HIV-1 RT enzyme	0.31 μM (IC_{50})	[157]
76	HIV-1	HIV-1 RT enzyme	0.57 μM (IC_{50})	[157]
77	HIV-1	HIV-1 RT enzyme	28.4 μM (IC_{50})	[157]
78	HIV-1	Protease-PR and HIV-1 RT enzyme	–	[158]
79	HIV-1	Protease-PR and HIV-1 RT enzyme	–	[158]
80	HIV-1	Protease-PR and HIV-1 RT enzyme	–	[158]
81	HIV-1	Protease-PR and HIV-1 RT enzyme	–	[158]
82	HIV-1	Protease-PR and HIV-1 RT enzyme	100 μM (63%)	[159]

and it also showed various systemic side effects. In New York alone, almost 1376 SARS-CoV-2 infected patients were treated with hydroxychloroquine, and they developed severe side effects with enhanced death of 32.3% [93]. The WHO has withdrawn hydroxychloroquine from the list of repurposed drugs for COVID-19.

IFN-α: IFN-α has popular antiviral activity against hepatitis. This drug was also used to prevent SARS-CoV viral infection in 2003. As per the latest guideline by WHO (6th edition) published that IFN-α can be used for the treatment of COVID-19 [94]. The specific dose in each person is 5 million U twice a day, and the method of administration is vapour inhalations [95].

EIDD-2801: EIDD-2801 is an antiviral drug for influenza which is typically administered orally. This oral drug participates in the genetic mutation of viral RNA and the RNA maker system of the virus, which makes the virus unable to infect healthy cells. This drug was found to be more effective for the treatment of the COVID-19 than remdesivir, a drug used in clinical trials against COVID-19 in March [96].

Baloxavir: Baloxavir is a common antiviral and anti-influenza oral drug. Recently it has shown remarkable antiviral activity against the SARS-CoV-2 virus in vitro. The specific dosage of baloxavir is 80 mg per day [97].

Favipiravir: Favipiravir is an antiviral drug that is used for a broad spectrum of viral diseases. In China and Japan, this drug is used primarily as anti-influenza. Very few clinical trial data are available against the treatment of SARS-CoV-2 diseases using favipiravir. The appropriate dose of favipiravir is 1600 mg twice daily on day 1, then 600 mg twice daily for 7–10 days [98].

Marboran/Methisazone: Methisazone is an inhibitor for smallpox. This drug inhibits mRNA and protein synthesis. This drug effectively eliminated the SARS-CoV-2 infection. This compound showed a very high binding affinity to several viral proteins (PDB ID 5R7Y, 5R7Z, 5R80, 5R81, and 5R82) [99].

Tenofovir: Tenofovir is an anti-HIV drug. This drug inhibits the HIV-1 reverse transcriptase and has shown potent antiviral activity against the SARS-CoV-2 virus [100].

6.2 The Combined Drugs

Lopinavir/ritonavir: Lopinavir/ritonavir combined drug is extensively used against HIV infection. This combined drug with ribavirin is also widely used against SARS and MERS-CoV viruses. In February 2020, several countries have used this combined drug for COVID-19 infection. The specific dose of this combo drug is 400 mg and 100 mg, respectively twice per day. After several clinical trials, it was observed that this combined drug did not have significant activity against COVID-19 infected patients. So, the use of this combo drug has been stopped from the first week of April 2020. The major drawback of this drug is a heart attack [101].

Nitazoxanide/Azithromycin: Nitazoxanide was used against SARS-CoV, MERS-CoV and other coronaviruses [102]. Azithromycin is a macrocyclic lactone antibiotic and was found active against zika and ebolaviruses. Therefore, the combined Nitazoxanide/Azithromycin could be the potential drug against COVID-19. This combo drug is still in use against SARS-CoV-2. Moreover, the use of a proper dose of this combo drug has proved the reduction of the symptoms in patients without any complication. It has a minor adverse effect [103].

Hydroxychloroquine/azithromycin: Ministry of Health and Family Welfare (MoHFW), Govt. of India has confirmed that the combined drug hydroxychloroquine/azithromycin was found potential to fight against COVID-19. As per the report in the New England Journal of Medicine, this combination drug has cured 100% of patients at the end of the sixth day [104]. The literature survey has also confirmed that only eight out of 14 patients or 57.1% have become cured by the application of hydroxychloroquine. This combo drug has been restricted to children and pregnant women. The adverse effect of the combined drug is an unstable heart rate. Therefore, WHO, MoHFW or ICMR have been directed to apply this combined drug in proper dose under medical observation to minimize the side effects [105].

6.3 The Triple Combine Drug

Lopinavir/ritonavir/ribavirin: The University of Hong Kong has confirmed that the triple combination drug (Lopinavir/ritonavir with ribavirin was found highly useful for COVID-19. These triple combination drugs are going to phase II trial. The patients treated with the triple-drug combination tested negative within five days but for those treated with lopinavir/ritonavir combo drug, it takes 12 days [106].

7 Transition Metal-Based Antiviral Agents

Therapeutic potential of metal-based compounds was known from ancient Indian, Egyptians, Greeks, and Chinese to modern societies as the remedy for a broad array of diseases. Discovery of bacteriostatic and anti-neoplastic properties of cis-dichlorodiaminoplatinum (II) (cisplatin) marked the new era of medicinal inorganic chemistry. Systematic therapeutic applications of metal-based compounds have advanced significantly and gained much attention and interest. Transition metal complexes are generally potent to modulate the reactivity of the organic substrates, or the role of metal ion can be modulated by the use of a suitable organic substance or the ligands [107, 108]. (i) Charge: The formal charge on a metal complex can vary from negative, neutral to positive depending upon the nature of the ligands. The formal charge on metal complexes becomes critical in determining the solubility and stability of the complex in a biological medium, cell permeability, and interaction with biological molecules. (ii) Structure and bonding: Metal complexes can possess various coordination numbers and geometry which enable the metal complexes to interact with biomolecules diversely. (iii) Reactivity: Metal complexes possess several types of metal–ligand interactions which allows tunable thermodynamic and kinetic properties of the complexes which enable the metal complexes in broadening the mechanistic possibilities to interact with various biomolecules of interest. (iv) Partially filled d orbitals: The electronic configuration of the metal ion is essential to dictate the magnetic and electronic properties of the transition metal complexes which enable the transition metal complexes to be used for diagnostic purposes (MRI contrast agents). (v) Redox activity: Ligand-induced tunability of

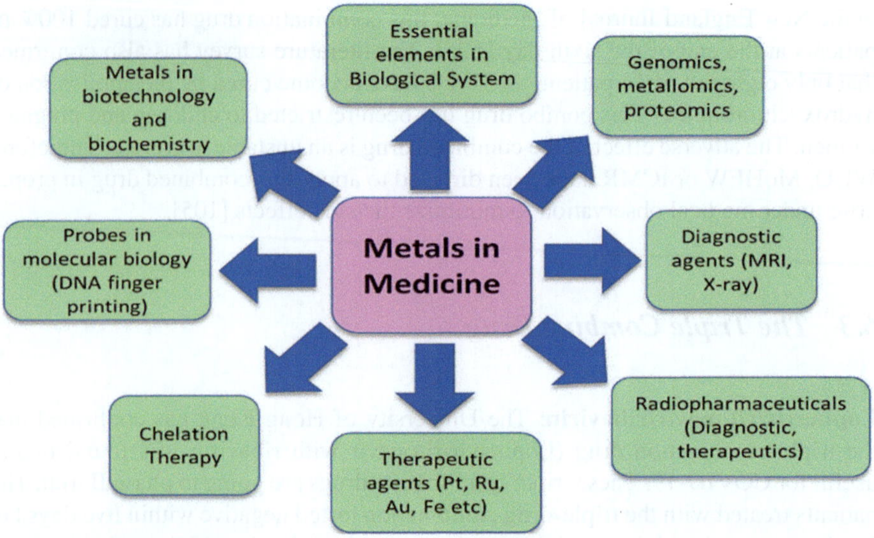

Fig. 1.3 This figure illustrates how metals play important role in medicinal chemistry from therapeutics to diagnostics

redox properties of transition metal complexes between two or more oxidation states could be important for determining the mechanism of reactivity of the transition metal complexes. Altering the redox activity of the metal ions by the ligands is an essential strategy for drug development. (vi) Lewis acidity: Lewis acid properties of Fe(III) or Zn(II) became very important for the strategic design of drugs for specific purposes.

In post cisplatin era, the medicinal inorganic chemistry is dominated by most of the transition metals in the periodic table for the treatment of cancer, bacterial and viral, microbial diseases, immunological disorder, inflammatory diseases, etc. (Fig. 1.3). Herein we discussed the metal-based antiviral agents primarily with the insight into the specifically targeted sites and the mechanisms of action [109].

7.1 *Vanadium (V)*

Vanadium and oxovanadium complexes are well known for cancer therapeutic and antibacterial applications [110]. Currently, several studies have shown that thiourea and polyoxotungstates based oxovanadium complexes have the potential to inhibit the Human Immunodeficiency (HIV) virus by targeting the infected immortalized T cells. Solubility and stability of vanadium (IV) complexes under physiological conditions were major concerns to their antiviral application. F. M. Uckun et al. employed porphyrinato ligand to stabilize VO^{2+}. All the water-soluble oxo-vanadium complexes (1–3) exhibited antiviral activities against HIV-1 RT. The aminosulfonyl

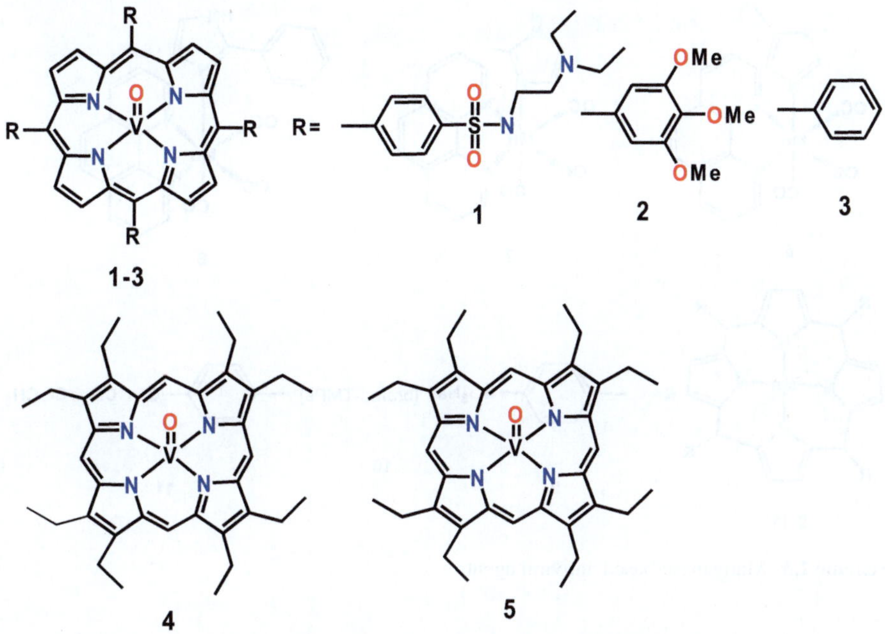

Scheme 1.2 Vanadium-based antiviral agents

VO^{2+} complex (1) targeted HIV-1 reverse transcriptase (RT) and exhibited 97% inhibition at 5 μM concentration [111, 112].

C. Che et al. reported the interaction between porphyrin-based oxovanadium complex (**1–5**) and various targeted proteins of HIV by computational simulation (Scheme 1.2). Complex **1** showed the selectively binds with the CD4 protein having binding energy -138.61 kcal mol^{-1}. The computational result suggested that complex **1** was highly active against HIV-1 as it blocks the infiltration of the virus into the host cells (e.g. Hut/CCR5) [113].

7.2 Manganese (Mn)

R. M. Carlos et al. synthesized a series of manganese carbonyl complexes (**6–8**) and those complexes were highly active against A. aegypti larvae and also strongly controlled mosquito proliferation (Scheme 1.3) [114]. The low lethal concentrations (LC_{99}) values of the complexes (6–8) had suggested that these complexes were highly effective against A. aegypti larvae and had shown ~90% mortality at 0.033–0.046 g/L after 96 h of incubation. These complexes strongly interacted with P450 enzyme. B Meunier et al. reported the carboxylate and sulphonate-functionalized Mn-porphyrin complexes (**9–11**) were more active and target the gp120 HIV protein but not bound

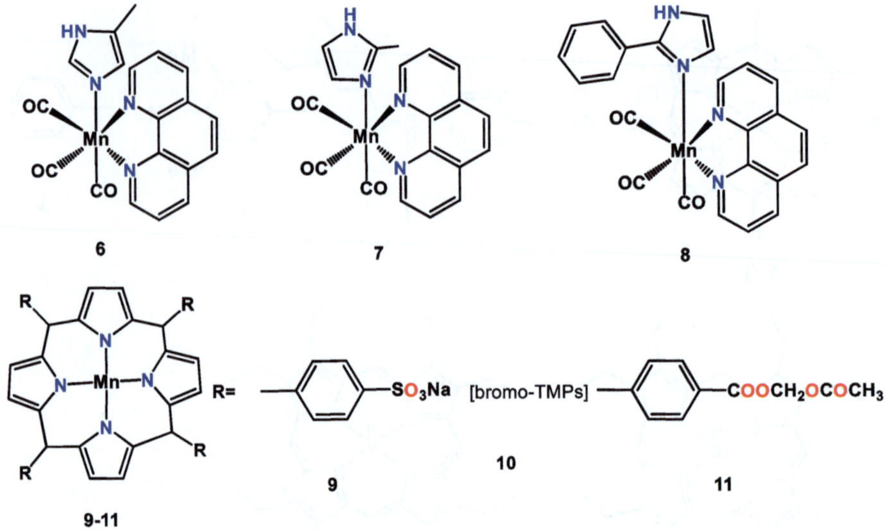

Scheme 1.3 Manganese-based antiviral agents

with the CD4 cellular receptor. The EC_{50} value of complex **9–11** were respectively >100 μg/mL, 70 μg/mL, >125 μg/mL and >100 μg/mL, 100 μg/mL, >125 μg/mL respectively against HUT-78/HIV-l and HUT-78/HIV-2 cells [115].

7.3 Iron (Fe)

B. Meunier et al. synthesized a new series of water-soluble porphyrin-based iron(III) complexes, and these complexes (**12–14**) exhibited antiviral activities against human immunodeficiency virus (HIV1, HIV-2) (Scheme 1.4). The carboxylate and sulphonate-functionalized Fe-porphyrin complexes efficiently targeted the gp120 HIV protein. The EC_{50} value of complex **12**, **13** and **14** were respectively 7 μg/mL, >20 μg/mL, 37 μg/mL and 40 μg/mL, >20 μg/mL, 60 μg/mL respectively against HUT-78/HIV-l and HUT-78/HIV-2 cells [114, 115]. Ferroquine (FQ) is a well-known antimalarial drug. J. Dubuisson et al. reported remarkable anti-HCV activity in hepatoma cell lines. The detailed mechanistic studies showed that FQ could prevent HCV RNA replication and could strongly bind to HCV envelope glycoproteins E1 and E2 [116]. However, FQ had no role in virion secretion and viral assembly. Most importantly, FQ inhibited the cell-to-cell spread of the Hepatitis C virus. Finally, FQ is an excellent anti-HCV molecule, and it was also used in combination with other direct-acting antivirals. The IC_{50} value of FQ against HCV was around 1 μM. E. D. Clercq et al. used FQ and its derivative (**16–18**) against SARS-CoV in Vero cell. It was reported that FQ inhibited the replication of the SARS-CoV virus (Scheme 1.4) [117]. G. Meng et al. reported that ferric ammonium nitrate (FAC) salt (**19**) has shown

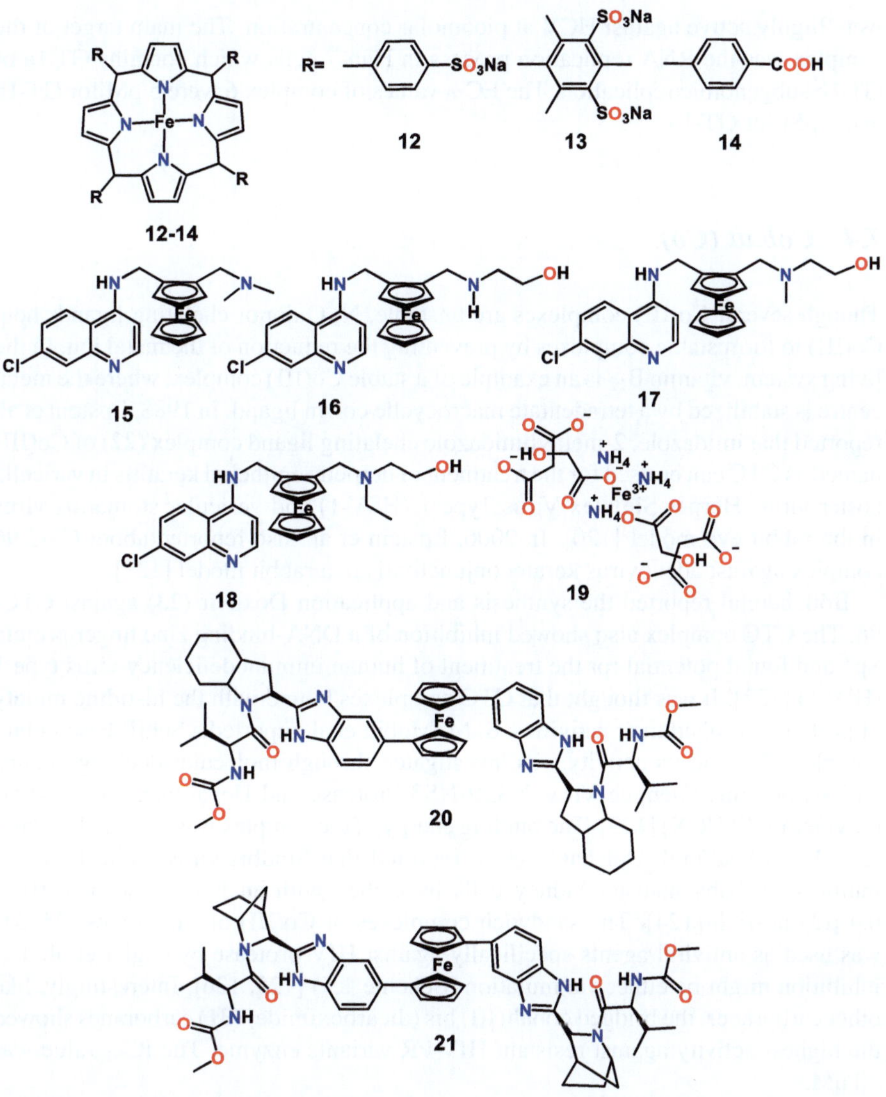

Scheme 1.4 Iron-based antiviral agents

excellent antiviral properties against various types of viral infections, e.g. Influenza A, HIV, Zika, and Enterovirus 71 (EV71) [118]. Mechanistically, it has been shown that ferric ammonium citrate (FAC) prevented several viral infections by blocking the endosomal viral release and viral fusion. In general, FAC targeted the virus and inhibits it in its early stage of viral replication. Recently in 2018, Jason A. Wiles et al. have synthesized a series of ferrocene-containing organometallic complexes (**20–21**) and their applications against the hepatitis C virus (HCV) [119]. These complexes were highly active against HCV at picomolar concentration. The main target of the complex was the RNA replication process in Huh-7 cells which contains GT-1a or GT-1b subgenomic replication. The EC_{50} values of complex **6** were 5 pM for GT-1b and 7 pM for GT-1a.

7.4 Cobalt (Co)

Though several Co(III) complexes are unstable, N, O donor chelating ligands help Co(III) to form stable complexes by preventing the reduction of the metal ion. In the living system, vitamin B_{12} is an example of a stable Co(III) complex, where the metal centre is stabilized by a tetradentate macrocyclic corrin ligand. In 1988, Epstein et al. reported that imidazole, 2-methylimidazole chelating ligand complex (**22**) of Co(III) named as CTC can be used for the treatment of herpetic epithelial keratitis in varicella zoster virus, Herpes Simplex Virus Type 1 (HSV-1) and vesicular stomatitis virus in the rabbit eye model [120]. In 2006, Epstein et al. also reported about CTC-96 complex against adenovirus keratoconjunctivitis in a rabbit model [121].

Böttcheretal reported the synthesis and application Doxovir (**23**) against CTC-96. The CTC complex also showed inhibition of a DNA-binding zinc finger protein Sp1 and found potential for the treatment of human immunodeficiency virus type 1 (HIV-1) [122]. It was thought that CTC complexes bound with the histidine moiety of protein, inhibited their activities. S. M. Mobin et al. reported a Schiff base-cobalt complex (24) and its activity was investigated through molecular docking against various proteins (Dengue Virus NS2B/NS3 Protease and Hepatitis C virus NS5B polymerase (2WCX) [123]. The binding energy of the complex was −11.21 kcal/mol and -10.88 kcal/mol. Delehanty et al. reported that Sindbis virus replication was inhibited in baby hamster kidney cells by Cohex with an IC_{50} value of $0.10 \pm 0.4\ \mu M$ at 48 h [124]. The sandwich complexes of Co(III) of carboranes (**25–31**) was used as antiviral agents specifically against HIV protease by Cigler et al. The inhibition might be either via mutation (Scheme 1.5) [125, 126]. Interestingly, like other carboranes, the bridged cobalt(III) bis (dicarboximide) (**31**) carboranes showed the highest activity against resistant HIV-PR variants enzyme. The IC_{50} value was 50 nM.

7.5 Nickel (Ni)

Tetra coordinated Ni complexes are most common, and various of them show antiviral activity. Gallego et al. in 2011 reported a complex $[\{(DES)MeN(CH_2)_2NMe(DES)\}Ni(H_2O)_2]\cdot 2H_2O$ (**32**) against HIV in MT-2 cells by inhibiting the virus replication process. At 5 μM concentration, viral replication was 50% inhibited [127]. Rogolino et al. reported that salicyl hydrazonic-based complex, $Ni(HL)_2\cdot 2H_2O$ (**33**) exhibited antiviral properties against the herpes

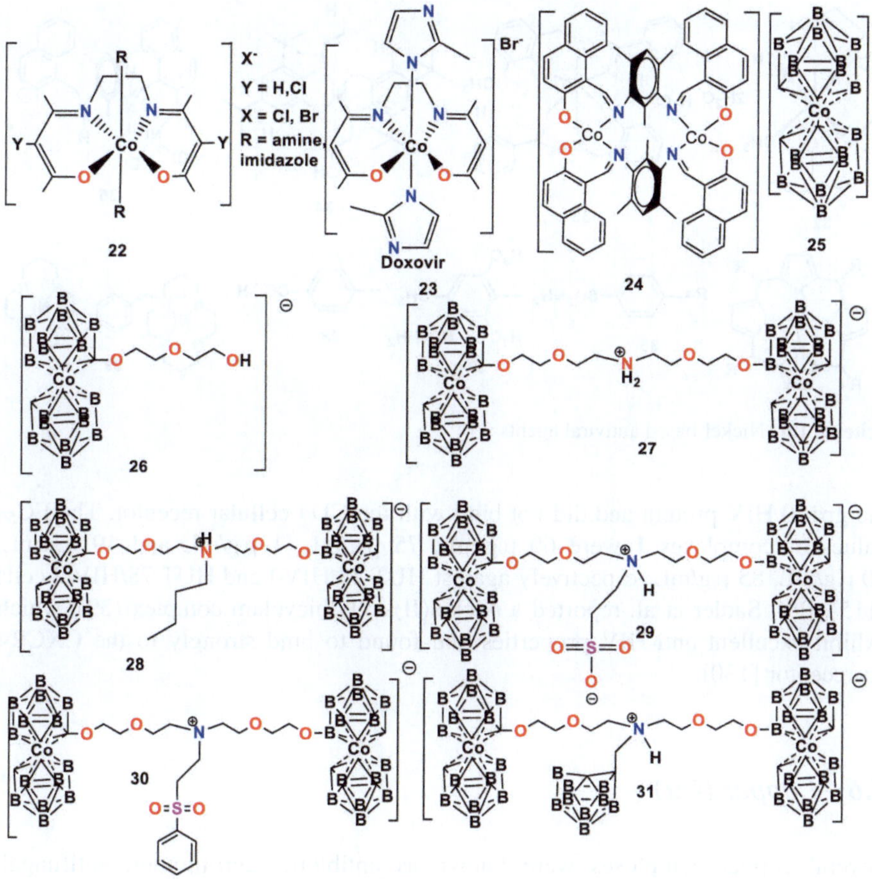

Scheme 1.5 Cobalt-based antiviral agents

simplex virus. This complex was highly active against HSV-1 and HSV-2, and the corresponding EC_{50} values were 1.8 and 4.3 μM [128]. S. M. Mobin et al. reported a Schiff base nickel complex (**34**) and their activity was investigated through molecular docking against various proteins (Dengue Virus NS2B/NS3 Protease and Hepatitis C virus NS5B polymerase (2WCX). The binding energy of the complex was −10.45 kcal/mol and −10.29 kcal/mol [123]. O. I. Shadyro et al. synthesized a new type of 4,6-di-tert-butyl-3-[(2-hydroxyethyl) thio]benzene-1,2-diol (L)-based Ni complex (**35**), and it was applied as the antifungal and anti-HIV agents. The complex exhibited moderate activity and EC_{50} value of the complex was 2.86 μM against HIV [129].

In 1997, B. Meunier et al. synthesized a new series of water-soluble porphyrin-based Ni (II) complexes (**36–38**), and these complexes exhibited antiviral activities against human immunodeficiency virus (HIV1, HIV-2) (Scheme 1.6). The carboxylate and sulphonate-functionalized Ni-porphyrin complexes were more active against

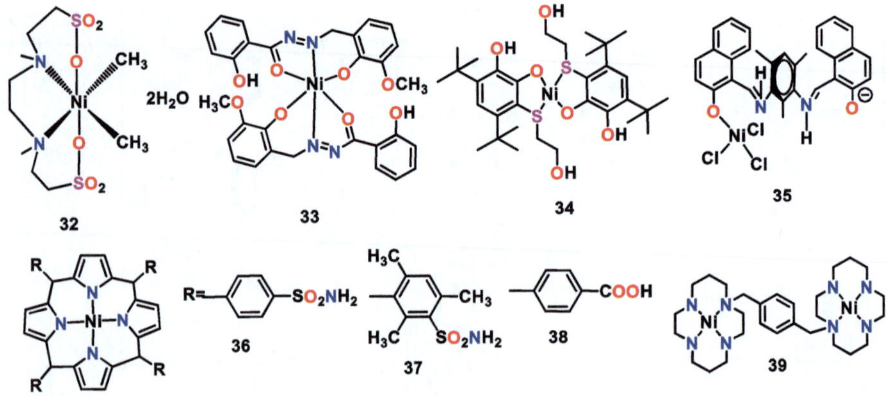

Scheme 1.6 Nickel-based antiviral agents

the gp120 HIV protein and did not bind with the CD4 cellular receptor. The EC_{50} values of complexes **1** were 60 μg/mL, 75 μg/mL, 1 μg/mL and 19 μg/mL, 70 μg/mL, 85 μg/mL respectively against HUT-78/HIV-1 and HUT-78/HIV-2 cells [115]. P. J. Sadler et al. reported a Nickel(II)-xylylbicyclam complex (**39**), which exhibit excellent anti-HIV properties and found to bind strongly to the CXCR4 co-receptor [130].

7.6 Copper (Cu)

Several copper complexes were known as antibiotic, anti-tumour, antifungal, antiviral, anti-inflammatory, and cytotoxic agents [131]. Several strategies had been used to prevent viral replication and the damaging of the viral surface. S. M. Goyel et al. reported in 1985 that viral surface was negatively charged so that it can attract the positively charged metal complex to interact very well [132]. The whole life cycle of the virus proceeded through few necessary steps such as binding, internalization, fusion and uncoating of the genome, replication, new particle formation and finally releasing of the new particle [133]. The Cu complex prevented the viral infection in several ways—(i) Endosome-lysosome degradation pathway. (ii) Redox reactions resulted in the oxidative degradation of the viral surface due to the production of different types of ROS such as $O2^-$, ·OH, OH^-, etc. (iii) Cu^+ ion in the presence of H_2O_2 played a significant role in viral inactivation process. (iv) Few Cu chelating complexes inhibited virus-induced apoptosis via RNA replication, and transcriptase enzyme inhibited the growth and replication of RNA tumour viruses [134].

E. Pilotti reported new thiosemicarbazones based Cu complexes [aqua (pyridoxalthio-semicarbazonato) copper(II)] (**40**) which inhibited significantly the Exo/Endogenous HIV-1 replication and also showed less antiviral activity against the retroviruses HTLV-1/2. The complex exhibited the value of CC_{50} was >10 μg/mL

[135]. The antiviral activity of curcumin-based Cu complex (**41**) exhibited remarkable activity against vesicular stomatitis virus, Coxsackievirus B^4, Respiratory syncytial virus and the EC_{50} value was ~4 μg/mL. This Cur-Cu complex (41) was also active against the feline coronavirus (fipv) [136]. F. J. D. L. Mata et al. in 2012, synthesized a new class of dmeddp and edtp based Cu complexes (**42–43**) and these complexes inhibited 50–60% HIV replication at a concentration of 5 μM [137]. Copper(II) complex containing fluoroquinolone and phenanthroline-based ligands (**47**) showed significant cytotoxicity against Vero cell line, and in vitro condition, it showed very less antiviral activity against gammaherpesvirus 68 (MHV-68) (Scheme 1.7) [138]. Complexes **44** and **45** were active against influenza virus, and they might target specifically His37 in influenza M2. These complexes blocked both the WT (wild type) and the amantadine-resistant M2S31N protein. The EC_{50} values were 3.7 and 1.1 μM in MDCK viral cells [139]. S. M. Mobin et al. reported a Schiff base containing copper complex (**48**) and their activity was investigated from molecular docking against several proteins involved in viral replication (Dengue Virus NS2B/NS3 Protease and Hepatitis C virus NS5B polymerase (2WCX) [136]. The binding energy of the complex was −11.19 kcal/mol and −11.11 kcal/mol [123]. O. I. Shadyro et al. synthesized a new type of 4,6-di-tert-butyl-3-[(2-hydroxyethyl)thio]benzene-1,2-diol (L)-based copper complex (**48**), and it was applied as antifungal and anti-HIV agents. The complex exhibited moderate activity and EC_{50} value of the complex was 8.36 μM

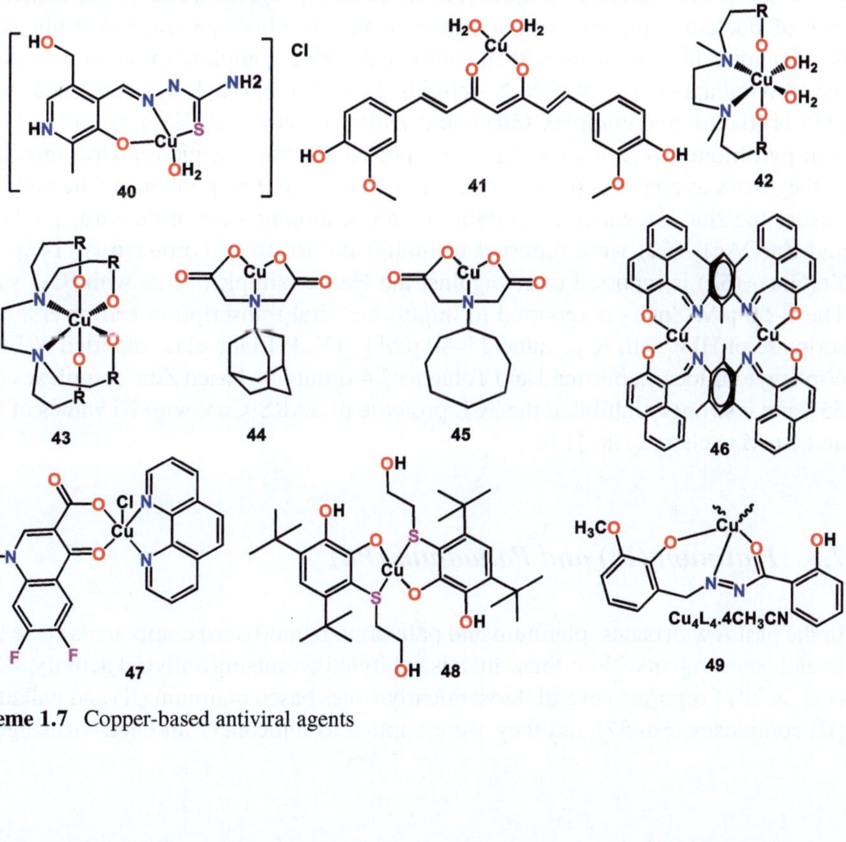

Scheme 1.7 Copper-based antiviral agents

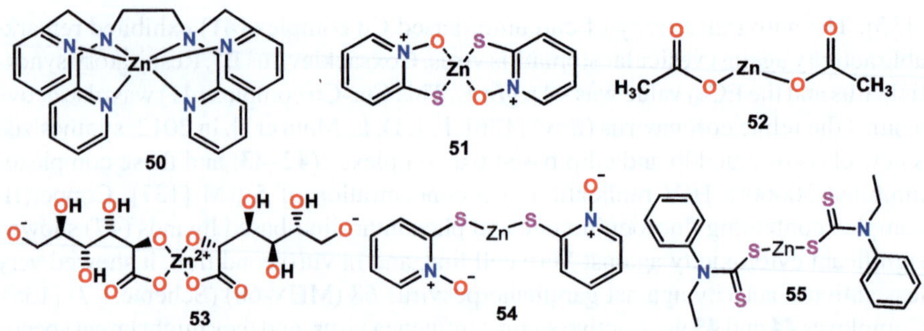

Scheme 1.8 Antiviral zinc complexes

against HIV [129]. L. Naesens et al. reported a new acyl hydrazones-based copper(II) complex (**49**), and it was applied against the herpes virus (HSV-1; HSV-2) [128].

7.7 *Zinc (Zn)*

Zinc is one of the essential elements in the biological system. It takes parts in processes like DNA replication, etc. The Lewis acid properties, catalyzing the hydrolysis of nucleic acid, proteins or peptides and its important role to bring structural rigidity in several metalloenzymes make zinc complexes potential to act as the antiviral agent. The N,N,N′,N′-tetrakis(2-pyridinylmethyl)-1,2-ethylenediamine (TPEN)-based zinc complex (**50**) could inhibit DENV and JEV replication [140]. Zinc pyrithione (PT) complex (**51**) was reported to inhibit the picornavirus infections as they work as zinc ionophore which can labialize the zinc pool inside the body and release the zinc ion which can inhibit the replication process of the virus [141]. PT and $Zn(OAc)_2$ (**52**) were reported to inhibit the RdRp of coronavirus [142]. The $Zn(Glu)_2$ (**53**) is reported to act against the Herpes simplex virus with IC_{50} value. The 1–50 μM $ZnCl_2$ is reported to inhibit the viral transcription and reverse transcriptase of HIV with IC_{50} value 25–80 μM [143]. P. Liang et al. reported *N*-Ethyl-*N*-phenyldithiocarbamic acid and Toluene-3,4-dithiolato based Zinc complexes (**54**–**55**), and it strongly inhibited the 3CL protease of SARS-CoV with Ki values of 0.17 and 1 μM (Scheme 1.8) [144].

7.8 *Platinum (Pt) and Palladium (Pd)*

In the past few decades, platinum and palladium metal-based compounds were used as anticancer agents. Now, these metals exhibited promising antiviral activity. B. Jan et al. in 2011 reported several thiosemicarbazones based platinum (II) and palladium (II) complexes (**56**–**57**) and they were applied as anticancer and anti-virus agents.

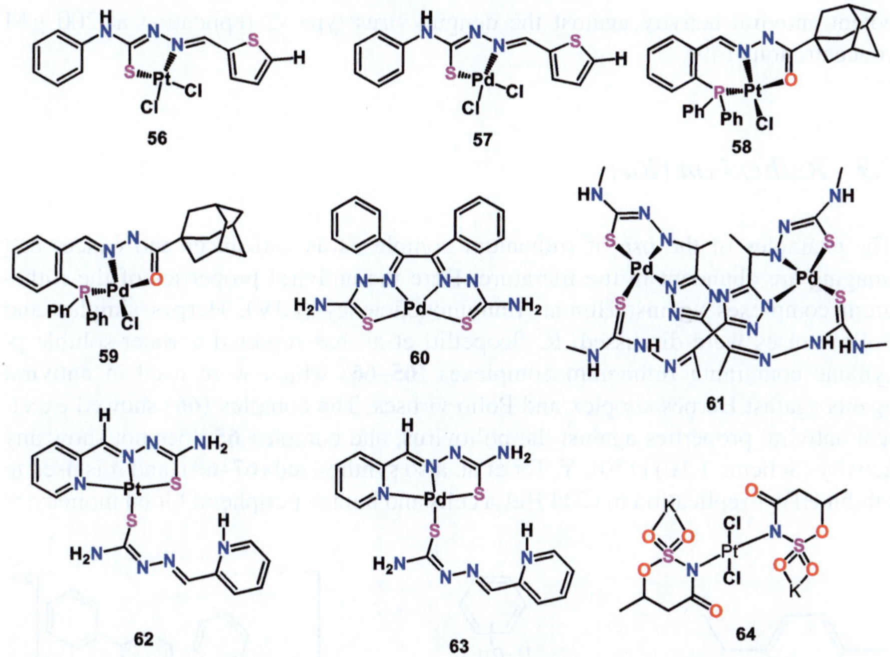

Scheme 1.9 Platinum and palladium-based complexes exhibiting in vitro antiviral activity

The compounds were slightly active against cytomegalovirus. Complex **56** was more active than complex **57**, and the EC_{50} values were 2.9 μM and 5 μM, respectively, against HCMV [145]. B. Vukovic-Gacic et al. also reported a new class of platinum and palladium complexes (**58–59**), and they were tested against type-1 poliovirus in Hep-2 cells. These complexes highly inhibited the virus receptors and prevented viral adsorption. The EC_{50} value of complex **59** was 32.5 ± 3.7 IU/mL (Scheme 1.9) [146].

P. Souza et al. synthesized a series of mono and dinuclear benzyl bis(thiosemicarbazone), and 3,5-diacyl-1,2,4-triazole bis(4-methylthiosemicarbazone)-based palladium (II) complexes (**60–61**) and their antiviral activity against herpes simplex virus type 1 (HSV 1) and type 2 (HSV 2). Complex **59** showed promising antiviral activity against acyclovir-resistant viruses R-100 and PU with IC50 values 0.01 μM and 0.01 μM, respectively [147]. Interestingly M. A. Demertzis et al. prepared pyridine-2-carbaldehyde thiosemicarbazone (HFoTsc) based platinum (II) and palladium (II) complexes (**62–63**) which were used in HSV replication. The assay was performed with MDBK cells. Complex **63** was found the most active and highly selective anti-HSV agent with an IC_{50} value of 0.01 μM (Scheme 1.9) [148]. In 2010, M. Cavicchioli et al. synthesized a new monomeric platinum (II) complex $K_2[PtCl_2(ace)_2]$ (**64**), and it was used to inhibit the dengue virus type 2 (New Guinea C strain) in Vero cells. The complex exhibited potent antiviral activity against the dengue virus type -2 replication at 200 μM concentration [149].

7.9 Ruthenium (Ru)

The examples of the use of ruthenium complexes as anticancer and cancer cell imaging are abundant in the literature. Here the antiviral properties of the ruthenium complexes against Human Immunodeficiency (HIV), Herpes simplex and Polio viruses were discussed. R. Scopelliti et al. has reported a water-soluble p-cymene containing ruthenium complexes (**65–66**) which were used in antiviral agents against Herpes simplex and Polio viruses. The complex (**66**) showed excellent antiviral properties against the poliovirus, and complex **65** does not show any activity (Scheme 1.10) [150]. Y. Tor et al. also synthesized (**67–69**), and it is used to inhibit HIV-1 replication in CD4 HeLa cells and human peripheral blood monocytes

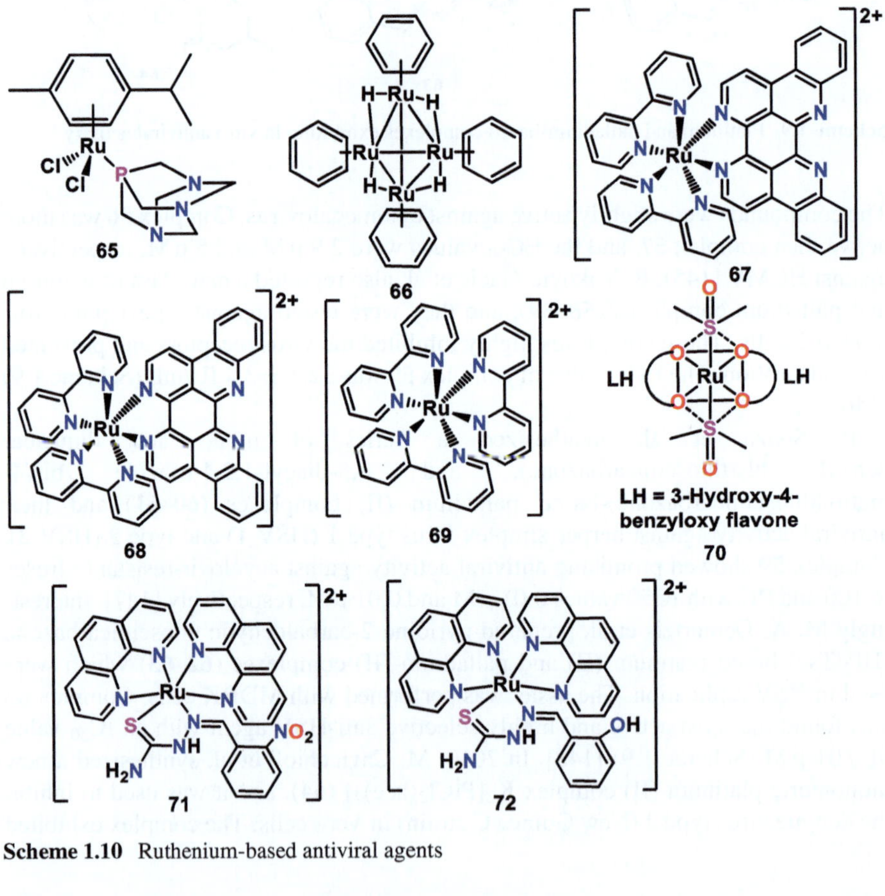

Scheme 1.10 Ruthenium-based antiviral agents

by using a plaque formation assay. The complexes **66–68** were firmly bound with nucleic acid with IC_{50} values of the complex were approximate 0.8 μM, 30 μM and >100 μM but it was also reported that without eilatin ruthenium complex showed very less activity (15 ± 100 fold lower) against HIV [151].

S M Pandey et al. reported a different type of octahedral 3-hydroxy flavone containing ruthenium complex (**70**) and their antiviral and anticancer properties against HIV and 11 cancer cell lines (MCF, CAKI, A549, PC3, HOS, KB-VIN, KB, SK-ME-L-2, U 87-MG, HCT-8, IA9) [152]. This complex was used in mice model. The EC_{50} value of the complex is 41.9 μg/ml against H9 cells. Ruthenium complexes (**71–72**) containing 1,10-phenanthroline and thiosemicarbazone were tested against several viruses, e.g. Cytomegalo-virus, HSV-1 and 2, Varicella Zoster Virus and Vaccinia Virus [153]. The EC50 value of complexes **71** and **72** were 11 μM and >20 μM against AD-169.

7.10 Gold (Au)

Medicinal properties of gold (Au) were known from ancient times. Various gold compounds were used to treat rheumatoid arthritis, AIDS, cancer. For rheumatoid arthritis, auranofin has been used clinically from the last century [154, 155]. In 1993, M. E. Gurney et al. reported two compounds (Aurothioglucose and aurothiomalate) (**73–74**) showed promising anti-HIV-1 activity in vitro due to Au(I) ligand exchange with cysteinyl thiol on the surface of the virus (Scheme 1.11) [156]. The target protein of the compound was human immune deficiency virus-1 (HIV-1) reverse transcriptase. In 2004, Chi-Ming Che et al. synthesized a new series of porphyrin-based gold (I) compounds (**75–77**) which were exhibited good anti-HIV activity [157]. Compounds were strongly effective inhibitors of HIV-1 reverse transcriptase with submicromolar level concentration (IC_{50} 0.31–28.4 μM). In 2009, D. Meyer et al. and J. Darkwa et al. reported 11 Au(I) phosphine (**78–81**) and ([3,5-Me2bpzaH$_2$][AuCl$_4$]Cl (**82**) compounds and they were used to inhibit HIV. The targeted proteins were reverse transcriptase-RT and protease-PR [158, 159]. Complex **82** at 100 μM concentration inhibited 63% HIV-1 protease.

8 Nanoparticles-Based Antiviral Agent

Recently nanoparticles-based antiviral agents are the most emerging area of research in the biomedical application because of their important physiochemical properties, including tunable size and shape, regulated surface charge, superparamagnetic nature, NIR light-absorbing surface plasmon resonance band, biocompatibility, highly luminescence properties, upconversion, and biodegradability [160]. Additionally, the nanoparticles can easily conjugate (via covalent or non-covalent bonding)

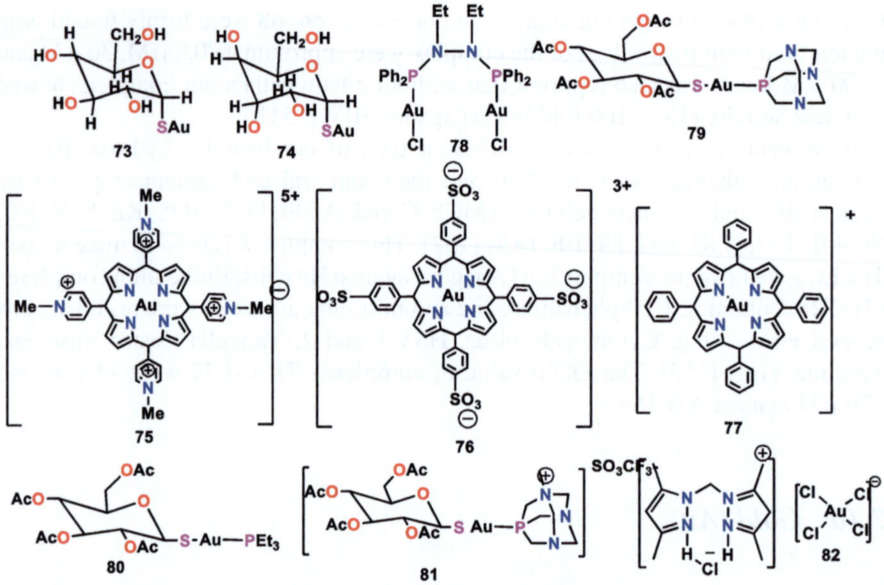

Scheme 1.11 Gold-based compounds with in vitro antiviral activity

with one or more bioactive molecules or functional groups. Furthermore, nanoparticles can penetrate through the blood-air barrier and blood–brain barrier. From the nineteenth century, Silver (Ag), Gold (Au), Copper (Cu), Titanium oxide (TiO_2), Silicon oxide (SiO_2) nanoparticles have been extensively used against various viruses including HSV, dengue virus type-2, hepatitis B (HBV) and vesicular stomatitis [161, 162]. Here we briefly discussed the nanomaterial as antiviral agents.

8.1 Gold Nanoparticles (AuNPs)

Gold nanoparticles (AuNPs), gold nanorod, and functionalized AuNPs exhibited excellent biological properties including the detection of DNA sequences, various proteins, and bacteria, and also primarily used in anticancer studies [163]. Such nanoconjugates have served also as an excellent drug delivery platform. Chakravarthy et al. reported gold nanorod strongly inhibited the replication of influenza virus (H1N1) by blocking the expression of IFN-β stimulated genes [164]. IFN complex and spherical shape hyaluronic acid conjugate AuNPs also inhibit the hepatitis C viral (HCV) infection reported by Lee and group [165]. Wen et al. reported a sialic acid conjugate of AuNPs which inhibit influenza viral infection by multivalent interactions [166]. The foot-and-mouth disease virus (FMDV) in its early stage and

HIV-1 virus has been strongly inhibited by AuNPs and peptide-functionalized triazoles (AuNP-PT) nanoparticles. Therefore, various AuNPs functionalized molecules showed excellent antiviral activity against HCV, HIV, and FMDA virus.

8.2 Silver Nanoparticles (AgNPs)

Silver nanoparticles (AgNPs) and functionalized silver nanoparticles (AgNPs) are already used in several biomedical fields such as anticancer, antiviral, anti-inflammatory, antibacterial, anti-angiogenesis, and antifungal agents [167]. Lara et al. synthesis a polyvinylpyrrolidone (PVP)-conjugated AgNPs which exhibited excellent antiviral activity against HIV-1 transmission [168]. The nanoparticles were strongly bound to HIV envelope glycoprotein gp120 in such a manner that inhibited CD4-dependent virion binding and stopped the replication of the virus. Yao et al. and Gaikwad et al. reported an AgNP functionalized polyurethane condoms (PUCs) nanoparticle and fungi-Mediated AgNPs which inhibit HIV-1/2 and parainfluenza virus effectively via inhibiting the interaction of the virus with the host cell [169]. Mori et al. also synthesized size-dependent chitosan based AgNPs which exhibited remarkable antiviral effects in influenza (H1N1) A virus [170]. In 2016, Huang etal reported a Curcumin-coated AgNPs which also showed excellent antiviral effect against the respiratory syncytial virus (RSV) by activating CD8+ T cell.

8.3 Graphene Oxide (GO) and Quantum Dots

The two-dimensional, hexagonal carbon lattice structure of graphene oxide (GO) and functionalized graphene oxide (GO) are great attention to biomedicinal researchers because of their excellent electronic and thermal properties. Graphene oxide (GO) is versatile used as an antiviral, antimicrobial, and anticancer agent [171]. In 2015, Tang et al. reported graphene oxide (GO) showed remarkable antiviral agents against influenza A (H9N2) virus, EV71, and FMDV through blocking the pathway of the pathogenic agent [172]. Sarid et al. also synthesized the reduced sulphonated GO which exhibited excellent antiviral activity in HSV-1 virus by preventing the virus from interacting with the host cells [173]. The curcumin-coated GO showed excellent antiviral activity and very good biocompatibility reported by Huang et al. The combination of curcumin and graphene oxide showed better activity rather than free curcumin or GO against RSV viral infection [174].

Quantum dots (QDs) is 2–10 nm size and have excellent electron transport properties and its widely used in cell imaging, anticancer, cell labelling and various other nanomedicine field. Basically, gold, carbon, silver, and CdSe quantum dots are mostly used in this field. Xiao et al. reported GHS functionalized CdTe quantum dots which showed excellent antiviral activity against pseudorabies virus by the interaction of surface protein of the virus [175]. Benzoxazine coated quantum dots exhibited remarkable antiviral activity against porcine parvovirus by the blocking of virus and host cell interaction [176].

8.4 Zinc Oxide

Zinc oxide nanoparticles also played a vital role in anticancer and microbial activity due to their partially negative charge on the surface. Antoine et al. reported micro-nano structures (MNSs) ZnO nanoparticles which showed remarkable antiviral activity against simplex virus type-2 (HSV-2) by blocking the entry of viruses to the host cell [177]. Tetrapod shaped Zinc oxide nanoparticles inhibit the HSV-1 viral infection and also the increase immune system against the HSV-2 virus [178].

9 Transition Metal Complexes and Nanocomposite from COVID-19 Perspective

The role of transition metal complexes as well as the nanocomposites as the antiviral agents are reviewed, and potential utility of the transition metal complexes and nanoparticles against SARS-CoV-2 in pandemic COVID-19 has initiated researches in this domain. There are only a few reports on in vitro applications of transition metal complexes as the antiCOVID-19 agents.

The gold metal-containing triethyl phosphine drug auranofin is the FDA-approved drug for rheumatoid arthritis since 1985. Auranofin has also proven its therapeutic potentials against various diseases like HIV, bacterial/parasitic infection, antimicrobial, anticancer, etc. [179, 180]. The actual mechanism of auranofin was the inhibition of oxidation/reduction with the enzymes that mainly generate different types of reactive oxygen species which leads to intercellular oxidative stress and apoptosis [181–183]. Moreover, anti-inflammatory properties of auranofin are expected to reduce the cytokine storms and thereby reduce the respiratory distress of alveolar adema.

Recently in 2020, L. Messori etal referred to auranofin (AF, **80**) as the potential inhibitor of SARS-COV-2 replication with utmost efficacy and urged the scientific community to explore auranofin against COVID-19 [184]. M. Kumar et al. also reported that auranofin remarkably inhibited SARS-COV-2 replication, and within 48 h infection was reduced by 95%. The EC_{50} value was 1.5 μM against SARS-CoV-2 infected Huh7 cells [185]. Ingo Ott et al. recently evaluated five different organometallic gold(I) complexes (**82–87**) and auranofin as inhibitors of SARS-CoVs virus (Scheme 1.12). The complexes can inhibit the interaction of the ACE2 receptor of the host cells and spike protein of the SARS-CoV-2 virus and could impede the entry of the virus into the host alveolar epithelial cells. The gold complexes also inhibited the SARS-CoV-2 papain-like protease (PLpro), which is an important enzyme

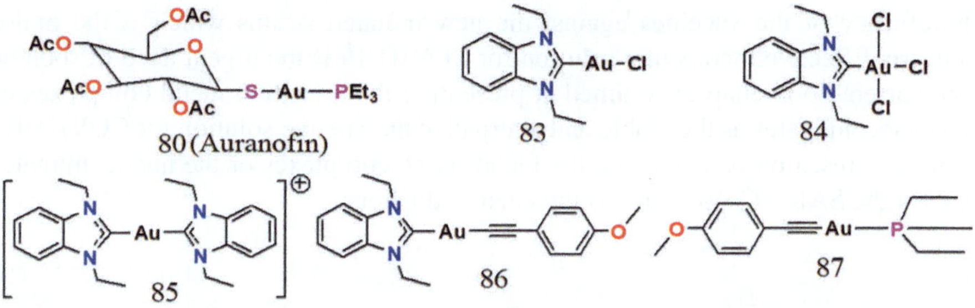

Scheme 1.12 Selected gold-based compounds showed in vitro activity against SARS-CoV-2 virus

for the replication of the virus. These complexes, overall, exhibited excellent IC_{50} values in the range of 16–25 μM [186]. Several other palladium-based complexes were explored recently for the therapeutic potential against SARS-CoV-2 in silico [187, 188]. Recently our group explored in silico the potential inhibitory potential of selected transition metal complexes against RNA-dependent RNA polymerase (RdRp) of SARS-CoV-2 virus [189]. Several metal complexes also have emerged as potential tools against SARS-CoV-2 [189–194]. Sportelli et al. (2020) stated that researchers need to explore the metal-based nanomaterial as the strategic tools to fight COVID-19 [195].

10 Summary and Conclusion

An outbreak of pandemic COVID-19 is the most challenging catastrophe of the present century. More than 173 million people across 222 countries are infected with SARS-CoV-2, and the death toll was raised to 3.7 million. The most critical challenge to the scientific community at present is the lack of therapeutic options for COVID-19. The attractive strategy for the treatment of COVID-19 is the repurposing of WHO-approved drugs or plasma therapy. Newly discovered vaccines are in the stage of the clinical trial. The unique properties like a broad spectrum of formal charge and oxidation states, wide range of coordination number and geometry, tunable kinetic, thermodynamic and redox properties, diverse reaction pathways inherited by transition metal complexes have emerged as the viable alternative in realm of medicinal chemistry and also could be potentially important in current pandemic COVID-19. However, the application of the transition metal-based compounds or the nanocomposites in COVID-19 treatment are virtually unexplored. The present chapter articulated the origin, transmission and biology of COVID-19 caused by SARS-CoV-2, repurposed drugs in the treatment of COVID-19, advances in metal-based complexes and nanoconjugates as the antiviral agents, in vitro applications of selected transition metal complexes against SARS-CoV-2 virus.

Vaccination, although, has emerged as the best strategy in preventing the SARS-CoV-2 infection, the outbreak of several mutated strains of the virus has reduced the efficacy of the vaccines against the new mutated strains which is the major concern. Effective therapeutic solution for COVID-19 is the urgent need of society. The current book chapter is aimed at presenting the transition metal complexes or the nanoconjugates as the viable and alternative therapeutic solution for COVID-19. Extensive research on exploring the metal-based complexes or the nanoconjugates against the SARS-CoV-2 virus is the need of the hour.

References

1. Lu R, Zhao X, Li J, Niu P, Yang B, Wu H, Wang W, Song H, Huang B, Zhu N, Bi Y, Ma X, Zhan F, Wang L, Hu T, Zhou H, Hu Z, Zhou W, Zhao L, Chen J, Meng Y, Wang J, Lin Y, Yuan J, Xie Z, Ma J, Liu WJ, Wang D, Xu W, Holmes EC, Gao GF, Wu G, Chen W, Shi W, Tan W (2020) Genomic characterisation and epidemiology of 2019 novel coronavirus: implications for virus origins and receptor binding. Lancet 395:565–574
2. https://www.who.int/emergencies/diseases/novel-coronavirus-2019
3. Zhu N, Zhang D, Wang W, Li X, Yang B, Song J, Zhao X, Huang NB, Shi W, Lu R, Niu P, Zhan F, Ma X, Wang D, Xu W, Wu G, Gao GF, Tan W (2020) A novel coronavirus from patients with pneumonia in China. Engl J Med 382:727–733
4. https://www.who.int/csr/media/sars_wha.pdf
5. Baig AM, Khaleeq A, Ali U, Syeda H (2020) Evidence of the COVID-19 virus targeting the CNS: tissue distribution, host-virus interaction, and proposed neurotropic mechanisms. ACS Chem Neurosci 11:995–998
6. Aminnejad R, Shafiee H (2020) Is regional anesthesia safe enough in suspected or confirmed COVID-19 patients? ACS Chem Neurosci 11:1371–1371
7. Xiong C, Jiang L, Chen Y, Jiang Q (2020) Evolution and variation of 2019-novel coronavirus. bioRxiv. https://doi.org/10.1101/2020.01.30.926477
8. Guerra S (2020) Oral mucosal lesions in a COVID-19 patient: New signs or secondary manifestations? Int J Infect Dis 97:326–328
9. https://en.wikipedia.org/wiki/National_responses_to_the_COVID-19_pandemic
10. https://en.wikipedia.org/wiki/COVID-19_pandemic_lockdowns
11. Godbole T (2020) Domestic violence rises amid coronavirus lockdowns in Asia 11 April, 2020
12. Mental health and psychosocial considerations during the COVID-19 outbreak. World Health Organization. 18 March 2020
13. https://en.m.wikipedia.org/wiki/Social_impact_of_the_COVID-19_pendemic
14. Opatz T, Senn-Bilfinger J, Richert C (2020) Thoughts on what chemists can contribute to fighting SARS-CoV-2—a short note on hand sanitizers, drug candidates and outreach. Angew Chem Int Ed 59:9236–9240
15. https://www.weforum.org/agenda/2020/02/coronavirus-economic-effects-global-economy-trade-travel/
16. Leng Z, Zhu R, Hou W, Feng Y, Liu H, Jin R, Jin K, Zhao RC (2020) Transplantation of ACE2—mesenchymal stem cells improves the outcome of patients with COVID-19 pneumonia. Aging Dis 11:216–228
17. https://www.who.int/emergencies/diseases/novel-coronavirus-2019/situation-reports/
18. Corman VM, Muth D, Niemeyer D, Drosten C (2018) Hosts and sources of endemic human coronaviruses. Adv Virus Res 100:163–188
19. Van Doremalen N, Bushmaker T, Morris DH, Holbrook MG, Gamble A, Williamson BN, Tamin A, Harcourt JL, Thornburg NJ, Gerber SI, Lloyd-Smith JO, Wit E, Munster VJ (2020) Aerosol and surface stability of SARS-CoV-2 as compared with SARS-CoV-1. N Engl J Med 382:1564–1567

20. https://www.worldometers.info/coronavirus/
21. Shalev D, Shapiro PA, Psychiatry E (2020) Epidemic psychiatry: the opportunities and challenges of COVID-19. Gen Hosp Psychiatry 64:68–71
22. Encinar JA, Menendez JA (2020) Potential drugs targeting early innate immune evasion of SARS-coronavirus 2 via 2′-O-methylation of viral RNA. Viruses 12:525
23. Tang X, Wu C, Li X, Song Y, Yao X, Wu X, Duan Y, Zhang H, Wang Y, Qian Z, Cui J, Lu J (2020) On the origin and continuing evolution of SARS-CoV-2. Natl Sci Rev 7:1012–1023
24. Lopez Bernal J, Andrews N, Gower C, Gallagher E, Simmons R, Thelwall S, Stowe J, Tessier E, Groves N, Dabrera G, Myers R, Campbell C, Amirthalingam G, Edmunds M, Zambon M, Brown K, Hopkins S, Chand M, Ramsay M (2021) Effectiveness of COVID-19 vaccines against the B.1.617.2 variant. https://doi.org/10.1101/2021.05.22.21257658. Zhou D, Dejnirattisai W, Supasa P, Liu C, Mentzer AJ, Ginn HM, Zhao Y, Duyvesteyn HME, Tuekprakhon A, Nutalai R, Wang B, Paesen GC, Lopez-Camacho C, Slon-Campos J, Hallis B, Coombes N, Bewley K, Charlton S, Walter TS, Skelly D, Lumley SF, Dold C, Levin R, Dong T, Pollard AJ, Knight JC, Crook D, Lambe T, Clutterbuck E, Bibi S, Flaxman A, Bittaye M, Belij-Rammerstorfer S, Gilbert S, James W, Carroll MW, Klenerman P, Barne E, Dunachie SJ, Fry EE, Mongkolsapaya J, Ren J, Stuart DI, Screaton GR (2021) Evidence of escape of SARS-CoV-2 variant B.1.351 from natural and vaccine-induced sera. Cell 198:2348–2361
25. Shang J, Ye G, Shi K, Wan Y, Luo C, Aihara H, Geng Q, Auerbach A, Li F (2020) Structural basis of receptor recognition by SARS-CoV-2. Nature 581:221–224. Chang EL, Simmers C, Knight DA (2010) Cobalt complexes as antiviral and antibacterial agents. Pharmaceuticals 3:1711–1728
26. Musib D, Raza MK, Kundu S, Roy M (2018) Modulating in vitro photodynamic activities of copper(II) complexes. Eur J Inorg Chem 2018:2011–2018
27. Lu R, Zhao X, Li J (2020) Genomic characterisation and epidemiology of 2019 novel coronavirus: implications for virus origins and receptor binding. Lancet 395:565–574
28. Walls AC, Park Y, Tortorici MA, Wall A, McGuire AT, Veesler D (2020) Structure, function, and antigenicity of the SARS-CoV-2 spike glycoprotein. Cell 181:281–292
29. Shereen MA, Khan S, Kazmi A, Bashir N, Siddique R (2020) COVID-19 infection: origin, transmission, and characteristics of human coronaviruses. J Adv Res 24:91–98
30. Hamre D, Procknow JJ (1966) A new virus isolated from the human respiratory tract. Proc Soc Exp Biol Med 121:190–193
31. McIntosh K, Dees JH, Becker WB, Kapikian AZ, Chanock RM (1967) Recovery in tracheal organ cultures of novel viruses from patients with respiratory disease. Proc Natl Acad Sci USA 57:933–940
32. Tyrrell DA, Cohen S, Schlarb JE (1993) Signs and symptoms in common colds. Epidemiol Infect 111:143–156
33. Abdul-Rasool S, Fielding BC (2010) Understanding human coronavirus HCoV-NL63. Open Virol J 4:76–84
34. Fouchier RA, Hartwig NG, Bestebroer TM, Niemeyer B, de Jong JC, Simon JH, Osterhaus ADME (2004) A previously undescribed coronavirus associated with respiratory disease in humans. Proc Natl Acad Sci USA 101:6212–6216
35. Lau SK, Woo PC, Yip CC, Tse H, Tsoi HW, Cheng VC, Lee P, Tang BSF, Cheung CHY, Lee RA, So L, Lau Y, Chan K, Yuen K (2006) Coronavirus HKU1 and other coronavirus Infections in Hong Kong. J Clin Microbiol 44:2063–2071
36. Cheng VC, Lau SK, Woo PC, Yuen KY (2007) Severe acute respiratory syndrome coronavirus as an agent of emerging and reemerging infection. Clin Microbiol Rev 20:660–694
37. Devi LR, Raza MK, Musib D, Ramu V, Devi J, Roy M (2019) Nucleus targeting anthraquinone-based copper (II) complexes as the potent PDT agents: synthesis, photo-physical and theoretical evaluation. Inorg Chim Acta 500:119208
38. Musib D, Raza MK, Martina K, Roy M (2019) Mn(I)-based photoCORMs for trackable, visible light-induced CO release and photocytotoxicity to cancer cells. Polyhedron 172:125–131

39. Hilgenfeld R, Peiris M (2013) From SARS to MERS: 10 years of research on highly pathogenic human coronaviruses. Antiviral Res 100:286–295
40. Gao H, Yao H, Yang S, Li L (2016) From SARS to MERS: evidence and speculation. Front Med 10:377–382
41. Coleman CM, Frieman MB (2013) Emergence of the Middle East respiratory syndrome coronavirus. PLoSPathog 9:e1003595
42. Huang C, Wang Y, Li X, Ren L, Zhao J, Hu Y, Zhang L, Fan G, Xu J, Gu X, Cheng Z, Yu T, Xia J, Wei Y, Wu W, Xie X, Yin W, Li H, Liu M, Xiao Y, Gao H, Guo L, Xie J, Wang G, Jiang R, Gao Z, Jin Q, Wang J, Cao B (2020) Clinical features of patients infected with 2019 novel coronavirus in Wuhan China. Lancet 395:497–506
43. Risku M, Lappalainen S, Rasanen S, Vesikari T (2010) Detection of human coronaviruses in children with acute gastroenteritis. J Clin Virol 48:27–30
44. Chan JF, Kok KH, Zhu Z, Chu H, To KK, Yuan S, Yuen K (2020) Genomic characterization of the 2019 novel human-pathogenic coronavirus isolated from a patient with atypical pneumonia after visiting Wuhan. Emerg Microbes Infect 9:221–236
45. Letko M, Marzi A, Munster V (2020) Functional assessment of cell entry and receptor usage for SARS-CoV-2 and other lineage B betacoronaviruses. Nat Microbiol 5:562–569
46. Tian X, Li C, Huang A, Xia S, Lu S, Shi Z, Lu L, Jiang S, Yang Z, Wu Y, Ying T (2020) Potent binding of 2019 novel coronavirus spike protein by a SARS coronavirus-specific human monoclonal antibody. Emerg Microbes Infect 9:382–385
47. Zhou P, Yang XL, Wang XG, Hu B, Zhang L, Zhang W, Si HR, Zhu Y, Li B, Huang CL, Chen H, Chen J, Luo Y, Guo H, Jiang R, Liu M, Chen Y, Shen X, Wang X, Zheng X, Zhao K, Chen Q, Deng F, Liu L, Yan B, Zhan F, Wang Y, Xiao G, Shi Z (2020) A pneumonia outbreak associated with a new coronavirus of probable bat origin. Nature 579:270–273
48. (a) http://www.nhc.gov.cn/jkj/s7915/202001/e4e2d5e6f01147e0a8df3f6701d49f33.shtml. (b) http://www.bigd.big.ac.cn/gwh/
49. Khailany RA, Safdar M, Ozaslan M (2020) Genomic characterization of a novel SARS-CoV-2. Gene Rep 19:100682
50. Yuen K, Ye Z, Fung S, Chan C, Jin D (2020) SARS-CoV-2 and COVID-19: the most important research questions. Cell Biosci 10:40
51. Wu F, Zhao S, Yu B, Chen YM, Wang W, Song ZG, Hu Y, Tao ZW, Tian JH, Pei YY, Yuan ML, Zhang YL, Dai FH, Liu Y, Wang QM, Zheng JJ, Xu L, Holmes EC, Zhang YZ (2020) A new coronavirus associated with human respiratory disease in China. Nature 579:265–269
52. Gorbalenya AE, Baker SC, Baric RS, de Groot RJ, Drosten C, Gulyaeva AA et al (2020) The species severe acute respiratory syndrome-related coronavirus: classifying 2019-nCoV and naming it SARS-CoV-2. Nat Microbiol 5:536–544
53. Paraskevis D, Kostaki EG, Magiorkinis G, Panayiotakopoulos G, Sourvinos G, Tsiodras S (2020) Full-genome evolutionary analysis of the novel corona virus (2019-nCoV) rejects the hypothesis of emergence as a result of a recent recombination event. Infect, Genet Evol 79:104212
54. National Microbiology Data Center. (http://nmdc.cn/coronavirus)
55. Leng Z, Zhu R, Hou W, Feng Y, Liu H, Jin R, Jin K, Zhao RC (2020) Transplantation of ACE2—mesenchymal stem cells improves the outcome of patients with COVID-19 Pneumonia. Aging Dis 11:216–228
56. Toljan K (2020) Letter to the editor regarding the viewpoint "evidence of the COVID-19 virus targeting the CNS: tissue distribution, host-virus interaction, and proposed neurotropic mechanism. ACS Chem Neurosci 11:1192–1194
57. Yan R, Zhang Y, Li Y, Xia L, Guo Y, Zhou Q (2020) Structural basis for the recognition of SARS-CoV-2 by full-length human ACE2. Science 367:1444–1448
58. Siu YL, Teoh KT, Lo J, Chan CM, Kien F, Escriou N, Tsao SW, Nicholls JM, Altmeyer R, Peiris JSM, Bruzzone R, Nal B (2008) The M, E, and N structural proteins of the severe acute respiratory syndrome coronavirus are required for efficient assembly, trafficking, and release of virus-like particles. J Virol 82:11318–11330

59. Shang J, Ye G, Shi K, Wan Y, Luo C, Aihara H, Geng Q, Auerbach A, Li F (2020) Structural basis of receptor recognition by SARS-CoV-2. Nature 581:221–224
60. Wan Y, Shang J, Graham R, Baric RS, Li F (2020) Receptor Recognition by the Novel Coronavirus from Wuhan: an Analysis Based on Decade-Long Structural Studies of SARS Coronavirus. J Virol 94:e00127-e220
61. Wrapp D, Wang N, Corbett KS, Goldsmith JA, Hsieh CL, Abiona O (2020) Cryo-EM structure of the 2019-nCoV spike in the prefusion conformation. Science 367:1260–1263
62. Walls AC, Park YJ, Tortorici MA, Wall A, McGuire AT, Veesler D (2020) Targets of T cell responses to SARS-CoV-2 coronavirus in humans with COVID-19 disease and unexposed individuals. Cell 20:S0092
63. Tan W, Zhao X, Ma X, Wang W, Niu P, Xu W, Gao GF, Wu G (2020) A novel coronavirus genome identified in a cluster of pneumonia cases—Wuhan, China 2019–2020. China CDC Weekly 2:61–62
64. Tian X, Li C, Huang A, Xia S, Lu S, Shi Z, Lu L, Jiang S, Yang Z, Wu Y, Ying T (2020) Potent binding of 2019 novel coronavirus spike protein by a SARS coronavirus-specific human monoclonal antibody. Emerg Microbes Infect 9:382–385
65. Andersen KG, Rambaut A, Lipkin WI, Holmes EC, Garry RF (2020) The proximal origin of SARS-CoV-2. Nat Med 26:450–452
66. Kikkert M (2020) Innate immune evasion by human respiratory RNA viruses. J Innate Immun 12:4–20
67. Kumar S, Maurya VK, Prasad AK, Bhatt MLB, Saxena SK (2020) Structural, glycosylation and antigenic variation between 2019 novel coronavirus (2019-nCoV) and SARS coronavirus (SARS-CoV). Virus Dis 31:13–21
68. Zhang Y, Geng X, Tan Y, Li Q, Xu C, Xu J, Hao L, Zeng Z, Luo X, Liu F, Wang H (2020) New understanding of the damage of SARS-CoV-2 infection outside the respiratory system. Biomed Pharmacother 127:110195
69. Mehta P, McAuley DF, Brown M, Sanchez E, Tattersall RS, Manson JJ (2020) COVID-19: consider cytokine storm syndromes and immunosuppression. Lancet 395:1033–1034
70. Walls AC, Park YJ, Tortorici MA, Wall A, McGuire AT, Veesler (2020) Structure, function, and antigenicity of the SARSCoV-2 Spike Glycoprotein. Cell 180:1–12
71. Wang Y, Wang Y, Chen Y, Qin Q (2020) Coronavirus infections and immune responses. J Med Virol 2020:1–9
72. Wang X, Xu W, Hu G, Xia S, Sun Z, Liu Z, Xie Y, Zhang R, Jiang S, Lu L (2020) SARS-CoV-2 infects T lymphocytes through its spike protein-mediated membrane fusion. Cell Mol Immunol, 1–3
73. Tang B, Bragazzi NL, Li Q, Tang S, Xiao Y, Wu J (2020) An updated estimation of the risk of transmission of the novel coronavirus (2019-nCov). Infect. Dis. Model 5:248–255
74. Wrapp D, Wang N, Corbett KS, Goldsmith JA, Hsieh CL, Abiona O, Graham BS, McLellan JS (2020) Cryo-EM structure of the 2019-nCoV spike in the prefusion conformation. Science 367:1260–1263
75. Guo D (2020) Old weapon for new enemy: drug repurposing for treatment of newly emerging viral diseases. Virol Sin 35:253–255
76. Maxmen A (2020) More than 80 clinical trials launch to test coronavirus treatments. Nature 578:347–348
77. Ali A, Pooya A, Simani L (2020) Central nervous system manifestations of COVID-19: A systematic review. J Neurol Sci 413:116832
78. Ge XY (2013) Isolation and characterization of a bat SARS-like coronavirus that uses the ACE2 receptor. Nature 503:535–538
79. Graham RL, Sparks JS, Eckerle LD, Sims AC, Denison MR (2008) SARS coronavirus replicase proteins in pathogenesis. Virus Res 133:88–10
80. Hurst KR, Koetzner CA, Masters PS (2013) Characterization of a critical interaction between the coronavirus nucleocapsid protein and nonstructural protein 3 of the viral replicase-transcriptase complex. J Virol 87:9159–9172

81. Koyama T, Platt D, Parida L (2020) Variant analysis of COVID-19 genomes, Bull. World Health Organ. Variant analysis of COVID-19 genomes. https://doi.org/10.2471/BLT.20.253591
82. Drosten C, Günther S, Preiser W, Werf S van der, Brodt HR, Becker S, Rabenau H, Panning M, Kolesnikova L, Fouchier RA, Berger A, Burguière A, Cinatl J, Eickmann M, Escriou N, Grywna K, Kramme S, Manuguerra J, Müller S, Rickerts V, Stürmer M, Vieth S, Klenk H, Osterhaus ADME, Schmitz H, Doerr HW (2003) Identification of a novel coronavirus in patients with severe acute respiratory syndrome. N Engl J Med 348:1967–1976
83. Cui J, Li F, Shi ZL (2019) Origin and evolution of pathogenic coronaviruses Nat. Rev Microbiol 17:181–192
84. (a) Morgenstern B, Michaelis M, Baer PC, Doerr HW, Cinatl Jr. J (2005) Ribavirin and interferon-beta synergistically inhibit SARS-associated coronavirus replication in animal and human cell lines. Biochem Biophys Res Commun 326:905–908. (b) Wang M, Cao R, Zhang L, Yang X, Liu J, Xu M, Shi Z, Hu Z, Zhong W, Xiao G (2020) Remdesivir and chloroquine effectively inhibit the recently emerged novel coronavirus (2019-nCoV) in vitro. Cell Res 30:269–271
85. Adhikari S, Meng S, Wu Y (2020) Epidemiology, causes, clinical manifestation and diagnosis, prevention and control of coronavirus disease (COVID-19) during the early outbreak period: a scoping review. Infect Dis Poverty 9:29
86. (a) Agostini ML, Andres EL, Sims AC, Graham RL, Sheahan TP, Lu X, Smith EC, Case JB, Feng JY, Jordan R (2020) Coronavirus susceptibility to the antiviral remdesivir (GS-5734) Is mediated by the viral polymerase and the proofreading exoribonuclease MBIO 9:00221. (b) Yamamoto N, Matsuyama S., Hoshino T, Yamamoto N (2020) Nelfinavir inhibits replication of severe acute respiratory syndrome coronavirus 2 in vitro bioRxiv. https://doi.org/10.1101/2020.04.06.026476
87. Wang M, Cao R, Zhang L, Yang X, Liu J, Xu M, Shi Z, Hu Z, Zhong W, Xiao G (2020) Remdesivir and chloroquine effectively inhibit the recently emerged novel coronavirus (2019-nCoV) in vitro. Cell Res 30:269–271
88. (a) Yamamoto N, Yang R, Yoshinaka Y, Amari S, Nakano T, Cinatl J, Rabenau H, Doerr HW, Hunsmann G, Otaka A, et al (2004) HIV protease inhibitor nelfinavir inhibits replication of SARS-associated coronavirus Biochem Biophys Res Commun 318:719–725. (b) Wang X, Cao R, Zhang H, Liu J, Xu M, Hu H, Li Y, Zhao L, Li W, Sun X, Yang X, Shi Z, Deng F, Hu Z, Zhong W, Wang M (2020) The anti-influenza virus drug, arbidol is an efficient inhibitor of SARS-CoV-2 in vitro. Cell Discov 6:28
89. http://www.sd.chinanews.com/2/2020/0205/70145.html
90. Uhlemann AC, Krishna S (2005) Antimalarial multi-drug resistance in Asia. Curr Top Microbiol Immunol 295:39–53
91. (a) Cortegiani A, Ingoglia G, Ippolito M, Giarratano A, Einav S (2020) A systematic review on the efficacy and safety of chloroquine for the treatment of COVID-19. J Crit Care 57:279–283. (b) Konig MF, Kim AH, Scheetz MH, Graef ER, Liew JW, Simard J, Machado PM, Gianfrancesco M, Yazdany J, Langguth D, Robinson PC (2020) Baseline use of hydroxychloroquine in systemic lupus erythematosus does not preclude SARS-CoV-2 infection and severe COVID-19. Ann Rheum Dis 79:1386–1388
92. Yao X, Ye F, Zhang M, Cui C, Huang B, Niu P, Liu X, Zhao L, Dong E, Song C, Zhan S, Lu R, Li H, Tan W, Liu D (2020) In vitro antiviral activity and projection of optimized dosing design of hydroxychloroquine for the treatment of severe acute respiratory syndrome coronavirus 2 (SARS-CoV-2). Clin Infect Dis 71:732–739. https://doi.org/10.1093/cid/ciaa237
93. Magagnoli J, Narendran S, Pereira F (2020) Outcomes of hydroxychloroquine usage in United States veterans hospitalized with Covid-19. medRxiv. (2020), DOI: https://doi.org/10.1101/2020.04.16.20065920.
94. Dong L, Hu S, Gao J (2020) Discovering drugs to treat coronavirus disease 2019 (COVID-19). Drug Discov Ther 14:58–60
95. Stockman LJ, Bellamy R, Garner P (2006) SARS: systematic review of treatment effects. PLos Med 3:e343

96. https://www.laboratoryequipment.com/563201-COVID-19-Treatment-Update-Remdesivir-Hydroxychloroquine-Leronlimab-Ivermectin-and-More/
97. Fang L, Karakiulakis G, Roth M (2020) Are patients with hypertension and diabetes mellitus at increased risk for COVID-19 infection? Lancet Respir Med 8:e21
98. Gurwitz D (2020) Angiotensin receptor blockers as tentative SARS-CoV-2 therapeutics. Drug Dev Res 81:537–540
99. Shah B, Modi P, Sagar SR (2020) In silico studies on therapeutic agents for COVID-19: Drug repurposing approach. Life Sci 252:117652
100. Richardson P, Griffin I, Tucker C, Smith D, Oechsle O, Phelan A, Stebbing J (2020) Baricitinib as potential treatment for 2019-nCoV acute respiratory disease. Lancet 395:e30–e31
101. Kim UJ, Won EJ, Kee SJ, Jung SI, Jang HC (2016) Combination therapy with lopinavir/ritonavir, ribavirin and interferon-α for Middle East respiratory syndrome. Antivir Ther 21:455–459
102. Kelleni MT (2020) Nitazoxanide/azithromycin combination for COVID-19: a suggested new protocol for early management. Pharmacol Res 157:104874
103. Haffizulla J, Hartman A, Hoppers M (2014) Effect of nitazoxanide in adults and adolescents with acute uncomplicated influenza: a double-blind, randomised, placebo-controlled, phase 2b/3 trial. Lancet Infect Dis 14:609–618
104. Cavalcanti AB, Zampieri FG, Rosa RG, Azevedo LCP, Veiga VC, Avezum A, Damiani LP, Marcadenti A, Kawano-Dourado L, Lisboa T, Junqueira DLM, de Barros e Silva PGM, Tramujas L, Abreu-Silva EO, Laranjeira LN, Soares AT, Echenique LS, Pereira AJ, Freitas FGR, Gebara OCE, Dantas VCS, Furtado RHM, Milan EP, Golin NA, Cardoso FF, Maia IS, Hoffmann Filho CR, Kormann APM, Amazonas RB, Bocchi de Oliveira MF, Serpa-Neto A, Falavigna M, Lopes RD, Machado FR, Berwanger O (2020) Hydroxychloroquine with or without Azithromycin in mild-to-moderate Covid-19. N Engl J Med NEJMoa2019014
105. Gautret P, Lagier JC, Parola P, Hoang VT, Meddeb L, Mailhe M, Doudier B, Courjon J, Giordanengo V, Vieira VE, Dupont HT, Honore S, Colson P, Chabriere E, Scola BL, Rolain JM, Brouqui P, Raoult D (2020) Hydroxychloroquine and azithromycin as a treatment of COVID-19: results of an open-label non-randomized clinical trial. Int J Antimicrob Agents 56:105949
106. Hung IF, Lung K, Tso EY, Liu R, Chung TW, Chu M, Ng Y, Lo J, Chan J, Tam AR, Shum H, Chan V, Wu AK, Sin K, Leung W, Law W, Lung DC, Sin S, Yeung P, Yip CC, Zhang RR, Fung AY, Yan EY, Leung K, Daniel J, Chu AW, Chan W, Ng AC, Lee R, Fung K, Yeung A, Wu T, Chan JW, Yan W, Chan W, Chan JF, Lie AK, Tsang OT, Cheng VC, Que T, Lau C, Chan K, To KK, Yuen K (2020) Triple combination of interferon beta-1b, lopinavir–ritonavir, and ribavirin in the treatment of patients admitted to hospital with COVID-19: an open-label, randomised, phase 2 trial. The Lancet 395:1695–1704. https://doi.org/10.1016/S0140-6736(20)31042-4
107 (a) Banerjee S, Chakravarty AR (2015) Metal complexes of curcumin for cellular imaging, targeting, and photoinduced anticancer activity. Acc Chem Res 48:2075–2083. (b) Sadler PJ, Li H, Sun H (1999) Coordination chemistry of metals in medicine: target sites for bismuth. Coord Chem Rev 185:689–709
108. Imberti C, Zhang P, Huang H, Sadler PJ (2020) New designs for phototherapeutic transition metal complexes. Angew Chem Int Ed 59:61–73
109. García-Gallego S, Serramía MJ, Arnaiz E, Díaz L, Muñoz-Fernández MA, Gómez-Sal P, Ottaviani MF, Gómez R, de la Mata FJ (2011) Transition-metal complexes based on a sulfonate-containing N-donor ligand and their use as HIV antiviral agents. Eur J Inorg Chem 2011:1657–1665
110. Thompson KH, Orvig C (2001) Coordination chemistry of vanadium in metallopharmaceutical candidate compounds. Coord Chem Rev 219:1033–1053
111. D'Cruz OJ, Dong Y, Uckun FM (2003) Potent dual anti-HIV and spermicidal activities of novel oxovanadium(V) complexes with thiourea non-nucleoside inhibitors of HIV-1 reverse transcriptase. Biochem Biophys Res Commun 302:253–264

112. Sun RW, Ma D, Wong EL, Che C (2007) Some uses of transition metal complexes as anticancer and anti-HIV agents. Dalton Trans 43:4884–4892
113. d Aguiar I, d Santos ER, Mafud AC, Annies V, Navarro-Silva MA, Malta VRdS, Gambardella MTdP, Marques FdA, Carlos RM (2017) Synthesis and characterization of Mn(I) complexes and their larvicidal activity against Aedes aegypti, vector of dengue fever. Inorg Chem Commun 84:49–55
114. Song R, Witvrouw M, Schols D, Robert A, Bolzorini J, Clercq E De, Bernodou J, Meunier B (1997) Anti-HIV activities of anionic metalloporphyrins and related compounds. Antivir Chem Chemother 8:85-97
115. Vausselin T, Calland N, Belouzard S, Descamps V, Douam F, Helle F, François C, Lavillette D, Duverlie G, Wahid A, Fénéant L, Cocquerel L, Guérardel Y, Wychowski C, Biot C, Dubuisson J (2013) The antimalarial ferroquine is an inhibitor of hepatitis C virus. Hepatology 58:86–97
116. Biot C, Daher W, Chavain N, Fandeur T, Khalife J, Dive D, Clercq ED (2006) Design and synthesis of hydroxyferroquine derivatives with antimalarial and antiviral activities. J Med Chem 49:2845–2849
117. Wang H, Li Z, Niu J, Xu Y, Ma L, Lu A, Wang X, Qian Z, Huang Z, Jin X, Leng Q, Wang J, Zhong J, Sun B, Meng G (2018) Antiviral effects of ferric ammonium citrate. Cell Discov 4:14
118. Gadhachanda VR, Eastman KJ, Wang Q, Phadke AS, Patel D, Yang W, Marlor CW, Deshpande M, Huang M, Wiles JA (2018) Ferrocene-based inhibitors of hepatitis C virus replication that target NS5A with low picomolar in vitro antiviral activity. Bioorg Med Chem Lett 28:3463–3471
119. Belema M, Nguyen VN, Bachand C, Deon DH, Goodrich JT, James CA, Lavoie R, Lopez OD, Martel A, Romine JL, Ruediger EH, Snyder LB, St. Laurent DR, Yang F, Zhu J, Wong HS, Langley DR, Adams SP, Cantor GH, Chimalakonda A, Fura A, Johnson BM, Knipe JO, Parker DD, Santone KS, Fridell RA, Lemm JA, O'Boyle DR, Colonno RJ, Gao M, Meanwell NA, Hamann LG. J Med Chem 28:2013–2032
120. Asbell PA, Epstein SP, Wallace JA, Epstein D, Stewart CC, Burger MR (1998) Efficacy of cobalt chelates in the rabbit eye model for epithelial herpetic keratitis. Cornea 17:550–557
121. Epstein SP, Pashinsky YY, Gershon D, Winicov I, Srivilasa C, Kristic KJ, Asbell PA (2006) Efficacy of topical cobalt chelate CTC-96 against adenovirus in a cell culture model and against adenovirus keratoconjunctivitis in a rabbit model. BMC Ophthalmol 6:22
122. Chang EL, Simmers C, Knight DA (2010) Cobalt Complexes as antiviral and antibacterial agents. Pharmaceuticals (Basel) 3:1711–1728
123. Saini AK, Kumari P, Sharma V, Mathur P, Mobin SM (2016) Varying structural motifs in the salen based metal complexes of Co(ii), Ni(ii) and Cu(ii): synthesis, crystal structures, molecular dynamics and biological activities Dalton Trans 45:19096–19108
124. Delehanty JB, Bongard JE, Thach DC, Knight DA, Hickey TE, Chang EL (2008) Antiviral properties of cobalt(III)-complexes. Bioorg Med Chem 16:830–837
125. Cígler P, Kožíšek M, Řezáčová P, Brynda J, Otwinowski Z, Pokorná J, Plešek J, Grüner B, Dolečková-Marešová L, Máša M, Sedláček J, Bodem J, Kräusslich HG, Král V, Konvalinka J (2005) From nonpeptide toward noncarbon protease inhibitors: Metallacarboranes as specific and potent inhibitors of HIV protease. Proc Natl Acad Sci USA 102:15394–15399
126. Řezáčová P, Pokorná J, Brynda J, Kožíšek M, Cígler P, Lepšík M, Fanfrlík J, Řezáč J, Šašková KG, Sieglová I, Plešek J, Šícha V, Grüner B, Oberwinkler H, Sedláček J, Kräusslich HG, Hobza P, Král V, Konvalinka J (2009) Design of HIV Protease Inhibitors Based on Inorganic Polyhedral Metallacarboranes. J Med Chem 52:7132–7141
127. García-Gallego S, Jesús Serramía M, Arnaiz E, Díaz L, Muñoz-Fernández MA, Gómez-Sal P, Ottaviani MF, Gómez R, Mata FJDL (2011) Transition-metal complexes based on a sulfonate-containing N-donor ligand and their use as HIV antiviral agents. Eur J Inorg Chem, 1657–1665
128. Rogolino D, Carcelli M, Bacchi A, Compari C, Contardi L, Fisicaro E, Gatti A, Sechi M, Stevaert A, Naesens L (2015) A versatile salicyl hydrazonic ligand and its metal complexes as antiviral agents. J Inorg Biochem 150:9–17

129. Loginova NV, Koval'chuk TV, Polozov GI, Osipovich NP, Rytik PG, Kucherov II, Chernyavskaya AA, Sorokin VL, Shadyro OI (2008) Synthesis, characterization, antifungal and anti-HIV activities of metal(II) complexes of 4,6-di-tert-butyl-3-[(2-hydroxyethyl)thio]benzene-1,2-diol. Eur J Med Chem 43:1536–1542
130. Hunter TM, McNae IW, Simpson DP, Smith AM, Moggach S, White F, Walkinshaw MD, Parsons S, Sadler PJ (2007) Configurations of Nickel-Cyclam antiviral complexes and protein recognition. Chem Eur J 13:40–50
131. Bitu MNA, Hossain MS, Zahid AASM, Zakaria CM, Kudrat-E-Zahan M (2019) Anti-pathogenic activity of Cu(II) complexes incorporating Schiff bases: a short review. Am J Heterocycl Chem 5:11–23
132. Zerda KS, Gerba CP, Goyel SM (1985) Adsorption of viruses to charge-modified silica. Appl Environ Microbial 49:91–95
133. Ishida T (2016) Bacteriolyses of Cu2+ solution on bacterial cell walls/cell membrane and DNA base-pairing damages. Biomed Res Trace Elemt 27:151–161
134. Srivastava A (2009) Antiviral activity of copper complexes of isoniazid against RNA tumor viruses. Resonance 14:754–760
135. Pelosi G, Bisceglie F, Bignami F, Ronzi P, Schiavone P, Carla RM, Casoli C, Pilotti E (2010) Antiretroviral activity of thiosemicarbazone metal complexes. J Med Chem 53:8765–8769
136. Chauhan G, Rath G, Goyal AK (2013) Non-invasive systemic drug delivery through mucosal routes. Nanomedicine, Biotechnol Int J 41:4
137. García-Gallego S, Sánchez Rodríguez J, Luis Jiménez J, Cangiotti M, Francesca Ottaviani M, Muñoz-Fernández MÁ, Gómez R, Mata FJDL (2012) Polyanionic N-donor ligands as chelating agents in transition metal complexes: synthesis, structural characterization and antiviral properties against HIV. Dalton Trans 41:6488–6499
138. Dorotíková S, Kožíšková J, Malček M, Jomová K, Herich P, Plevová K, Briestenská K, Chalupková A, Mistríková J, Milata V, Dvoranová D, Bučinský L (2015) Tetracarboxylato-platinum(IV) complexes featuring monodentate leaving groups—a rational approach toward exploiting the platinum(IV) prodrug strategy. J Inorg Biochem 150:160–173
139. McGuire KL, Hogge J, Hintze A, Liddle N, Nelson N, Pollock J, Brown A, Facer S, Walker S, Lynch J, Harrison RG, Busath DD (2019) Copper complexes as influenza antivirals: reduced zebrafish toxicity. Eng Nanomater - Health Saf. https://doi.org/10.5772/intechopen.88786
140. Kar M, Khan NA, Panwar A, Bais SS, Basak S, Goel R, Sopory S, Medigeshi GR (2019) Zinc chelation specifically inhibits early stages of dengue virus replication by activation of NF-κB and induction of antiviral response in epithelial cells. Zinc chelation specifically inhibits early stages of dengue virus replication by activation of NF-κB and induction of antiviral response in epithelial cells. Front Immunol 10:2347
141. Krenn BM, Gaudernak E, Holzer B, Lanke K, Van Kuppeveld FJM, Seipelt J (2009) Antiviral activity of the zinc ionophores pyrithione and hinokitiol against picornavirus infections. J Virol 83:58–64
142. Velthuis AJ, van den Worm SH, Sims AC, Baric RS, Snijder EJ, van Hemert MJ (2010) Zn2+ inhibits coronavirus and arterivirus RNA polymerase activity in vitro and zinc ionophores block the replication of these viruses in cell culture. PLoS Pathog 6:1001176
143. Arens M, Travis S (2000) Detection of Legionella pneumophila using a real-time PCR hybridization assay. J Clin Microbiol 38:1758–1762
144. Hsu JTA, Kuo C, Hsieh H, Wang Y, Huang K, Lin CPC, Huang P, Chen X, Liang P (2004) Evaluation of metal-conjugated compounds as inhibitors of 3CL protease of SARS-CoV. FEBS Lett 574:116–120
145. Karaküçük-Iyidogan A, Tasdemir D, Oruç-Emre EE, Jan B (2011) Novel platinum(II) and palladium(II) complexes of thiosemicarbazones derived from 5-substitutedthiophene-2-carboxaldehydes and their antiviral and cytotoxic activities. Eur J Med Chem 46:5616–5624
146. Simic V, Kolarevic S, Brceki I, Jeremic D, Vukovic-gacic B (2016) Cytotoxicity and antiviral activity of palladium(II) and platinum(II) complexes with 2-(diphenylphosphino)benzaldehyde 1-adamantoylhydrazone. Turk J Biol 40:661–669

147. Genova P, Varadinova T, Matesanz AI, Marinova D, Souza P (2004) Toxic effects of bis(thiosemicarbazone) compounds and its palladium(II) complexes on herpes simplex virus growth. Toxicol Appl Pharmacol 197:107–112
148. Kovala-Demertzi D, Varadinova T, Genova P, Souza P, Demertzis MA (2007) Platinum(II) and palladium(II) complexes of pyridine-2-carbaldehyde thiosemicarbazone as alternative antiherpes simplex virus agents. Bioinorg Chem Appl 2007:56165
149. Cavicchioli M, Massabni AC, Heinrich TA, Costa-Neto CM, Abrão EP, Fonseca AL, Castellano EE, Corbi PP, Lustri WR, Leite CQF (2010) Pt(II) and Ag(I) complexes with acesulfame: crystal structure and a study of their antitumoral, antimicrobial and antiviral activities. J Inorg Biochem 104:533–540
150. Allardyce CS, Dyson PJ, Ellis DJ, Salter PA, Scopelliti R (2003) Synthesis and characterisation of some water soluble ruthenium(II)–arene complexes and an investigation of their antibiotic and antiviral properties. J Organomet Chem 668:35–42
151. Luedtke NW, Hwang JS, Glazer EC, Gut D, Kol M, Tor Y (2002) Eilatin Ru(II) complexes display anti-HIV activity and enantiomeric diversity in the binding of RNA. Chem Bio Chem 3:766–771
152. Mishra L, Singh AK, Trigun SK, Singh SK, Pandey SM (2004) J Exp Biol 42:660–666
153. Anchuri SS, Gangarapu K, Thota S, Karki SS, Clercq ED, Andrei G, Snoeck R, Balzarini J (2016) Evaluation of hepatoprotective and antioxidant activity of newly synthesized Ho(III) complex. Biointerface Res Appl Chem 6:1491–1496
154. Gunatilleke SS, Barrios AM (2006) Inhibition of lysosomal cysteine proteases by a series of Au(I) complexes: a detailed mechanistic investigation. J Med Chem 49:3933–3937
155. Gunatilleke SS, de Oliveira CAF, McCammon JA, Barrios AM (2008) The ruthenium(II)-arene compound RAPTA-C induces apoptosis in EAC cells through mitochondrial and p53-JNK pathways. J Biol Inorg Chem 13:555–561
156. Okada T, Patterson BK, Ye SQ, Gurney ME (1993) Silencing the alarms: Innate immune antagonism by rotavirus NSP1 and VP3. Virology 192:631–642
157. Sun RWY, Yu WY, Sun H, Che CM (2004) In vitro inhibition of human immunodeficiency virus type-1 (HIV-1) reverse transcriptase by gold(III) porphyrins. Chem Bio Chem 5:1293–1298
158. Fonteh P, Meyer D (2009) Novel gold(i) phosphine compounds inhibit HIV-1 enzymes. Metallomics 1:427–433
159. Fonteh PN, Keter FK, Meyer D (2010) Cellular mechanisms of cadmium toxicity related to the homeostasis of essential metals. Bio. Metals 23:185–196
160. Qasim M, Baipaywad P, Udomluck N, Na D, Park H (2014) Enhanced therapeutic efficacy of lipophilic amphotericin B against Candida albicans with amphiphilic poly(N-isopropylacrylamide) nanogels. Macromol Res 22:1125–1131
161. Gurunathan S, Qasim M, Park C, Yoo H, Kim JH, Hong K (2018) Cytotoxic potential and molecular pathway analysis of silver nanoparticles in human colon cancer cells HCT116. Int J Mol Sci 19:2269
162. Jeyaraj M, Gurunathan S, Qasim M, Kang MH, Kim JH (2019) A comprehensive review on the synthesis, characterization, and biomedical application of platinum nanoparticles. Nanomaterials 9:1719
163. Chakravarthy KV, Bonoiu AC, Davis WG, Ranjan P, Ding H, Hu R, Bowzard JB, Bergey EJ, Katz JM, Knight PR (2010) Gold nanorod delivery of an ssRNA immune activator inhibits pandemic H1N1 influenza viral replication. Proc Natl Acad Sci USA 107:10172–10177
164. Lee MY, Yang JA, Jung HS, Beack S, Choi JE, Hur W, Koo H, Kim K, Yoon SK, Hahn SK (2012) Hyaluronic acid-gold nanoparticle/interferon α complex for targeted treatment of hepatitis C virus infection. ACS Nano 6:9522–9531
165. Halder A, Das S, Ojha D, Chattopadhyay D, Mukherjee A (2018) Highly monodispersed gold nanoparticles synthesis and inhibition of herpes simplex virus infections. Mater Sci Eng C 89:413–421
166. Papp I, Sieben C, Ludwig K, Roskamp M, Böttcher C, Schlecht S, Herrmann A, Haag R (2010) Inhibition of influenza virus infection by multivalent sialic-acid-functionalized gold nanoparticles. Small 6:2900–2906

167. Hu R, Li S, Kong F, Hou R, Guan X, Guo F (2014) Inhibition effect of silver nanoparticles on herpes simplex virus 2. Genet Mol Res 13:7022–7028
168. Lara HH, Ayala-Nuñez NV, Ixtepan-Turrent L, Rodriguez-Padilla C (2010) Mode of antiviral action of silver nanoparticles against HIV-1. J Nanobiotechnol 8:1–10
169. Mohammed Fayaz A, Ao Z, Girilal M, Chen L, Xiao X, Kalaichelvan PT, Yao X (2012) Inactivation of microbial infectiousness by silver nanoparticles-coated condom: a new approach to inhibit HIV- and HSV-transmitted infection. Int J Nanomed 7:5007–5018
170. Mori Y, Ono T, Miyahira Y, Nguyen VQ, Matsui T, Ishihara M (2013) Antiviral activity of silver nanoparticle/chitosan composites against H1N1 influenza a virus. Nanoscale Res Lett 8:93
171. Ghosal K, Sarkar K (2018) Biomedical applications of graphene nanomaterials and beyond. ACS Biomater Sci Eng 4:2653–2703
172. Song Z, Wang X, Zhu G, Nian Q, Zhou H, Yang D, Qin C, Tang R (2015) Virus capture and destruction by label-free graphene oxide for detection and disinfection applications. Small 11:1171–1176
173. Sametband M, Kalt I, Gedanken A, Sarid R (2014) Herpes simplex virus type-1 attachment inhibition by functionalized graphene oxide. ACS Appl Mater Interfaces 6:1228–1235
174. Yang XX, Li CM, Li YF, Wang J, Huang CZ (2017) Synergistic antiviral effect of curcumin functionalized graphene oxide against respiratory syncytial virus infection. Nanoscale 9:16086–16092
175. Du T, Cai K, Han H, Fang L, Liang J, Xiao S (2015) Probing the interactions of CdTe quantum dots with pseudorabies virus. Sci Rep 5:1–10
176. Huang S, Gu J, Ye J, Fang B, Wan S, Wang C, Ashraf U, Li Q, Wang X, Shao L (2019) Benzoxazine monomer derived carbon dots as a broad-spectrum agent to block viral infectivity. J Colloid Interface Sci 542:198–206
177. Antoine TE, Mishra YK, Trigilio J, Tiwari V, Adelung R, Shukla D (2012) Prophylactic, therapeutic and neutralizing effects of zinc oxide tetrapod structures against herpes simplex virus type-2 infection. Antivir Res 96:363–375
178. Antoine TE, Hadigal SR, Yakoub AM, Mishra YK, Bhattacharya P, Haddad C, Valyi-Nagy T, Adelung R, Prabhakar BS, Shukla D (2016) Intravaginal zinc oxide tetrapod nanoparticles as novel immunoprotective agents against genital herpes. J Immunol 196:4566–4575
179. Tejman-Yarden N, Miyamoto Y, Leitsch D, Santini J, Debnath A (2013) A reprofiled drug, auranofin, is effective against metronidazole-resistant Giardia lamblia. Chemother 57:2029–2035
180. Harbut MB, Vilcheze C, Luo X, Hensler ME, Guo H, Yang B, Chatterjee AK, Nizet V, Jacobs JWR, Schultz PG, Wang F (2015) Auranofin exerts broad-spectrum bactericidal activities by targeting thiol-redox homeostasis. Proc Natl Acad Sci USA 112:4453–4458
181. Musib D, Raza MK, Pal M, Roy M (2021) A red light-activable $Mn^{I}(CO)_3$-functionalized gold nanocomposite as the anticancer prodrug with theranostic potential. Appl Organomet 35:e6110
182. Musib D, Banerjee S, Garai A, Soraisam U, Roy M (2018) Synthesis, theory and in vitro photodynamic activities of new copper(II)-histidinito complexes. Chemistry Select 3:2767–2775
183. Musib D, Pal M, Raza MK, Roy M (2020) Photo-physical, theoretical and photo-cytotoxic evaluation of a new class of lanthanide(iii)–curcumin/diketone complexes for PDT application. Dalton Trans 49:10786–10798
184. Marzo T, Messori L, Marzo T, Messori L (2020) A role for metal-based drugs in fighting COVID-19 infection? The case of auranofin. ACS Med Chem Lett 11:1067–1068
185. Rothan HR, Stone S, Natekar J, Kumari P, Arora K, Kumar M (2020) The FDA-approved gold drug Auranofin inhibits novel coronavirus (SARS-COV-2) replication and attenuates inflammation in human cells. bioRxiv preprint. https://doi.org/10.1101/2020.04.14.041228
186. Gil-Moles M, Basu U, Büssing R, Hoffmeister H, Türck S, Varchmin A, Ott I (2020) Gold metallodrugs to target coronavirus proteins: inhibitory effects on the spike-ACE2 interaction and on PLpro protease activity by auranofin and gold organometallics. Chem Eur J 26:15140–15144

187. Milenković DA, Dimić DS, Avdovićac EH, Marković ZS (2020) Several coumarin derivatives and their Pd(II) complexes as potential inhibitors of the main protease of SARS-CoV-2, an in silico approach. RSC Adv 10:5099–35108
188. Haribabu J, Srividya S, Mahendiran D, Gayathri D, Venkatramu V, Bhuvanesh N, Karvembu R (2020) Synthesis of palladium(II) complexes via Michael addition: antiproliferative effects through ROS-mediated mitochondrial apoptosis and docking with SARS-CoV-2. Inorg Chem 59:17109–17122
189. Pal M, Musib D, Roy M (2021) Transition metal complexes as potential tools against SARS-CoV-2: an in silico approach. New J Chem 45:1924–1933
190. https://www.rcsb.org/
191. http://www.rcsb.org/pdb/workbench/workbench.do
192. Yadav M, Dhagat S, Eswari JS (2020) Emerging strategies on in silico drug development against COVID-19: challenges and opportunities. Eur J Pharm Sci 155:105522
193. El Hawary SS, Khattab AR, Marzouk HS, El Senousy AS, Alex MGA, Aly OM, Telebe M, Abdelmohsen UR (2020) In silico identification of SARS-CoV-2 spike (S) protein–ACE2 complex inhibitors from eight Tecoma species and cultivars analyzed by LC-MS. RSC Adv 10:43103–43108
194. Maffucci I, Contini A (2020) In silico drug repurposing for SARS-CoV-2 main proteinase and spike proteins. J Proteome Res 19:4637–4648
195. Sportelli MC, Izzi M, Kukushkina EA, Hossain SI, Picca RA, Ditaranto N, Cioffi N (2020) can nanotechnology and materials science help the fight against SARS-CoV-2? Nanomaterials (Basel) 4:802

Chapter 2
Nanocarriers as Drug Delivery Vectors

Joydeep Biswas and Bandita Datta

1 Introduction

The emerging field of nanobiotechnology attempts to bring together the advantages of nanotechnology and biotechnology to produce better materials for applications in biology [1, 2]. Most of the fundamental constituents of the biological systems may be classified as nanomaterials since at least one dimension falls in the 1–100 nm regimes that fall in the size domains of nanomaterials. Thus, ribonucleic acid (RNA) or deoxyribonucleic acid (DNA)—the data storage/translation systems in cells—is about a nanometer thick and can extend to many microns. Similarly, most of the proteins that carry out the metabolic functions of the cells are of the nanometer size range. So, these systems are bonafide nanomaterials.

Natural biological systems come with a lot of desirable qualities. These include high specificity of intra- and intermolecular interactions between the biomolecules (e.g. interactions between nucleotide base-pairs or the same between avidin and biotin proteins). These attractive interactions of such a magnitude can be utilized for developing appropriate nanoparticle assembly. Another advantage is the uniformity of biological systems. Thus, bacteria from the same colony are of nearly uniform size. Biological systems also often have anisotropic shapes. For instance, the cilia or flagella of microorganisms adopt a tubular body. These characteristics of biological specimens and the anisotropic forms of many living systems were indeed exploited for templating the syntheses of nanomaterials and achieving their self-assembly [3–8].

J. Biswas (✉)
Department of Chemistry, Sikkim Manipal Institute of Technology, Sikkim Manipal University, East Sikkim, Majitar 737136, India
e-mail: joydeep.biswas@smit.smu.edu.in

B. Datta
Department of Chemistry, Amity Institute of Applied Science, Amity University, Kolkata 700156, India

B. P. Swain (ed.), *Nanostructured Biomaterials*, Materials Horizons: From Nature to Nanomaterials, https://doi.org/10.1007/978-981-16-8399-2_2

Recently, there has been a strong urge to design and develop nanomaterials for utilization in biological applications [9, 10]. Progress in research based on polymers has guided the fabrication of numerous vectors for drug delivery. Supramolecular assemblies based on polymers and dendritic molecules are thoroughly investigated for drug and gene delivery [11]. Solid lipid-based nanocarriers are emerging delivery vectors owing to the ease of fabrications, scale-up ability, biodegradability and biocompatibility of preparation ingredients. Solid lipid-based nanoparticles are appropriate nanocarriers for both hydrophobic and hydrophilic therapeutic agents. [12]. Solid lipid-based nanocarriers are efficient drug delivery vectors for delivering several therapeutically essential active components from small drug molecules to macromolecules (e.g. polypeptides and even nucleic acids) [13].

The drug must be attached to the nanocarrier first for the nanocarriers to perform as a drug delivery vector. This step is essential and there are numerous techniques of achieving this: the drugs can be encapsulated in the nanocarriers, or the drug perhaps bonded covalently or adsorbed to the nanocarrier surface. There are definite benefits to each technique. For example, in covalent bonding, the pharmaceuticals are linked to the nanocarriers employing recognition ligands. In addition, the implementation of covalent bonding provides the capability to regulate the number of pharmaceutical agents connected to the nanocarrier.

2 Nanocarriers Delivery to the Specific Site

Nanocarriers-based drugs delivery to the specific site is frequently accomplished through active or passive strategies. Active strategies comprise altering the physical conditions, i.e. pH, magnetism and temperature, to acquire the nanocarriers to specific regions [14]. Passive strategies include changing the enhanced permeability and retention (EPR) effect [15]. For instance, minute nanocarrier specially assembles in tumours because of the EPR effect of tumours. Therefore, the nanocarriers must not gather inside the biological cells for a longer time, as they might influence natural physiological activities. Nevertheless, minute nanocarriers could also be increased reactive because of their elevated surface area and possibly enhanced cytotoxic.

3 Classifications of Nanocarriers

3.1 Liposomes

Liposomes are the first kind of nanocarriers and are about 20 nm to several micrometres in dimension. Liposomes are sphere-shaped and comprise mostly steroids and phospholipids. Liposomes are tailored readily by dispersing lipid molecules in highly

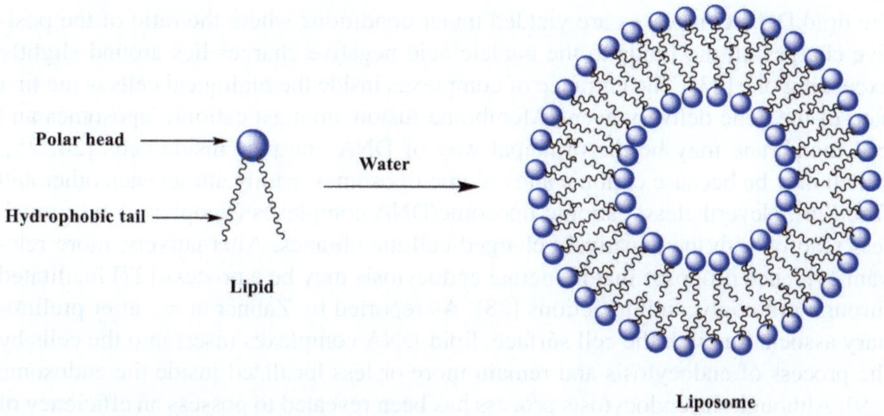

Fig. 1 Schematic of the formation of liposome

pure water (Fig. 1). A pharmaceutical agent is often encapsulated within the liposome. It is often successively released from the pharmaceutical agent by altering physical factors, viz., exterior surroundings, osmotic gradient and pH.

Moreover, various surface alterations increase the half-life of the liposomes. For instance, the addition of polyethylene glycol (PEG) enhances the longevity of liposomes by inhibiting recognition by phagosomes. Correspondingly, PEG is covalently bonded with phosphatidylethanolamine (PE) biological lipid to yield PEG-PE conjugates (Fig. 2), which have been reported to be non-toxic and potentially employed to precisely target the nanocarrier to the mitochondria [16].

The utility of nanocarriers in physiological applications has been recently revealed. For example, appropriately designed nanoparticles exhibited effective DNA recognition behaviour [17]. Liposomes also exhibited efficient DNA transfection ability [18, 19]. Further, these have been employed in the suppression of DNA transcription [20]. Issues on nanoparticle toxicity have also been addressed [21]. The variation of surface charge on nanoparticle interaction with biosystems has also been developed very recently [22].

The mechanism of cationic liposome-mediated gene delivery is not understood at a molecular level. Gene delivery seems to be generally extreme efficient when

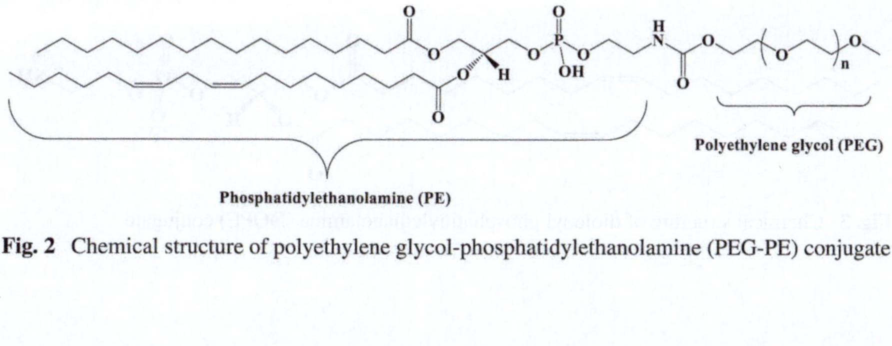

Fig. 2 Chemical structure of polyethylene glycol-phosphatidylethanolamine (PEG-PE) conjugate

the lipid:DNA complexes are yielded under conditions where the ratio of the positive charges on the lipids to the nucleic acid negative charges lies around slightly exceeding one [23]. The entrance of complexes inside the biological cells is the first step of the gene delivery event. Membrane fusion amongst cationic liposomes and cell membrane may be the principal way of DNA entrance inside cells [24, 25], which may be because cationic and anionic liposomes readily attract each other and fuse [26]. Nevertheless, cationic liposome/DNA complexes (lipoplexes) fuse much less readily with the negatively charged cell membranes. Alternatively, more relevant evidence proposes that moderate endocytosis may be a process [27] facilitated through proteoglycan interactions [28]. As reported by Zabner et al., after preliminary association with the cell surface, lipid:DNA complexes insert into the cells by the process of endocytosis and remain more or less localized inside the endosome [29]. Although the endocytosis process has been revealed to possess an efficiency of up to 80%, the observed lipid-mediated gene expression is often not more than 50%, which shows the inefficiency of the escape of the nucleic acid from the endosomal compartment. When nucleic acids succeed in escaping from the endosome, they do so almost immediately after the endocytosis has taken place (i.e. the early endosome escape) [30, 31].

Dioleoyl phosphatidylethanolamine (DOPE) (Fig. 3) is mixed with a cationic liposome as a helper lipid to form the transfecting recipe. DOPE helps release DNA from the lipoplexes encapsulated in the endosome by inducing polymorphic changes in the lipid phase below biological conditions [32]. It is well known that DOPE containing mixed lipid assemblies can convert to an inverted hexagonal phase at room temperature and under biological pH (~7.4) from the lamellar liquid crystalline characteristics of most biological membranes [32]. Furthermore, it is generally understood that when two membranes fuse, the lipids in the adjoining bilayer membranes adopt a hexagonal phase [33]. Therefore, the inclusion of DOPE in the transfecting recipe provides a means for endosome disruption, thereby triggering DNA release from the lipoplexes.

Furthermore, coulombic interactions amongst the cationic liposome and endosomal membranes prompt the flip-flop of the anionic lipids from the monolayer of the endosome membranes that faces the cytoplasm [34]. As a result, these membranes laterally diffuse into the complex, forming charge neutralized ion-pairs of lipids [35]. Consequently, ionic interactions between the DNA and liposomes are interrupted and permit the DNA to diffuse spontaneously into the cytoplasm (Fig. 4) [36]. Finally,

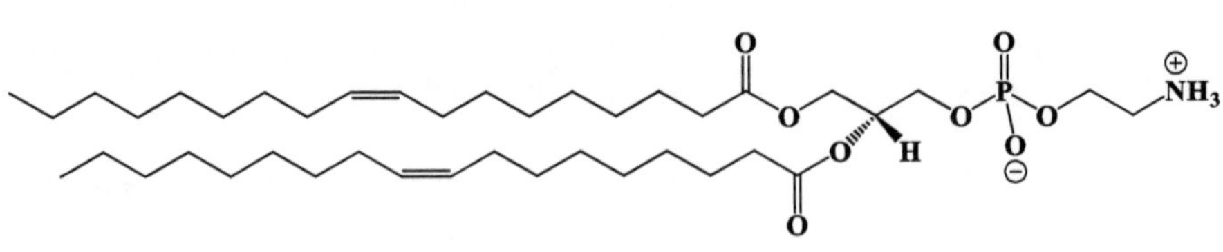

Fig. 3 Chemical structure of dioleoyl phosphatidylethanolamine (DOPE) conjugate

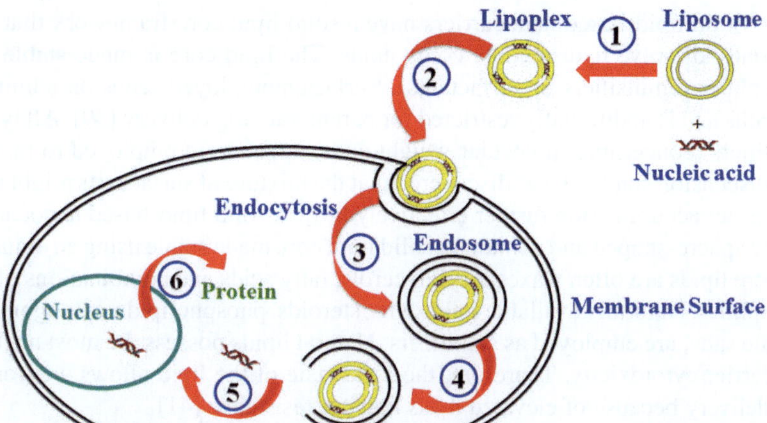

Fig. 4 Schematic of the primary mechanism of the nucleic acid delivery into the cells with lipid suspensions. Steps involved: **(1)** liposome-nucleic acid complexation; **(2)** fusion with the cell membrane; **(3)** cellular uptake of the lipoplexes; **(4)** the nucleic acid decomplexes from the lipoplexes and the nucleic acid is unconfined into the cytoplasm; **(5)** nuclear uptake of the nucleic acid and **(6)** transcription expression of the protein

after releasing the nucleic acid into the cytoplasm from the lipoplexes, the nucleic acid sequence enters into the nucleus initiating the protein synthesis.

Researchers investigated the effect of DOPE in the membrane of cholesterol derived lipids and observed that the cholesterol derived lipids stabilize DOPE in the Lα phase [37]. In addition, DOPE facilitated cholesterol derived lipids from surviving in a bilayer state. Therefore, it was imagined that probably ice was locally formed during the freezing phase. Still, the surroundings of phosphate groups of DOPE were influenced with roughly hydration states [38], which gave rise to a structure of the mixed liposomes, increasing the transfection to a significant level.

3.2 *Solid Lipid-Based Nanocarriers*

Lipid-based nanocarriers comprise solid lipid-based nanocarriers, lipid–drug conjugates and nanostructured lipid carriers. The solid lipid-based nanocarriers are based on solid lipids and afford decent physical stability and tolerability. On the other hand, lipid-drug conjugates and nanostructured lipid carriers amalgam solid and liquid lipids with enhanced load capacity and diminished pharmaceutical agent expulsion characteristics.

A solid lipid-based nanocarrier is usually sphere-shaped, possessing a size range of 10–1000 nm. The idiom lipid is utilized herein in a greater context, comprising steroids, fatty acids, monoglycerides, diglycerides, triglycerides and waxes. Solid lipid-based nanocarriers have a solid lipid core framework that could potentially dissolve hydrophobic compounds. The lipid core is made stable using amphiphiles (emulsifiers or surfactants). Surfactantemployed relies on administration paths and is additionally restricted for parenteral drug delivery [39]. All types of emulsifiers (concerning molecular weight and charge) were employed to make the lipid suspension stable. It was discovered that the mixture of surfactants might inhibit nanocarrier accumulation further effectively [40]. A solid lipid-based nanocarrier is usually sphere-shaped and contains a solid lipid core made stable using an emulsifier. The core lipids are often waxes, acylglycerols, fatty acids and combinations of these amphiphiles. Naturally available lipids, viz., steroids, phospholipids, sphingomyelins and bile salts, are employed as stabilizers. Natural lipids possess the most negligible nanocarrier cytotoxicity. Therefore, the solid state of the lipid allows well-ordered drug delivery because of elevated mass transfer resistance [41].

Solid lipid-based nanocarriers can play a role as the fundamental for parenteral and oral drug delivery vectors. Solid lipid-based nanocarriers amalgamate the benefits of lipid emulsion and polymer-based nanoparticle entities, disabling the temporal and in vivo durability problems that trouble the traditional and polymeric nanoparticles drug delivery methodologies [42]. It was suggested that solid lipid-based nanocarriers combine several benefits over the traditional colloidal nanocarriers, i.e. an amalgamation of hydrophilic and hydrophobic pharmaceutical agents achievable, avoiding organic solvents, no cytotoxicity of the nanocarrier, probability of controlled drug release, targeting, amplified stability and no issues concerning large scale fabrication [13]. The latest research has revealed the usage of these nanocarriers as a stage for the delivery of micronutrient iron orally by integrating$FeSO_4$salt in a lipid framework comprised of octadecanoic acid [43]. In addition, carvedilol-filled solid lipid-based nanocarriers have been fabricated employing a mechanical procedure, i.e. hot high-pressure homogenization for oral delivery utilizing a surfactant (i.e. poloxamer 188) and a lipid (i.e. compritol 888) (Fig. 5) [44]. An alternative instance of drug delivery employing solid lipid-based nanocarrier could have been solid lipid suspended in aqueous media fabricated to trap pharmaceutical agents inside its assembly. However, upon indigestion, the solid lipid-based nanocarriers

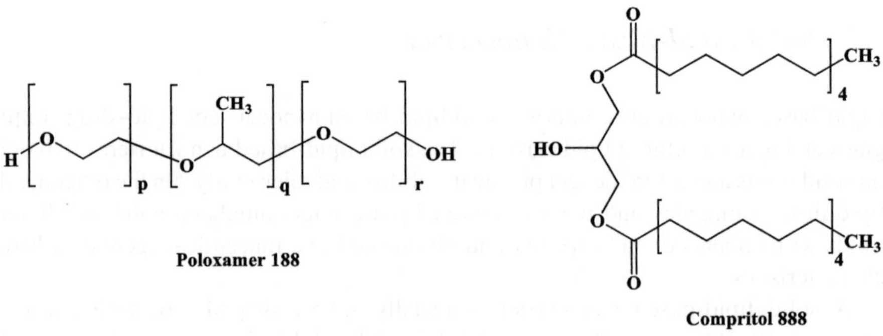

Fig. 5 Chemical structure of poloxamer 188 and compritol 888

are susceptible to gastric and intestinal acids that solubilize the solid lipid-based nanocarriers and release the pharmaceutical agents into the system [45].

Numerous nanostructured formulations were employed for ocular (related to the eyes or vision) drug delivery and generated precise encouraging outcomes. Solid lipid-based nanocarriers have been observed as a prospective drug nanocarrier formulation ever since the decade 1990s. Solid lipid-based nanocarriers do not demonstrate cytotoxicity as they are tailored from biological lipids. Therefore, solid lipid-based nanocarriers are exclusively beneficial in ocular drug delivery. Solid lipid-based nanocarriers could boost pharmaceutical agents' corneal absorption and enhance the ocular pharmacokinetics (mainly bioavailability) of hydrophobic and hydrophilic pharmaceuticals [46]. In addition, they benefit from allowing autoclave sterilization, an essential step en route for designing ocular preparations [47].

Benefits of solid lipid-based nanocarriers comprise biological lipids use (which reduces the risk of severe and long-lasting cytotoxicity), avoiding organic solvents, a potentially broad utilization range (e.g. intravenous, oral and dermal administrations) and high-pressure homogenization as a conventional fabrication technique. Furthermore, enhanced bioavailability, shielding of sensitive pharmaceutical agents from the light, water, exterior surroundings and surprisingly controlled release features have been affirmed via incorporating inadequately water-soluble drug molecules in the solid lipid-based nanocarrier assembly. Additionally, solid lipid-based nanocarriers can transport equally hydrophilic and hydrophobic pharmaceutical agents and are more reasonable than surfactant or polymer-based nanocarriers [48].

3.3 *Polymer-Based Nanocarriers*

Polymer-based nanocarriers are developed from artificial polymers, which range from 10 to 100 nm. Polymer-based nanocarriers have been classified into biodegradable and non-biodegradable, which is because pharmaceutical agents are often adjoined on the exterior surface of the nanocarriers through the polymerization reactions and are frequently released by diffusion or desorption inside the specific target tissue. Biodegradable nanocarriers could perhaps experience breakdown inside the body to offer glycolic acid and lactic acid. Polymer-based nanocarriers are additionally stable in blood, non-thrombogenic and most importantly, non-toxic.

Functionalized amphiphilic polymers currently play a fundamental role in drug delivery skill progression by furnishing the controlled release of pharmaceuticals [49]. The release of drugs could possibly occur in several means: continuous dosing over an extended time, cyclic dosage or tunable release of the pharmaceutical agent [50]. Frequently, such polymers spontaneously fabricate nanocarriers wrapping hydrophilic as well as hydrophobic pharmaceutical agents. Furthermore, these polymers are typically employed as nanocarriers to stabilize and transport water-insoluble or sensitive therapeutic agents (e.g. short single strands of synthetic DNA or RNA) to their specific targets.

Furthermore, therapeutic utilization, chemically modified amphiphilic polymers, e.g. hyperbranched bowl-shape polymers, can function as blocking agents to diminish non-specific binding in diagnostic assays [51]. Other suitable chemically modified polymers function as non-covalent binders for definite solids, liquids or gases.

Maximum drugs require a suitable preparation for their fruitful utilization as medicines. The motives for preparing could also be the pharmaceutical agents' aqueous insolubility, sensitivity, instability, fast clearance or undefined distribution within the human body. Therefore, mainly amphiphilic polymers are frequently employed as critical additives to prepare the drug delivery vectors properly. For instance, these polymers are polyglycolic acids (PGA) or polylactic acids (PLA) or their mixtures and also polyethylene glycol (PEG) block-copolymers with PLA, PGA and poly(amino acid)s (PAA) [52]. Alternate appropriate polymers are linear or branched polyamines, viz., linear poly(ethylene imine)s (LPEI) or branched poly(ethylene imine)s (BPEI) (Fig. 6). Currently, bowl-shaped dendrimer-like polymers, viz. hyperbranched polyglycerols (HPG), are incorporated into the class of amphiphilic functional polymers for drug delivery application [53].

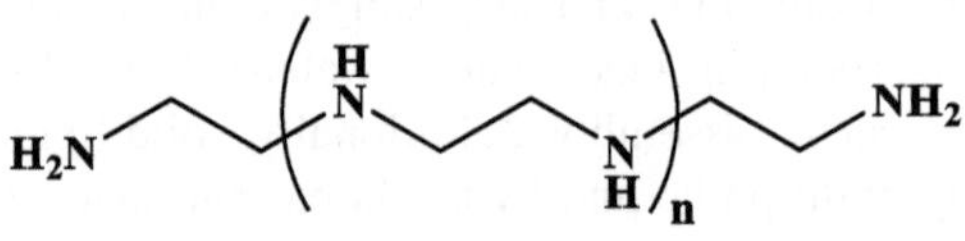

Linear poly(ethylene imine)s (LPEI)

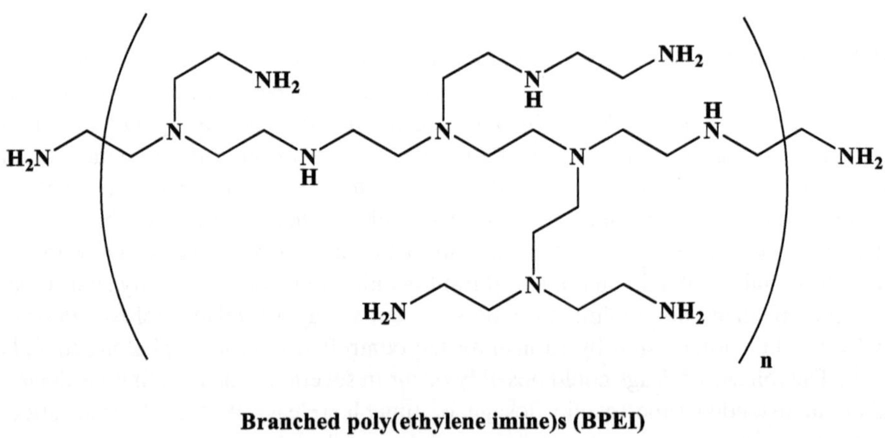

Branched poly(ethylene imine)s (BPEI)

Fig. 6 Chemical structure of linear poly(ethylene imine)s (LPEI) or branched poly(ethylene imine)s (BPEI)

3.4 Dendrimer-Based Nanocarriers

Dendrimer-based nanocarriers comprise the following structural characteristics: core, dendrons (dendritic structure) and hydrophilic surface-active ligands. The dendrimers are linked to the centre and the sort of surface-active groups determines the properties of the nanocarriers. In addition, various ligands can connect to the exterior circumference of dendrimers, viz., vitamins, polyethylene glycols, peptides, antimicrobial agents or antibodies. These inclusions alter the physical and chemical characteristics of the dendritic molecules.

Usages of dendrimers typically comprise adjoining other molecular moieties to the dendrimer exterior circumference, serving as pharmaceutically active molecules, targeting components, affinity ligands, radioligands, sensing agents (viz., dye molecules) or imaging agents. Dendrimers possess the revolutionary prospects for these accomplishments due to their structure, resulting in a multivalent entity. Specifically, one dendrimer molecule contains numerous potential sites to adjoin to an active ligand. Therefore, scientists targeted to employ the lyophilic exterior surface of the dendritic molecules to perform reactions (primarily photochemistry) that yield surface activated derivatives, which are very difficult to synthesize. Furthermore, dendrimers containing groups, viz., carboxylic acid and phenol at the exterior surface were reported to be water-soluble and fabricated to incorporate their function in drug delivery and perform chemical reactions in their interiors [54]. Hence, it allows scientists to adjoin both targeting ligands and pharmaceutical agents to the same dendritic molecule, reducing the adverse side effects of drugs on robust biological cells [55].

The novel category of dendritic molecules was considered a significant contender for host–guest interaction since their entry in the middle of the 1980s [56]. Dendritic molecules with a hydrophilic exterior surface and hydrophobic core demonstrate micelle-like features and act as nanocontainers for hydrophobic pharmaceutical agents [57]. Moreover, dendritic molecules are potentially employed as solubilizing agents. In 1985, Newkome et al. suggested the utilization of dendritic molecule, [27]-arborol (Fig. 7) as unimolecular micelles [58]. This resemblance emphasized the usefulness of dendrimers as solubilizing agents [59]. However, most drug molecules accessible in pharmaceutical manufacturing are hydrophobic and this characteristic specifically generates significant preparation complications. This disadvantage of pharmaceutical agents could be enhanced by the dendritic skeleton, which could perhaps be employed to encapsulate on dissolving the medicines due to the potential of such structures to join in wide-ranging hydrogen bonding with water molecules [60–65].

Dendrimer-specialized research laboratories worldwide are tirelessly attempting to employ dendrimer's dissolving characteristic in their approach to discover dendritic molecules as target specific nanocarrier and drug delivery vectors [66–70]. However, for dendrimers to be employed in medicinal applications, they need to overcome the critical regulatory obstacles to succeed in the market. One dendritic structure tailored to accomplish this is often poly(ethoxyethylglycinamide) (PEE-G)

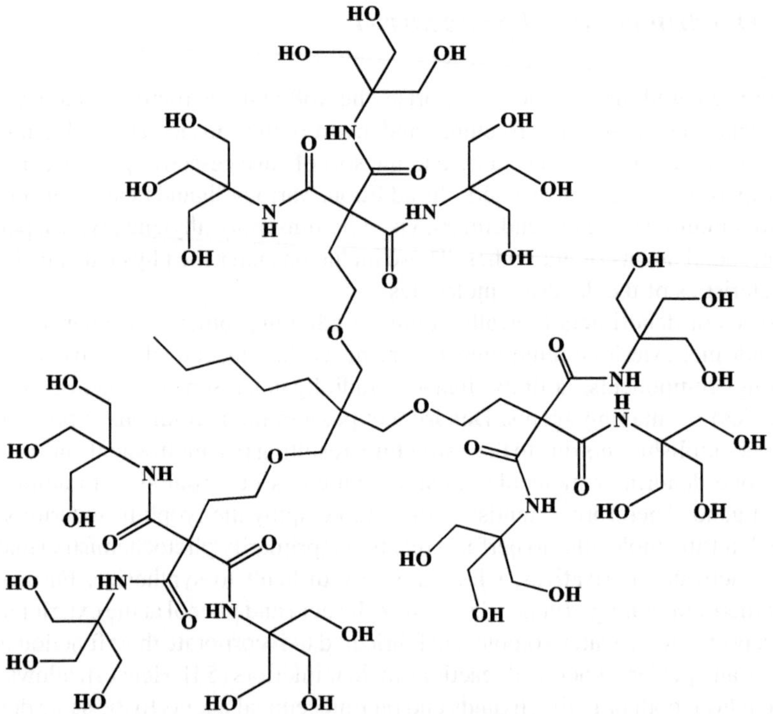

Fig. 7 Chemical structure of dendritic molecule, [27]-arborol

[71, 72]. This dendritic structure was tailored and revealed to possess greater HPLC purity, solubility in water, high stability and low intrinsic cytotoxicity.

4 Conclusion

Herein, we described concisely various nanocarriers for drug delivery to a specific site. Liposomes, solid lipids, polymers and dendrimers are potential nanocarriers for drug delivery described here. Nanocarriers-based drugs delivery to the specific site is frequently accomplished through active or passive strategies. Liposomes are the first kind of nanocarriers. A drug is often encapsulated within the liposome and is often successively released from the drug by altering physical factors. Liposomes are also efficient to deliver genetic materials to a specific site. Solid lipid-based nanocarriers can transport equally hydrophilic and hydrophobic drugs and are more reasonable than surfactant or polymer-based nanocarriers. Solid lipid-based nanocarriers are efficient drug delivery vectors for delivering several therapeutically essential active components from small drug molecules to macromolecules (e.g. polypeptides and even nucleic acids). Supramolecular assemblies based on polymers and dendritic molecules are thoroughly investigated for drug and gene delivery. Functionalized amphiphilic polymers currently play a fundamental role in drug delivery skill progression by furnishing the controlled release of drugs. Polymers are typically employed as nanocarriers to stabilize and transport water-insoluble or sensitive therapeutic agents (e.g. short single strands of synthetic DNA or RNA) to their specific targets. Dendritic molecules with a hydrophilic exterior surface and hydrophobic core demonstrate micelle-like features and act as nanocontainers for hydrophobic drugs. In future, it is assumed that the fabrication of suitable nanocarriers for efficient drug delivery with minimal cytotoxicity will enhance the use of nanocarriers in modern medicine.

References

1. Niemeyer CM (2001) Nanoparticles, proteins, and nucleic acids: biotechnology meets materials science. Angew Chem Int Ed 40:4128–4158
2. Niemeyer CM (2004) Semi-synthetic DNA–protein conjugates: novel tools in analytics and nanobiotechnology. Biochem Soc Trans 32:51–53
3. Blum AS, Soto CM, Wilson CD, Cole JD, Kim M, Gnade B, Chatterji A, Ochoa WF, Lin T, Johnson JE, Ratna BR (2004) Cowpea mosaic virus as a scaffold for 3-D patterning of gold nanoparticles. Nano Lett 4:867–870
4. Losic D, Mitchell JG, Voelcker NH (2005) Complex gold nanostructures derived by templating from diatom frustules. Chem Commun, 4905–4907
5. Petty JT, Zheng J, Hud NV, Dickson RM (2004) DNA-templated Ag nanocluster formation. J Am Chem Soc 126:5207–5212
6. Fu A, Micheel CM, Cha J, Chang H, Yang H, Alivisatos AP (2004) Discrete nanostructures of quantum dots/Au with DNA. J Am Chem Soc 126:10832–10833
7. Behrens S, Rahn K, Habicht W, Bohm K-J, Rosner H, Dinjus E, Unger E (2002) Nanoscale particle arrays induced by highly ordered protein assemblies. Adv Mater 14:1621–1625
8. Mirkin CA, Storhoff JJ (1999) Programmed materials synthesis with DNA. Chem Rev 99:1849–1862
9. Wang EC, Wang AZ (2014) Nanoparticles and their applications in cell and molecular biology. Integr Biol 6:9–26
10. Siddique S, Chow JCL (2020) Application of nanomaterials in biomedical imaging and cancer therapy. Nanomaterials 10:1700(1)–1700(40)
11. Pillai O, Panchagnula R (2001) Polymers in drug delivery. Curr Opin Chem Biol 5:447–451
12. Ghasemiyeh P, Mohammadi-Samani S (2018) Solid lipid nanoparticles and nanostructured lipid carriers as novel drug delivery systems: applications, advantages and disadvantages. Res. Pharm. Sci. 13:288–303
13. Puri A, Loomis K, Smith B, Lee J-H, Yavlovich A, Heldman E, Blumenthal R (2009) Lipid-based nanoparticles as pharmaceutical drug carriers: from concepts to clinic. Crit Rev Ther Drug Carrier Syst 26:523–580
14. Patra JK, Das G, Fraceto LF, Campos EVR, Rodriguez-Torres MDP, Acosta-Torres LS, Diaz-Torres LA., Grillo R, Swamy MK, Sharma S, Habtemariam S, Shin H-S (2018) Nano based drug delivery systems: recent developments and future prospects. J Nanobiotechnol 16:71(1)–71(33)
15. Golombek SK, May J-N, Theek B, Appold L, Drude N, Kiessling F, Lammers T (2018) Tumor targeting via EPR: strategies to enhance patient responses. Adv Drug Deliv Rev 130:17–38
16. Biswas S, Dodwadkar NS, Deshpande PP, Torchilin VP (2012) Liposomes loaded with paclitaxel and modified with novel triphenylphosphonium-PEG-PE conjugate possess low toxicity, target mitochondria and demonstrate enhanced antitumor effects *in vitro* and *in vivo*. J Control Release 159:393–402

17. Mahtab R, Harden HH, Murphy CJ (2000) Temperature- and salt-dependent binding of long DNA to protein-sized quantum dots: thermodynamics of "inorganic protein"–DNA interactions. J Am Chem Soc 122:14–17
18. Balazs DA, Godbey WT (2011) Liposomes for use in gene delivery. J Drug Deliv, 326497(1)–326497(12)
19. Rodrigues BDS, Banerjee A, Kanekiyo T, Singh J (2019) Functionalized liposomal nanoparticles for efficient gene delivery system to neuronal cell transfection. Int J Pharm 566:717–730
20. McIntosh CM, Esposito EA, Boal AK, Simard JM, Martin CT, Rotello VM (2001) Inhibition of DNA transcription using cationic mixed monolayer protected gold clusters. J Am Chem Soc 123:7626–7629
21. Goodman CM, McCusker CD, Yilmaz T, Rotello VM (2004) Toxicity of gold nanoparticles functionalized with cationic and anionic side chains. Bioconjugate Chem 15:897–900
22. Hong R, Emrick T, Rotello VM (2004) Monolayer-controlled substrate selectivity using noncovalent enzyme–nanoparticle conjugates. J Am Chem Soc 126:13572–13573
23. Gershon H, Ghirlando R, Guttman SB, Minsky A (1993) Mode of formation and structural features of DNA-cationic liposome complexes used for transfection. Biochemistry 32:7143–7151
24. Struck DK, Hoekstra D, Pagano RE (1981) Use of resonance energy transfer to monitor membrane fusion. Biochemistry 20:4093–4099
25. Smith JG, Walzem RL, German JB (1993) Liposomes as agents of DNA transfer. Biochim Biophys Acta 1154:327–340
26. Duzgunes N, Goldstein JA, Friend DS, Felgner PL (1989) Fusion of liposomes containing a novel cationic lipid, N-[2,3-(dioleyloxy)propyl]-N, N, N-trimethylammonium: induction by multivalent anions and asymmetric fusion with acidic phospholipid vesicles. Biochemistry 28:9179–9184
27. Farhood H, Serbina N, Huang L (1995) The role of dioleoylphosphatidylethanolamine in cationic liposome mediated gene transfer. Biochim Biophys Acta 1235:289–295
28. Mislick KA, Baldeschweiler JD (1996) Evidence for the role of proteoglycans in cation-mediated gene transfer. Proc Natl Acad Sci USA 93:12349–12354
29. Zabner J, Fasbender AJ, Moninger T, Poellinger KA, Welsh MJ (1995) Cellular and molecular barriers to gene transfer by a cationic lipid. J Biol Chem 270:18997–19007
30. Zelphati O, Szoka FC Jr (1996) Intracellular distribution and mechanism of delivery of oligonucleotides mediated by cationic lipids. Pharm Res 13:1367–1372
31. Friend DS, Papahadjopoulos D, Debs RJ (1996) Endocytosis and intracellular processing accompanying transfection mediated by cationic liposomes. Biochim Biophys Acta 1278:41–50
32. Verkleij AJ (1984) Lipidic intramembranous particles. Biochim Biophys Acta 779:43–63
33. Allen TM, Hong K, Papahadjopoulos D (1990) Membrane contact, fusion and hexagonal (HII) transitions in phosphatidylethanolamine liposomes. Biochemistry 29:2976–2985
34. Pires P, Simões S, Nir S, Gaspar R, Düzgünes N, de Lima MCP (1999) Interaction of cationic liposomes and their DNA complexes with monocytic leukemia cells. Biochim Biophys Acta 1418:71–84
35. Pöyry S, Vattulainen I (2016) Role of charged lipids in membrane structures - Insight given by simulations. Biochim Biophys Acta 1858:2322–2333
36. Xu Y, Szoka FC Jr (1996) Mechanism of DNA Release from Cationic Liposome/DNA Complexes Used in Cell Transfection. Biochemistry 35:5616–5623
37. Fenske DB, Cullis PR (1992) Chemical exchange between lamellar and non-lamellar lipid phases. a one- and two-dimensional ^{31}P-NMR study. Biochim Biophys Acta 1108:201–209
38. Sanderson PW, Williams WP, Cunningham BA, Wolfe DH, Lis LJ (1993) The effect of ice on membrane lipid phase behaviour. Biochim Biophys Acta 1148:278–284
39. Hippalgaonkar K, Majumdar S, Kansara V (2010) Injectable Lipid Emulsions-Advancements, Opportunities and Challenges. AAPS PharmSciTech 11:1526–1540
40. Raya SA, Saaid IM, Ahmed AA, Umar AA (2020) A critical review of development and demulsification mechanisms of crude oil emulsion in the petroleum industry. J Pet Explor Prod Technol 10:1711–1728

41. ud Din F, Aman W, Ullah I, Qureshi OS, Mustapha O, Shafique S, Zeb A (2017) Effective use of nanocarriers as drug delivery systems for the treatment of selected tumors. Int J Nanomedicine 12:7291–7309
42. García-Pinel B, Porras-Alcalá C, Ortega-Rodríguez A, Sarabia F, Prados J, Melguizo C, López-Romero JM (2019) Lipid-based nanoparticles: application and recent advances in cancer treatment. Nanomaterials 9:638(1)–638(23).
43. Zariwala MG, Elsaid N, Jackson TL, López FC, Farnaud S, Somavarapu S, Renshaw D (2013) A novel approach to oral iron delivery using ferrous sulphate loaded solid lipid nanoparticles. Int J Pharm 456:400–407
44. Shah MK, Madan P, Lin S (2014) Preparation, evaluation and statistical optimization of carvedilol-loaded solid lipid nanoparticle for lymphatic absorption via oral administration. Pharm Dev Technol 19:475–485
45. Pandey R, Sharma S, Khuller GK (2005) Oral solid lipid nanoparticle-based antitubercular chemotherapy. Tuberculosis 85:415–420
46. Arana L, Salado C, Vega S, Aizpurua-Olaizola O, de la Arada I, Suarez T, Usobiaga A, Arrondo JLR, Alonso A, Goñi FM, Alkorta I (2015) Solid lipid nanoparticles for delivery of Calendula officinalis extract. Colloids Surf B Biointerfaces 135:18–26
47. Seyfoddin A, Shaw J, Al-Kassas R (2010) Solid lipid nanoparticles for ocular drug delivery. Drug Deliv 17:467–489
48. Mukherjee S, Ray S, Thakur RS (2009) Solid lipid nanoparticles: a modern formulation approach in drug delivery system. Indian J Pharm Sci 71:349–358
49. Patra JK, Das G, Fraceto LF, Campos EVR, Rodriguez-Torres MDP, Acosta-Torres LS, Diaz-Torres LA, Grillo R, Swamy MK, Sharma S, Habtemariam S, Shin HS (2018) Nano based drug delivery systems: recent developments and future prospects. J Nanobiotechnol 16:71(1)–71(33)
50. Kamaly N, Yameen B, Wu J, Farokhzad OC (2016) Degradable controlled-release polymers and polymeric nanoparticles: mechanisms of controlling drug release. Chem Rev 116:2602–2663
51. Chen S, Zhang XZ, Cheng SX, Zhuo RX, Gu ZW (2008) Functionalized amphiphilic hyperbranched polymers for targeted drug delivery. Biomacromol 9:2578–2585
52. Morita T, Horikiri Y, Suzuki T, Yoshino H (2001) Applicability of various amphiphilic polymers to the modification of protein release kinetics from biodegradable reservoir-type microspheres. Eur J Pharm Biopharm 51:45–53
53. Abbina S, Vappala S, Kumar P, Siren EMJ, La CC, Abbasi U, Brooks DE, Kizhakkedathu JN (2017) Hyperbranchedpolyglycerols: recent advances in synthesis, biocompatibility and biomedical applications. J Mater Chem B 5:9249–9277
54. Kaanumalle LS, Ramesh R, Maddipatla VSNM, Nithyanandhan J, Jayaraman N, Ramamurthy V (2005) Dendrimers as photochemical reaction media. Photochemical behavior of unimolecular and bimolecular reactions in water-soluble dendrimers. J Org Chem 70:5062–5069
55. Madaan K, Kumar S, Poonia N, Lather V, Pandita D (2014) Dendrimers in drug delivery and targeting: Drug-dendrimer interactions and toxicity issues. J. Pharm. Bioallied Sci. 6:139–150
56. Tomalia DA, Naylor AM, Goddard WA III (1990) Starburst dendrimers: molecular-level control of size, shape, surface chemistry, topology, and flexibility from atoms to macroscopic matter. Angew Chem Int Ed 29:138–175
57. Fréchet JM (1994) Functional polymers and dendrimers: reactivity, molecular architecture, and interfacial energy. Science 263:1710–1715
58. Newkome GR, Yao Z, Baker GR, Gupta VK (1985) Micelles Part 1. Cascade molecules: a new approach to micelles, A-arborol. J Org Chem 50:155–158
59. Choudhary S, Gupta L, Rani S, Dave K, Gupta U (2017) Impact of dendrimers on solubility of hydrophobic drug molecules. Front Pharmacol 8:261(1)–261(23)
60. Stevelmens S, Hest JC, Jansen JF, Boxtel DA, de Bravander-van den B, Miejer EW (1996) Synthesis, characterization and guest-host properties of inverted unimolecular micelles. J Am Chem Soc 118:7398–7399
61. Gupta U, Agashe HB, Asthana A, Jain NK (2006) Dendrimers: novel polymeric nanoarchitectures for solubility enhancement. Biomacromol 7:649–658

62. Thomas TP, Majoros IJ, Kotlyar A, Kukowska-Latallo JF, Bielinska A, Myc A, Baker JR (2005) Targeting and inhibition of cell growth by an engineered dendritic nanodevice. J Med Chem 48:3729–3735
63. Bhadra D, Bhadra S, Jain P, Jain NK (2002) Pegnology: a review of PEG-ylated systems. Pharmazie 57:5–29
64. Asthana A, Chauhan AS, Diwan PV, Jain NK (2005) Poly(amidoamine) (PAMAM) dendritic nanostructures for controlled site-specific delivery of acidic anti-inflammatory active ingredient. AAPS PharmSciTech 6:E536–E542
65. Bhadra D, Bhadra S, Jain S, Jain NK (2003) A PEGylated dendritic nanoparticulate carrier of fluorouracil. Int J Pharm 257:111–124
66. Khopade AJ, Caruso F, Tripathi P, Nagaich S, Jain NK (2002) Effect of dendrimer on entrapment and release of bioactive from liposomes. Int J Pharm 232:157–162
67. Prajapati RN, Tekade RK, Gupta U, Gajbhiye V, Jain NK (2009) Dendrimer-mediated solubilization, formulation development and in vitro-in vivo assessment of piroxicam. Mol Pharm 6:940–950
68. Chauhan AS, Sridevi S, Chalasani KB, Jain AK, Jain SK, Jain NK, Diwan PV (2003) Dendrimer-mediated transdermal delivery: enhanced bioavailability of indomethacin. J Control Release 90:335–343
69. Kukowska-Latallo JF, Candido KA, Cao Z, Nigavekar SS, Majoros IJ, Thomas TP, Balogh LP, Khan MK, Baker JR Jr (2005) Nanoparticle targeting of anticancer drug improves therapeutic response in animal model of human epithelial cancer. Cancer Res 65:5317–5324
70. Quintana A, Raczka E, Piehler L, Lee I, Myc A, Majoros I, Patri AK, Thomas T, Mulé J, Baker JR Jr (2002) Design and function of a dendrimer-based therapeutic nanodevice targeted to tumor cells through the folate receptor. Pharm Res 19:1310–1316
71. Toms S, Carnachan SM, Hermans IF, Johnson KD, Khan AA, O'Hagan SE, Tang C-W, Rendle PM (2016) Poly ethoxy ethyl glycinamide (PEE-G) dendrimers: dendrimers specifically designed for pharmaceutical applications. ChemMedChem 11:1583–1586
72. Shrestha R, Teesdale-Spittle PH, Lewis AR, Rendle PM (2020) Gadolinium complexes attached to poly ethoxy ethyl glycinamide (PEE-G) dendrons: magnetic resonance imaging contrast agents with increased relaxivity. ChemPlusChem 85:1881–1892

Chapter 3
Cocrystals and Their Induced Activity of Drugs

Bandita Datta and Joydeep Biswas

1 Introduction

In modern times science has obtained a divergent branch, i.e. drug discovery. It comprises of two stages, namely 'lead structure' and 'drug candidate'. The 'lead structure' stage leads to selecting an active pharmaceutical compound (APC), whereas the drug candidate involves the formulation of the APC into an easily absorbed solid form. Although the process starts with thousands of candidates, around 5–10 candidates pass to the 'drug candidate' stage. These candidates are further scrutinized and finally, one candidate is selected, called the Active Pharmaceutical Ingredient (API). This API, once obtained, is further converted into the solid drug form by formulations. The drug formulation includes optimization of the properties of the solid drug. Solubility, dissolution rate and permeability are the essential properties of an API required for the API's bioavailability in the drug form. The API may be available in various solid forms. Drug discovery helps in the selection of the optimal API from all the available forms [1].

In the last decade, excellent progress has been made in the biomedical field. Some of the factors are a rise in the number of therapeutic targets, which helps design the drugs according to their structure and discover interactions of gene networks and their products. Sequencing of the genome of humans has revealed that more than 20,000 protein-coding genes are available. The discovery of such a high number of proteins based on disease genes requires a comprehensive, workable strategy for screening them [2].

B. Datta (✉)
Department of Chemistry, Amity Institute of Applied Science, Amity University, Kolkata 700156, India
e-mail: bdatta@kol.amity.edu

J. Biswas
Department of Chemistry, Sikkim Manipal Institute of Technology, Sikkim Manipal University, East Sikkim, Majitar 737136, India

B. P. Swain (ed.), *Nanostructured Biomaterials*, Materials Horizons: From Nature to Nanomaterials, https://doi.org/10.1007/978-981-16-8399-2_3

Stability, solubility, bioavailability and mechanical properties of active pharmaceutical ingredients (API) have become the main snag in the pharmaceutical industry [3]. The enhancement of solubility shows its impact on the pharmacokinetic profile. The increased solubility facilitates better absorption of the API in the gastrointestinal tract and reduces the subsequent dosage of the drug [4]. The predicament of 'solubility of new drugs' which severely affect bioavailability, has led to pharmaceutical crystals being prioritized because of their higher stability, reproducibility comparison to amorphous and other solid solutions such as partially crystalline forms, subcooled liquid and the different types of crystal forms that have variable dissolution rate and intrinsic solubility [5, 6]. To enhance the aqueous solubility of drugs, various methods have been designated, such as salt formation, micro-ionization, emulsification, polymer-drug vehicles [7].

Crystalline materials have gained significance in the pharmaceutical industry due to their inherent stability. They also have an ingrained effect on the purification and isolation of chemical substances [8]. Chemists and engineers in the pharmaceutical industry are paying increasing attention to the influence of such materials on drug discovery and initial stages of development [9]. APIs can be developed in various crystalline forms like polymorphs, solvates, hydrates, salts, cocrystals, etc. Each form shows distinct physicochemical properties that influence the bioavailability, purification, stability and other properties of a drug [10].

2 Pharmaceutical Cocrystals

A crystalline phase consists of supramolecular networks, where the molecules are inter-linked through synthons with the help of non-covalent interactions like H-bonding, van der Waals, π-π interactions, etc. [11]. Synthons are those units that help in connecting the supramolecular compounds, for example, O–H–H bonds in carboxylic acids and N–H–O bonds in amides which play an essential role in biochemical reactions. A cocrystal is a crystalline system consisting of two or more components compounded in a stoichiometric ratio. The constituents of pharmaceutical cocrystal are an API and one or more cocrystal formers which are compounded to form crystalline solids at room temperature [12]. They have the components which are bonded to one another through various non-covalent bonds in various stoichiometric ratios (1:1, 1:2. 1:3) of the API and the cocrystal former [11]. Their synthesis is based on the principle of the synthesis of supramolecule molecules. Hence understanding the chemistry of supramolecular compounds is a necessity. A study on suitable cocrystal former which forms a non-covalent bond with the functional groups present in API is also significant. Cocrystals can be further into different forms like cocrystal anhydrases, cocrystal hydrates (solvates), anhydrases of cocrystals of salts and hydrates (solvates) of cocrystals of salts [13].

Cocrystals can be synthesized both in the presence of solvent and solvent-free methods. Preparation of cocrystal in the presence of solvent involves slurry formation, solvent-evaporation, cooled crystallization and precipitation. The solvent-free method involves neat grinding or sonication. Although solvent-assisted grinding and wet sonication (use of solvent) has also been employed for the synthesis [14]. Vigorous research is being further carried out to design various cocrystals and understand their mechanistic pathway of formation [15].

Co-crystallizations in solutions are based on slow evaporation, reaction crystallization and cooling crystallization. In evaporation cocrystallization, solvent or solvent mixtures are used for obtaining the solubility of components. The reaction cocrystallization method provides the cocrystals stability region in non-conformable saturating solvents using different reactant concentrations [16]. The cooling cocrystallization process is based on heating the solvent and reactants in a reactor to a high temperature and cooling back down. The precipitation happens after the solution becomes supersaturated [17]. Disadvantages of solution methods are solubility and solvent problems such as the severe interaction with solvents and the broad solubility difference between drugs and cocrystal formers [18].

3 Mechanochemical Cocrystallization

A mechanochemical or co-grinding method is an economical, environmentally friendly and simple method in which the absence of organic solvent makes this method more preferable than the others [19]. Grinding for many years has been used in synthesis, but only in the 1970s have applications of co-grinding been revealed and, since then, has been used in the formation of metallic alloys and polymeric composites [20]. In this process, such variables as grinding rate and duration time, grinding pot volume, the drug and excipient ratio must be controlled by experimental design [19].

4 Mechanisms of Mechanical Cocrystallization

The two primary methodologies for cocrystal synthesis via grinding in pestle & mortar or ball mills mechanically. Firstly, the method involved is renowned dry grinding, which involves mixing the components (API and cocrystal former) and grinding them manually or mechanically using a 'ball mill'. The second technique employed is 'liquid-assisted grinding' (LAG) or 'solvent-drop grinding'. Here a solvent is added in a minimal amount to the mixture. The latter methodology benefits over the former method because it produces higher yield, morphology, higher crystallinity of the product and significantly extends the opportunity to use various reactants and products [21]. The solid-state cocrystallization by neat grinding and liquid-assisted grinding method might follow several mechanisms, including molecular diffusion, eutectic formation, or cocrystallization.

Rastogi et al. were the first group to explain the importance of molecular diffusion in the formulation of naphthalene and picric acid into a cocrystal. This diffusion occurred mainly when both the reactants showed significant vapour pressures in the solid state [22]. Kuroda et al. prepared cocrystal by passing over the vapours of p-benzoquinone over crystals of 2,2′-biphenol and 4,4′-biphenol. Mechanochemical grinding enhances the reaction rate by creating a new surface in which the solid reactants mix well. A new lattice strain and defects are introduced [23]. Davey et al. reported the cocrystallization of benzophenone and diphenylamine. A linear N–H/O H-bond is formed by the solvent-free grinding of the reactants, which first forms a metastable eutectic. The eutectic-mediated cocrystallization can be very efficiently carried out during mechanical grinding. During grinding, new reactant surfaces are generated for eutectic formation, which subsequently induces nucleation followed by solidification in the eutectic phase [24] (Fig. 1).

Nguyen et al. investigated the neat grinding of phenazine and mesaconic acid, which involved an amorphous phase as an intermediate. PXRD and Terahertz 'time-domain spectroscopy' (THz-TDS) studies revealed that after grinding for 60 min. It afforded complete conversion, but only a 60% yield of the cocrystal was obtained. It was conclusive that the remaining 40% of the product must be present in the amorphous state [25]. The microscopic level of understanding of the mechanism was addressed by Cinčić et al. who carried out the synthesis of cocrystals between thiomorpholine (tmo) and tetrafluoro-1,4-diiodobenzene (1,4-tfib). A 1:1 mixture of 1,4-tfib and tmo was ground for 30 min, which resulted in the formation of linear

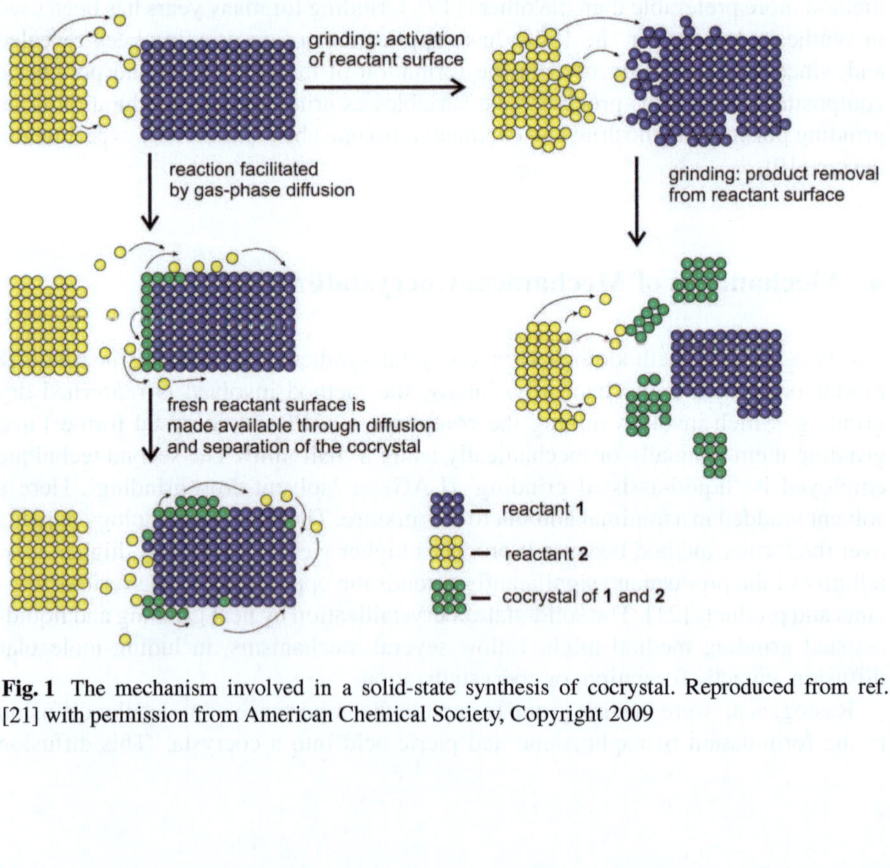

Fig. 1 The mechanism involved in a solid-state synthesis of cocrystal. Reproduced from ref. [21] with permission from American Chemical Society, Copyright 2009

bonded chains. When the mixture was initially ground for four mins, (1,4-tfib).(tmo) crystallized out with a conformer as shown by the PXRD pattern. Further studies were carried out using single-crystal analysis, which gave us an idea of the formation of intermediate consisting of finite assemblies of (1,4-tfib).$(tmo)^2$. The mechanical grinding of this mixture leads to the formation of finite assemblies joined by the strong N-I bonds. The grinding is further continued with 1,4-tfib resulting in the polymerization of these finite assemblies through the weak S-I bonds [26].

Till now, coordination complexes of pharmaceutical molecules have been scarcely explored. This pharmaceutical form is of two kinds; one is metallodrugs in which the metal ion acts as a cofactor and is an essential active component; whereas the other one is metallo pharmaceuticals in which the metal ion acts as a carrier for the active pharmaceutical ingredient [27]. Some examples of metallodrugs include bismuth subsalicylate complex used as Pepto-Bismol [28] and platinum complexes such as cisplatin, carboplatin or oxaliplatin, showing applications in cancer treatment [29]. Moulton et al. exploited the copper (II) carboxylate motifs to increase the lipophilicity of carboxylate APIs [30].

Braga et al. have reported coordination complexes between zinc and copper (II) chlorides and neuroleptic drug gabapentin [31]. They also extended this work by generating metal–organic complexes with silver nitrate and APIs like 4-aminosalicyclic acid [32]. This pioneering work showed the importance of coordination chemistry for the generation of new solid forms of API. Further probing into the structure revealed that in the solution phase, complexation produced hydrates. Although in the neat grinding, it was found that an anhydrous product was obtained. It was found that 2-dimensional sheets were formed, which were held together via Ag–O and Ag–N coordination bonds, as well as hydrogen bonds between O and H. Although this was a high yielding reaction, the disadvantage was the use of toxic transition metals. To avoid this, Chow et al. came up with an alternative of synthesizing API forming coordination complexes with biologically magnesium ions directly from magnesium oxide [33]. An efficient and environment-friendly methodology was employed to synthesize the metallodrug bismuth subsalicylate, which is the API for the drug Pepto-Bismol for gastrointestinal problems. The synthesis was performed by solvent-mediated mechanochemical grinding [34].

5 Neat Grinding Versus Liquid-Assisted Grinding

Braga et al. observed that the course of the mechanochemical reaction varied with the polarity of the solvent. A mixture of pimelic acid and 1,1′-bis(4-pyridyl)ferrocene was ground in solvent-free conditions, which did not yield any cocrystal formation, whereas the presence of the vapours of solvents like CH_2Cl_2, $CHCl_3$, Et_2O, CH_3NO_2 and ethyl lactate provided the cocrystal of stoichiometric 1:1 ratio. A pyridinium salt of stoichiometric 1:2 ratio was obtained in isopropyl alcohol, methanol, water or ethanol [35]. The above observation indicated the profound effect of the microscopic amounts of the liquid on the course of mechanochemical cocrystallization.

The disadvantages of neat grinding compared to solution crystallization led to the systematic development of LAG to form cocrystal. Jones et al. tried to synthesize cocrystal from cyclohexane-1,3-cis,5-cis-tricarboxylic acid (CTA) and 4,7-phenanthroline(47P) by neat grinding but was obtained in partial yield as compared to solution crystallization. While the grinding was carried out in a few drops of methanol, a good yield of cocrystal formation was attained within minutes [36]. Several experiments were conducted for solvent-mediated grinding of API and coformer. Even if one of the reactants is soluble in the solvent, it increases the rate of cocrystal formation.

Liquid-assisted grinding was also shown to form cocrystal polymorph following green chemistry protocols by Jones et al. Caffeine and Glutaric acid (GA) was crushed in a ball mill in the presence of chloroform solution produced cocrystals of two different morphologies, i.e. rods and blocks (Fig. 2). Single X-ray diffraction studies revealed that the rods formed have monoclinic (form I) and the blocks have a triclinic (form II) crystal system in 1.3:1 caffeine: GA stoichiometry. Equal stoichiometry of caffeine and GA was employed to prepare cocrystals mechanochemically both in the presence and in the absence of solvents. The solvent-free method produced cocrystal form I, while the presence of few drops of a solvent such as n-heptane, cyclohexane, or hexane while grinding also yielded cocrystal in form I. On the contrary, when polar solvents were added like $CHCl_3$, CH_2Cl_2, CH_3CN and water during grinding produced cocrystal in form II. Hence we can conclude that the choice of solvent

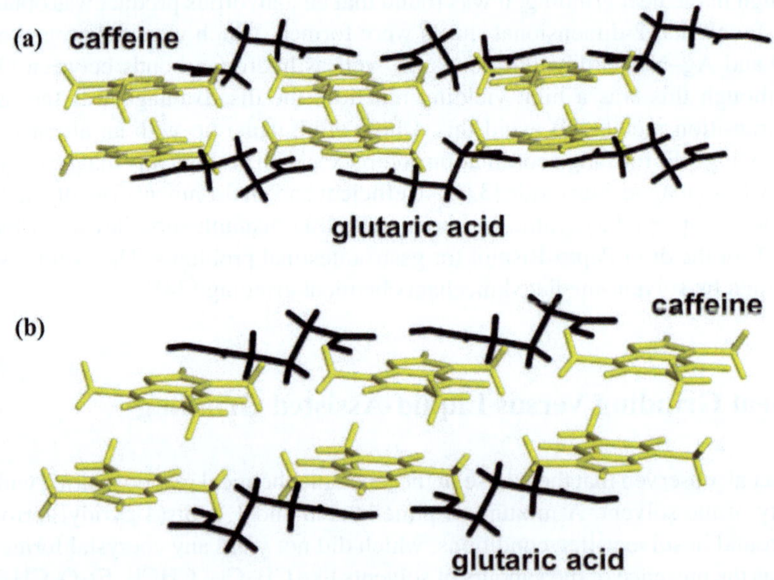

Fig. 2 Representation of cocrystal formation by solvent-drop method **a** Production of cocrystal in cyclohexane as solvent, **b** Production of cocrystal in chloroform solvent. Reproduced from ref. [21] with permission from American Chemical Society, Copyright 2009

in solvent-assisted grinding can produce different cocrystal polymorphs following green chemistry protocols [37]. Further to understand the influence of solvent phase, Friščić et al. introduced a parameter η, which is the ratio of the solvent volume to the weight of the combination of API and conformer [38].

6 A Brief Overview of Mechanochemistry in Drug Discovery

In recent times, the time-consuming early stage of drug discovery and more extended clinical development has increased the expense and time to acquire a new lead compound. Using mechanical grinding can be traced back to the 1890s when Ling and Baker synthesized tetrachloroquihydrone from grinding of meta-dichloroquinone and meta-dichloroquinol [39]. The grinding of naphthalene and picric acid resulted in the formation of cocrystal by vapour diffusion was shown by Rastogi et al. [22]. Whitesides et al. were the first group who had applied crystal engineering for the formation of pharmaceutical cocrystals. Whitesides et al. used substituted barbituric acids and melamines or cyanuric acids to generate supramolecular 'linear tape,' 'crinkled tape,' and 'rosette' motifs. The diverse forms of the cocrystals of an API were demonstrated using these studies [40, 41]. Cocrystal of 4-aminobenzoic acid and3,5-dinitrobenzoic acid was used to prepare cocrystal mechanochemically by Etter et al. [42]. Both grinding and crystallization from solution methods were employed to formulate the cocrystals of anthracene as conformer with 4-methyl-3,5-dinitrobenzoic acid and 4-chloro-3,5-dinitrobenzoic acid with anthracene by Pedireddi et al. Although they were unable to form cocrystals of 3,5-dinitrobenzoic acid and anthracene by grinding methods, the use of benzene as a solvent allowed them to obtain the cocrystal from solution method [43]. Caira prepared various cocrystals using the drug sulfadimidine with various carboxylic acids like benzoic, anthranilic, salicylic acid and acetylsalicylic (aspirin) [44]. Procedures such as LAG, solution, vapour-digestion and grinding cocrystallization were applied by Braga et al. to form cocrystals of various API molecules and precursors like piracetam, nicotinamide, diacetamide, malonamide and barbituric acid with $CaCl_2$ [45, 46]. The formation of ionic cocrystal (cocrystals between allowed them to obtain crystals by sol organic molecule and ionic salts) was first reported by Beevers and Cochran in 1946 between sucrose and sodium bromide [47]. Braga et al. synthesized ionic cocrystals of piracetam and nicotinamide, which have a lower intrinsic dissolution rate than the pure APIs [46]. Organic-organometallic [48] and inorganic-organometallic [49] hybrid materials were also produced by grinding and kneading techniques. A minute amount of solvent vapour showed inflation of the rate of the covalent formation and enhancement of the yield, as noted by Toda et al. [50].

Aneef et al. investigated the formation of a cocrystal of curcumin with resorcinol. Cocrystal was formed when the mixture in 1:1 mixture was ground together in the LAG method. The 'in vitro' dissolution studies showed that while a pure drug showed 39.45% of bioavailability, the cocrystal form showed 97.24% drug release. The 'in vivo' studies for the antiulcer activity of curcumin were evaluated. The studies revealed that the cocrystal thus obtained with the conformer resorcinol, at a dose of 100 mg/kg body weight, demonstrated 76.63% antiulcer activity [51]. Such examples showcase the importance of cocrystals for biomedical applications.

Similarly, a green mechanical grinding technique was employed for the formation of Lomefloxacin-Salicylic acid and Lomefloxacin-Urea cocrystals. Lomefloxacin-Salicylic acid cocrystal was found to have shown higher solubility than individual lomefloxacin as a drug. Both LAG and neat grinding methodologies were used, but neat grinding gave better yields and purity results. This study further establishes the importance of neat grinding as an alternative method for producing cocrystals with better pharmacological properties [52].

Theophylline is one of the vital drugs used to treat chronic obstructive pulmonary disease (COPD) and asthma [53]. An equimolar quantity of theophylline and nicotinamide was ground with the help of pestle and mortar to form a cocrystal by Lu and Rohani. They obtained a 1:1 molar ratio of TP:NCT, which was found to possess unique properties like thermal, spectroscopic and X-ray diffraction patterns. Compared to pure anhydrous theophylline, the cocrystal showed better solubility and hygroscopicity, an important property to be possessed by a drug for better action [54].

Recently, Tan et al. have reported the formation of non-steroidal anti-inflammatory cocrystals. The API chosen was flufenic acid (FFA) with theophylline (TP) along with 2-pyridone and 4,4′-bipyridine utilizing the solvent-drop grinding method. A 'mixer mill' having stainless steel grinding jars and grinding balls with the speed of 20 Hz/30 min was used for carrying out the cocrystal formation for 100 mg (0.36 mmol) of FFA. 2 drops of methanol was the choice of solvent which was used for mechanochemical grinding of FFA with TP. The physicochemical properties of the cocrystal obtained with TP were investigated, one of the essential requirements for the pharmaceutical industry compared with the existing cocrystal of FFA:Nicotinamide. The cocrystal they obtained showed better solubility property which is essential for bioavailability. It also took care of the hygroscopic property of TP, which generally causes the problem [55].

Caira et al. worked on the cocrystal of 3-(6-Methoxypyridin-3-yl)-5-(4-methylsulfonyl phenyl)-pyridin-2-amine (MMP), which is an essential component of antimalarial drugs that can be administered orally. This API is found to show potency against plasmodial and antimalarial activity in Plasmodium berghei-infected mice. Hence the physicochemical properties must be improved for this potent drug. They prepared five salts and one cocrystal with different coformers like suberic acid, oxalic acid, fumaric acid, saccharin, salicylic acid and adipic acid. The salt formed with saccharin and fumaric acid along with the cocrystal formed proved to show more remarkable solubility properties than MMP itself [56].

Emmerling et al. also attempted and successfully prepared a cocrystal of equal quantity of Carbamazepine and indomethacin in a 1:1 stoichiometric ratio. A conventional ball mill was employed, which yielded the cocrystal in a few seconds itself. They also attempted to prepare cocrystals of different stoichiometric quantities by prolonged grinding, resulting in the loss of crystallinity of the mixture. The cocrystal formed showed better physicochemical properties as compared to the individual API [57].

Nangia et al. studied the importance of sulfacetamide, a prescribed antibiotic for treating ocular infections and prepared its cocrystals to improve the therapeutic action. From the earlier examples, we have understood that the synthesis of cocrystal improves the bioavailability, solubility and dissolution rate, which improves the therapeutic action. Hence these properties are the most studied concerning the cocrystals. Novel cocrystals were synthesized with coformers like caffeine, isonicotinamide, theophylline, bipyridine and 4-aminopyridine. In the environment of pH 7 maintained with the help of phosphate buffer media, the solubility measurement was carried out. They observed that sulfacetamide–caffeine cocrystal had low solubility than the drug, but it had stability for 24 h in the pH 7 slurry medium. The low solubility and good stability of the sulfacetamide–caffeine cocrystal was helpful to tackle the insufficient habitation time and faster removal issues of the API [58]. In pharmaceutical industries, we have seen numerous examples where only one of the enantiomers is pharmacologically active and is used to treat an illness. Hence the separation of one of the enantiomers from the racemic mixtures has gained much importance lately, as the usage of the pure enantiomer as the dosage is more advantageous than the racemic mixture. The racemic mixture sometimes shows low bioavailability and prolonged toxicity [59].

In this context, Jones et al. have exhibited the racemic mixture of malic acid can be transformed into separate diastereomeric cocrystals by mechanochemical grinding with pure enantiomer form of tartaric acid [60]. Working on a similar principle, Leyssens et al. also investigated the cocrystal formation with amino acids. They have shown an affinity towards compounds containing carboxyl groups and amino groups in the terminals, forming a strong charge assisted hydrogen bonding. They also observed that enantiospecific cocrystal of S-naproxen with D- but not with L-tyrosine was formed [61]. Nangia et al. have recently prepared different cocrystals of Voriconazole with m-nitrobenzoic acid, p-aminobenzoic acid and p-hydroxybenzoic acid coformers by solid-state grinding to enhance the therapeutic properties [62].

7 Characterization of Cocrystals

Characterization of cocrystals is a significant factor in understanding the structural and physical properties of the cocrystals [63–66]. The techniques employed include Infrared spectroscopy, differential scanning calorimeter, powder X-ray diffraction, thermogravimetric analysis and single-crystal X-ray crystallography [63, 64]. XD methods were best used for determining the structures of the cocrystals. To understand the thermogravimetric properties, differential thermal analysis and thermogravimetric analysis and differential scanning colorimetry were employed. NMR can also be used for the determination of the detailed structural analysis of the pharmaceutical cocrystals [67, 68]. Desai et al. prepared carbamazepine cocrystals and characterized them with the help of X-ray diffractometry, differential scanning calorimetry, visual morphology, infrared spectroscopy, etc. [69]. Cocrystals of gentisic acid and piracetam were characterized by PXRD, melting point, IR, DSC and single-crystal X-ray diffraction [70].

8 Benefits of Cocrystals

Cocrystals are more advantageous over amorphous forms as it is more thermodynamically stable. The mechanochemical cocrystallization techniques are applicable for all nonionic, ionic and weakly ionic active pharmaceutical compounds. As we have seen in various examples given before, the cocrystallization technique enhances the properties of API, for example, dissolution properties, oral bioavailability, stability, compressibility, flowability and hygroscopic properties. In recent years, much advancement in better physicochemical modification and increased efficacy of API has been observed, which is apparent by the upsurge in the amount of patent grants and their applications.

9 Conclusion

The above studies have taught us the importance of the physiological properties of a potential API, such as thermal stability and bioavailability in the initial stages of drug discovery.

These properties decide the fate of the final drug candidate, storage, dissolution properties and dosage form. Different forms such as hydrates, cocrystals, amorphous, solvates, salts and polymorphs solids exist for an API. Recent progress in crystal engineering technology and advances made in supramolecular chemistry has helped us by increasing our understanding of the concept of cocrystals and factors affecting them, such as stability and solubility. Analysis of various case studies on cocrystal formation using different techniques shows the competencies and limitations of crystal engineering strategies. A summarized literature survey provides countless opportunities for designing cocrystals for biomedical applications and clinical relevance. Further, innovative methods for the characterization of cocrystals should be developed, such as structure determination through single crystals X-ray, Infrared, NMR crystallography or powder X-ray diffraction.

10 Future Research

It is the responsibility of a drug discovery scientist to formulate a drug that has physical and chemical stability as well as can be manufactured at a commercial scale. Discoveries and the development of new methodologies have constantly been coming out in the academic environment, making it necessary for drug discovery scientists to frequently interact with their colleagues in industries and academia. The drug discovery process in the preliminary stages is divided into four phases, namely D1, D2, D3 and D4, followed by four clinical phases. All the phases are interrelated for drug discovery. D3 phase involves lead optimization, which involves changing the structure of a compound to obtain different properties. This work is carried out by biologists and chemists. The former checks the efficiency of the drug candidate on biological systems and the latter uses this information to carry out some more modifications which are further retested by the biologists. The altered compounds thus formed is the drug candidate or API in the solid form. Extensive work is being carried out to understand the properties of all solid forms of a new chemical entity in the pharmaceutical industry. Solid forms and polymorphs were the most common forms found, but recently more explorations are carried out on cocrystals. Synthesis of cocrystals can be categorized into liquid-based and solvent-free methods. Solvent-based cocrystallization faces challenges in terms of scale-up processes. Moreover, in this method, drying of the solvent to acceptable levels from the products must be carried out. Numerous research articles are available for neat grinding of the compounds in mills have been reported. Recently, solvent-assisted grinding methodology has also been established as a more efficient method to prepare cocrystals. This mechanochemical method of cocrystal synthesis avoids challenges associated with kinetic crystallization and solubility differences between API and the coformer. Even though there has been a surge in cocrystals' preparation, its applications in the pharmaceutical industries have been limited. The reason behind this may be due to the problem in scaling up the synthesis to industrial levels.

Mechanochemical synthesis of APIs can be further developed, combined with mechanochemical ball milling for the formation of cocrystals supported by a photochemical or thermal reaction. Eco-friendly metals should be explored to synthesize metallodrugs and metallo pharmaceuticals for their use in novel drug discovery.

References

1. Drews J (2000) Drug discovery: a historical perspective. Science 287:1960–1964
2. Herrling PL (2005) The drug discovery process. In: Rudin M (ed) Progress in drug research
3. Dhumal RS, Kelly AL, York P, Coates PD, Paradkar A (2010) Cocrystalization and simultaneous agglomeration using hot melt extrusion. Springer Science & Business Media, pp 2726–2733
4. Chadha R, Saini A, Arora P, Jain DS, Dasgupta A, Guru Row TN (2011) Multicomponent solids of lamotrigine with some selected coformers and their characterization by thermoanalytical, spectroscopic and X-ray diffraction methods. CrystEngComm 13:6271–6284

5. Shan N, Zaworotko MJ (2008) The role of cocrystals in pharmaceutical science. Drug Discovery Today 13:440–446
6. Rohani S (2010) Applications of the crystallization process in the pharmaceutical industry. Front Chem Eng 4:2–9
7. Qiao N, Li M, Schlindwein W, Malek N, Davies A, Trappitt A (2011) Pharmaceutical cocrystals: an overview. Int J Pharma 419:1–11
8. Paul EL, Tung HH, Midler M (2005) Organic crystallization processes. Powder Technol 150:133–143
9. Gardner CR, Walsh CT, Almarsson Ö (2004) Drugs as materials: valuing physical form in drug discovery. Nat Rev Drug Disc 3:926–934
10. Rodrıguez-Hornedo N, Nehm SJ, Jayasankar A (2006) Cocrystals: design, properties and formation mechanisms. In: Swarbrick, Boylan (eds) Encyclopedia of pharmaceutical technology. Informa Health Care, London, pp 615–635
11. Byrn SR, Pfeiffer RR, Stowell JG (1999) Solid-state chemistry of drugs. West Lafayette, SSCI
12. Yadav AV, Shete AS, Dabke AP, Kulkarni PV, Sakhare SS (2009) Cocrystals: a novel approach to modify physicochemical properties of active pharmaceutical ingredients. Indian J Pharm Sci 71(4):359–370
13. Ning S, Michael JZ (2008) The role of cocrystals in pharmaceutical science. Drug Discov Today 13:440–446
14. Zaworotko M (2005) Polymorphism in cocrystals and pharmaceutical cocrystals. Congress of the International Union of Crystallography, Florence; Pasha MA, Datta B (2014) Mechanochemistry: an efficient method of solvent-free synthesis of 3-amino-2,4-dicarbonitrile-5-methylbiphenyls. J Saudi Chem Soc 18(1):47–51
15. Aakeröy CB, Beatty AM, Helfrich BA, Nieuwenhuyzen M (2003) Do polymorphic compounds make good cocrystallising agents? A structural case study that demonstrates the importance of synthon flexibility. Cryst Growth Des 3:159–165
16. Kumar S, Nanda S (2017) Pharmaceutical cocrystals: an overview. Indian J Pharm Sci 79(6):858–871
17. Yang P, Zhao D, Margolese DI, Chmelka BF, Stucky GD (1999) Block copolymer templating syntheses of mesoporous metal oxides with large ordering lengths and semicrystalline framework. Chem Mater 11(10):2813–2826
18. Yamamoto K, Tsutsumi S, Ikeda Y (2012) Establishment of cocrystal cocktail grinding method for rational screening of pharmaceutical cocrystals. Int J Pharm 437(1–2):162–171 (2012); Desale PK (2013) A novel method: co-crystallisation. Int J Pharm Invent 3(1):19–26
19. Kürti L, Kukovecz Á, Kozma G, Ambrus R, Deli MA, Szabó-Révész P (2011) Study of the parameters influencing the co-grinding process for the production of meloxicam nanoparticles. Powder Technol 212(1):210–217
20. Tadiera S, Bulay NL, Rey C, Combes C (2011) Co-grinding significance for calcium carbonate–calcium phosphate mixed cement. Part I: effect of particle size and mixing on solid phase reactivity. Acta Biomater 7(4):1817–1826
21. Friščić T, Jones W (2009) Recent advances in understanding the mechanism of cocrystal formation via grinding. Cryst Growth Des 9(3):1621–1637
22. Rastogi RP, Bassi PS, Chadha SL (1962) Kinetics of reaction between naphthalene and picric acid in the solid state. J Phys Chem 66:2707–2708
23. Kuroda R, Higashiguchi K, Hasebe S, Imai Y (2004) Crystal to crystal transformation in the solid state. CrystEngComm 6:464–468
24. Chadwick K, Davey R, Cross W (2007) How does grinding produce cocrystals? Insights from the case of benzophenone and diphenylamine. CrystEngComm 9:732–734
25. Nguyen KL, Friščić T, Day GM, Gladden MF, Jones W (2007) Terahertz time-domain spectroscopy and the quantitative monitoring of mechanochemical cocrystal formation. Nat Mater 6:206–209
26. Cinčić D, FriščićT, Jones W (2008) Stepwise mechanism for the mechanochemical synthesis of halogen-bonded cocrystal architectures. J Am Chem Soc 130:7524–7525

27. Delori A, Friščić T, Jones W (2012) The role of mechanochemistry and supramolecular design in the development of pharmaceutical materials. CrystEngComm 14:2350–2362
28. Ge R, Sun H (2007) Bioinorganic chemistry of bismuth and antimony: target sites of metallodrugs. Acc Chem Res 40:267–274
29. Guo Z, Sadler PJ (1999) Metals in medicine. Angew Chem Int Ed 38:1512–1531
30. Ma Z, Moulton B (2007) Mixed-ligand coordination species: a promising approach for "second-generation" drug development. Cryst Growth Des 7:196–198; Ma Z, Moulton B (2007) Supramolecular medicinal chemistry: mixed-ligand coordination complexes. Mol Pharm 4:373–385
31. Braga D, Grepioni F, Maini L, Brescello R, Cotarca L (2008) Simple and quantitative mechanochemical preparation of the first zinc and copper complexes of the neuroleptic drug gabapentin. CrystEngComm 10:469–471
32. Braga D, Grepioni F, Andre V, Duarte MT (2009) Drug-containing coordination and hydrogen bonding networks obtained mechanochemically. CrystEngComm 11:2618–2621
33. Chow EHH, Strobridge FC, Friščić T (2010) Mechanochemistry of magnesium oxide revisited: facile derivatisation of pharmaceuticals using coordination and supramolecular chemistry. Chem Commun 46:6368–6370
34. Andre V, Hardeman A, Halasz I, Stein RS, Jackson GJ, Reid DG, Duer MJ, Curfs C, Duarte MT, Friščić T (2011) Mechanosynthesis of the metallodrug bismuth subsalicylate from Bi_2O_3 and structure of bismuth salicylate without auxiliary organic ligands. Angew Chem Int Ed 50:7858–7861
35. Braga D, Giaffreda SL, Grepioni F, Chierotti MR, Gobetto R, Palladino G, Polito M (2007) Solvent effect in a "solvent free" reaction. CrystEngComm 9:879–881
36. Shan N, Toda F, Jones W (2002) Mechanochemistry and cocrystal formation: effect of solvent on reaction kinetics. Chem Commun 2372–2373
37. Trask AV, Motherwell WDS, Jones W (2004) Solvent-drop grinding: green polymorph control of cocrystallization. Chem Commun 8:890–891
38. Friščić T, Childs SL, Rizvi SAA, Jones W (2009) The role of solvent in mechanochemical and sonochemical cocrystal formation: a solubility-based approach for predicting cocrystallisation outcome. CrystEngComm 11:418–426
39. Ling AR, Baker JL (1893) XCVI—Halogen derivatives of quinone. Part III. Derivatives of quinhydrone. J Chem Soc Trans 63:1314–1327
40. Zerkowski JA, Seto CT, Wierda DA, Whitesides GM (1990) Design of organic structures in the solid state: hydrogen-bonded molecular "Tapes". J Am Chem Soc 112:9025–9026; Zerkowski JA, Seto CT, Whitesides GM (1992) Solid-state structures of "Rosette" and "Crinkled Tape" motifs derived from the cyanuric acid melamine lattice. J Am Chem Soc 114:5473–5475; Zerkowski JA, Whitesides GM (1994) Steric control of secondary, solid-state architecture in 1:1 complexes of melamines and barbiturates that crystallize as crinkled tapes. J Am Chem Soc 116;4298–4304; Zerkowski JA, MacDonald JC, Seto CT, Wierda DA, Whitesides GM (1994) Design of organic structures in the solid state: molecular tapes based on the network of hydrogen bonds present in the cyanuric acid—melamine complex. J Am Chem Soc 116:2382–2391; Zerkowski JA, Mathias JP, Whitesides GM (1994) New varieties of crystalline archeticture produced by small changes in molecular structure in tape complexes of melamines and barbiturates. J Am Chem Soc 116:4305–4315
41. Vishweshwar P, McMahon JA, Bis JA, Zaworotko MJ (2006) Pharmaceutical cocrystals. J Pharm Sci 95:499–516
42. Etter MC, Frankenbach GM (1989) Hydrogen-bond directed cocrystallization as a tool for designing acentric organic solids. Chem Mater 1:10–12
43. Pedireddi VR, Jones W, Chorlton AP, Docherty R (1996) Creation of crystalline supramolecular arrays: a comparison of cocrystal formation from solution and by solid-state grinding. Chem Commun 987–988
44. Caira MR, Nassimbeni LR, Wildervanck AF (1995) Selective formation of hydrogen bonded cocrystals between a sulfonamide and aromatic carboxylic acids in the solid state. J Chem Soc Perkin Trans 2:2213–2216

45. Braga D, Grepioni F, Lampronti GI, Maini L, Turrina A (2011) Ionic cocrystals of organic molecules with metal halides: a new prospect in the solid formulation of active pharmaceutical ingredients. Cryst Growth Des 11:5621–5627; Tilborg A, Michaux C, Norberg B, Wouters J (2010) Advantages of cocrystallization in the field of solid-state pharmaceutical chemistry: l-Proline and $MnCl_2$. Eur J Med Chem 45:3511–3517
46. Braga D, Grepioni F, Maini L, Prosperi S, Gobetto R, Chierotti MR (2010) From unexpected reactions to a new family of ionic cocrystals: the case of barbituric acid with alkali bromides and caesium iodide. Chem Commun 46:7715–7717
47. Beevers CA, Cochran W (1946) X-ray examination of sucrose. Nature 157:872
48. Braga D, Giaffreda SL, Grepioni F, Palladino G, Polito M (2008) Mechanochemical assembly of hybrid organic–organometallic materials. Solid–solid reactions of 1,1′-di-pyridyl-ferrocene with organic acids. New J Chem 32:820–828
49. Braga D, Maini L, Giaffreda SL, Grepioni F, Chierotti MR, Gobetto R (2004) Supramolecular-complexation of alkali cations through mechanochemical reactions between crystalline solids. Chem–Eur J 10:3261–3269
50. Nakamatsu S, Toyota S, Jones W, Toda F (2005) The important role of solvent vapor in an organic solid state reaction. Chem Commun 3808–3810
51. Mohammed Y (2012) Solubility enhancement of cocrystal based solid dosage forms. Semantic Scholar, Corpus ID. 104284655
52. Nanjwade VK (2012) Development and characterization of novel pharmaceutical crystalline complex of lomefloxacin. Int J Drug Dev Res 4(1):227–233
53. Van Andel AE, Reisner C, Menjoge SS, Witek TJ (1999) Analysis of inhaled corticosteroid and oral theophylline use among patients with stable COPD from 1987 to 1995. Chest 115:703–707
54. Lu J, Rohani S (2009) Preparation and characterization of theophylline-nicotinamide cocrystal. Org Process Res Dev 13:1269–1275; Li P, Chu Y, Wang L, Wenslow Jr. RM, Yu K, Zhang H, Deng Z (2014) Structure determination of the theophylline–nicotinamide cocrystal: a combined powder XRD, 1D solid-state NMR, and theoretical calculation study. CrystEngComm 16:3141–3147
55. Aitipamula S, Wong ABH, Chowa PS, Tan RBH (2014) Cocrystallization with flufenamic acid: comparison of physicochemical properties of two pharmaceutical cocrystals. CrystEngComm 16:5793–5801
56. Cruickshank DL, Younis Y, Njuguna NM, Ongarora DSB, Chibale K, Caira MR (2014) Alternative solid-state forms of a potent antimalarial aminopyridine: X-ray crystallographic, thermal and solubility aspects. CrystEngComm 16:5781–5792
57. Tröbs L, Emmerling F (2014) Mechanochemical synthesis and characterisation of cocrystals and metal organic compounds. Faraday Discuss 170:109–119
58. Goud NR, Khan RA, Nangia A (2014) Modulating the solubility of sulfacetamide by means of cocrystals. CrystEngComm 16:5859–5863
59. McConathy J, Owens MJ (2003) Prim. Care Companion J Clin Psychiatr 5:70–73; Ariens EJ (1984) Eur J Clin Pharmacol 26:663–668
60. Eddleston MD, Arhangelskis M, Friščić T, Jones W (2012) Solid state grinding as a tool to aid enantiomeric resolution by cocrystallization. Chem Commun 48:11340–11342
61. Tumanova N, Tumanov N, Robeyns K, Filinchuk Y, Wouters J, Leyssens T (2014) Structural insight into cocrystallization with zwitterionic co-formers: cocrystals of S-naproxen. CrystEngComm 16:8185–8196
62. Kumar SS, Thakuria R, Nangia A (2014) Pharmaceutical cocrystals and a nitrate salt of voriconazole. CrystEngComm 16:4722–4731
63. Basavoju S, Bostrom D, Velaga SP (2008) Indomethacin-saccharin cocrystal: design, synthesis and preliminary pharmaceutical characterization. Pharm Res 25:530–541
64. Lu J, Rohani S (2009) Preparation and characterization of theophylline−nicotinamide cocrystal. Org Process Res Dev 13(6):1269–1275
65. Zoghbi A, Wang B (2015) Carvedilol solubility enhancement by inclusion complexation and solid dispersion: review. J Drug Deliv Therapeutics 5(2):1–8

66. Wenger M, Bernstein J (2008) An alternate crystal form of gabapentin: a cocrystal with oxalic acid. Cryst Growth Des 8:1595–1598
67. Vogt FG, Clawson JS, Strohmeier M, Edwards AJ (2009) Solid-state NMR analysis of organic cocrystals and complexes. Cryst Growth Des 9:921–937
68. Pindelska E, Sokal A, Kolodziejski W (2017) Pharmaceutical cocrystals, slats and polymorphs: advanced characterization techniques. Adv Drug Deliv Rev 1:111–146
69. Desai H, Rao L, Amin P (2014) Carbamazepine cocrystals by solvent evaporation technique: formulation and characterization studies. Am J Pharmatech Res 4(2):479–493
70. Fleischman SG, Kuduva SS, McMahon JA, Moulton B, Walsh RB, Rodriguez-Hornedo N, Zaworotko MJ (2003) Crystal engineering of the composition of pharmaceutical phases. 2. Multiple component crystalline solids involving carbamazepine. Cryst Growth Des 3:909–919

Chapter 4
An Introduction on Evolution of Azole Derivatives in Medicinal Chemistry

Arup K. Kabi, Sattu Sravani, Raghuram Gujjarappa, Aakriti Garg, Nagaraju Vodnala, Ujjawal Tyagi, Dhananjaya Kaldhi, Ravichandiran Velayutham, Sreya Gupta, and Chandi C. Malakar

Abbreviations

U.S. FDA	United States food and drug administration
HIV	Human immunodeficiency virus
SAR	Structure-activity relationship
DNA	Deoxyribonucleic acid
CNS	Central nervous system
HCl	Hydrochloric acid
CYP450	Cytochromes P450
NSAID	Non-steroidal anti-inflammatory drugs
ACE	Angiotensin-converting enzyme
BCRP	Breast cancer resistance protein
P-gp	P-glycoprotein
RNA	Ribonucleic acid
CML	Chronic myelogenous leukemia
COX	Cyclooxygenase
PGH2	Prostaglandin H2
PGG_2	Prostaglandin G2
POX	Peroxidase
PGI_2	Prostaglandin I2
PGE_2	Prostaglandin E2
PGD_2	Prostaglandin D2

A. K. Kabi · R. Gujjarappa · N. Vodnala · U. Tyagi · D. Kaldhi · C. C. Malakar (✉)
Department of Chemistry, National Institute of Technology Manipur, Langol, Manipur Imphal-795004, India
e-mail: cmalakar@nitmanipur.ac.in

S. Sravani · A. Garg · R. Velayutham · S. Gupta (✉)
Department of Medicinal Chemistry, National Institute of Pharmaceutical Education and Research (NIPER), Chunilal Bhawan, 168, Maniktala Road, Kolkata, West Bengal 700054, India

B. P. Swain (ed.), *Nanostructured Biomaterials*, Materials Horizons: From Nature to Nanomaterials, https://doi.org/10.1007/978-981-16-8399-2_4

$PGF_{2\alpha}$ Prostaglandin F2alpha
TxA_2 Thromboxane A2

1 Introduction

The pharmaceutically valued scaffolds reported in the literature generally consist of a heterocyclic moiety as an essential content of their structure. Heterocyclic molecules, such as indole [1–4], benzothiazole [5, 6], triazole [7–13], pyrimidine [14, 15], morpholine [16, 17], thiazole [18–27], pyrrolidine [28, 29], benzoxazole [30, 31], pyridine [32–37], benzimidazoles [38–42], thiophene [43–50], tetrazole [51], quinoline [52–57], isoquinoline [58, 59], pyrrole [60, 61] and imidazole [62–67] are imperative ingredients of various bioactive scaffolds. Among these heterocyclic moieties, the azole and its relative derivatives are one of the most crucial scaffolds found in the pharmacologically active molecules. Azoles occupy a domain of interest in natural and synthetic chemistry [68–73]. After the development of azole derivatives in the 1840s, the research activities based upon azoles have become a rapidly developing and increasingly effective area of research because of their efficiency as drug molecules and extensive scope in supramolecular fields, agrochemicals, smart materials, ligands, artificial acceptors and biomimetic catalysis, etc. [74–78]. Especially, the applications of azole derivatives in medicinal chemistry have accomplished great breakthroughs. In this chapter, we aim to discuss the biological and pharmacological activity of some azole derivatives and analyze the azole-related heterocycle architecture, density and structural diversity among U.S. FDA-approved small-molecule drug construction as well as current developments on azole analogues.

On Investigating of U.S. FDA-approved drug database, we determined that the best magnitude of the abundance of *N*-heterocyclic molecules would be to spotlight entirely on architecturally diverse drugs based on small molecules. The entire number of particular drugs embedded with a minimum of one nitrogen heterocycle rises from 613 to 640 (59%). Interestingly, for the small-molecule drugs, the amount of nitrogen-atoms per drug are 2.3 N/drug; whereas, that is 35% greater in those consist of a heterocycle embedded with nitrogen (3.1 N/drug) [79].

Having assembled and classified all of the 139 pharmaceuticals consist of an azole-related heterocycle, we were determined to take up the most common ones first (Fig. 1) [79]. The above-mentioned figure explains the top azole-related molecules in order of diminishing frequency expressed by a down-to-scale solid coloured bar, which features the assorted ring system of azoles. The most frequent azole-associated molecules that occurred in a sum of 30 exclusive small-compound drugs is thiazole followed by imidazole and tetrazole with 24 and 16 drugs respectively.

Significantly, benzimidazoles were found in 13 approved drugs and 1,2,4-triazole about 11 drugs. Consequently, imidazoline, tetrazole, pyrazole, thiadiazole and benzisoxazole rounding off the top 10with almost equal representation. Aromatic

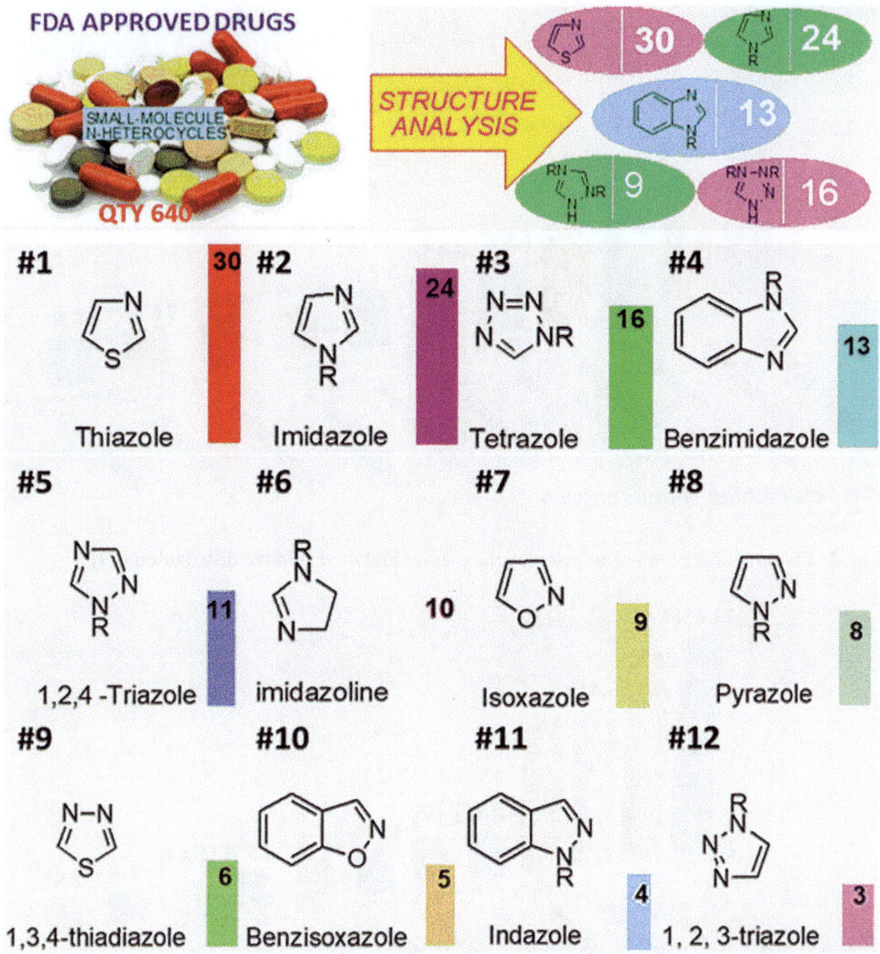

Fig. 1 Topmost frequent used azole-related *N*-heterocyclic moieties in U.S. FDA recognized drugs [79]

moieties are frequent structural units of many authorized pharmaceuticals and azole-containing heterocycles, which belong to the maximum of the top 12 motifs, is no exception.

As it is evident from (Fig. 2) the corresponding predominance of a broad range of nitrogen heterocyclic scaffolds differs significantly. Six-membered and five-membered rings are the most commonly employed followed by fused rings. Given the significance of six and five-membered heterocyclic rings, we determined to allocate these investigations and presented these subsections as non-aromatic and aromatic nitrogen heterocyclic scaffolds. The Given data [79] in (Figs. 2 and 3) admits exceptional discrepancies amongst two ring sizes, with 62.4% of the 5-membered

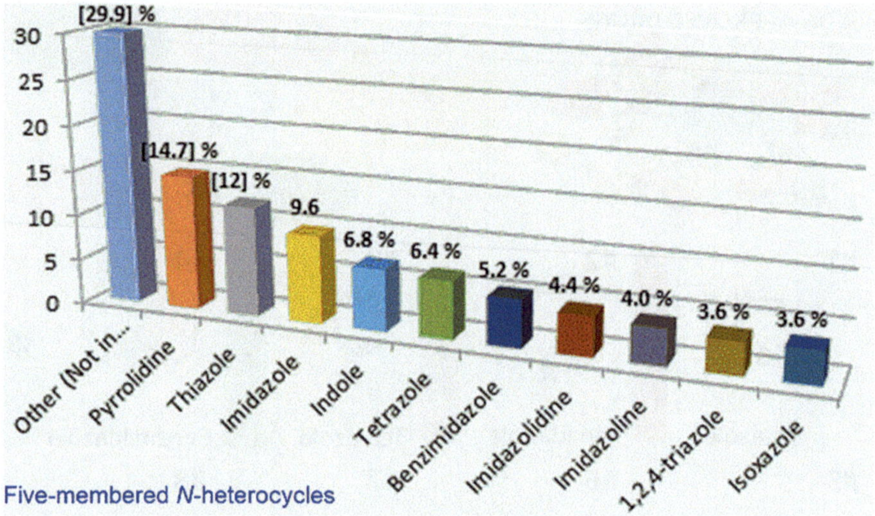

Fig. 2 Five-membered nitrogen heterocyclic classes and their relative distribution [79]

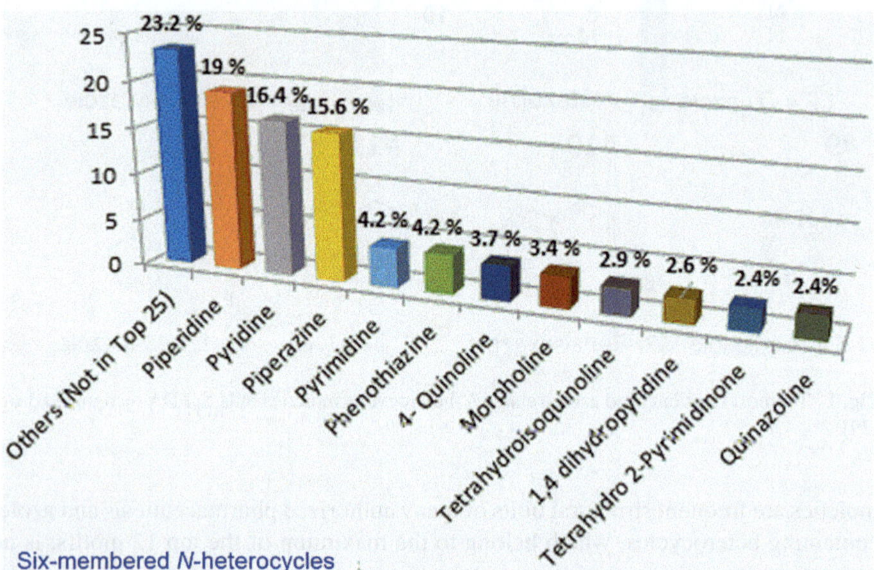

Fig. 3 Six-membered nitrogen heterocyclic classes and their relative distribution [79]

ring *N*-heterocyclic molecules being aromatically correlated to only 27.9% of 6-membered rings. The associated bar graph showcases the respective value of heterocyclic compounds from these vital classes and represents their architecture among the topmost of all five-membered nitrogen-containing heterocycles.

Azole cores having electron-deficient nature and good electron-accepting capacity as well as thermal and chemical stability are broadly applied in optics and electrochemistry [80]. The heterocycles having electron-rich properties may not only cheerfully donate or accept a proton but also readily form various weak bonding interactions. Therefore, the distinct framework aspects of azole-related rings are advantageous for its related compounds to eagerly interact with a range of receptors and enzymes in biological arrangements via coordination, ion–dipole, hydrogen bonds, cation-, hydrophobic effects, van der Waals interactions, π-π stacking and thereby representational immense bioactivities [81]. Azole ring is present in naturally widespread molecules and a number of bioactive entities in human metabolism [82]. To wonder, nature has selected these unique types of azole rings to be presented in most abundant biological molecules such as vitamin B12, histamine, haemoglobin and deoxyribonucleic acid (DNA) for exerting miscellaneous biological functions. This shows that azole-related rings could be crucial for the physiological response of imperative biological activities [83–85]. These specific physiological properties and extremely important appearance in essential processes have been engaging exceptional importance in azole-based biological and pharmaceutical chemistry. The presence of azole ring in amusing compounds may be agreeable for developing solubility property to certain extension due to its hetero atoms efficiently leading to the construction of hydrogen bonds. More predominantly, azole ring with manifold binding sites is adequate for coordinating with diverse nature of inorganic metal ions or collaborating with organic scaffolds through non-covalent bond to produce drugs of supramolecular character, which might have not only biological functions of azole molecules themselves, however also the assets of diverse drugs of supramolecular activity, probably exerting mechanisms using dual actions these are accessible to overwhelm resistances of drugs [83–87]. On the other hand, an azole ring as an attractive binding site could collaborate with distinct cations and anions as well as biomolecules within the human body. Therefore, azole rings have been intermittently incorporated into fluorescent frames to achieve fictitious fluorescent derivatives as pathologic probes and diagnostic agents to audit the biochemical development of biologically valuable ions and molecules in the living system for accepting biological phenomena. All the above-mentioned properties of azoles exhibit the enormous aptitude of azole-embedded molecules in pharmaceutical chemistry and a lot of growing work has been conducted toward their possible productive utilization in diverse areas.

In this chapter, an overview of the biological activity of the azole-related heterocyclic rings and its derivatives as well as pharmacological importance has been given. Owing to huge diversity in the biological field, the azole family has focused the consideration of scientists to study its skeleton chemically and biologically.

This chapter highlights the different pharmacological properties of azole derivatives. Studies on the biological activity of azole derivatives developed by many scientists around the globe are reported. For the first time, we are reporting the biological and pharmacological activities of the entire azole family at a glance.

2 Biological Activities of Azole Derivatives

The azole-containing heterocycle ring is omnipresent in nature and azole functionality plays a decisive role in many substrates as well as the human body also. The azole-related ring has been found in several naturally occurring important commodities which include the α-amino acid histidine, general constituents of most proteins, histamine, purine and biotin [88–98]. They are normally encountered in drugs that display diversification of pharmacological activities such as anti-inflammatory, histamine-H3 antagonist, antioxidant, gastro-protective, antitumoral, antiparasitic, antiviral, antibacterial, antitubercular as well as their potential activities in diagnostics and pathology [99–106]. The biological applicability of these kinds of heteroaromatic groups is because of their excellent bioisosteres of biomolecules. Particularly a huge number of azole-based compounds as clinical drugs such as anticancer [107, 108]. (dacarbazine, azathioprine, tipifarnib, zoledronic acid), antifungal [109, 110] (miconazole, ketoconazole, clotrimazole, oxiconazole), Antiparasitic [111, 112] (metronidazole, ornidazole, benznidazole, secnidazole), antihistaminic [113] (imetit, immepip, cimetidine, thioperamide), antineuropathic [114] (fipamezole, nafimidone, dexmedetomidine), antihypertensive [115] (eprosartan, olmesartan, losartan), drugs have been broadly use to treat different types of diseases having high therapeutic efficiency, which have shown the enormous development value. This has been energetically promoting much effort to spotlight on azole-embedded pharmaceutical agents and the enlarging research and developments have become an increasingly active and developing topic and almost extended to the entire range of medicinal field. Arrangement of azole drugs according to pharmacological actions.

- Adrenergic Agents: Clonidine, Phentolamine, Tolazoline
- Cholinergic Agents: Pilocarpine HCl
- Sulphonamides: Sulphamethizole, Sulfisoxazole, Sulfamethoxazole.
- Cephalosporins: Cefazolin sodium, Cefonicid sodium, Ceforanide, Cefoperazone, Cefmetazole, Cefotetan sodium, Cefixime, Cefotaxime sodium
- Diuretics: Muzolimine, Acetazolamide, Methazolamide.
- Cardiovascular Agents: Saralasin, Methimazole
- Antihistaminic Agents: Famotidine, Cimetidine
- Analgesics: Etonitazene
- Anti-inflammatory Analgesics: Tolmetin, Phenylbutazone
- Steroids: Imazodan, Pimobendan.
- Amino Acids: Histidine, Tryptophan

- Antiviral Agents: Ribavirin, Ritonavir, Vidarabin, Acyclovir, Valacyclovir, Ganacyclovir, Famciclovir, Penciclovir.
- Antineoplastic Agents: Dacarbazine, Mercaptopurine
- Immunotherapy: Levamisole
- CNS depressants: Etomidate, Alprazolam, Midazolam, Triazolam.
- Antipsychotics: Ondansetron
- CNS stimulants: Pentylene tetrazole, Methylxanthines, Pemoline, Isocarboxazid, Mazindol, Trazodone HCl
- Vitamins: Vitamin B_1, Vitamin B_{12}.

2.1 *Mechanism of Action of Azole Antifungals*

They act by decreasing the CYP450 membrane-bound, i.e. "Lanosterol-14-α-demethylase" which is the enzyme involved in the early stages of the synthesis of ergosterol inhibits oxidative demethylation and retains excess of lanosterol which destabilizes the cell (Fig. 4) [116].

2.2 *SAR of Azole Antifungals*

- They are characterized by the presence of a weakly basic imidazole or 1,2,4-triazole nucleus for antifungal activity.
- The amidines "N" atom at the 3rd position of imidazole and 4th position of triazole is responsible for the decreased concentration of heme iron due to

$$\text{Complexation} \rightarrow \text{Decrease } Fe^{+2} \rightarrow \text{Decrease cell respiration} \rightarrow \text{Hypoxia}$$

- All azole consists of 2 or 3 bulkier substituents like phenyl or substituted phenyl group enrich imparts lipophilicity to disturb cell membrane.

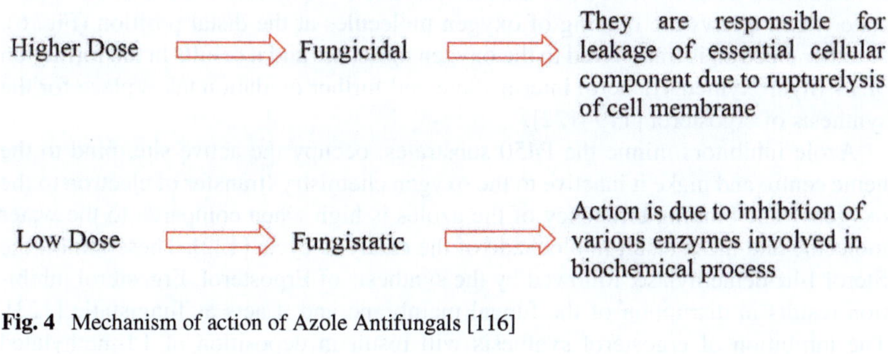

Fig. 4 Mechanism of action of Azole Antifungals [116]

Azole has a unique structure that makes it easily bind to the active site through various non-covalent forces such as electrostatic, Vander Waals, coordination bonds, hydrophobic interactions and hydrogen bonds. Azole and its derivatives have a broad spectrum of activity in various categories. It can be proved by seeing the entries of the azoles in the Drugbank database (approved: 202; illicit: 09; nutraceutical: 09; experimental 756 and investigational: 94) [117].

The list of FDA-approved *N*-heterocycle-containing drugs has been portrayed in Table 1 [117].

2.3 Molecular Mechanism of Azole Antifungals

Azole and triazole-based antifungals are cytochrome P450-dependent enzymes, i.e. Lanosterol 14α-demethylase inhibitors. The prevention of this enzyme results in the suppression of biosynthesis of ergosterol at the C-14 demethylation stage and it is responsible for its antifungal activity [118–120]. These azole derivatives are used to treat various fungal infections both systemically and superficially. The advantage of these compounds is high selectivity towards fungi over human ergosterol [118, 119]. Ergosterol is a steroid moiety in fungi and parasites. It is important for the viability of the parasites and their growth as it regulates the fluidity of the plasma membrane, its biogenesis and function [119–121]. The source of the Ergosterol in the fungi is by a sterol synthetic pathway, i.e. from Lanosterol, a precursor for various steroids. It is transformed into Ergosterol by Sterol 14α-demethylase as one of the enzymes. Sterol14α-demethylase is a Cytochrome P450 (CYP51) family enzyme (Fig. 5) [119].

P450 enzymes have an active site that consists of a heme moiety at the centre and there is a Fe-S linkage between the Cysteinate residue and the backbone of the protein. The Cysteine ligand is present on the axial side of the heme and its distal end of the heme is occupied by the water molecule in the resting state. The active site of the Sterol 14α-demethylase is occupied by Sterol substrate such that the 14α-methyl group is 5A^0 above the heme iron (Fe^{+3}) of the enzyme. An electron is transferred from the CYP51 to the heme iron (Fe^{+3}) and it will be reduced to Fe^{+2} state which allows the binding of oxygen molecules at the distal position (Fig. 6). Another electron is transferred to the oxygen molecule and it results in the formation of 14-(hydroxymethyl) sterol intermediate and further oxidation takes place for the synthesis of ergosterol [119–122].

Azole inhibitors mimic the P450 substrates, occupy the active site, bind to the heme centre and make it inactive to the oxygen chemistry (transfer of electron to the oxygen). The binding efficiency of the azoles is high when compared to the water molecule and these result in blockade of the catalytic cycle [122]. These inhibit the Sterol 14α-demethylase, followed by the synthesis of Ergosterol. Ergosterol inhibition results in disruption of the fungal membrane and it acts as fungistatic [123]. The inhibition of ergosterol synthesis will result in deposition of 14-methylated sterols and lanosterol. This accumulation results in disruption of the structure of

Table 1 Azole-containing drugs and their mode of action [117]

Azoles	Drugs	Mechanism of action	Category
Pyrrole	Ramipril	ACE inhibitor	Congestive heart failure, Hypertension, Myocardial function, Anti-tumour
	Tolmetin	NSAID	Anticancer, Rheumatoid arthritis, Osteoarthritis
	Telaprevir	Directly acting antiviral agent against Hepatitis-C virus. Breast cancer resistance protein (BCRP), Inhibition of P-glycoprotein(P-gp)	Antiviral, Anticancer
Pyrazole	Betazole	Histamine (H2) agonist	Zollinger-Ellison syndrome
Imidazole	Bifonazole	Inhibition of Ergosterol biosynthesis. It induces the detachment of hexokinase from mitochondria related to the B16 melanoma cells. It results in the reduction of viability of cells	Antifungal, Skin cancer treatments, In the treatment of melanoma
	Clotrimazole	Phosphoglyceride and triglyceride biosynthesis inhibition and blockade of ion channel pathway	Antimycotic activity
	Miconazole	14α-demethylase inhibitor	Antifungal, Used in the treatment of breast cancer
	Tetryzoline		Used in eye drops and nasal drops
	Econazole	14α-demethylase inhibitor	Treatment of prostate cancer
	Dacarbazine	Inhibition of DNA synthesis by acting as a purine analogue and interaction with the thiol group	Cytotoxic effects, Anticancer
	Tolazoline	Interacts with adrenergic, cholinergic and histaminic receptors	Persistent pulmonary hypertension
	Butoconazole	Presumed to inhibit the synthesis of steroid	Actively used against vaginal infections mediated by Candida albicans
	Carbimazole	Decreases the uptake of inorganic Iodine by thyroid	Antithyroid activity

(continued)

Table 1 (continued)

Azoles	Drugs	Mechanism of action	Category
	Losartan	Angiotensin receptor blocker	Systolic dysfunction, heart failure, Coronary artery disease, myocardial infarction, hypertension, etc. Coronary artery disease, diabetic neuropathy, diabetic neuropathy and systolic dysfunction, hypertension
	Eprosartan	Angiotensin-II receptor antagonist	Treatment of high blood pressure
Triazole	Ribavirin	RNA dependent RNA polymerase inhibitor	Hepatitis-C, Respiratory Syncytial respiratory disease
Isoxazole	Sulfisoxazole	Inhibits folic acid synthesis	Treatment of infections by Gram-positive and Gram-negative bacteria
Thiazole	Sulfathiazole		Antimicrobial and active for both Gram-negative and Gram-positive
	Thiabendazole	Chelating agent	Vermicidal and or Vermifugal agent
	Dasatinib	Suppress the actions of Src-family tyrosine kinase, BCR/ABL and other oncogene kinases	Chronic myelogenous leukemia (CML)

plasma membrane and it also disturbs the activities of various other membrane-bound enzymes which are involved in chitin synthesis and nutrient transport. It also affects the cell growth and proliferation of the fungi [121]. Ergosterol inhibition results in various secondary effects like the morphogenic transformation of yeast to mycelia are inhibited, fungal adherence is decreased and their will be toxic effects on membrane phospholipids [123].

Azole antifungals exhibit interaction between the N4 or N3 atom in the triazole or imidazole and the heme iron associated with the CYP51 enzyme. The other atoms in the molecule exhibit aromatic stacking, hydrophobic and various van der Waals interactions with the amino acid component of the enzyme [119]. Miconazole is the topical antifungal. Ketoconazole, Itraconazole, Fluconazole are the systemic antifungals [120].

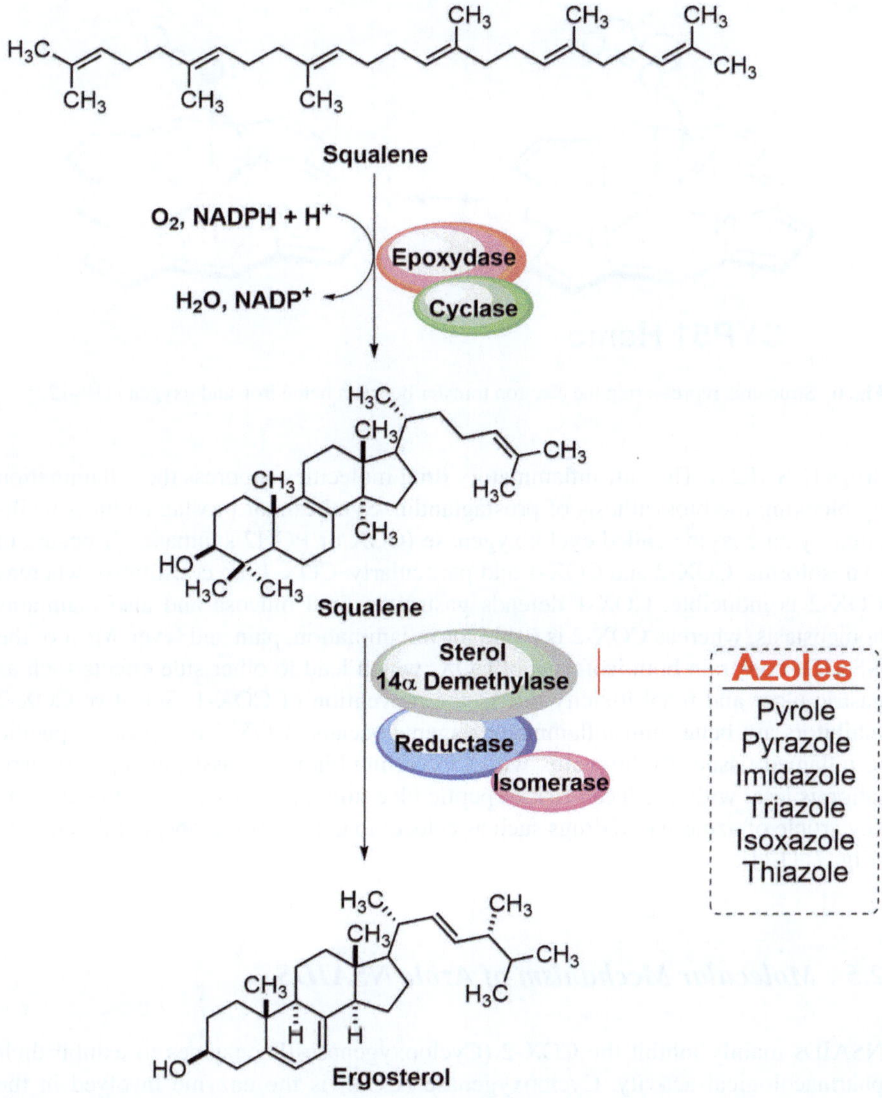

Fig. 5 Molecular mechanism of action of azole antifungals [119]

2.4 *Mechanism of Action of Azole Non-Steroidal Anti-Inflammatory Drugs (NSAIDs)*

NSAIDs are known as non-narcotic drugs which act as pain relievers and lower body temperature and azoles are among the imperative classes of scaffolds known as NSAIDs. However, these drugs do not contain steroidal nuclei and also have an anti-inflammatory action, they are recognized as non-steroidal anti-inflammatory

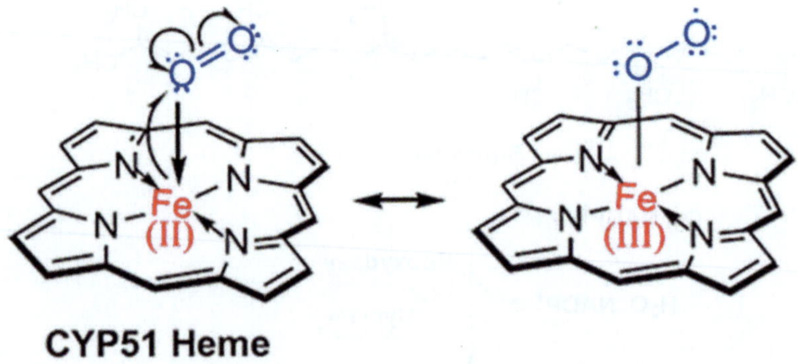

Fig. 6 Structures representing the electron transfer between heme iron and oxygen [119–122]

drugs (NSAIDs). The anti-inflammatory drug molecules suppress the inflammation by blocking the biosynthesis of prostaglandin. Synthesis of prostaglandin is facilitated by an enzyme called cyclooxygenase (COX or PGH2 synthase). It occurs in two isoforms, COX-2 and COX-1 and particularly, COX-1 are constitutive whereas COX-2 is inducible. COX-1 defends gastrointestinal mucosa and also maintains homeostasis, whereas COX-2 is liable for inflammation, pain and fever. Most of the NSAIDs constrain both isoforms of COX, which lead to other side effects such as gastric ulcer and renal toxicity due to the prevention of COX-1. Selective COX-2 inhibitors are better anti-inflammatory agents because COX-2 is typically specific to inflamed tissue. Additionally, with COX-2 inhibitors the associated gastric irritation is less, with a reduced risk of peptic ulceration. Selectivity for COX-2 is the key article of azole-based drugs such as celecoxib and other members of this family (Fig. 7) [124].

2.5 *Molecular Mechanism of Azole NSAIDS*

NSAIDs mainly inhibit the COX-2 (Cyclooxygenase-II) enzymes to exhibit their pharmacological activity. Cyclooxygenase (COX) is the enzyme involved in the synthesis of inflammatory mediators. This enzyme converts the Arachidonic acid into PGG_2 by the addition of two oxygen molecules. The PGG_2 enters into the active site of the Peroxidase (POX) and leads in the generation of PGH_2 after two-electron reduction; various inflammatory mediators like PGI_2, PGE_2, PGD_2, $PGF_{2\alpha}$, TxA_2 are formed by the action of some enzymes like synthases (Fig. 8) [124].

Cyclooxygenase is embedded in three domains namely a C-terminal catalytic domain, a membrane-binding domain and a short N-terminal epidermal growth factor domain which is large and globular in shape. The catalytic consists of POX and COX active sites in the reverse sides and a heme prosthetic group are present at the base

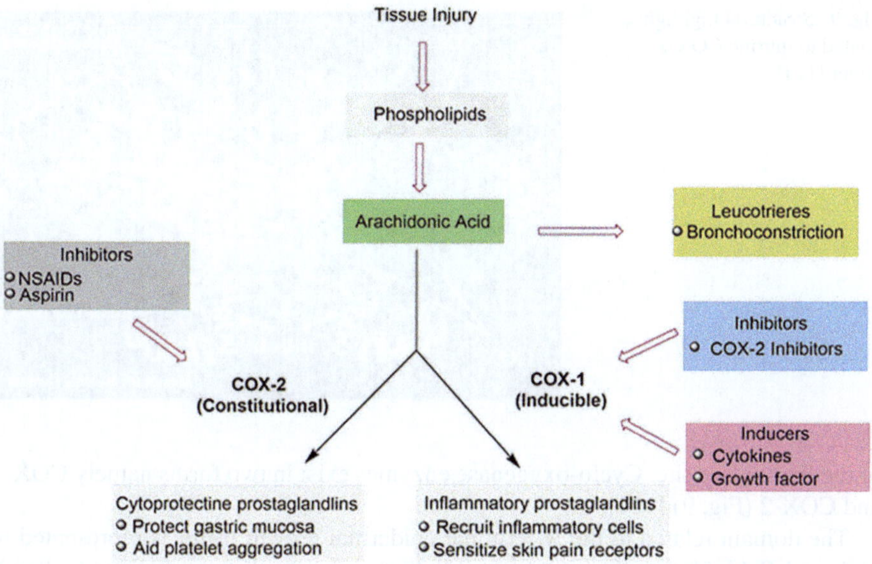

Fig. 7 Mechanism of action of azole non-steroidal anti-inflammatory drugs [124]

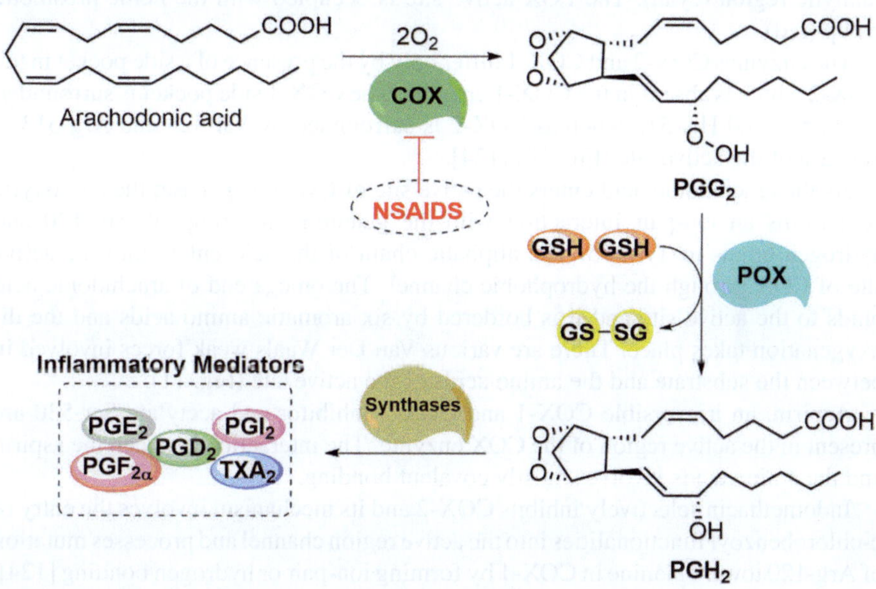

Fig. 8 Pathway for inflammatory mediator's production and NSAIDS showing its site of inhibition [124]

Fig. 9 Structural highlights related to murine COX-2 dimer [124]

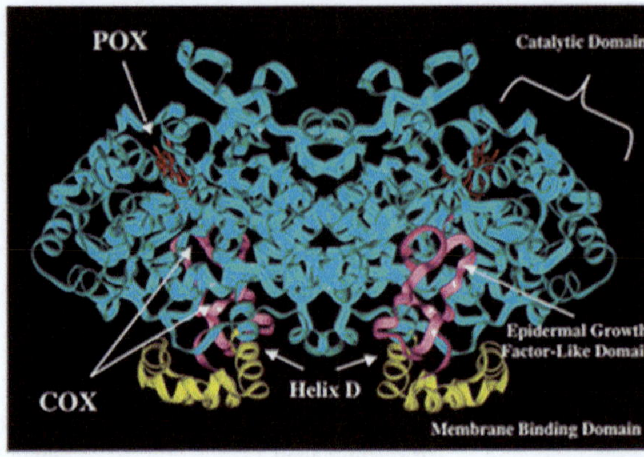

of the Peroxidase site. Cyclo-oxygenase enzymes exist in two forms namely COX-1 and COX-2 (Fig. 9) [124].

The domain related to the *N*-terminal epidermal growth factor is represented in pink and R highlights the membrane-binding region with four -helices (yellow). Helix D shows up in the COX active region, which is situated in the large globular catalytic region (cyan). The POX active site is occupied with the heme prosthetic group (red).

The enzymes COX-2 and COX-1 differ only by the presence of a side pocket in the COX-2 which is absent in the COX-1 enzyme. The COX-1 side pocket is surrounded by Ile-523 and His-513 whereas COX-2 is surrounded by Val-523 and Arg-513 at the base of the active site (Fig. 10) [124].

As the arachidonic acid enters the active site of Cyclooxygenase; the carboxylic acid forms an ion-pair interaction with the guanidinium group of Arg-120 and hydrogen bonds to Tyr-355. The aliphatic chain of the acid enters into the active site of COX through the hydrophobic channel. The omega end of arachidonic acid binds to the active site and it is bordered by six aromatic amino acids and the di-oxygenation takes place. There are various Van Der Waals weak forces involved in between the substrate and the amino acids of the active site (Fig. 11).

Aspirin, an irreversible COX-1 and COX-2 inhibitor and acetylate Ser-530 are present in the active region of the COX enzyme. The interaction between the aspirin and the amino acids involves mostly covalent bonding.

Indomethacin selectively inhibits COX-2 and its mechanism involves the entry of p-chlorobenzoyl functionalities into the active region channel and processes mutation of Arg-120 toward alanine in COX-1 by forming ion-pair or hydrogen bonding [124].

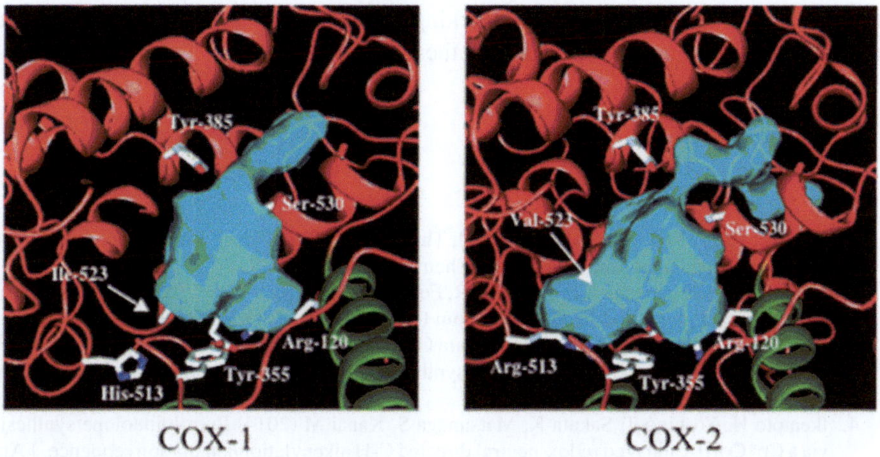

Fig. 10 Catalytic regions of COX proteins are highlighted as red in a ribbon diagram with membrane-binding regions (predominantly helix D) represented in green. Residues lining the COX active region of these proteins are highlighted in white with solvent reachable surfaces in the active region indicated as translucent light-blue [124]

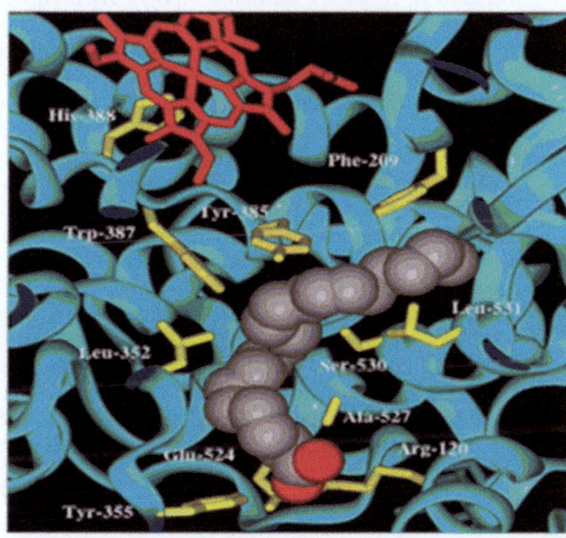

Fig. 11 Arachidonic acid is bound to the active site of the COX-2 enzyme [124]

3 Summary/Conclusion

This chapter emphasizes the medicinal importance of azole derivatives in various fields of medicinal chemistry. The biological significance and mode of action of azole-containing drugs have been discussed in detail for the benefit of readers. The application of azole derivatives as antifungals, NSAIDs has been elucidated in brief to captivate the interest of chemists working in the field of medicinal chemistry and drug discovery to broaden the utility of these molecules.

References

1. Galm U, Hager MH, Van Lanen SG, Ju J, Thorson JS, Shen B (2005) Antitumor antibiotics: bleomycin, enediynes, and mitomycin. Chem Rev 105:739–758
2. Fernandez LS, Buchanan MS, Carroll AR, Feng YJ, Quinn RJ, Avery VM (2009) Flinderoles A-C: antimalarial bis-indole alkaloids from flindersia species. Org Lett 11:329–332
3. Elmegeed GA, Baiuomy AR, Abdel- Salam OME (2007) Evaluation of the anti-inflammatory and anti-nociceptive activities of novel synthesized melatonin analogues. Eur J Med Chem 42:1285–1292
4. Ikemoto H, Yoshino T, Sakata K, Matsunaga S, Kanai M (2014) Pyrroloindolone synthesis via a Cp*CoIII-catalyzed redox-neutral directed C-H alkenylation/annulation sequence. J Am Chem Soc 136:5424–5431
5. Kok SHL, Gambari R, Chui CH, Yuen MCW, Lin E, Wong RSM, Lau FY, Cheng GYM, Lam WS, Chan SH, Lam KH, Cheng CH, Lai PBS, Yu MWY, Chueng F, Tang JCO, Chan ASC (2008) Synthesis and anti-cancer activity of benzothiazole containing phthalimide on human carcinoma cell lines. Bioorg Med Chem 16:3626–3631
6. Liu C, Lin J, Pitt S, Zhang RF, Sack JS, Kiefer SE, Kisah K, Doweyko AM, Zhang H, Marathe PH, Trzaskos J, Mckinnon M, Dodd JH, Barrish JC, Schieven GL, Leftheris K (2008) Benzothiazole based inhibitors of p38α MAP kinase. Bioorg Med Chem Lett 18:1874–1879
7. Shneine JK, Alaraji YH (2016) Chemistry of 1, 2, 4-triazole: a review article. Int J Sci Res 5:2319–7064
8. Kapkoti DS, Singh S, Luqman S, Bhakuni RS (2018) Synthesis of novel 1,2,3-triazole based artemisinin derivatives and their antiproliferative activity. New J Chem 42:5978–5995
9. Kuang J, Chen B, Ma S (2014) Copper-mediated efficient three-component synthesis of 1,2,4-triazoles from amines and nitriles. Org Chem Front 1:186–189
10. Sun GX, Yang MY, Shi YX, Sun ZH, Liu XH, Wu HK, Li BJ, Zhang YG (2014) Microwave assistant synthesis, antifungal activity and DFT theoretical study of some novel 1,2,4-triazole derivatives containing pyridine moiety. Int J Mol Sci 15:8075–8090
11. Tang R, Jin L, Mou C, Yin J, Bai S, Hu D, Wu J, Yang S, Song B (2013) Synthesis, antifungal and antibacterial activity for novel amide derivatives containing a triazole moiety. Chem Cent J 7:30–36
12. Ötvös SB, Fülöp F (2015) Flow chemistry as a versatile tool for the synthesis of triazoles. Catal Sci Technol 5:4926–4941
13. Xu K, Huang L, Xu Z, Wang Y, Bai G, Wu Q, Wang X, Yu S, Jiang Y (2015) Design, synthesis, and antifungal activities of novel triazole derivatives containing the benzyl group. Drug Des Dev Ther 9:1459–1467
14. Bello AM, Konforte D, Poduch E, Furlonger C, Wei L, Liu Y, Lewis M, Pai EF, Paige CJ, Kotra LP (2009) Structure-activity relationships of orotidine-5′-monophosphate decarboxylase inhibitors as anticancer agents. J Med Chem 52:1648–1658
15. Krajewska E, Shugar D (1982) Pyrimidine ribonucleoside phosphorylase activity vs 5-and/or 6-substituted uracil and uridine analogues, including conformational aspects. Biochem Pharmacol 31:1097–1102
16. Pal'chikov VA (2013) Morpholines. synthesis and biological activity. Russ J Org Chem 49:787–814
17. Naim MJ, Alam O, Alam MJ, Alam P, Shrivastava N (2015) A review on pharmacological profile of morpholine derivatives. Int J Pharmacol Pharm Sci 1:40–51
18. Gentles RG, Grant-Young K, Hu S, Huang Y, Poss MA, Andres C, Fiedler T, Knox R, Lodge N, Weaver CD, Harden DG (2008) Initial SAR studies on apamin-displacing 2-aminothiazole

blockers of calcium-activated small conductance potassium channels. Bioorg Med Chem Lett 18:5316–5319
19. Zhao M, Yin J, Huffman MA, McNamara JM (2006) A very concise synthesis of a potent N-(1,3-thiazol-2-yl)pyridin-2-amine KDR kinase inhibitor. Tetrahedron 62:1110–1115
20. Heng S, Gryncel KR, Kantrowitz ER (2009) A library of novel allosteric inhibitors against fructose 1,6-bisphosphatase. Bioorg Med Chem 17:3916–3922
21. Gallardo-Godoy A, Gever J, Fife KL, Silber BM, Prusiner SB, Renslo AR (2011) 2-aminothiazoles as therapeutic leads for prion diseases. J Med Chem 54:1010–1021
22. Yeh VSC (2004) Recent advances in the total syntheses of oxazole-containing natural products. Tetrahedron 60:11995–12042
23. Fu RG, Wang Y, Xia F, Zhang HL, Sun Y, Yang DW, Wang YW, Yin P (2019) Synthesis of 2-Amino-5-acylthiazoles by a tertiary amine-promoted one-pot three-component cascade cyclization using elemental sulfur as a sulfur source. J Org Chem 84:12237–12245
24. Miura T, Funakoshi Y, Fujimoto Y, Nakahashi J, Murakami M (2015) Facile synthesis of 2,5-disubstituted thiazoles from terminal alkynes, sulfonyl azides, and thionoesters. Org Lett 17:2454–2457
25. Zificsak CA, Hlasta DJ (2004) Current methods for the synthesis of 2-substituted azoles. Tetrahedron 60:8991–9016
26. Turchi IJ, Dewar MJS (1975) Chemistry of oxazoles. Chem Rev 75:389–437
27. Esposito G, Teta R, Miceli R, Ceccarelli LS, Sala GD, Camerlingo R, Irollo E, Mangoni A, Pirozzi G, Costantino V (2015) Isolation and assessment of the in vitro anti-tumor activity of smenothiazole A and B, chlorinated thiazole-containing peptide/polyketides from the caribbean sponge, smenospongia aurea. Mar Drugs 13:444–459
28. Bellina F, Rossi R (2006) Synthesis and biological activity of pyrrole, pyrroline and pyrrolidine derivatives with two aryl groups on adjacent positions. Tetrahedron 62:7213–7256
29. Saraswat P, Jeyabalan G, Hassan MZ, Rahman MU, Nyola NK (2016) Review of synthesis and various biological activities of spiro heterocyclic compounds comprising oxindole and pyrrolidine moieties. Synth Commun 46:1643–1664
30. Neumann KT, Lindhardt AT, Bang-Andersen B, Skrydstrup T (2015) Palladium-catalyzed carbonylative sonogashira coupling of aryl bromides using near stoichiometric carbon monoxide. Org Lett 17:2094–2097
31. Saha P, Ramana T, Purkait N, Ali MA, Paul R, Punniyamurthy T (2009) J Org Chem 74:8719–8725
32. Roth HJ, Kleemann A (1988) Drug synthesis in pharmaceutical chemistry, vol 1, pp 88–114. Wiley, New York
33. Joule JA, Mills K (2000) Heterocyclic chemistry, 4th edn. Blackwell Science, Cambridge, pp 63–120
34. Olbe L, Carlsson E, Lindberg P (2003) A Proton-pump inhibitor expedition: the case histories of omeprazole and esomeprazole. Nat Rev Drug Discovery 2:132–139
35. Michael JP (2005) Quinoline, quinazoline and acridone alkaloids. Nat Prod Rep 22:627–646
36. Katritzky AR, Ramsden CA, Scriven EFV, Taylor RJK (2008) Comprehensive heterocyclic chemistry. Elsevier, Oxford 1:1–13718
37. Henry GD (2004) De novo synthesis of substituted pyridines. Tetrahedron 60:6043–6061
38. Alamgir M, Black DSC, Kumar N (2007) Synthesis, reactivity and biological activity of benzimidazoles. In: Khan MTH (ed) Bioactive heterocycles III, vol 9. Topics in Heterocyclic Chemistry. Springer, Berlin, Heidelberg, pp 87–118
39. Horton DA, Bourne GT, Smythe ML (2003) The combinatorial synthesis of bicyclic privileged structures or privileged substructures. Chem Rev 103:893–930
40. Skalitzky DJ, Marakovits JT, Maegley KA, Ekker A, Yu X-H, Hostomsky Z, Webber SE, Eastman BW, Almassy R, Li J, Curtin NJ, Newell DR, Calvert AH, Griffin RJ, Golding BT (2003) Tricyclic benzimidazoles as potent poly(ADP-ribose) polymerase-1 inhibitors. J Med Chem 46:210–213
41. Rao A, Chimirri A, De Clercq E, Monforte AM, Monforte P, Pannecouque C, Zappalà M (2002) Synthesis and anti-HIV activity of 1-(2,6-difluorophenyl)-1H,3Hthiazolo[3,4-a]benzimidazole structurally-related 1,2-substituted benzimidazoles. Il Farmco. 57:819–823

42. Sun Q, Wu R, Cai S, Lin Y, Sellers L, Sakamoto K, He B, Peterson BR (2011) Synthesis and biological evaluation of analogues of AKT (protein kinase B) inhibitor-IV. J Med Chem 54:1126–1139
43. Ueda K, Amaike K, Maceiczyk RM, Itami K, Yamaguchi J (2014) β-selective C-H arylation of pyrroles leading to concise syntheses of lamellarins C and I. J Am Chem Soc 136:13226–13232
44. Young IS, Thornton PD, Thompson A (2010) Synthesis of natural products containing the pyrrolic ring. Nat Prod Rep 27:1801–1839
45. Roth BD (2002) The discovery and development of atorvastatin, a potent novel hypolipidemic agent. Prog Med Chem 40:1–22
46. Fürstner A, Domostoj MM, Scheiper B (2005) Total synthesis of dictyodendrinB. J Am Chem Soc 127:11620–11621
47. Zhang W, Liu Z, Li S, Yang T, Zhang Q, Ma L, Tian X, Zhang H, Huang C, Zhang S, Ju J, Shen Y, Zhang C (2012) Spiroindimicins A-D: new bisindole alkaloids from a deep-sea-derived actinomycete. Org Lett 14:3364–3367
48. Beck EM, Hatley R, Gaunt MJ (2008) Synthesis of rhazinicine by a metal-catalyzed C-H bond functionalization strategy. Angew Chem Int Ed 47:3004–3007
49. Chen L, Min H, Zeng W, Zhu X, Liang Y, Deng G, Yang Y (2018) Transition-metal-free sulfuration/annulation of alkenes: economical access to thiophenes enabled by the cleavage of multiple C-H bonds. Org Lett 20:7392–7395
50. Davis RA, Buchanan MS, Duffy S, Avery VM, Charman SA, Charman WN, White KL, Shackleford DM, Edstein MD, Andrews KT, Camp D, Quinn RJ (2012) Antimalarial activity of pyrroloiminoquinones from the australian marine sponge Zyzzya sp. J Med Chem 55:5851–5858
51. Nichick MN, Voitekhovich SV, Lesnyak V, Matulis VE, Zheldakova RA, Lesnikovich AI, Ivashkevich OA (2011) 1-Substituted tetrazole-5-thiol-capped noble metal nanoparticles. J Phys Chem C 115:16928–16933
52. Michael JP (2008) Quinoline, quinazoline and acridone alkaloids. Nat Prod Rep 25:166–187
53. Barluenga J, Rodriguez F, Fananas FJ (2009) Recent advances in the synthesis of indole and quinoline derivatives through cascade reactions. Chem Asian J 4:1036–1048
54. Michael JP (2007) Quinoline, quinazoline and acridone alkaloids. Nat Prod Rep 24:223–246
55. Koen A, Peter V, Jerome G, Hinrich WHG, Marc NJ, Hans W, Van GJ, Philip T, Min Z, Ennis L, Peter W, de Didier C, Emma H, Sven H, Emmanuelle C, Chantal TP, Nacer L, Vincent J (2005) A Diarylquinoline drug active on the ATP synthesis of synthase of mycobacterium tuberculosis. Science 307:223–227
56. Rouffet M, de Oliveira CAF, Udi Y, Agrawal A, Sagi I, McCammon JA, Cohen SM (2010) From sensors to silencers: quinoline- and benzimidazole-sulfonamides as inhibitors for zinc proteases. J Am Chem Soc 132:8232–8233
57. Bax BD, Chan PF, Eggleston DS, Fosberry A, Gentry DR, Gorrec F, Giordano I, Hann MM, Hennessy A, Hibbs M, Huang J, Jones E, Jones J, Brown KK, Lewis CJ, May EW, Saunders MR, Singh O, Spitzfaden CE, Shen C, Shillings A, Theobald AJ, Wohlkonig A, Pearson ND, Gwynn MN (2010) Type IIA topoisomerase inhibition by a new class of antibacterial agents. Nature 466:935–940
58. Alajarín R, Burgos C (2011) Modern heterocyclic chemistry. In: Alvarez- Builla J, Vaquero JJ, Barluenga J (eds) Wiley-VCH, Weinheim, vol 3, pp 1527–1629
59. Kletsas D, Li W, Han Z, Papadopoulos V (2004) Peripheral-type benzodiazepine receptor (PBR) and PBR drug ligands in fibroblast and fibrosarcoma cell proliferation: role of ERK, c-Jun and ligand-activated PBR-independent pathways. Biochem Pharmacol 67:1927–1932
60. Walsh CT, Garneau-Tsodikova S, Howard-Jones AR (2006) Biological formation of pyrroles: nature's logic and enzymatic machinery. Nat Prod Rep 23:517–531
61. Fürstner A (2003) Chemistry and biology of roseophilin and the prodigiosin alkaloids: a survey of the last 2500 years. Angew Chem Int Ed 42:3582–3603
62. Khann IK, Yu Y, Huff RM, Weier RM, Xu X, Koszyk FJ, Collins PW, Cogburn JN, Lsakson PC, Koboldt CM, Masferrer JL, Perkins WE, Seibert K, Veenhuizen AW, Yuan J, Yang D, Zhang YY (2000) Selective cyclooxygenase-2 inhibitors: Heteroaryl modified 1,2-diarylimidazoles are potent, orally active antiinflammatory agents. J Med Chem 43:3168–3185

63. Shin J, Rho JR, Seo Y, Lee HS, Cho KW, Kwon HJ, Sim CJ (2001) Wondonins A and B, new bis (dihydroxystyryl) imidazoles from a two-sponge association. Tetrahedron Lett 42:1965–1968
64. Newman DJ, Cragg GM, Snader KM (2003) Natural products as sources of new drugs over the period 1981–2002. J Nat Prod 66:1022–1037
65. Pastor IM, Yus M (2009) Bioactive N-phenylimidazole derivatives. Curr Chem Bio 3:385–408
66. Kon Y, Kubota T, Shibazaki A, Gonoi T, Kobayashi J (2010) Ceratinadins A-C, new bromotyrosine alkaloids from an Okinawan marine sponge Pseudoceratina sp. Bioorg Med Chem Lett 20:4569–4572
67. Molina P, Tárraga A, Otón F (2012) Imidazole derivatives: a comprehensive survey of their recognition properties. Org Biomol Chem 10:1711–1724
68. El-Garhy OH (2014) An overview of the azoles of interest. Int J Curr Pharm Res 7:1–6
69. Gao H, Shreeve JM (2011) Azole-based energetic salts. Chem Rev 111:7377–7436
70. Nfor EN, Asobo PF, Nenwa J, Nfor ON, Njapba JN, Njong RN, Offiong OE (2013) Nickel (II) and iron (II) complexes with azole derivatives: synthesis, crystal structures and antifungal activities. Int J Inorg Chem. https://doi.org/10.1155/2013/987574
71. Sandhu SS, Shukla H, Aharwal R, Kumar S, Shukla S (2014) Antifungal azole derivatives and their pharmacological potential: prospects & retrospect. The Nat Prod J 4:140–152
72. Maertens JA (2004) History of the development of azole derivatives. Clin Microbiol Infect 1:1–10
73. Dwoskin LP, Jewell AL, Cassis LA (1992) DuP 753, a nonpeptide angiotensin II-1 receptor antagonist, alters dopaminergic function in rat striatum. Arch Pharmacol 345:153–159
74. Hill RA (2009) Marine natural products. Annu Rep Prog Chem, Sect B Org Chem 105:150–166
75. Forte B, Malgesini B, Piutti C, Quartieri F, Scolaro A, Papeo G (2009) A submarine journey: the pyrroleimidazole alkaloids. Mar Drugs 7:705–753
76. Jin Z (2011) Muscarine, imidazole, oxazole, and thiazole alkaloids. Nat Prod Rep 28:1143–1191
77. Gao G, Xiao R, Yuan Y, Zhou CH, You JS, Xie RG (2002) Efficient imidazolium catalysts for the benzoin condensation. J Chem Res 2002:262–263
78. Jiang HY, Zhou CH, Luo K, Chen H, Lan JB, Xie RG (2006) Chiral imidazole metalloenzyme models: Synthesis and enantioselective hydrolysis for a-amino acid esters. J Mol Catal A Chem 260:288–294
79. Vitaku E, Smith DT, Njardarson JT (2014) Analysis of the structural diversity, substitution patterns, and frequency of nitrogen heterocycles among U.S. FDA approved pharmaceuticals. J Med Chem 57:10257–10274
80. Miller RD, Lee VY, Moylan CR (1994) Substituted azole derivatives as nonlinear optical chromophores. Chem Mater 6:1023–1032
81. Zhang L, Peng X-M, Damu GLV, Geng R-X, Zhou C-H (2014) Comprehensive review in current developments of imidazole-based medicinal chemistry. Med Res Rev 34:340–437
82. Aleksandrova EV, Kravchenko AN, Kochergin PM (2011) Properties of haloimidazoles. Chem Heterocycl Compd 47:261–289
83. Narasimhan B, Sharma D, Kumar P (2011) Biological importance of imidazole nucleus in the new millennium. Med Chem Res 20:1119–1140
84. Shalini K, Sharma PK, Kumar N (2010) Imidazole and its biological activities: a review. Chem Sinica 1:36–47
85. Bhatnagar A, Sharma PK, Kumar N (2011) A review on "Imidazoles": their chemistry and pharmacological potentials. Int J PharmTech Res 3:268–282
86. Zhou CH, Gan LL, Zhang YY, Zhang FF, Wang GZ, Jin L, Geng RX (2009) Review on supermolecules as chemical drugs. Sci China Ser B 52:415–458
87. Zhou CH, Zhang YY, Yan CY, Wan K, Gan LL, Shi Y (2010) Recent researches in metal supramolecular complexes as anticancer agents. Anticancer Agents Med Chem 10:371–395
88. Faulkner DJ (2000) Marine natural products. Nat Prod Rep 17:7–55

89. Ho JZ, Mohareb RM, Ahn JH, Sim TB, Rapoport H (2003) Enantiospecific synthesis of carbapentostatins. J Org Chem 68:109–114
90. Lombardino JG, Wiseman EH (1974) Preparation and anti-inflammatory activity of some nonacidic trisubstituted imidazoles. J Med Chem 17:1182–1188
91. Lo YS, Nolan JC, Maren TH, Welstead WJ Jr, Gripshover DF, Shamblee DA (1992) Synthesis and physiochemical properties of sulfamate derivatives as topical antiglaucoma agents. J Med Chem 35:4790–4794
92. Bertini I, Sigel A, Sigel H (2001) Handbook on metalloproteins. Marcel Dekker, New York, J Am Chem Soc 123:12748–12748
93. El khoury Y, Hellwig P (2009) Infrared spectroscopic characterization of copper-polyhistidine from 1,800 to 50 cm^{-1}: model systems for copper coordination. J Biol Inorg Chem 14:23–24
94. Schmiedekamp A, Nanda V (2009) Metal-activated histidine carbon donor hydrogen bonds contribute to metalloprotein folding and function. J Inorg Biochem 103:1054–1060
95. Abdelhamid RF, Obara Y, Uchida Y, Kohzuma T, Dooley DM, Brown DE, Hori H (2007) π–π interaction between aromatic ring and copper-coordinated His81 imidazole regulates the blue copper active-site structure. J Biol Inorg Chem 12:165–173
96. Greiner BA, Marshall NM, NarducciSarjeant AA, McLauchlan CC (2007) Imidazole-based Nickel(II) and Cobalt(II) coordination complexes for potential use as models for histidine containing metalloproteins. Inorg Chim Acta 360:3132–3140
97. Patel DP, Swink SM, Castelo-Soccio L (2017) A review of the use of biotin for hair loss. Skin Appendage Disord 3:166–169
98. Sarper O, Bulak E, Kaim W, Varnali T (2006) Modeling bis-(1-methylimidazol-2-yl)glyoxal, 'big', in complex ions [M(big)2]+n, M=Au, Pt, Cu, Re, Zn: a DFT study. Mol Phys 104:833–838
99. Showalter HDH, Johnson JL, Hoftiezer JM (1987) Anthrapyrazole anticancer agents. Synthesis and structureactivity relationships against murine leukemias. J Med Chem 30:121–131
100. Tewari AK, Mishra A (2001) Synthesisandanti-inflammatory activities of N4, N5-disubstituted-3-methyl-1H-pyrazolo[3,4c]pyridazines. Bioorg Med Chem 9:715–718
101. Larsen JS, Zahran MA, Pedersen EB, Nielsen C (1999) Synthesis of triazenopyrazole derivatives as potential inhibitors of HIV-1. Monatshefte für Chemie 130:1167–1173
102. Michon V, Penhoat CH, Tombret F, Gillardin JM, Lepage F, Erthon L (1995) Preparation, structural analysis and anticonvulsant activity of 3- and 5-aminopyrazole N-benzoyl derivatives. Eur J Med Chem 30:147–155
103. Yıldırım I, Özdemir N, Akçamur Y, Dinçer M, Andaç O (2005) 4-Benzoyl-1,5-diphenyl-1H-pyrazole-3-carboxylic acid methanol solvate. Acta Crystallogr 61:256–258
104. Bailey DM, Hansen PE, Hlavac AG (1985) 3,4-Di phenyl 1H-pyrazole-1-propanamine antidepressants. J Med Chem 28:256–260
105. Chu CK, Cutler SJ (1986) Chemistry and antiviral activities of acyclonucleosides. J Heterocycl Chem 23:289–319
106. Bahal SC, Dubey BL, Nath N, Srivastava JK (1984) Synthesis, characterisation and fungitoxicity of the complexes of Hg(II), Cd(II), Cu(II) and Ag(I) with 3-o-tolyoxymethyl-4aryl-5-mercapto-1,2,4-triazole. Inorg Chim Acta 91:43–45
107. Joukhadar C, Klein N, Mader RM, Schrolnberger C, Rizovski B, Heere-Ress E, Pehamberger H, Strauchmann N, Jansen B, Műller M (2001) Penetration of dacarbazine and its active metabolite 5-aminoimidazole-4-carboxamide into cutaneous metastases of human malignant melanoma. Cancer 92:2190–2196
108. Hamdy RC (2010) Zoledronic acid: clinical utility and patient considerations in osteoporosis and low bone mass. Drug Des Devel Ther 4:321–335
109. Erick Cuevas Yañez EC, Sánchez AC, Becerra JMS, Muchowski JM, Almanza RC (2004) Synthesis of miconazole and analogs through a carbenoid intermediate. Rev Soc Quím Méx 48:49–52
110. Rossello A, Bertini S, Lapucci A, Macchia M, Martinelli A, Rapposelli S, Herreros E, Macchia B (2002) Synthesis, antifungal activity, and molecular modeling studies of new inverted oxime ethers of oxiconazole. J Med Chem 45:4903–4912

111. Atia AJ (2009) Synthesis and antibacterial activities of new metronidazole and imidazole derivatives. Molecules 14:2431–2446
112. Huang X-S, Liu K, Yin Y, Li W-M, Ran W, Duan M, Wang L-S, Zhu H-L (2011) The synthesis, structure and activity evaluation of secnidazole derivatives as helicobacter pylori urease inhibitors. Curr Bioact Compd 7:268–280
113. Balraj C, Satheshkumar A, Ganesh K, El-Mossalamy EH, Elango KP (2013) Synthesis and characterization of molecular complexes of cimetidine with water soluble 1,4-benzoquinones. J Mol Struct 1050:166–173
114. Gerlach AT, Dasta JF (2007) Dexmedetomidine: an updated review. Ann Pharmacother 41:245–252
115. Burnier M, Wuerzner G (2011) Pharmacokinetic evaluation of losartan. Expert Opin Drug Metab Toxicol 7:643–649
116. Ghannoum MA, Rice LB (1999) Antifungal agents: mode of action, mechanisms of resistance, and correlation of these mechanisms with bacterial resistance. Clin Microbiol Rev 12:501–517
117. Wishart DS, Feunang YD, Guo AC, Lo EJ, Marcu A, Grant JR, Sajed T, Johnson D, Li C, Sayeeda Z, Assempour N, Iynkkaran I, Liu Y, Maciejewski A, Gale N, Wilson A, Chin L, Cummings R, Le D, Pon A, Knox C, Wilson M (2018) DrugBank 5.0: a major update to the DrugBank database for 2018. Nucleic Acids Res 46:D1074–D1082
118. Lamb D, Kelly D, Kelly S (1999) Molecular aspects of azole antifungal action and resistance. Drug Resist Updates 2:390–402
119. Emami S, Tavangar P, Keighobadi M (2017) An overview of azoles targeting sterol 14α-demethylase for antileishmanial therapy. Eur J Med Chem 135:241–259
120. Georgopapadakou NH, Walsh TJ (1996) Antifungal agents: chemotherapeutic targets and immunologic strategies. Antimicrob Agents Chemother 40:279–291
121. Yang H, Tong J, Lee CW, Ha S, Eom SH, Im YJ (2015) Structural mechanism of ergosterol regulation by fungal sterol transcription factor Upc2. Nat Commun 6:6129
122. Balding PR, Porro CS, McLean KJ, Sutcliffe MJ, Maréchal JD, Munro AW, Visser SPD (2008) How do azoles inhibit cytochrome P450 enzymes? A density functional study. J Phys Chem A 112:12911–12918
123. Gupta VK, Sharma AK, Sharma R, Diwan S, Saini S (2014) Azoles as effective antifungal agents: trends, scope and relevance. The Nat Prod J 4:82–92
124. Blobaum AL, Marnett LJ (2007) Structural and functional basis of cyclooxygenase inhibition. J Med Chem 50:1425–1441

Chapter 5
Overview on Biological Activities of Thiazole Derivatives

Raghuram Gujjarappa, Arup K. Kabi, Sattu Sravani, Aakriti Garg, Nagaraju Vodnala, Ujjawal Tyagi, Dhananjaya Kaldhi, Sreya Gupta, and Chandi C. Malakar

Abbreviations

U.S. FDA	United States Food and Drug Administration
WHO	World Health Organization
LEDs	light-emitting diode
DNA	Deoxyribonucleic Acid
CNS	Central Nervous System
PBP3	penicillin-binding protein 3
NTZ	Nitazoxanide
NSAID	Non-steroidal Anti-inflammatory Drugs
PFOR	Pyruvate: Ferredoxin/Flavodoxinoxidoreductase
OAB	Overactive Bladder Symptoms
HER2	Human Epidermal Growth Factor Receptor 2
RNA	Ribonucleic Acid
HCV	Hepatitis C Virus
COX	Cyclooxygenase
CYP3A	Cytochrome P450 3A
AD	Alzheimer's Disease
GERD	Gastroesophageal Reflux Disease
PI3K-α	Phosphotidylinositol-3-kinase-α
TPOR	Thrombopoietin Receptor
PARP	poly(ADP-ribose)polymerase

R. Gujjarappa · A. K. Kabi · N. Vodnala · U. Tyagi · D. Kaldhi · C. C. Malakar (✉)
Department of Chemistry, National Institute of Technology Manipur, Langol, Imphal Manipur 795004, India
e-mail: cmalakar@nitmanipur.ac.in

S. Sravani · A. Garg · S. Gupta (✉)
Department of Medicinal Chemistry, National Institute of Pharmaceutical Education and Research (NIPER), Kolkata, Chunilal Bhawan 168, Maniktala Main Road, Kolkata 700054, India

B. P. Swain (ed.), *Nanostructured Biomaterials*, Materials Horizons: From Nature to Nanomaterials, https://doi.org/10.1007/978-981-16-8399-2_5

IMPDH	Inosine-5′-monophosphate dehydrogenase
TNF-α	Tumour necrosis factor alpha
XMP	Xanthosine monophosphate

1 Introduction

Thiazole is a five-membered planar aromatic heterocyclic ring with a pyridine-like odour and pale yellow colour. Its chemical formula is C_3H_3NS. Hantzch and Weber were the first to characterise thiazole in 1887 and Prop rooted its structure in 1889 [1]. Thiazole has a higher π–electron delocalization than equivalent oxazoles, indicating superior aromaticity, as attested by the chemical shift of aromatic protons in NMR spectroscopy (between 8.77 and 7.27 ppm), displaying a strong diamagnetic ring current.

Thiazole is the most widely utilised heterocyclic molecule, including antiviral, antimicrobial, anti-inflammatory, antitumour, anti-HIV, anti-oxidant and other biological actions (Fig. 1) [2–7].

According to a review of the literature, thiazole ring alterations are particularly effective at increasing efficacy while lowering toxicity. Vitamins, pigments and alkaloids are all examples of naturally occurring nitrogen and sulphur-containing heterocycles with medicinal activity. In the same way, Penicillin contains thiazolidine ring

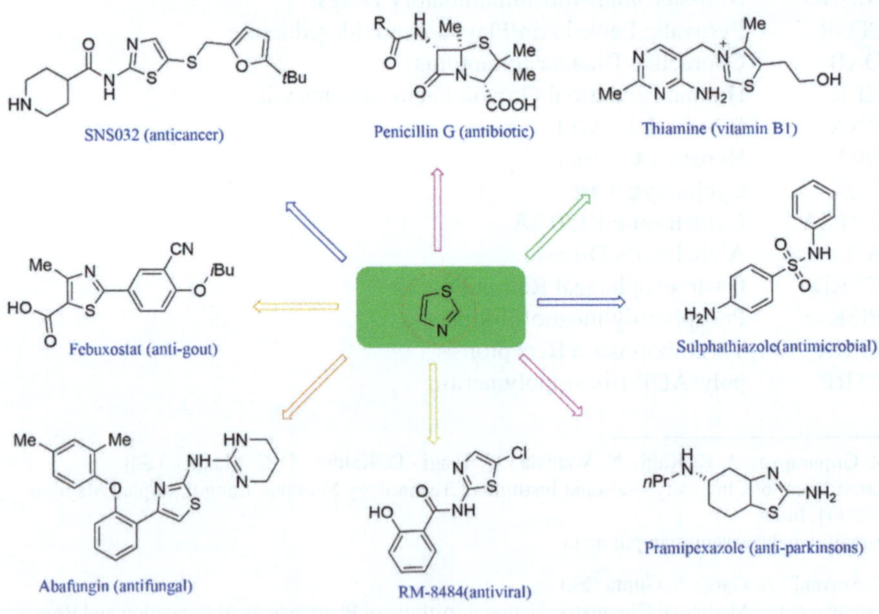

Fig. 1 Different pharmacological activities of thiazole analogues [2–7]

system and Thiamine (Vitamin B1) contains a thiazole system both occur naturally. It also has a wide range of applications in medication development for radiofonic, age-related, mental detention in children and clinical situations for neurodegenerative brain damage in Parkinson and Alzheimer's disease. Ritonavir (antiretroviral), Sulfathiazole (antimicrobial), Abafungin (antifungal) with brand names Bleomycin, abasol cream and Tiazofurin (antineuroplastic) are only a handful of the potent biologically active medicinal compounds that contain these scaffolds [8]. Thiazole-derived compounds have a lot of applications in agriculture, cosmetics, catalysis, light harvesting, mass manufacture of light-emitting diodes (LEDs), photo chromes and molecular switches as well as nonlinear optical materials [9]. Recently, thiazoles were found in drug development as fibrinogen receptor antagonists for the medication of pain, possessing antithrombotic properties and as novel bacterial DNA gyrase B inhibitors [10].

One of the reasons for the enormous regularity among five-membered azole-based aromatic nitrogen heterocycles, according to our analysis of exclusive U.S. FDA-approved drugs with a thiazole group (Fig. 2), is that it has appeared as an extensively used functional group for the massive class of β-lactam antibiotics. Unfortunately, this valuable family of medications accounts for 67 percent of all

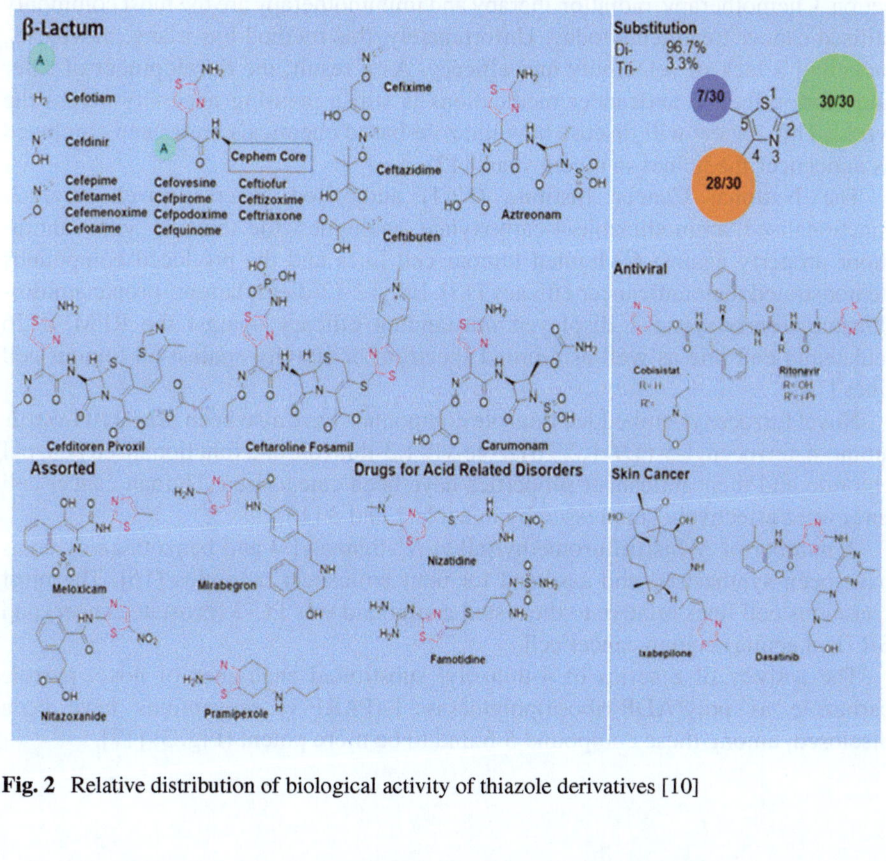

Fig. 2 Relative distribution of biological activity of thiazole derivatives [10]

thiazole-containing medications. Each thiazole drug has a substitution in the C2-position, with the majority of them additionally having a supplemental substituent in the C4 position. Even though there is no approved mono-substituted thiazole pharmaceutical, pramipexole is the only licensed trisubstituted thiazole drug.

The anti-HIV medications ritonavir and cobicistat are unique in that they have two distinct thiazole groups, as well as cefditoren pivoxil and are structurally identical. Nizatidine and famotidine, both used to treat peptic ulcers, are structurally fascinating since they include thioethers as well as intriguing nitro and sulphonamide groups.

The list of FDA-approved thiazole-containing drugs has been portrayed in Table 1 [10].

2 Antitumour Activity of Thiazole Derivatives

Cancer is the prime reason for death worldwide, which is witnessed for millions of deaths each year. Global cancer rates could rise by 50% to 29.5 million by 2040, according to WHO statistics [11].

As a result, the number of cancer patients continues to rise, necessitating quick action. Chemotherapy, radiation therapy and immunotherapy are the most commonly utilised cancer treatments today. Unfortunately, this method has many drawbacks, including a lack of selectivity and efficacy. As a result, the development of safer and more effective anticancer medications is still a pressing necessity across the world. This review will discuss how thiazole-based chemicals have been employed as anticancer medicines in recent years [12].

The National Cancer Institute (NCI) authorised several novel ethyl 2-functionalised aminothiazole-4-carboxylate analogues for their in vitro antitumour property against 60 human tumour cell lines and the produced compounds demonstrated their anticancer efficacy [13]. Ethyl 2-[3(diethylamino-propanamido]-thiazole-4-carboxylate 1 displayed outstanding efficacy against the RPMI-8226 leukaemia cell line as well as a broad spectrum of activity against all tumour cell lines [14].

Novel ferrocenyl-embedded thiazole compounds have also been prepared from 2-amino-4-ferrocenyl-5-(1*H*-1,2,3 triazole-yl)-1,3-thiazole and functionalised benzoyl chloride and their anticancer properties have been categorised. Human cancer cell lines were effectively suppressed by thiazole **2** and **3** [15].

A number of *N*-bis(trifluromethyl)alkyl-*N*'-thiazolyl **4** and benzothiazolylureas have been synthesised and assessed for their anticancer activities [16]. The most conscious cell lines relative to the tested compound was PC-3 (prostate cancer) and SR (leukaemia) human cancer cell.

The activity of a series of 4-thiazolyl substituted analogues of novel pyrrole carbazole as poly(ADP-ribose)polymerase-1-(PARP-1) preventions have been disclosed, among these compound 5 found to be more potent (Fig. 3) [17].

Table 1 FDA-approved drugs containing thiazole moiety [10]

Structure of the drug name of the drug (Drug bank id)	Category/Indication	Mechanism of action
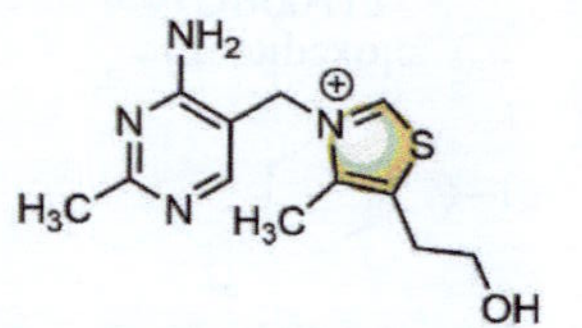 Thiamine (DB00152)	For Korsakov's alcoholic psychosis, niacin and thiamine deficiency, Wernicke-Korsakov syndrome, delirium and peripheral neuritis	Redirecting glycolytic flow to reduce intracellular protein glycation
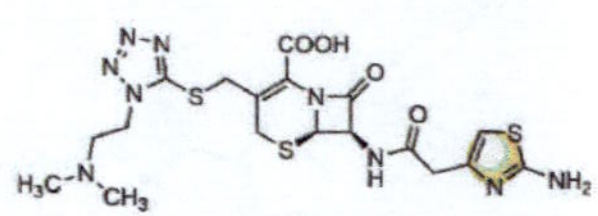 Cefotiam (DB00229)	For the cure of severe infections provoked by bacteria that are sensitive to antibiotics	Cefotiam's bactericidal effect is due to its affinity for penicillin-binding proteins (PBPs), which inhibits cell wall formation
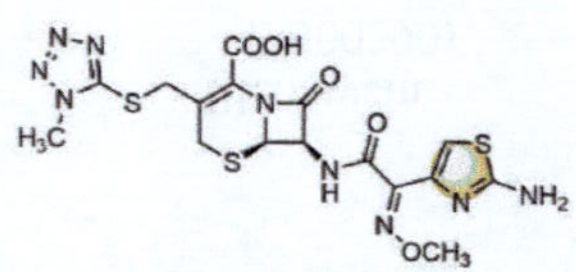 Cefmenoxime (DB00267)	Used for the treatment of female obstetric and gynecologic infections induced by vulnerable anaerobic and aerobic bacteria along with the gonococcus	Cefmenoxime's bactericidal effect is due to its affinity for PBPs, which inhibits cell wall formation

(continued)

Table 1 (continued)

Structure of the drug name of the drug (Drug bank id)	Category/Indication	Mechanism of action
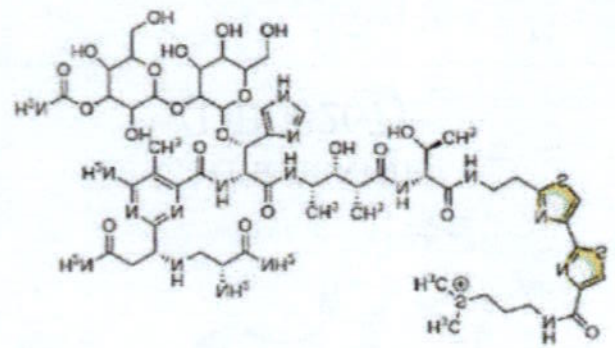 Bleomycin (DB00290)	For palliative care of malignant neoplasms (bronchus, trachea, lung), lymphomas and squamous cell carcinomas	The primary method of action appears to be the blockage of DNA formation, with confirmation of minor suppression of RNA and generation of protein
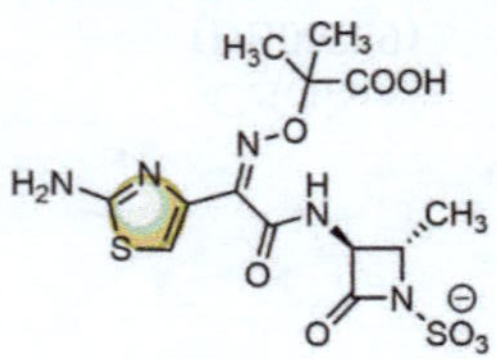 Aztreonam (DB00355)	For the medication of infections in the Urinary tract, infections of septicemia, diseases in the lower respiratory tract, infections in intra-abdominal and skin and skin-structure irregularities and gynecologic infections which are induced by gram-negative germs	Aztreonam's bactericidal effect is attributed to its strong affinity for penicillin-binding protein 3 (PBP3), which inhibits bacterial cell wall production. Aztreonam suppresses the third and last stage of bacterial cell wall generation by binding to PBP3
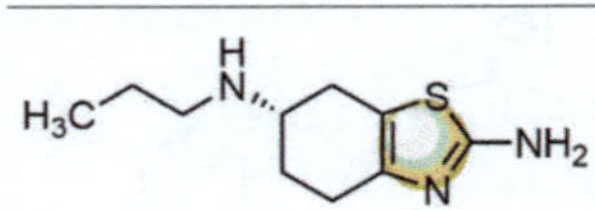 Pramipexole (DB00413)	This medication is recommended to cure the indications of Parkinson's disease. This medication can be used as immunotherapy or in combination with levodopa. It is also employed for the treatment of the mild and severe symptoms of primary Restless Legs Syndrome (RLS).	The dopamine receptors in the striatum of the brain, which receives a wide range of neurological input and is responsible for a wide range of processes, are stimulated by pramipexole

(continued)

Table 1 (continued)

Structure of the drug name of the drug (Drug bank id)	Category/Indication	Mechanism of action
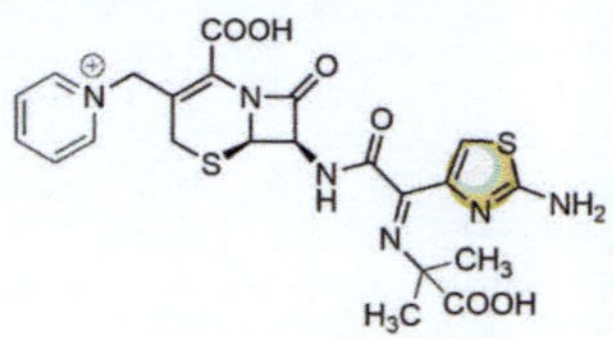 Ceftazidime (DB00438)	Infections lower respiratory tract, infections in the skin, urinary tract, bones and joints. Moreover, it is recommended for bacterial septicemia, gynecologic inflammation, infections in the central nervous system along with meningitis and intra-abdomen including peritonitis. These are among the health issues lead by the susceptible strains of organisms	Ceftazidime's bactericidal effect is due to its attraction toward penicillin-binding proteins (PBPs), which inhibits cell wall formation
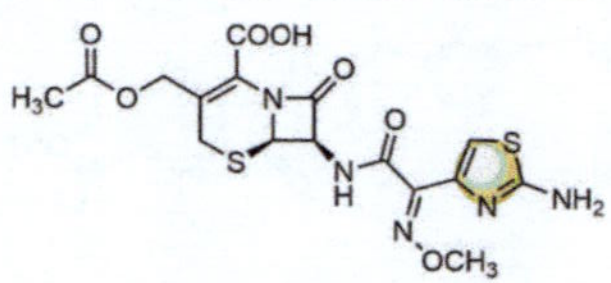 Cefotaxime	Recommended for the medication of meningitis, gonorrhoea and severe infections in the urine system and kidney (pyelonephritis). It is used for post-surgery infections	Cefotaxime's bactericidal effect is due to its affinity for penicillin-binding proteins (PBPs), which inhibits cell wall formation.
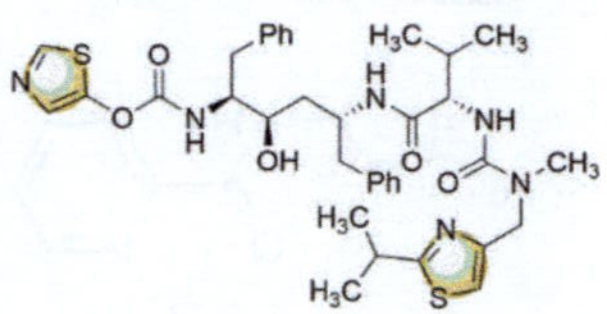 Ritonavir (DB00503)	For the cure of infections related to HIV-1, it is used in conjunction with other antiretroviral drugs	Ritonavic works by preventing the HIV viral proteinase enzyme from cleaving the structural and replicative proteins that result from important HIV genes including gag and pol

(continued)

Table 1 (continued)

Structure of the drug name of the drug (Drug bank id)	Category/Indication	Mechanism of action
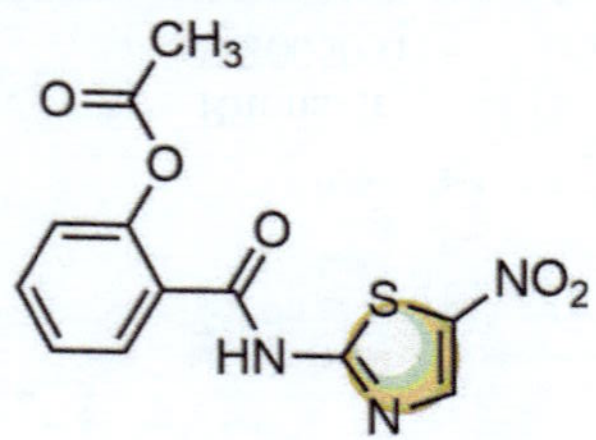 Nitazoxanide (NTZ) (DB00507)	For the treatment of diarrhoea produced by the protozoan Giardia lamblia in adults and children, as well as diarrhoea produced by the protozoan Cryptosporidium parvum in children In patients with HIV/immunodeficiency, nitazoxanide remains inferior to placebo in the medication of Cryptosporidium parvum-related diarrhoea	The most widely accepted mechanism of NTZ is that it interferes with anaerobic microorganisms' energy metabolism by inhibiting the pyruvate: ferredoxin/flavodoxinoxidoreductase (PFOR) cycle
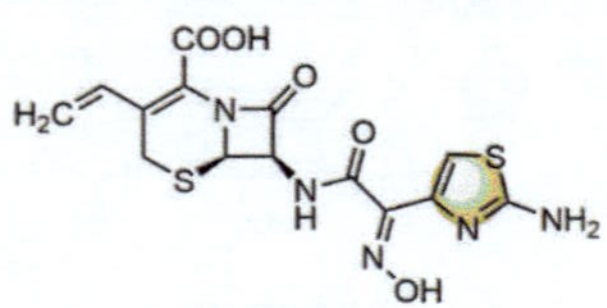 Cefdinir (DB00535)	NTZ is thought to affect anaerobic microorganisms' energy metabolism by inhibiting the pyruvate: ferredoxin/flavodoxin oxidoreductase cycle	In cephalosporins, the five-member thiazolidine rings that make up penicillin is replaced by a six-member dihydrothiazine ring, which confers enhanced bactericidal action. Thus Cefdinir and other cephalosporins can withstand inactivation by certain bacterial enzymes because of their 6-member ring

(continued)

Table 1 (continued)

Structure of the drug name of the drug (Drug bank id)	Category/Indication	Mechanism of action
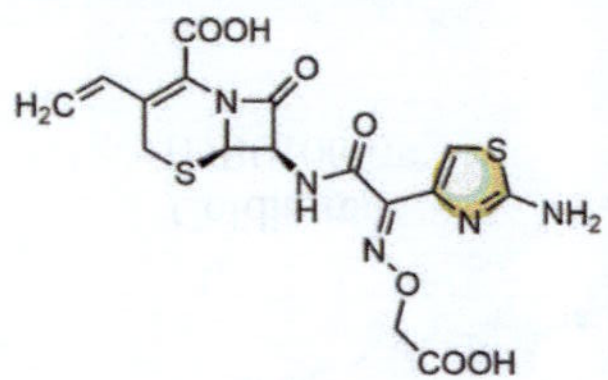 Cefixime (DB00671)	To be used in the treatment of infections such as (1) *Escherichia coli* and Proteus mirabilis, such as simple urinary tract infections, (2) *Haemophilusinfluenzae* (negative and positive strains of beta-lactamase), beta-lactamase positive Moraxella catarrhalis and otitis media (Streptococcus pyogenes) (3) used for tonsillitis and pharyngitis occurred by *S. pyogenes*, (4) acute exacerbations of chronic bronchitis and acute bronchitis originated by *Haemophilusinfluenzae* (negative and positive strains of beta-lactamase) and *Streptococcus pneumonia and* (5) recommended for urethral/cervical arisen due to *Neisseria gonorrhoeae* (non-penicillinase- and penicillinase-producing strains)	Affinity toward specific penicillin-binding proteins (PBPs) which are inside the bacterial cell wall, inhibiting the bacterial cell wall's third and final stages of formation. Cell lysis is then influenced by autolytic enzymes found in the cell wall of bacteria, such as autolysins; cefixime may interfere with an autolysin inhibitor
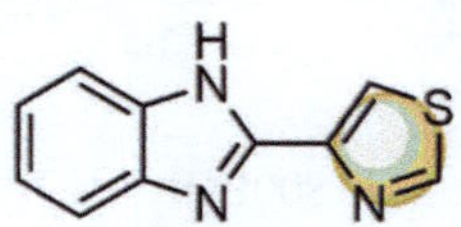 Thiabendazole (DB00730)	Used for the treatment of threadworm (Strongyloidiasis), visceral larva migrans, creeping eruption (cutaneous larva migrans) and trichinosis	Thiabendazole is a parasiticide and fungicide. Thiabendazole acts as a chelating agent in case of metal poisonings like mercury, lead and antimony
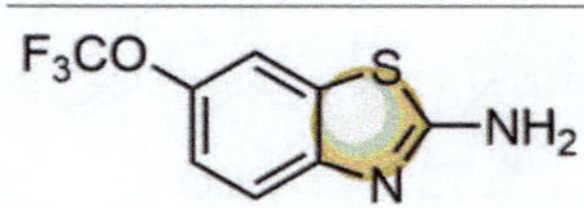 Riluzole (DB00740)	To treat amyotrophic lateral sclerosis (ALS, often known as Lou Gehrig's disease)	Riluzole's mechanism of action is uncertain. It has the following pharmacological properties: (1) glutamate release suppressor by the activation of glutamate reuptake, (2) voltage-dependent sodium channels inactivation and (3) capable of hindering the intracellular events caused by the binding of the transmitter at excitatory amino acid receptors

(continued)

Table 1 (continued)

Structure of the drug name of the drug (Drug bank id)	Category/Indication	Mechanism of action
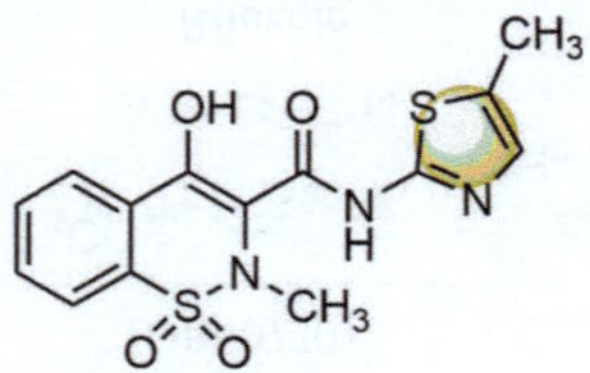 Meloxicam (DB00814)	Arthritis and osteoarthritis symptomatic therapy	Meloxicam's anti-inflammatory actions are thought to be related to the inhibition of prostaglandin synthetase (cyclo-oxygenase), which inhibits prostaglandin production
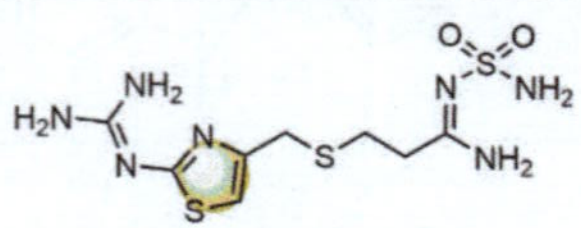 Famotidine	Treatment of Gastroesophageal reflux disease (GERD) and Peptic ulcer disease (PUD)	It binds aggressively to H2-receptors on the parietal cell's basolateral membrane, hinders the histamine effects. Diminish the secretion of nocturnal and basal gastric acid. It also participates to decrease the amounts, acidity and volume of gastric acid discharge in reply to stimuli such as coffee, food, betazole, insulin, or pentagastrin
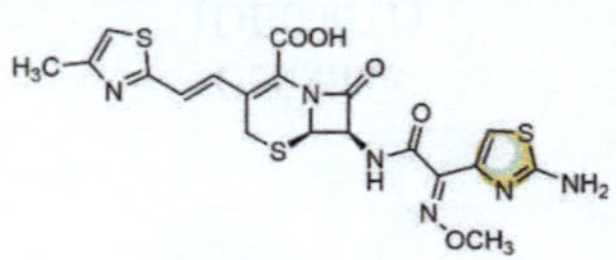 Cefditoren (DB01066)	Recommended for the adolescents and adult patients with mild to severe infections led by vulnerable microorganisms in an acute bacterial increase of chronic bronchitis, pharyngitis/tonsillitis, community-acquired pneumonia and infections of uncomplicated skin and skin structure	Cefditoren's bactericidal effect is due to its affinity for penicillin-binding proteins (PBPs), which inhibits the generation of cell walls

(continued)

Table 1 (continued)

Structure of the drug name of the drug (Drug bank id)	Category/Indication	Mechanism of action
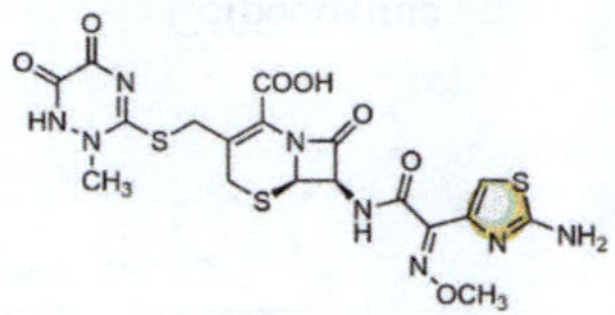 Ceftriaxone (DB01212)	For the cure of inflammations occurred by H. influenzae, S. pneumoniae, staphylococci, E. coli, S. pyogenes (group A beta-hemolytic streptococci), P. mirabilis, coagulase-negative staph diseases (respiratory, cutanecus, soft tissue, UTI, ENT) and Klebsiella sp.	Ceftriaxone works by preventing the bacterial cell wall from producing mucopeptides. Beta-lactam unit of Ceftriaxone binds to bacterial endopeptidases, carboxypeptidases and transpeptidases in the cytoplasmic membrane
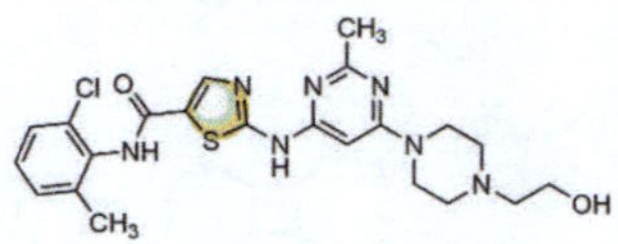 Dasatinib (DB01254)	Adult patients having rapid, chronic, or lymphoid or myeloid blast phase chronic myeloid leukaemia, who have developed intolerance or resistance to previous therapy. Patients with Philadelphia chromosome-positive acute lymphoblastic leukaemia who have developed intolerance or resistance to earlier therapy	SRC, BCR-ABL family (LCK, SRC, FYN, YES), EPHA2, c-KIT and PDGFR are suppressed by nanomolar concentrations of Dasatinib. Modelling research revealed that the Dasatinib is also assumed to correlate to numerous conformations of the ABL kinase
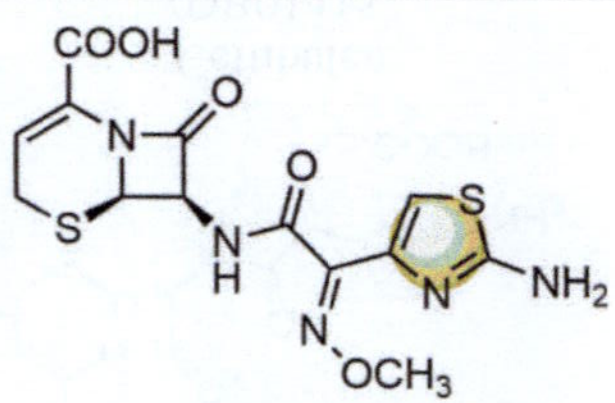 Ceftizoxime (DB01332)	For the treatment of diseases caused by sensitive microorganism strains	It inhibits the third and final steps of cell wall generation of bacteria by attaching to certain penicillin-binding proteins (PBPs) that exist inside the bacterial cell wall. Cell lysis is then induced by autolytic enzymes found in bacterial cell walls, such as autolysins; ceftizoxime may interfere with an autolysin inhibitor

(continued)

Table 1 (continued)

Structure of the drug name of the drug (Drug bank id)	Category/Indication	Mechanism of action
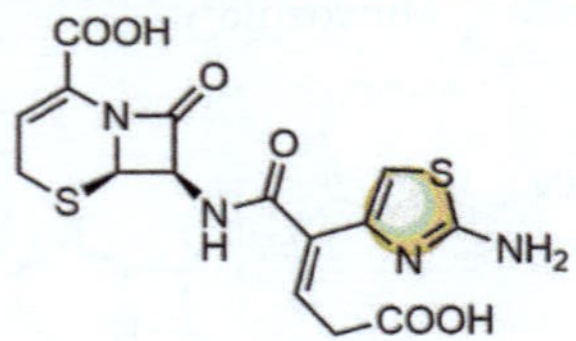 Ceftibuten (DB01415)	Acute bacterial aggravations of chronic bronchitis (ABECB), acute bacterial otitis media, pharyngitis and tonsillitis are among the conditions for which it is prescribed	Ceftibuten works as a bactericide by binding to key bacterial cell wall target proteins. The suppression of cell wall production occurs as a result of this interaction
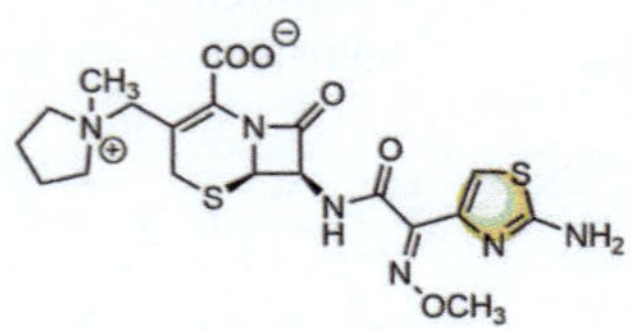 Cefepime (DB01413)	For the cure of Streptococcus pneumoniae pneumonia (moderate to severe), including instances with concurrent bacteremia, Pseudomonas aeruginosa, Klebsiella pneumonia, or Enterobacter species. Also used to treat febrile neutropenic individuals, as well as mild and serious inflammations in the urinary tract as well as pyelonephritis lead by Escherichia coli or Klebsiella pneumonia	Cephalosporins prevent the peptidoglycan layer of bacterial cell walls from forming. The peptidoglycan layer is critical for Gram-positive species' cell wall structural integrity
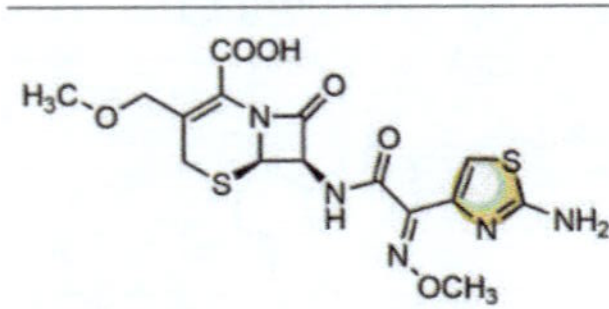 Cefpodoxime (DB01416)	Patients suffering from mild to moderate inflammation lead by the vulnerable strains of the indicated bacteria should use this medication	Cefpodoxime is an antibiotic that works against both Gram-negative and Gram-positive bacteria. In the appearance of beta-lactamase enzymes, cefpodoxime remains stable. Subsequently, many organisms unsusceptible to cephalosporins and penicillin may be sensitive to cefpodoxime due to their synthesis of beta-lactamase

(continued)

Table 1 (continued)

Structure of the drug name of the drug (Drug bank id)	Category/Indication	Mechanism of action
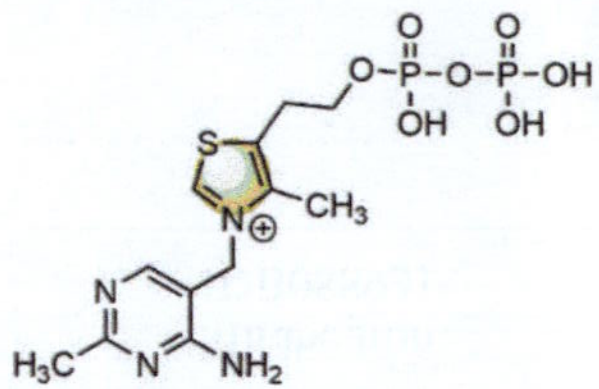 Cocarboxylase (DB01987)	NA	NA
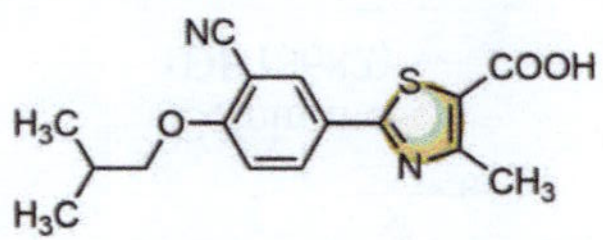 Febuxostat (DB04854)	Hyperuricemia in gout patients is treated with this drug	Febuxostat, a xanthine oxidase inhibitor, works by lowering uric acid levels in the blood. At therapeutic dosages, Febuxostat is unlikely to inhibit other enzymes involved in purine and pyrimidine production and metabolism
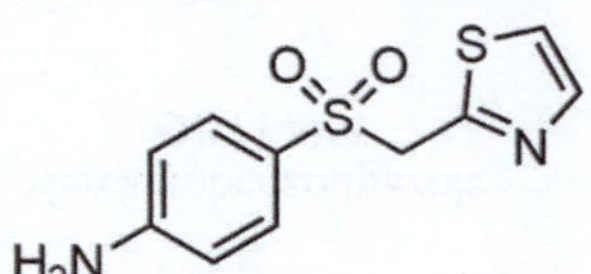 Sulfathiazole (DB06147)	Sulfathiazole is efficient against a wide spectrum of pathogenic bacteria, both Gram-negative and Gram-positive. It is still utilised in cattle, even though it is no longer used in people	Dihydropteroate production is inhibited by sulfathiazole

(continued)

Table 1 (continued)

Structure of the drug name of the drug (Drug bank id)	Category/Indication	Mechanism of action
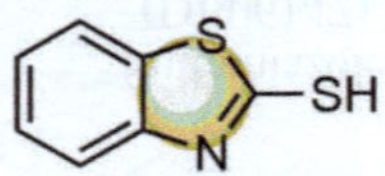 Mercaptobenzothiazole (DB11496)	NA	Thyroid peroxide is inhibited by 2-Mercaptobenzothiazole
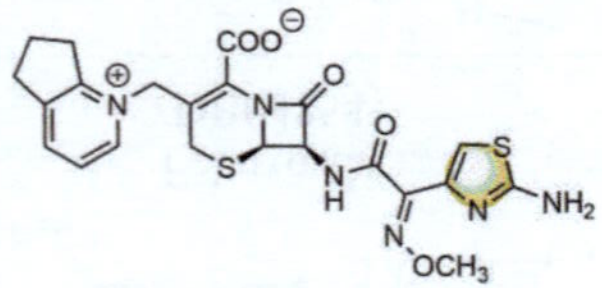 Cefpirome (DB13682)	NA	NA
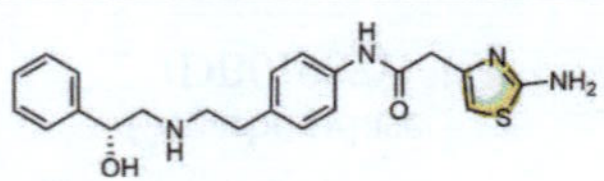 Mirabegron (DB08893)	It is a beta-3 adrenergic agonist recommended for the treatment of overactive bladder (OAB) symptoms which includes urge urine incontinence, urgency and frequency	Mirabegron is a beta-3 adrenergic receptor agonist that is both strong and selective. The detrusor smooth muscle relaxes when beta-3 receptors are engaged, allowing for a higher bladder capacity. Mirabegron has the potential to activate beta-1 and beta-2 adrenergic receptors at higher doses (200 mg)

(continued)

Table 1 (continued)

Structure of the drug name of the drug (Drug bank id)	Category/Indication	Mechanism of action
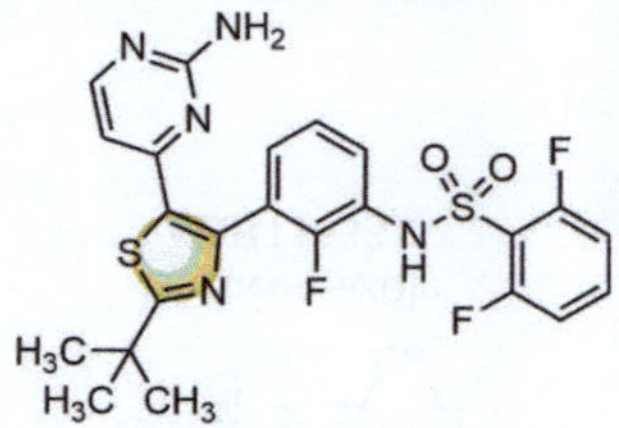 Dabrafenib (DB08912)	It is known as a kinase inhibitor which was licensed as a single medication for the cure of individuals with unresectable or metastatic melanoma who tested positive for the BRAF V600E mutation using an FDA-authorised test	Dabrafenib is an antineoplastic B-raf (BRAF) protein inhibitor that can be used orally. Dabrafenib preferentially binds to and suppresses the mechanism of B-raf, which may prevent tumour cells with a mutant BRAF gene from proliferating. B-raf is a member of the raf/mil family of serine/threonine protein kinases which monitors the MAP kinase/Extracellular Signal-Regulated Kinases signalling pathway, which can be constitutively active as a result of BRAF gene mutations
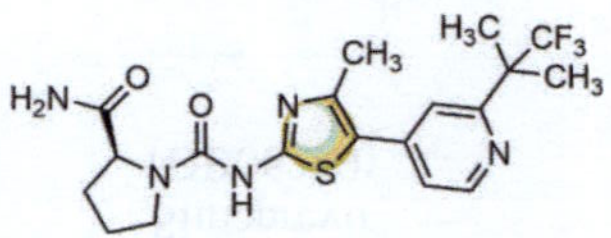 Alpelisib (DB12015)	Alpelisib is used in combination with fulvestrant to treat advanced or metastatic breast cancer in postmenopausal women and men. Hormone receptor (HR)-positive, human epidermal growth factor receptor 2 (HER2)-negative and PIK3CA-mutated cancers are required. Following advancement on or after an endocrine-based therapy, cancer must be diagnosed by an FDA-approved test	In response to the activation of the growth factor-tyrosine kinase pathway, phosphatidyl inositol-3-kinase-α (PI3K-α) is responsible for cell proliferation. The p110 catalytic subunit of PI3K is altered in some tumours, making it hyperactive. With the highest specificity for PI3Kα, Alpelisib inhibits it

(continued)

Table 1 (continued)

Structure of the drug name of the drug (Drug bank id)	Category/Indication	Mechanism of action
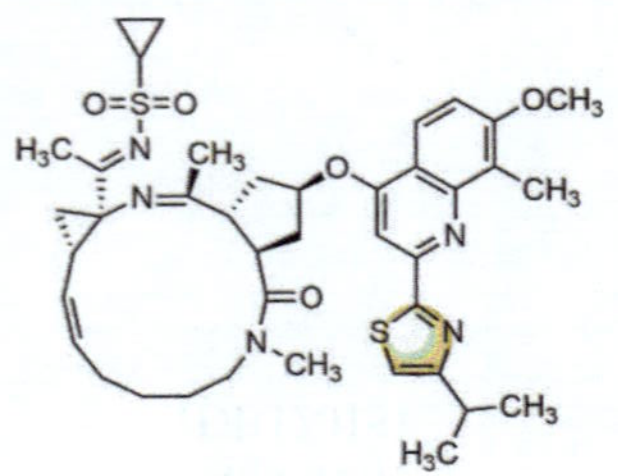 Simeprevir (DB06290)	It is used for the cure of adult patients having chronic hepatitis C virus (HCV) inflammation. It is normally recommended in addition with sofosbuvir in patients having HCV genotype 1 without compensated cirrhosis or cirrhosis. This is used in addition with ribavirin (RBV) and peginterferonalfa (Peg-IFN-alfa) in patients suffering from HCV genotype 1 or 4 without cirrhosis or compensated cirrhosis	Simeprevir accumulates in the liver after being taken up by OATP1B1/3 in hepatocytes. The cofactor N4A subunit and the proteolytic site-containing N3 subunit make up the NS3/4A heterodimeric complex. The NS3/4A protease separates the HCV polyprotein downstream of the NS3 site, forming the non-structural viral proteins NS4A, NS3, NS5A, NS4B and NS5A and then develops proteins. Simeprevir hinders the cleavage of HCV polyprotein by correlating to an expanded S2 subsite at the catalytic site of NS3 via induced-fit binding. Because NS3/4A inhibitors rely on a small number of contacts in the viral serine protease's substrate-binding groove, they are vulnerable to resistance and treatment failure due to a few crucial changes in these regions
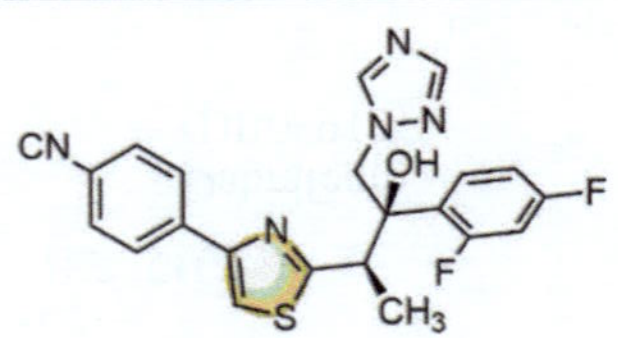 Isavucon-azole (DB11633)	For the medication of invasive aspergillosis in patients whose age is 18 years or above. Also, it is recommended for the cure of invasive mucormycosis in patients in patients whose age is 18 years or above and especially for the patients for whom amphotericin B is ineffective	Isavuconazole has fungicidal properties because it prevents the formation of ergosterol, a critical part of the cell membrane of the fungi. It prevents the transformation of lanosterol into ergosterol by hindering the cytochrome P-450 enzyme lanosterol 14-alpha-demethylase

(continued)

Table 1 (continued)

Structure of the drug name of the drug (Drug bank id)	Category/Indication	Mechanism of action
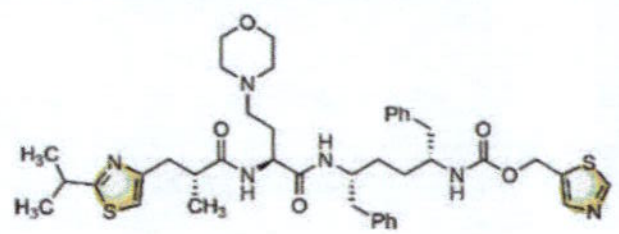 Cobicistat (DB09065)	Cobicistat is a CYP3A inhibitor used to boost atazanavir or darunavir (one daily dosing schedule) systemic exposure when used with other antiretroviral drugs to treat HIV-1 infection	Cobicistat is a cytochrome P450 3A (CYP3A) isoform inhibitor with a mechanism of action. Cobicistat's suppression of CYP3A-induced metabolism increases the systemic exposure of CYP3A substrates darunavir and atazanavir enhances the antiviral activity at lower doses. Cobicistat does not pose anti-HIV properties
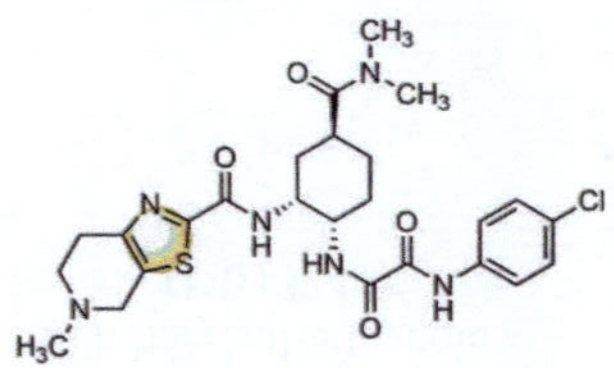 Edoxaban (DB09075)	Edoxaban is recommended to diminish the danger of stroke and systemic embolism (SE) (NVAF) in individuals having nonvalvular atrial fibrillation. In comparison to the warfarin at the highest dose, it was verified that it is not recommended for individuals having creatinine clearance (CrCL) > 95 mL/min due to the boosting risk of ischemic stroke (60 mg)	Edoxaban is a selective inhibitor of factor Xa, a serine endopeptidase involved in the conversion of prothrombin to thrombin in the clotting cascade
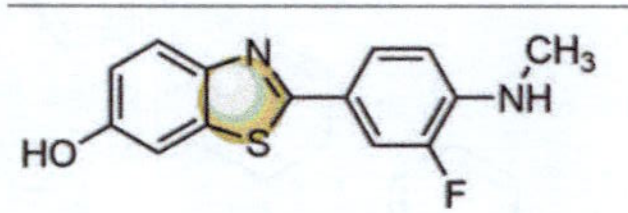 Flutemetamol (DB09151)	Flutemetamol F18 is used for adult patients to quantify amyloid neuritic plaque density with cognitive impairment who are being investigated for Alzheimer's disease or other diseases of cognitive decline employing Positron Emission Tomography (PET) imaging of the brain	Fluorine-18 (F 18) is a cyclotron-generated radionuclide that decays to produce stable oxygen-18 having the physical half-life of 109.8 min via orbital electron capture (3.3 percent) and positron emission (β + decay, 96.7 percent). The positron can eradicate an electron to deliver two gamma rays, each of which has an energy of 511 keV. Flumetamol F18 acquires in beta-amyloid plaques in the brain after being injected intravenously, making it detectable on positron emission tomography (PET).

(continued)

Table 1 (continued)

Structure of the drug name of the drug (Drug bank id)	Category/Indication	Mechanism of action
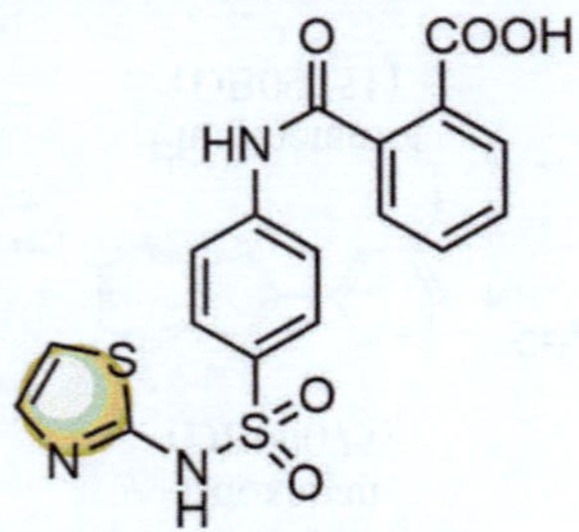 Phthalylsulfathiazole (DB13248)	NA	NA
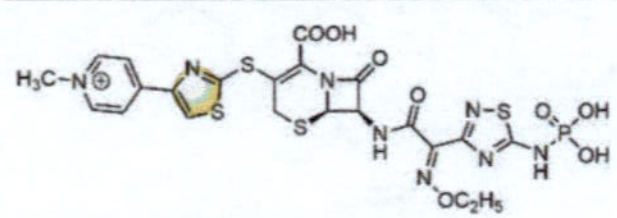 Ceftarolinefosamil (DB06590)	Ceftarolinefosamil is recommended for the medication of the inflammations induced by vulnerable isolates of the assigned bacteria	Ceftarolinefosamil is a bacterial antibiotic

(continued)

Table 1 (continued)

Structure of the drug name of the drug (Drug bank id)	Category/Indication	Mechanism of action
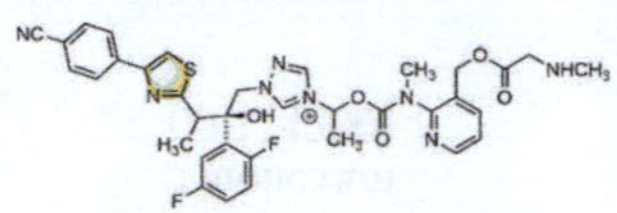 Isavuconazonium (DB06636)	Invasive aspergillosis and invasive mucormycosis are also treated with this drug	Antifungals in the triazole class, such as isavuconazonium, target and inhibit sterol 14–demethylase (Erg11p), a critical player in the ergosterol biosynthesis pathway's demethylation step. This inhibits the formation of ergosterol, a chemical abundant in the membranes of fungi such as Aspergillus, Candida and Mucorales and is involved in membrane integrity, fluidity and permeability regulation. Inhibition of Erg11p results in the accumulation of ergosterol precursors, which are toxic and lead to cell death
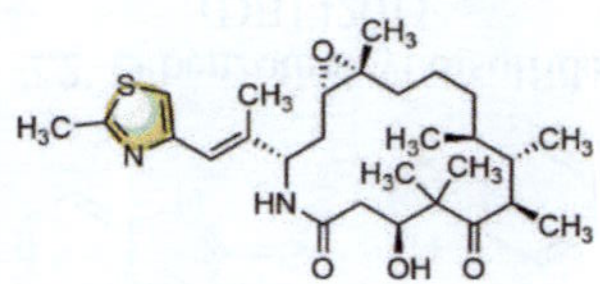 Ixabepilone (DB04845)	It is studied for head, breast, lung, prostate and neck cancer. It is also investigated for the treatment of melanoma, lymphoma (non- Hodgkin's), renal cell carcinoma and other cancers/tumours	Ixabepilone stabilises microtubules via binding to beta-tubulins (e.g. beta-III tubulin). Because microtubules are required for cell division and epothilones hinder cells from correctly dividing. Like taxol, Ixabepilone correlates to the αβ-tubulin heterodimer subunit. The rate of αβ-tubulin cleavage decreases after binding, stabilizing microtubules
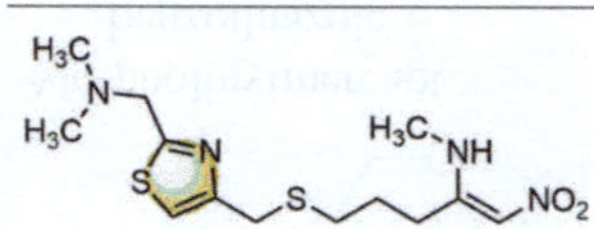 Nizatidine (DB00585)	Used for acid reflux disease (GERD), active benign gastric ulcer, peptic ulcer disease and active duodenal ulcer	At the H2-receptors on the gastric basolateral membrane of parietal cells, izatidine competes with histamine for binding. Basal and nocturnal stomach acid productions are reduced as a result of competitive inhibition. Food, coffee, insulin, betazole and pentagastrin all lower the gastric acid reaction to the medication

(continued)

Table 1 (continued)

Structure of the drug name of the drug (Drug bank id)	Category/Indication	Mechanism of action
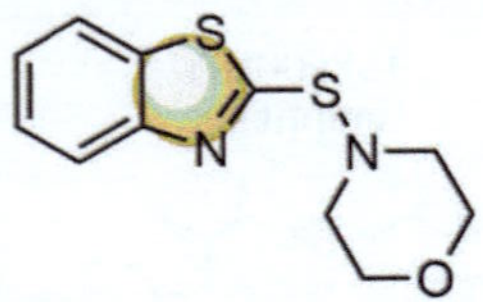 Morpholinylmercapto benzothiazole (DB14202)	NA	NA
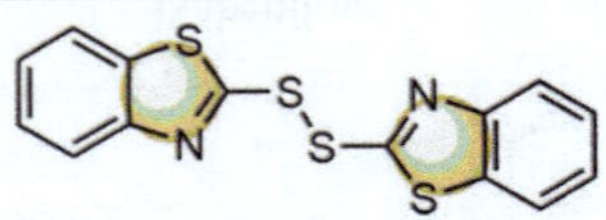 2,2'-Dibenzothiazyl disulfide (DB14201)	It is employed in the allergenic epicutaneous patch tests, which are used to benefit in the investigation of allergic contact dermatitis (ACD) in the age group of six and above	NA
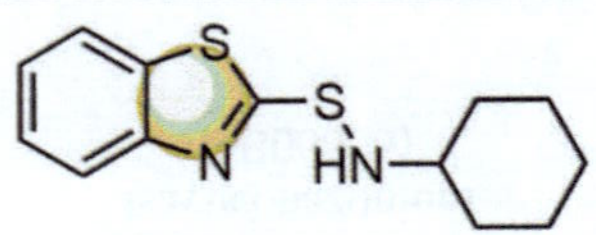 Thiohexam (DB14200)	Thiohexam has been licensed for the utilization in allergenic epicutaneous patch tests, which are employed to benefit in the investigation of allergic contact dermatitis (ACD) in the age group of six and above	NA

(continued)

Table 1 (continued)

Structure of the drug name of the drug (Drug bank id)	Category/Indication	Mechanism of action
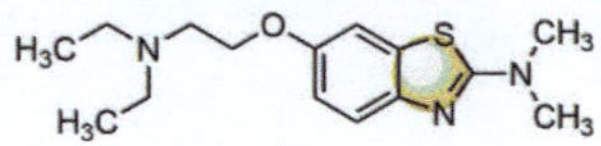 Dimazole (DB13858)	NA	NA
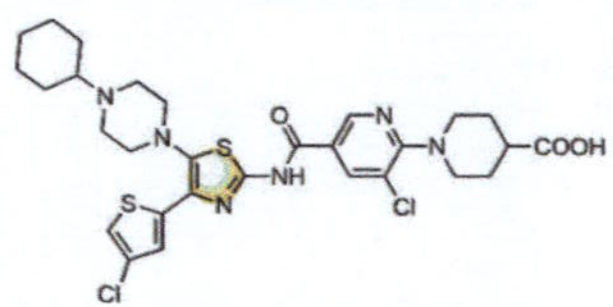 Avatrombopag (DB11995)	It's used to treat thrombocytopenia in adult patients suffering from chronic liver disease and requires surgery	Vatrombopag is small chemical thrombopoietin (TPO) receptor agonist that increases megakaryocyte reproduction and separation from the bone marrow progenitor cells, leading to enhanced platelet production. Avatrombopag does not compete with thrombopoietin for binding to the TPO
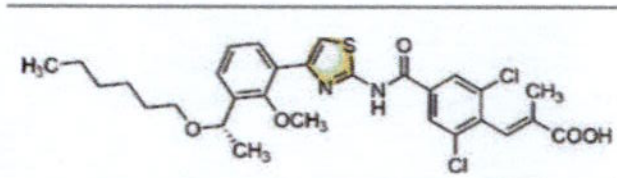 Lusutrombopag (DB13125)	Lusutrombopag is used to treat thrombocytopenia in people with chronic liver disease who are about to have a dental or medical procedure done	Lusutrombopag acts as an agonist for the thrombopoietin receptor (TPOR) expressed on megakaryocytes, mimicking the biological activities of endogenous thrombopoietin (TPO). It binds to the receptor's transmembrane domain and causes thrombocytopoiesis by activating the same signal transduction mechanism as endogenous TPO, which involves the JAK and STAT pathways

(continued)

Table 1 (continued)

Structure of the drug name of the drug (Drug bank id)	Category/Indication	Mechanism of action
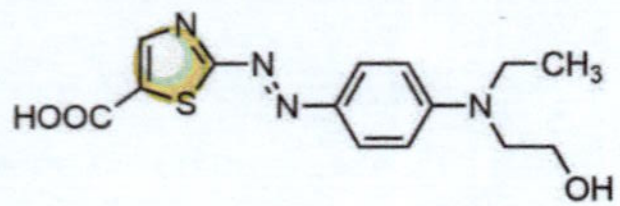 Disperse Blue 106 (DB14203)	NA	NA

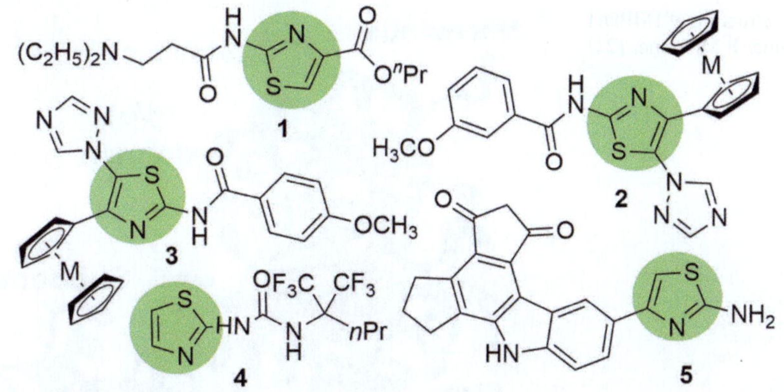

Fig. 3 Thiazole derivatives with antitumour activity [15–17]

2.1 Mechanism of Antitumour Activity of Thiazole Derivatives

The antitumour activity of thiazole and fused thiazole derivatives is mediated by interactions with several cancer bio targets such as TNF-α, IMPDH (inosine monophosphate dehydrogenase) and apoptosis inducers, kinase inhibitors, promatrix metalloprotein activation inhibition, inhibitors of signal transducer and activator of transcription 3, Bcl-2 family inhibitors, modulators of high expression in cancer 1 activity, histone deacetylases inhibitors [18, 19].

2.1.1 Thiazole Derivatives Acting on IMPDH

Nucleotides act as building blocks of nucleic acid, co-substrates and co-enzymes in the various metabolisms. Generally, cells will synthesise the nucleotides in two different ways namely de novo pathway and salvage pathway. The enzymes that are involved in the nucleotide biosynthesis pathway are vital for cell proliferation. IMPDH catalyses the initial step in the production of Guanine nucleotides and it is abundant in tumours and rapidly growing tissues [20].

Monomers of IMPDH have two domains: a catalytic domain and a subdomain. The catalytic domain consists of $(\beta/\alpha)_8$ barrel and the subdomain consists of 2 CBS domains. The transition from the catalytic domain to the subdomain is adaptable. The active site is located on the loops at the C-terminal ends of the catalytic domain's β-sheets (Fig. 4) [21].

IMPDH catalyses two dissimilar chemical transformations. One of the transformations involves a dehydrogenase reaction and results in the formation of NADH and an intermediate E-XMP. The other transformation involves hydrolysis of the intermediate E-XMP into XMP. IMPDH catalyses two different transformations by the single active site. It is possible as it acquires two different transition states for

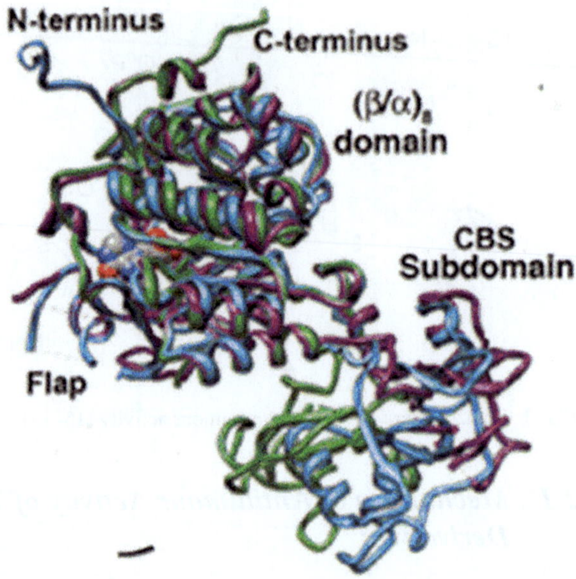

Fig. 4 Structure of IMPDH **a** Tetramer **b** Monomer [21]

catalysing two different reactions. The transformation of the enzyme during the reaction is as follows. NADH exits the enzyme, the moveable flap of the enzyme moves into the vacant dinucleotide site and the conserved Arg418-Tyr419 is carried into the active site. As a result, IMPDH has two distinct conformations: an open state for catalysing the dehydrogenase process and a closed state for catalysing the E-XMP hydrolysis process [22].

Thiazofurin, a thiazole derivative is a reversible inhibitor of IMPDH [23]. If numerous inhibitors are applied, thiazofurin is transformed into its dinucleotide form, TAD, which may bind to either the free enzyme or the intermediate forms (E-IMP, E-XMP) and subsequently block both the open and closed states for the redox and hydrolysis reactions (Fig. 5) [22].

2.1.2 Thiazoles as Kinase Inhibitors

Phosphatidylinositol-3-kinases (PI3Ks) are a kind of lipid kinase enzyme. It is involved in a number of biological activities, including cell survival, proliferation and differentiation. These kinases are involved in transducing signals for activation of serine/threonine kinases. The AKT pathway, which catalyses the phosphorylation event that transforms Phosphatidylinositol-4,5-bisphosphate to Phosphatidylinositol-3,4,5-triphosphate, activates different growth factors and cytokines as intracellular signals [23].

It has three classes of enzymes based on the regulatory subunits and the structural features. There are three types of classes: class-I, class-II and class-III. IA and IB

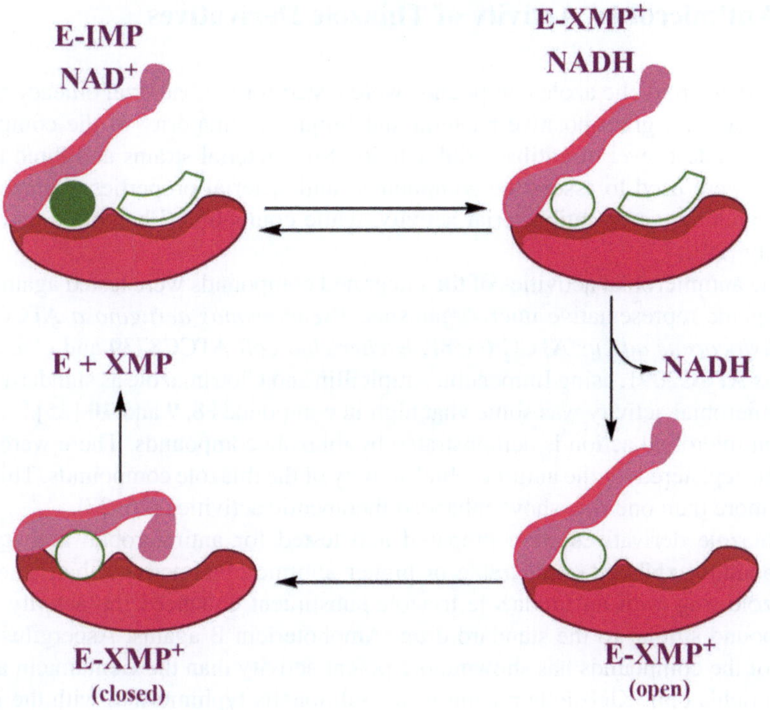

Fig. 5 Transition states of IMPDH [22]

are the two components that makeup class-I. The class-II contains three isoforms (PI3KC2α, PI3KC2β and PI3KC2ϒ) and the class-III contains only one subunit named Vsp34. Class-I is activated by the cell surface receptors and class-II enzymes are activated by RTKs, cytokine receptors and integrins (Fig. 6).

Several thiazole derivatives were synthesised which contain 2-carboxamide pyrrolidine urea and it has been reported for the patent by Novartis AG. These compounds have shown selective preferability for PI3Ks isoforms [20].

Fig. 6 Classification of PI3Ks into classes and sub-classes [23]

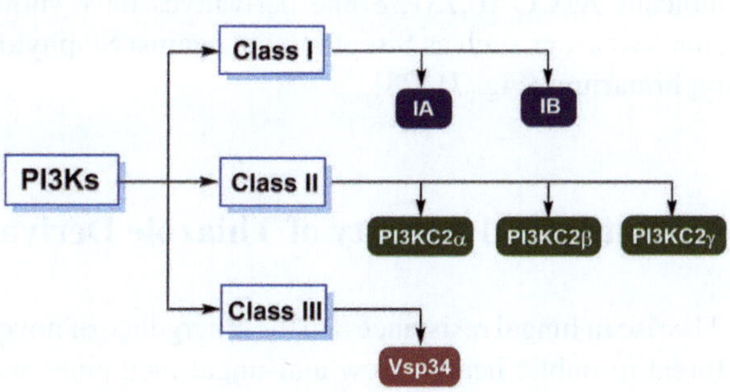

3 Antimicrobial Activity of Thiazole Derivatives

Isolated heterocyclic azole compounds were tested for antibacterial efficacy against gram-negative, gram-positive bacteria and fungi. The majority of the compounds have a modest level of antibacterial activity. Six bacterial strains and three fungus strains were used to assess the compounds' antibacterial properties. Compound **6** has a wide range of antibacterial activity, while compound **7** had good antifungal activity [24].

The antimicrobial activities of the integrated compounds were tested against four pathogenic representative microorganisms: *Pseudomonas aeruginosa* ATCC9027, *Staphylococcus aureus* ATCC 6538P, *Escherichia coli* ATCC8739 and *Candidaalbicans* ATCC2091, using Imipenem, Ampicillin and Clotrimazole as standard drugs. Antimicrobial activity was somewhat high in compounds **8**, **9** and **10** [25].

Antimicrobial action is demonstrated by thiazole compounds. There were many patents registered for the antimicrobial activity of the thiazole compounds. Thiazoles with more than one ring show enhanced therapeutic activities [26, 27].

Thiazole derivatives were prepared and tested for antimicrobial activity. The compounds exhibited comparable or higher antimicrobial activity than standards. Thiazole ring with an imidazole triazole substituent enhanced the activity of the compound similar to the standard drug, Amphotericin B against Aspergilus niger. One of the compounds has shown more potent activity than the Gentamicin against Escherichia coli, Klebsiella pneumoniae, Salmonella typhimurium with the inhibition zone of 26.7 mm, 25.3 mm and 26.33 mm respectively whereas the values of Gentamicin was 25.4, 22.6 and 23.3 respectively. This shows that the presence of the Thiazole ring with various substituents exhibits an efficient antimicrobial activity.

Diverse interactions of the thiazole ring with the amino acids present in the active region of the protein were observed during docking studies of thiazole substituents with various proteins. The thiazole ring has shown interactions with LysB97 in protein 17a-1ydo, Trp55 in 19f-1ydo, LysA297 in protein 24-1ydo and Asn36, Ala35, Ser38 & Ser41 in 19f-3k4p [26].

The 2-phenylamino thiazole derivatives were prepared and evaluated for its antimicrobial activity. Some derivatives have shown comparable activity with the standard drug Fluconazole against Enterococcus faecalis ATCC 29,212 and Candida albicans ATCC 10,231. Some derivatives have shown increased activity than the standard drugs such as Streptomycin against Staphylococcus aureus and Salmonella typhimurium (Fig. 7) [28].

4 Antifungal Activity of Thiazole Derivatives

The rise in fungal resistance and the emergence of novel infections has posed a serious threat to public health. New antifungal medicines with mechanisms separate from well-known families of therapeutic medicines are being developed with increasing

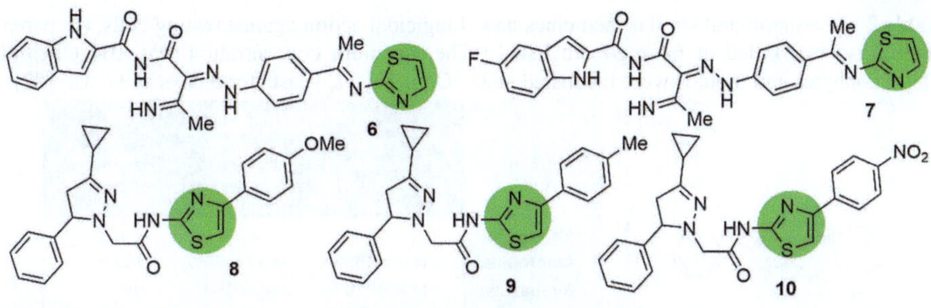

Fig. 7 Thiazole derivatives with antimicrobial activity [24–28]

vigour. The widespread clinical use of azole-based medical medications has attracted a lot of attention and their research and development has been a very busy and fast-moving key issue with a limitless amount of space. Antibacterial activity was investigated using a variety of compounds, including (*Staphylococcus aureus* ATCC 9144, *Pseudomonas aeruginosa* ATCC 2853, *Micrococcus luteus* ATCC 4698, *Staphylococcus Epidermidis* ATCC 155, *Bacillus cereus* ATCC 11,778, *Escherichia coli* ATCC 25,922, *Klebsiella Pneumoniae* ATCC 11,298) and antifungal (*Aspergillus fumigatus* ATCC 46,645 and *Aspergillus niger* ATCC 9029) activities by paper disc diffusion method. The majority of the compounds have antibacterial and antifungal properties. Compound **11** was found to show highest antibacterial activity and compound **12** exhibited maximum antifungal activity (Fig. 8) [29].

Thiazole derivatives exhibit a great antifungal activity with low cytotoxicity to the human cells [30]. Bis-Thiazole derivatives have been synthesised and evaluated for antifungal activity against *Aspergillus niger*, *Aspergillus flavus*, *Penicillium chrysogenum* and *Fusarium moniliforme* [31].

Abafungin, a thiazole ring containing drug has shown better activity than all the standard drugs used against Candida and Aspergillus with the MIC range of 0.5 μg/ml to 16 μg/ml and 0.1 μg/ml to 4 μg/ml respectively whereas the standards exhibited values greater than MIC of 2 μg/ml to > 64 μg/ml and 1 μg/ml to 16 μg/ml respectively. This indicates that thiazoles exhibit good antifungal activity.

Fig. 8 Thiazole derivatives with Antifungal activity [29]

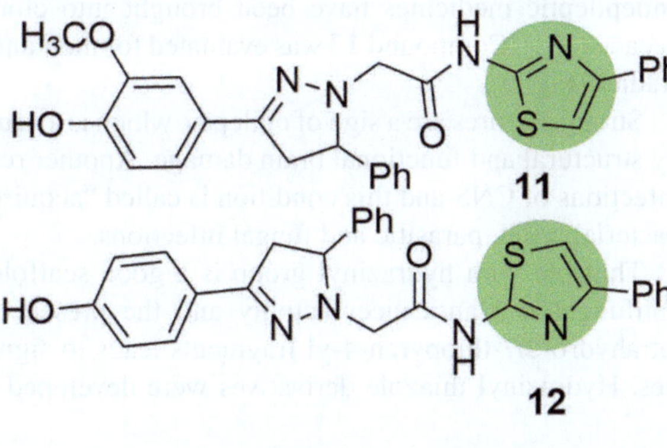

Table 2 Abafungin and similar medicines have fungicidal action against resting cells. (In parentheses, isolates killed at 64 mg/l are listed.) The inoculum concentration was 104 CFU/ml. Dermatophytes and moulds were incubated at 28 °C for 5 days, yeasts for 3 days at 37 °C.) [32]

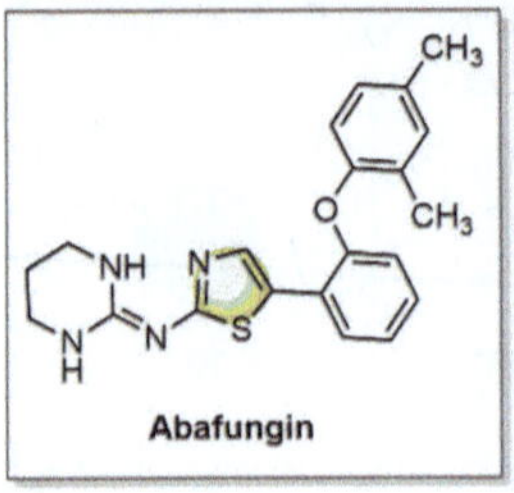

Antifungal Agent	MFC range, g/ml		
	Dermatographytes 43 isolates in demineralized water	Molds 20 isolates in saline	Yeast 38 isolates in saline
Abafungin	1-16	4-16	4-32
Amorolfine	1 to >64 (85%)	4 to >64 (63%)	>64 (0%)
Bifonazole	8 to >64 (18%)	>64 (0%)	Not done
Clotrimazole	8 to >64 (51%)	>64 (74%)	64 to >64 (5%)
Terbinafine	1-32	>64 (0%)	Not done

Abafungin causes ion leakage, ATP leakage and degradation of the fungal resting cells. It also showed the same effect in the growing cells but in a slower manner. In resting cells the ATP degradation was observed directly whereas in growing cells; a short increase of ATP synthesis followed by ATP degradation was observed. Abafungin has two mechanisms of action: it inhibits ergosterol manufacture at low doses and damages the cell membrane at higher quantities (Table 2) [32].

5 Anticonvulsant Activity of Thiazole Derivatives

Thiazole derivatives are one of the most investigated scaffolds of heterocyclic molecules due to their diverse pharmacological properties. Efforts have also been made to develop new anticonvulsant medicines with distinct mechanisms derived from well-known classes of scientific medications. The widespread use of azole-based medicines in medicine has drawn a lot of attention. Epilepsy is a frequent neurological syndrome that refers to a series of patterns in the brain that involve spontaneous, intermittent and aberrant electrical activity. During the last decade, epilepsy medication has been mostly forgotten. The maximal electroshock (MES) test and subcutaneous pentylene tetrazole (ScPTZ) test are the most explored animal models of epilepsy to evaluate anticonvulsant properties, despite the fact that new antiepileptic medicines have been brought into clinical use during the past two decades [33]. Compound **13** was evaluated for their anticonvulsant and neurotoxicity studies (Fig. 9).

Strong seizures are a sign of epilepsy, which is a neurological illness characterised by structural and functional brain damage. Another reason for epilepsy is due to the infections of CNS and this condition is called "acquired epilepsy". It may be due to bacterial, viral, parasitic and fungal infections.

Thiazole with hydrazinyl group is a good scaffold for exhibiting antibacterial, antifungal and anticancer activity and the presence of cyclopentylethylene and tetrahydro-2*H*-thiopyran-4-yl fragments leads to significant anticonvulsant properties. Hydrazinyl thiazole derivatives were developed and tested for antifungal and

Fig. 9 Thiazole derivatives with anticonvulsant activity [33]

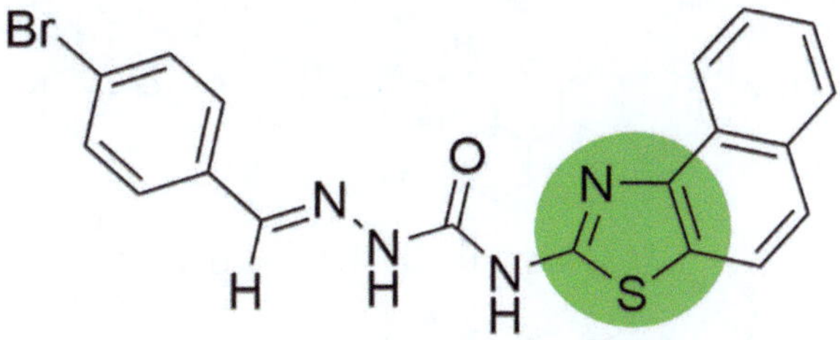

13

anticonvulsant activity. Some of them have shown potent antifungal activity. This indicates that by using these compounds, acquired epilepsy due to fungal infections can be cured. The anticonvulsant efficacy of these compounds was investigated and two compounds showed the maximum anticonvulsant effect at a dosage of 100 mg/kg, inhibiting seizures by 75 and 63 percent in the PTZ model, respectively. Four drugs showed action in the PTZ, MES and 6-Hz models, with no evidence of impaired motor coordination [34].

Anticonvulsant action was investigated using a new series of 2-substituted thiazolidin-4-ones. These compounds exhibited potent activity with the ED_{50} value of 18.5 mg/kg and 15.3 mg/kg in the MES and Sc-PTZ test and the protective index ($PI = TD_{50}/ED_{50}$) values 10.6 and 12.8 respectively. It is highly safer than the reference drug, Carbamazepine whose ED_{50} value is 11.8 mg/kg and 11.2 mg/kg in MES and Sc-PTZ and PI is 6.4 and 6.7 respectively [35].

6 Antibacterial Activity of Thiazole Derivatives

Pathogenic bacteria have caused serious diseases and a lot of mortality in many nations, especially in developing countries. These agents usually spread instantly and the most susceptibilities to them have been assigned to the immunocompromised persons, pregnant mothers and children and older individuals. Along with the dynamic resistance of bacteria to the current antibiotics as a result of irregular antibiotics utilization in medicine, the health and general hygiene of people are strongly at risk and therefore, to prevent this threat, identification and utilizing novel antibacterial compounds are necessary. In recent years, experimental researchers have popularised some thiazole derivatives as multi-therapeutic effect compounds including anticancer, anti-inflammatory and inhibitors of the parasites like *Leishmania* and the fungi such as *Candida.* Furthermore, the antibacterial effects of these compounds have been proven on a broad range of pathogens like *Bacillus subtilis, Staphylococcus aureus, Pseudomonas aeruginosa, Klebsiella aerogenes* that their potency and immense spectrum activity have promised the researchers to replace them with outdated drugs to which the bacteria are resisting. Maximum antibacterial activity was detected in the compounds **14**, **15**, **16**, **17** and **18**. Compounds **15**, **17**

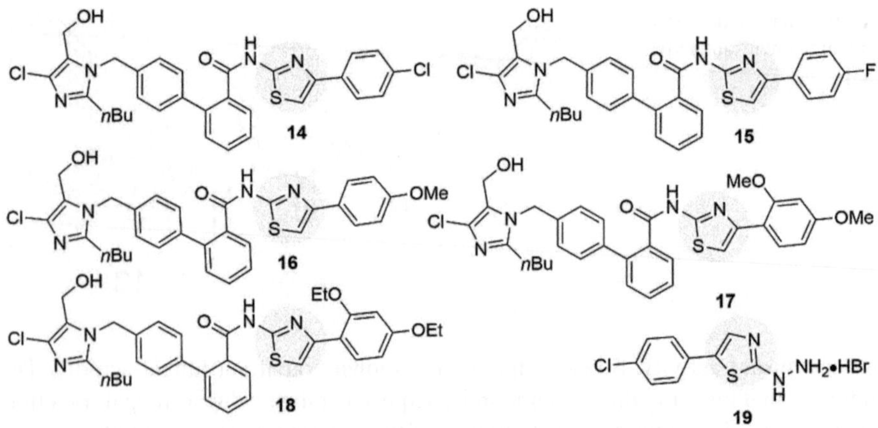

Fig.10 Thiazole derivatives with antibacterial activity [36]

and **18** demonstrated the most fungicidal efficacy, according to fungicidal screening data [36].

The compound **19** was tested for antibacterial activity against ten distinct gram-positive and gram-negative bacteria using the agar well diffusion method and it exhibited significant efficacy against *Bacillus subtilis* and *Staphylococcus aureus*, respectively, when compared to the standard antibiotic Levofloxacin (Fig. 10).

Thiazole derivatives exhibit antimicrobial activity. 1,3-thiazole derivatives with di or tri-substitution have exhibited various pharmacological activities. Febuxostat, Fatostatin and Nizatidine are drugs with thiazole with substitution. Febuxostat is xanthine oxidase inhibitor and urate-lowering drug, Fatostatin is an SREBP inhibitor and Nizatidine is used in GERD and peptic ulcer treatment.

C_2 position of thiazole ring requires the moieties like substituted phenyl ring which enhances the hydrophilic properties and it increases antibacterial effect [31]. Thiazole compounds exhibit a very good antibacterial effect and their Schiff bases exhibit higher antibacterial effects when compared to simple thiazoles. The ligands and the metal complexes were tested for antibacterial activity. The ligands have shown the inhibition zone of 27% to 45% in *Staphylococcus aureus* and *Pseudomonas aeruginosa*, 45% to 67% in *Escherichia coli* and *Klebsiella pneumoniae* whereas the metal complexes exhibited the activity of 64% to 100% in all the four species [37].

The 2-pyrazol-1-yl-thiazole derivatives were developed and screened for antibacterial activity. The compounds' MIC was determined to be between 0.03 and 0.08 μg/ml. One of the compounds has robust action compared to Erythromycin, with a MIC of 2.5 μg/ml and activity comparable to Levofloxacin, having a MIC of 0.016 μg/ml [38].

7 Anthelmintic and Insecticidal Activity of Thiazole Derivatives

Thiazole nucleus normally occurs in natural products. The most significant naturally occurring azole derivatives thiamine and penicillin contain thiazole ring systems. Moreover, a number of thiazole derivatives exhibit good anthelmintic [39] and insecticidal activity. Anthelmintic effects against earthworms (*Eudrilus Eugeniae*) and insecticidal activities against termites were performed on compounds **20** and **21**. (*Coptotermes formosanus*) (Fig. 11).

N-methylated derivatives of thiazolyl amino acids and peptides were synthesised and screened for the anthelmintic activity by Garg's method on *Eudrilus Eugeniae* and insecticidal activity by Morita et al. method on *Coptotermes formosanus.* Three compounds were found to have anthelmintic action and their activity was compared to that of Mebendazole, a conventional medicine. The paralyzing time of the compounds was found to be 50, 45 and 55 min. respectively whereas the Mebendazole paralyzing time was found to be 55 min. The death time of the worm of the three compounds was 55 min, 50 min. and 60 min. and the Mebendazole exhibited 60 min. Two compounds exhibited higher activity than the Mebendazole and one compound exhibited comparable activity with the Mebendazole.

The insecticidal activity was screened and compared with the standard drug, Chlorpyrifos and its death time was 2.40 h. The death time of the compounds exhibiting the activity was found to be 2.10 h, 1.45 min and 2.30 h respectively [40].

7.1 *Thiazoles as Acetylcholinesterase Inhibitors*

Thiamine inhibits the enzyme-Acetylcholinesterase and blocks the nerve impulse transmission in the body. Acetylcholinesterase consists of an anionic binding site. So, any molecule containing positive charge can bind to it and act as a reversible inhibitor at the physiological pH. Thiazole alone doesn't act as an inhibitor, but the addition of a methyl group at the fourth position makes it an inhibitor. As the carbon chain increases, the capability of the molecule to inhibit also increases. The

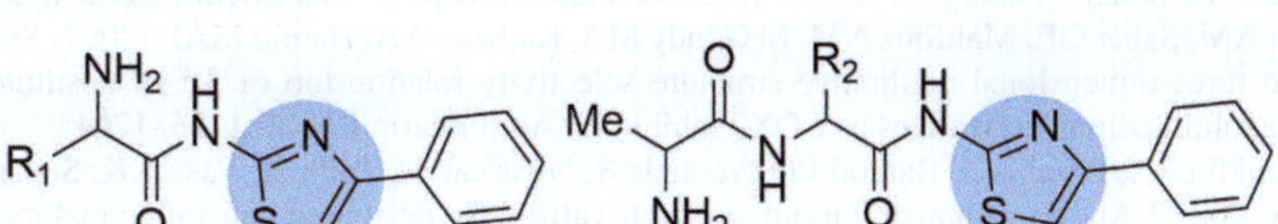

20 **21**

Fig.11 Thiazole derivatives with anthelmintic and insecticidal activity [39]

attachment of the polar groups like hydroxyl groups decreases the inhibitory activity of the molecule.

Thiazole in this molecule has the quaternary nitrogen atom which has a permanent charge and the alkyl chain (mostly ethanolic) on the thiazole ring are the molecular features responsible for the thiamine's inhibition of Acetylcholinesterase [41]. Acetylcholinesterase reversible inhibitors will be commonly used in the neurodegenerative disorder's treatment especially Alzheimer's disease [42].

8 Summary/Conclusion

This chapter highlights the medicinal significance of azole derivatives in various fields of medicinal chemistry and drug discovery. The biological importance and mode of action of thiazole-containing drugs have been discussed in detail for the benefit of the scientific community. To stimulate the interest of chemists working on drug discovery and medicinal chemistry research, the usage of thiazole derivatives as antimicrobials, antitumour, antivirals, NSAIDs, antifungals and anticancer agents has been briefly explained.

References

1. Cascioferro S, Parrino B, Carbone D, Schillaci D, Giovannetti E, Cirrincione G, Diana P (2020) Thiazoles, their benzofused systems, and thiazolidinone derivatives: versatile and promising tools to combat antibiotic resistance. J Med Chem 63:7923–7956
2. Beno BR, Yeung KS, Bartberger MD, Pennington LD, Meanwell NA (2015) A survey of the role of noncovalent sulfur interactions in drug design. J Med Chem 58:4383–4438
3. Fu RG, Wang Y, Xia F, Zhang HL, Sun Y, Yang DW, Wang YW, Yin P (2019) Synthesis of 2-amino-5-acylthiazoles by a tertiary amine-promoted one-pot three-component cascade cyclization using elemental sulfur as a sulfur source. J Org Chem 84:12237–12245
4. Miura T, Funakoshi Y, Fujimoto Y, Nakahashi J, Murakami M (2015) Facile synthesis of 2,5-disubstituted thiazoles from terminal alkynes, sulfonyl azides, and thionoesters. Org Lett 17:2454–2457
5. Rostom SAF, El-Ashmay IM, Abd El Razik HA, Badr MH, Ashour HMA (2009) Design and synthesis of some thiazolyl and thiadiazolyl derivatives of antipyrine as potential non-acidic anti-inflammatory, analgesic and antimicrobial agents. Bioorg Med Chem 17:882–895
6. Ali AM, Saber GE, Mahfouz NM, El Gendy MA, Radwan AA, Hamid MAE (2007) Synthesis and three-dimensional qualitative structure selectivity relationship of 3,5-disubstituted-2,4-thiazolidinedione derivatives as COX2 inhibitors. Arch Pharm Res 30:1186–1204
7. Franklin PX, Pillai AD, Rathod PD, Yerande S, Nivarkar M, Padh H, Vasu KK, Sudarsanam V (2008) 2-Amino-5-thiazolyl motif: a novel scaffold for designing anti-inflammatory agents of diverse structures. Eur J Med Chem 43:129–134
8. Roy RS, Gehring AM, Milne JC, Belshaw PJ, Walsh CT (1999) Thiazole and oxazole peptides: biosynthesis and molecular machinery. Nat Prod Rep 16:249–263
9. Frijaa LMT, Pombeiro AJL, Kopylovich MN (2016) Coordination chemistry of thiazoles, isothiazoles and thiadiazoles. Coord Chem Rev 308:32–55

10. Wishart DS, Feunang YD, Guo AC, Lo EJ, Marcu A, Grant JR, Sajed T, Johnson D, Li C, Sayeeda Z, Assempour N, Iynkkaran I, Liu Y, Maciejewski A, Gale N, Wilson A, Chin L, Cummings R, Le D, Pon A, Knox C, Wilson M (2018) DrugBank 5.0: a major update to the DrugBank database for 2018. Nucleic Acids Res 46:D1074–D1082
11. Ali I, Lone MN, Al-Othman ZA, Al-Warthan A, Sanagi MM (2015) Heterocyclic scaffolds: centrality in anticancer drug development. Curr Drug Targets 16:711–734
12. Huang ST, Hsei IJ, Chen C (2006) Synthesis and anticancer evaluation of bis(- benzimidazoles), bis(benzoxazoles), and benzothiazoles. Bioorg Med Chem 14:6106–6119
13. Kayagil I, Demirayak S (2009) Synthesis and anticancer activities of some thiazole derivatives. Phosphorus Sulfur Silicon 184:2197–2207
14. El-Subbagha HI, Abadi AH, Lehmann J (1999) Synthesis and antitumour activity of ethyl 2 - Substituted-aminothiazole-4-carboxylate analogs. Arch Pharm Pharm Med Chem 332:137–142
15. Shao L, Zhou X, Hu Y, Jin Z, Liu J, Fang J-X (2006) Synthesis and evaluation of novel ferrocenyl thiazole derivatives as anticancer agents. Synth React Inorg Met-Org Nano-Met Chem 36:325–330
16. Luzina EL, Popov AV (2009) Synthesis and anticancer activity of N-bis(trifluoromethyl)alkyl-NKthiazolyl and N-bis(trifluoromethyl)alkyl-NKbenzothiazolylureas. Eur J Med Chem 44:4944–4953
17. Dunn D, Husten J, Ator MA, Chatterjee S (2007) Novel poly(ADP-ribose polymerase-1inhibitors. Bioorg Med Chem 17:542–545
18. Kaur H, Goyal A (2018) A review on thiazole as anticancer agents. Int J Pharm Drug Anal 6:509–522
19. Sintchak MD, Nimmesgern E (1999) The structure of inosine 5^1-monophosphate dehydrogenase and the design of novel inhibitors. Immunopharmacology 47:163–184
20. Liu P, Cheng H, Roberts TM, Zhao JJ (2009) Targeting the phosphoinositide 3-kinase (PI3K) pathway in Cancer. Nat Rev Drug Discov 8:627–644
21. Morigi R, Locatelli A, Leoni A, Rambaldi M (2015) Recent patents on thiazole derivatives endowed with antitumour activity. Recent Pat Anticancer Drug Discov 10:280–297
22. Hedstrom L (2009) IMP dehydrogenase: structure, mechanism, and inhibition. Chem Rev 109:2903–2928
23. Cooney DA, Jayaram HN, Glazer RI, Kelley JA, Marquez VE, Gebeyehu G, Van Cott AC, Zwelling LA, Johns DG (1983) Studies on the mechanism of action of Thiazofurin metabolism to an analog of NAD with potent IMP-dehydrogenase-inhibitory activity. Adv Enzyme Regul 21:271–303
24. Mostafa MS, Abd El-Salam NM (2013) Synthesis and biological evaluation of 3-methyl-2pyrazolin-5-one derivatives containing thiazole and indole moieties. Der PharmaChemica 5:1–7
25. Qureshi A, Pradhan DA (2016) Antimicrobial thiazoles: a short review. Curr Res Bio Phar Sci 5:6–13
26. Althagafi I, El-Metwaly N, Farghaly TA (2019) New series of thiazole derivatives: synthesis, structural elucidation, antimicrobial activity molecular modeling and MOE docking. Molecules 24:1741
27. Sadek B, Al-Tabakha MM, Fahelelbom KMS (2011) Antimicrobial prospect of newly synthesized 1,3-thiazole derivatives. Molecules 16:9386–9396
28. Bikobo DS, Vodnar DC, Tiperciue ASB, Nastasa C, Douchet M, Oniga O (2017) Synthesis of 2-phenylamino-thiazole derivatives as antimicrobial agents. J Saudi Chem Soc 21:861–868
29. Saravanan G, Alagarsamy V, Pavitra TGV, Kumar GC, Savithri Y, Naresh L, Avinash P (2010) Synthesis, characterization and anti-microbial activities of novel thiazole derivatives. Int J Pharma Bio Sci 1:1–8
30. Naryana B, Raj KKV, Ashalatha BV, Kumari SN (2006) Antibacterial and Antifungal studies on some new acetylcinnolines&cinnolinyl thiazole derivatives. Indian J Chem 45B:1704–1709
31. Borde RM, Jadhav SB, Dhavse RR, Munde AS (2018) Design, synthesis, and pharmacological evaluation of some novel bis-thiazole derivatives. Asian J Pharm Clin Res 11:164–168

32. Borelli C, Schaller M, Niewerth M, Nocker K, Baasner B, Berg D, Tiemann R, Tietjen K, Fugmann B, Fugmann SL, Korting HC (2008) Modes of action of the new arylguanidine abafungin beyond interference with ergosterol biosynthesis and in vitro activity against medically important fungi. Chemotherapy 54:245–259
33. Łączkowski KZ, Sałat K, Misiura K, Podkowa A, Malikowska N (2016) Synthesis and anticonvulsant activities of novel 2-(cyclopentylmethylene)hydrazinyl-1,3-thiazoles in mouse models of seizures. J Enzyme Inhib Med Chem 31:1576–1582
34. Łączkowski KZ, Konklewska N, Biernasiuk A, Malm A, Sałat K, Furgała A, Dzitko K, Bekier K, Łączkowska AB, Paneth A (2018) Thiazoles with cyclopropyl fragment as antifungal, anticonvulsant, and anti-Toxoplasma gondii agents: synthesis, toxicity evaluation, and molecular docking study. Med Chem Res. https://doi.org/10.1007/s00044-018-2221-x
35. Gong GH, Wang D, Zhang JF, Wei CX, Quan ZS (2014) Anticonvulsant activity of 2-(Substituted-imino) thiazolidin-4-ones. Drug Res 64:5–9
36. Khan I, Ibrar A, Waqas M, White JM (2013) Synthesis, X-ray crystallographic studies and antibacterial screening of 1-(5-(4-Chlorophenyl) thiazol-2-yl) hydrazine hydrobromide. Phys Rev Res Int 3:10–17
37. Chohan ZH, Kausar S (2000) Synthesis, characterization and biological properties of tridentate nno, nns and nnn donor thiazole-derived furanyl, thiophenyl and pyrrolylschiff bases and their Co(II), Cu(II), Ni(II) and Zn(II) metal chelates. Met Based Drugs 7:17–22
38. Ivanenkov YA, Yamidanov RS, Osterman IA, Sergiev PV, Aladinskiy VA, Aladinskaya AV, Terentiev VA, Veselov MS, Ayginin AA, Skvortsov DA, Komarova KS, Sadovnikov SV, Matniyazov R, Sofronova AS, Malyshev AS, Machulkin AE, Petrov RA, Lukianov D, Iarovenko S, Bezrukov DS, Baymiev AK, Dontsova OA (2019) 2-Pyrazol-1-yl-thiazole derivatives as novel highly potent antibacterials. J Antibiot 72:827
39. Lunkad AS, Kothawade SN, Darkunde KK, Priya B, Bagmar UR, Bhandari DS (2013) Synthesis and screening anthelmintic activity of some thiazole derivatives. Int J Chem Sci 11:1146–2115
40. Himaja M, Nitesh G, Munirajesekhar D, Asif K, Mukesh SS (2012) Synthesis and Biological evaluation of some N-methylated derivatives of Thiazolylamino acids and peptides. J Pharm Sci Innovation 1:33–36
41. Alspach JD, Ingraham LL (1977) Inhibition of acetylcholinesterase by thiamine a structure-function study. J Med Chem 20:161–164
42. Colovic MB, Krstic DZ, Pasti TDL, Bondzic AM, Vasic VM (2013) Acetylcholinesterase inhibitors: pharmacology and toxicology. Curr Neuropharmacol 11:315–335

Chapter 6
Overview on Biological Activities of Imidazole Derivatives

Raghuram Gujjarappa, Arup K. Kabi, Sattu Sravani, Aakriti Garg, Nagaraju Vodnala, Ujjawal Tyagi, Dhananjaya Kaldhi, Ravichandiran Velayutham, Virender Singh, Sreya Gupta, and Chandi C. Malakar

Abbreviations

U.S. FDA	United States Food and Drug Administration
ZE	Zollinger-Ellison syndrome
ARB	Angiotensin II receptor blocker
DNA	Deoxyribonucleic Acid
CNS	Central Nervous System
ACE	Angiotensin-converting enzyme
GABA	Gamma-Aminobutyric Acid
NSAID	Non-steroidal Anti-inflammatory Drugs
MTIC	3-Methyl-(triazen-1-yl)imidazole-4-carboxamide
TIMP	Thioinosine monophosphate
HER2	Human Epidermal Growth Factor Receptor 2
RNA	Ribonucleic Acid
HCV	Hepatitis C Virus
IBS-D	Irritable Bowel Syndrome with Diarrhoea
CYP3A	Cytochrome P450 3A
ara-GTP	Ara-G triphosphate
GERD	Gastroesophageal Reflux Disease
DPP-4	Dipeptidyl peptidase-4

R. Gujjarappa · A. K. Kabi · N. Vodnala · U. Tyagi · D. Kaldhi · C. C. Malakar (✉)
Department of Chemistry, National Institute of Technology Manipur, Langol, Imphal, Manipur 795004, India
e-mail: cmalakar@nitmanipur.ac.in

S. Sravani · A. Garg · R. Velayutham · S. Gupta (✉)
Department of Medicinal Chemistry, National Institute of Pharmaceutical Education and Research (NIPER), Chunilal Bhawan, 168, Maniktala Main Road, KolkataKolkata 70005, India

V. Singh
Department of Chemistry, Central University of Punjab, Bathinda, Punjab 151001, India

B. P. Swain (ed.), *Nanostructured Biomaterials*, Materials Horizons: From Nature to Nanomaterials, https://doi.org/10.1007/978-981-16-8399-2_6

GLP-1	Glucagon-like peptide 1
GIP	Gastric inhibitory polypeptide
TK	Thymidine kinase
NS5A	Nonstructural protein 5A
XMP	Xanthosine monophosphate

1 Introduction

Imidazole is classified as a five-membered heterocyclic aromatic ring with three carbon and two nitrogen atoms clustered at positions 1 and 3. Gluoxaline was the initial name given to it (synthesis was primarily reported with ammonia and glyoxal). Amphoteric character is exposed in terms of electrophilic and nucleophilic attack and it is highly stable to acids, bases, heat, oxidative and reductive conditions. It's possible that it will produce a lot of intramolecular hydrogen bonds because of delocalization of hydrogen atoms on any of the two nitrogen atoms and exists in two tautomeric forms that are equal. The four atoms present in the ring contributes an electron towards the sextet of electrons and the remaining two electron were donated by the protonated nitrogen atom due to its persistent aromaticity. The property of imidazole to act as a base or an acid marks them as amphoteric in nature. Having a pK_a of 14.5, imidazole is more acidic than carboxylic acid, imides and phenols, but less acidic than alcohols. Because conjugate acid has a pK_a of about 7, imidazole is almost 60 times more basic than pyridine. These assets are characterised by resonance interactions that strengthen the basicity of the 3-nitrogen atom (Fig. 1) [1].

Imidazoles are important in medical chemistry, and their activity affects various structures in the human body, including histamine, histidine, biotin, alkaloids, and nucleic acid. A number of imidazole compounds have been proposed for use in medicine. Furthermore, marine sponges create a variety of intriguing, structurally diverse secondary metabolites, most of which include the imidazole moiety. Hundreds of similar compounds have been identified since the discovery of the first alkaloid 'oroidin' of this family in 1971 [2–5]. Imidazole compounds are commonly found in medications with a wide range of pharmacological properties, including anticancer, antioxidant, anti-inflammatory, gastroprotective, histamine-H3 antagonist, and antiparasitic properties. These heteroaromatic groups have biological applications because they are good bioisosteres of biomolecules. It has recently been reported

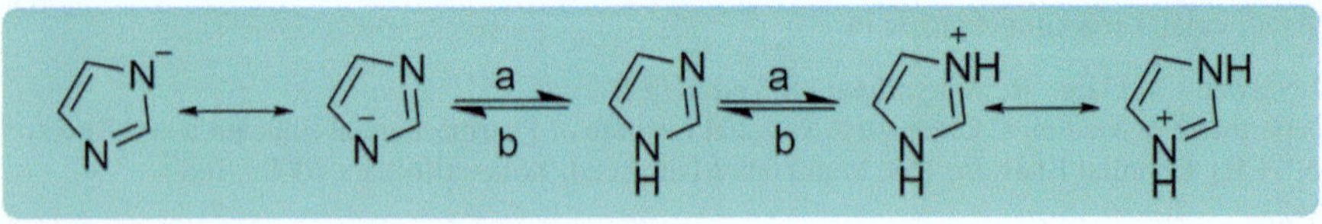

Fig. 1 Resonance structures of Imidazole [1]

that farnesyl protein transferase inhibitors were developed using cysteinyl moiety features that are similar to imidazole functions. Nevertheless, plenty of imidazole derivatives are used as medicinal medications such as antifungal agents (clotrimazole, ketoconazole, miconazole, and oxiconazole), antihypertensive (losartan, eprosartan, and olmesartan), anticancer (dacarbazine, zoledronic acid, azathioprine, and tipifarnib), antihistaminic (imetit, cimetidine, immepip, and thioperamide), antiparasitic (ornidazole, metronidazole, benznidazole, and secnidazole) and antineuropathic (nafimidone, dexmedetomidine, and fipamezole) drugs with high therapeutic potency have been extensively used to treat several types of diseases (Fig. 2) [6–14]. This has drawn a lot of interest to imidazole-based therapeutic compounds, and the growing research and advancements have become a hot topic. In addition, imidazole rings are broadly active as spin trapping classes, which makes them a promising property in the development of neuroprotective medications.

Among the database's five-membered aromatic nitrogen heterocycles, imidazoles are the most abundant. (Fig. 2) [15]. Eight (33%) of the 24 imidazole-containing frameworks belong to an antifungal agent class. In all eight drugs, a chlorinated aromatic ring and a mono-substituted imidazole group are present. All

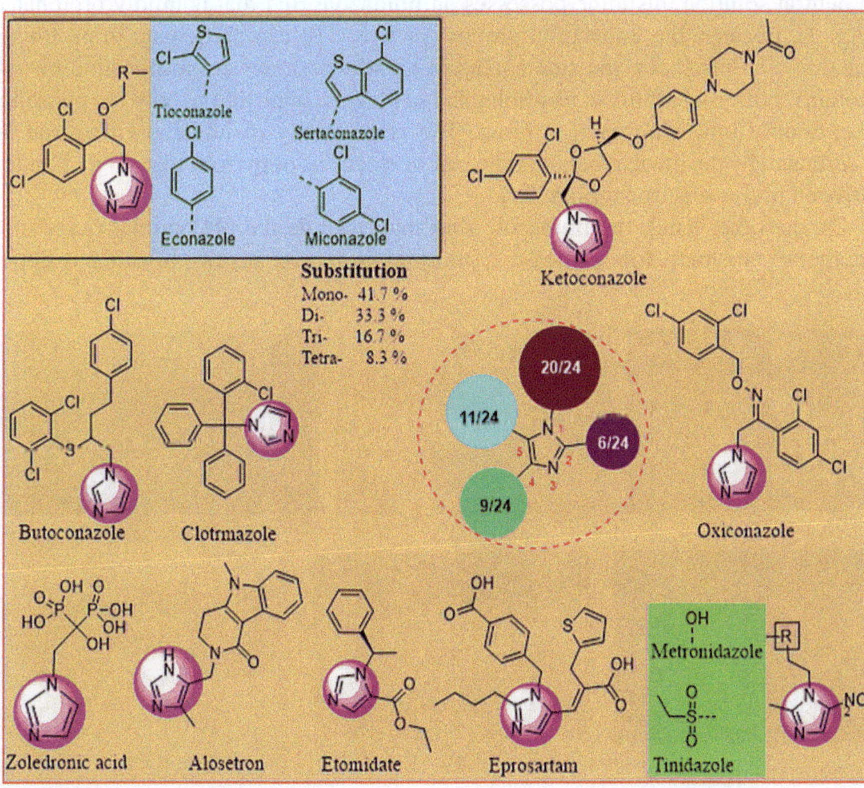

Fig. 2 Pharmaceuticals containing imidazoles [6–15]

chiral compounds in this antifungal category are available in racemic form except ketoconazole. Because of their small size and the presence of nitro substituents on the imidazole ring, metronidazole and tinidazole are essential antibacterial drugs. According to a study of substitution patterns, 42% of imidazole is mono-substituted and 33% is di-substituted. There are no evident replacement options among the remaining imidazole medicines, which are tri- (17%) and tetra-substituted (8%).

2 Naturally Occurring Imidazole and Their Activity

Imidazoles are a popular family of N-heterocycles that encompass a wide range of biological and chemically important compounds (Fig. 3) [16, 17]. They are found in a variety of biomolecules, including the essential amino-acid histidine as well as related chemicals like biotin and histamine. Several histaminergic ligands for histamine H1, H2, and H3 receptors have an imidazole structure [13]. These are now being employed in pharmacological research.

Histidine-(2-amino-3-(4-imidazolyl)-propanoic acid is an important amino acid. In neutral solution, histidine possesses an imidazole ring that is mildly protonated (Fig. 4). Because the imidazolyl groups ($pK_B = 8.0$) can participate in protolytic reaction circumstances, the side chains of histidyl residues are responsible for the proteins' buffering ability at physiological pH levels. In histidine, only the pyridine-like, doubly linked nitrogen is basic. The non-basicity of imidazole ring can be ascertained by the involvement of lone pair of electrons of pyrrole-like singly bonded nitrogen in 6 π-electron aromaticity.

On the other hand, the imidazole ring may include the pH optimal range of a number of enzymes. Ribonuclease, a phosphodiesterase, has two histidine residues

Fig. 3 Naturally occurring sources of imidazoles [16, 17]

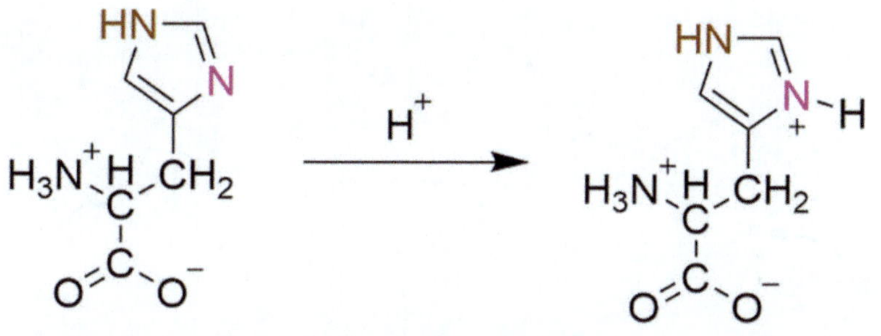

Fig. 4 Biological importance of imidazoles [18]

that are necessary for catalytic activity which is also responsible for the hydrolysis of pyrimidine-2,3-cyclic phosphoric acids. Cytidine-2,3-cyclic phosphoric acid is sandwiched between two imidazole groups in the active site's binding locus (Fig. 5) [18, 19].

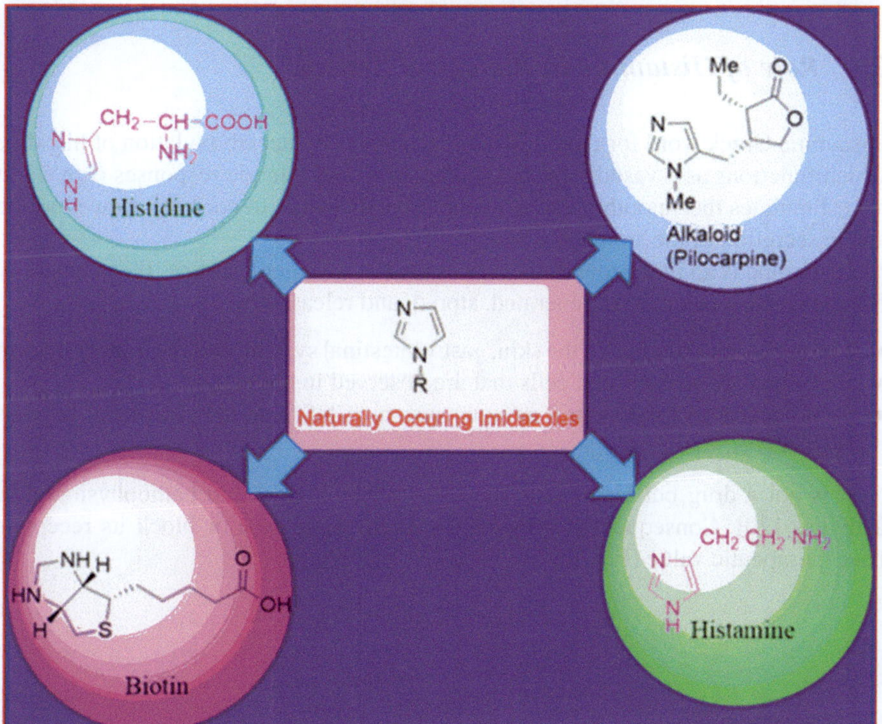

Fig. 5 Biological importance of imidazoles [18, 19]

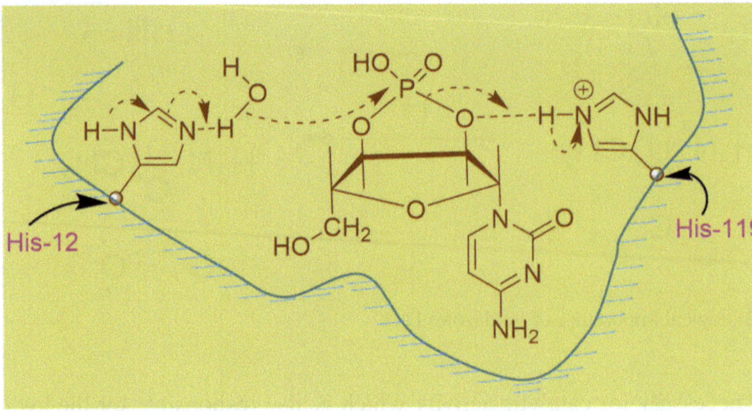

Fig. 6 Hydrolysis of cytidine-2,3-phosphate by ribonuclease [18, 19]

By removing the proton from a water molecule, His-12 acts as a common base throughout the ribonuclease chain (Fig. 6). The electrophilic phosphate group is then attacked nucleophilically by the intermediate OH-ions.

2.1 Role of Histamine in Biological System

Histamine Greek word for tissue (histos), originates by decarboxylation of histidine which functions as a vasodilator in inflammation and allergic responses (Fig. 7). It also stimulates the smooth muscles of the bronchial stem to contract, as well as the acidic secretion of the stomach.

Histamine is an imidazole derivative as well as a biogenic amine. It is a naturally occurring substance that is generated, stored, and released by

- Mast cells are abundant in the skin, gastrointestinal system, and respiratory system
- Basophils are white blood cells that are observed in blood, and
- Some neurons in the central nervous system (CNS) and the peripheral nervous system (PNS) (Fig. 8)

It is not a drug but significant due to its physiological and pathophysiological activities [20]. Consequently, drugs that prevent its release or block its receptors have therapeutic value (Fig. 9).

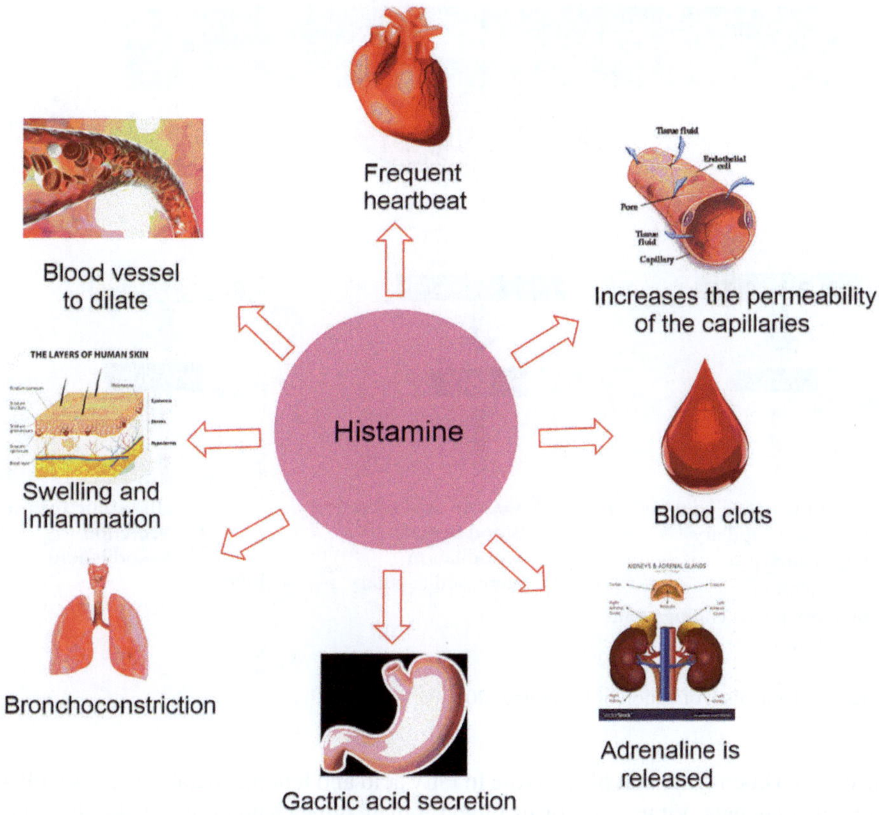

Fig. 7 Histamine and its biological actions [20]

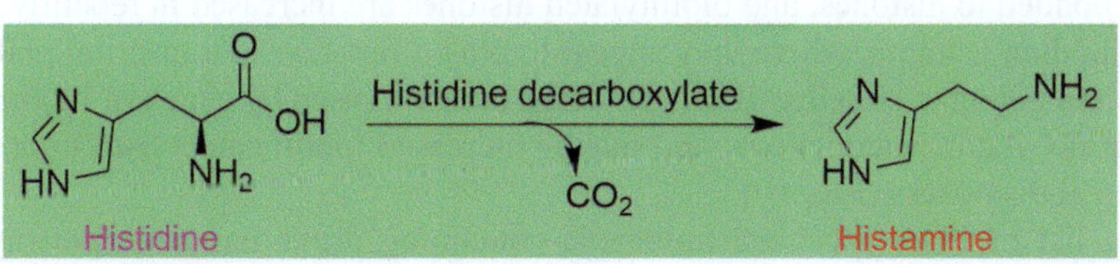

Fig. 8 Precursor for histamine [20]

2.2 *Biological Importance of Biotin*

Biotin is a B-complex vitamin that is water soluble (vitamin B_7). This sulfur-containing heterocyclic monocarboxylic acid acts as a coenzyme in the carboxylation process [21]. The structure is made up of imidazole (tetrahydro imidazole) and thiophene (tetrahydrothiophene) rings that are fused together. A valeric acid substituent is connected to one of the carbon atoms in the tetrahydrothiophene ring.

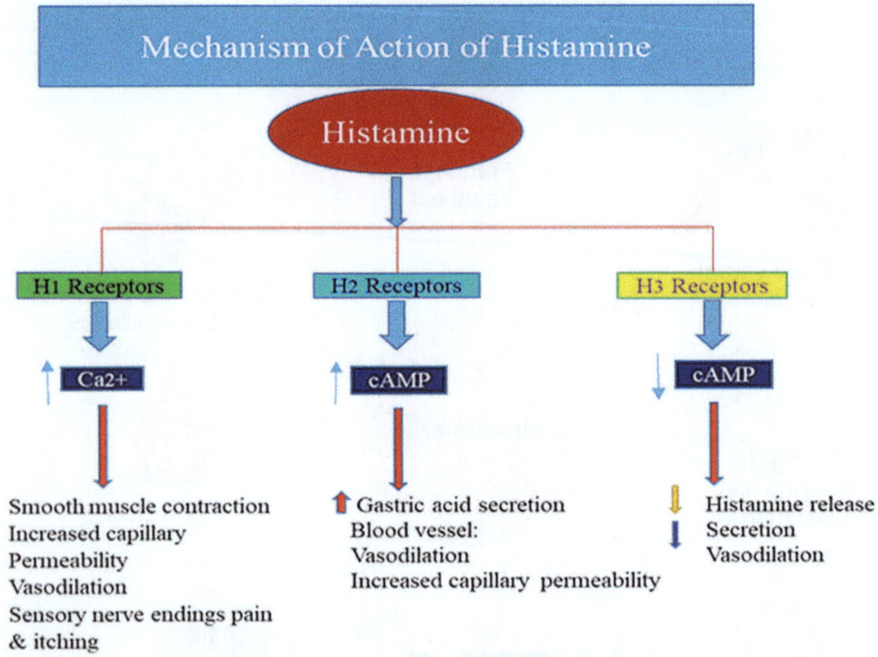

Fig. 9 Mechanism of action of histamine [20]

Biotin is a coenzyme that plays a role in fatty acid and leucine metabolism, as well as gluconeogenesis. Vitamin H forms biocytin in enzymes when it is covalently linked to the amino group of lysine. Biocytin is a coenzyme that is found in biotin. In response to a lack of biotin, carboxylase activity drops dramatically. Biotin is covalently bonded to histones, and biotinylated histones are increased in repetitive areas of the human genome, where they appear to play a function in transcriptional regulation and genomic stability. Biotin shortage can be caused by a lack of biotin in the diet, drug-vitamin interactions, and maybe increased biotin catabolism in pregnant women and smokers (Fig. 10).

The list of FDA approved imidazole-containing drugs has been portrayed in Table 1 [22].

3 Anticancer Activity of Imidazole Derivatives

Cancer is one of the top most significant health threats to humans, although it receives little attention worldwide. Cancer is a genetic disease that spreads by a multistep carcinogenesis process involving various physiological systems in the human body, including cell signalling and death, making it exceedingly difficult to treat [23]. New tumours, irregular bleeding, persistent coughs, unexpected weight loss, and changes

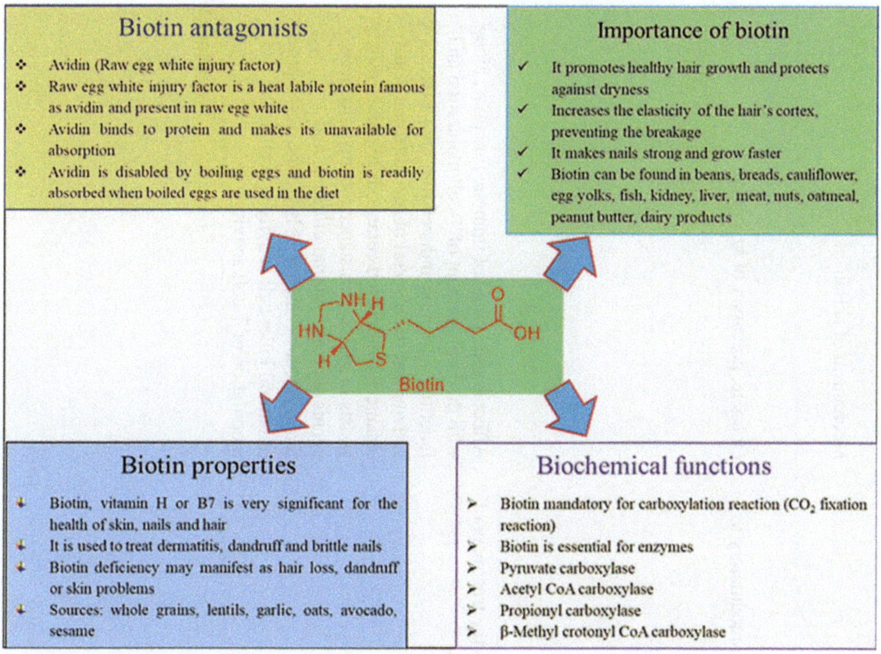

Fig. 10 Biological importance of Biotin [21]

in bowel motions are all possible signs and symptoms of cancer [24, 25]. Cancer is the main cause of death in the United States and stands second in the causes of death worldwide [26–28]. Anticancer drugs, radiation, surgical treatments, and chemotherapy have all been the subject of much research. In 2012, the World Health Organization (WHO) anticipated that in the upcoming two decades, the number of patients affected with cancer will increase to 22 million from 14 million [29, 30]. According to a UN DESA assessment from 2014, the present global population of 7.3 billion people will grow to 8.5 billion by 2030, 9.7 billion by 2050, and 11.2 billion by 2100. This group is thought to be responsible for 1.5 billion cancer cases and 1.2 billion deaths worldwide [31]. In addition, with an estimated 8.5 billion people, total fatalities are predicted to reach 2.14 billion by 2030 [32, 33]. Lung and breast cancers are the most frequent causes of mortality in women and men, respectively, with 522,000 deaths in 2012. In about 140 nations, these two cancer forms are the most frequent [34].

N-heterocyclic compounds have been the principal molecules in organic chemistry for more than a decade due to their superior pharmacological actions, notably their anticancer properties [35]. Researchers have been studying and creating imidazole-based molecules since the invention of imidazole in the 1840s [36–40]. Imidazoles have been employed as effective anticancer medicines in a variety of therapeutic applications. Antidotes for different malignancies, such as zoledronic acid, dacarbazine, tipifarnib, mercaptopurine and others, are now being used in clinics

Table 1 FDA approved imidazole-containing drugs [22]

Structure of the drug Name of the drug (drug bank ID)	Category/indication	Mechanism of action
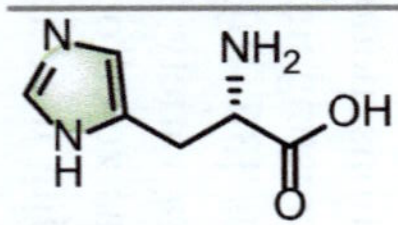 Histidine DB00117	Supplemental L-histidine, immunomodulatory as well as antioxidant activity	Obligate, precursor of histamine
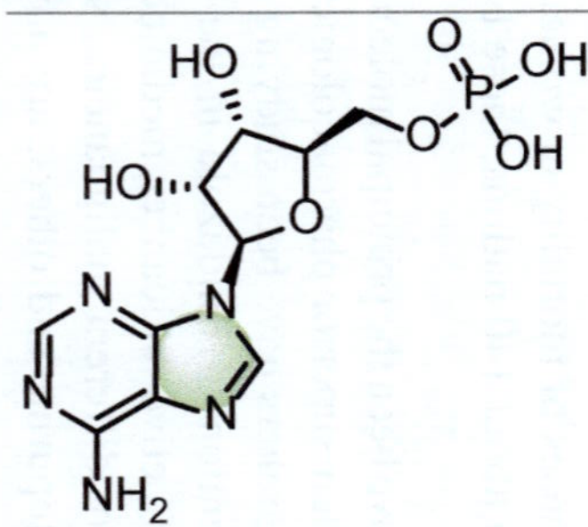 Adenosine phosphate DB00131	Supplemental nutrition, as well as the treatment of dietary deficiencies or imbalances	Affect a wide range of immune functions, along with the improvement of T-cell maturation and function, starvation-induced immunosuppressive activity and the reversal of malnutrition. It also enhances the delayed cutaneous hypersensitivity, assists in gaining resistance against infectious agents such as *Candida albicans*, and Staphylococcus aureus. Also in the development of natural killer cell activity as well as the modulation of T-cell activity

(continued)

Table 1 (continued)

Structure of the drug Name of the drug (drug bank ID)	Category/indication	Mechanism of action
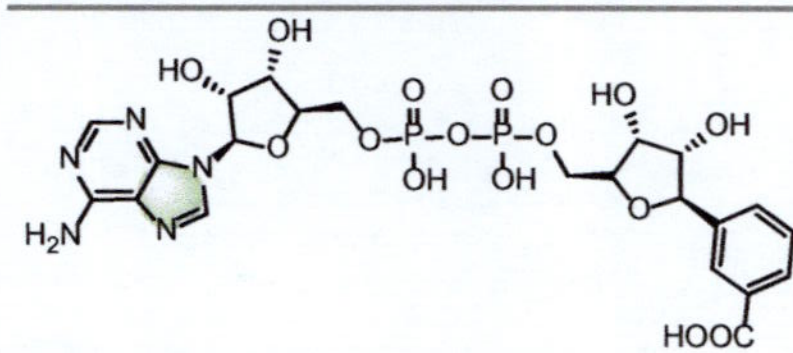 NADH DB00157	Used to cure chronic fatigue syndrome, Alzheimer's disease, Parkinson's disease, and cardiovascular disease	Role in energy generation and the electron-transport chain in mitochondria
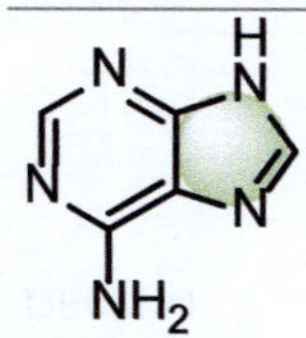 Adenine DB00173	For nutritional replenishment as well as the treatment of dietary deficiencies or imbalances	Transfers chemical energy between reactions to power various cellular metabolic activities

(continued)

Table 1 (continued)

Structure of the drug Name of the drug (drug bank ID)	Category/indication	Mechanism of action
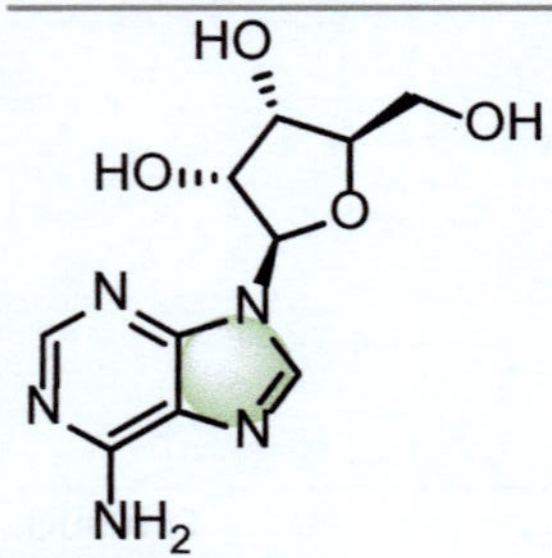 Vidarabine DB00194	Herpes zoster, herpes simplex, and chickenpox-varicella therapy	Herpes virus DNA replication takes place in two ways: (1) incorporation followed by the termination of viral DNA chain growth. (2) Inhibitory action on viral DNA polymerase by acting as a competitive inhibitor

(continued)

Table 1 (continued)

Structure of the drug Name of the drug (drug bank ID)	Category/indication	Mechanism of action
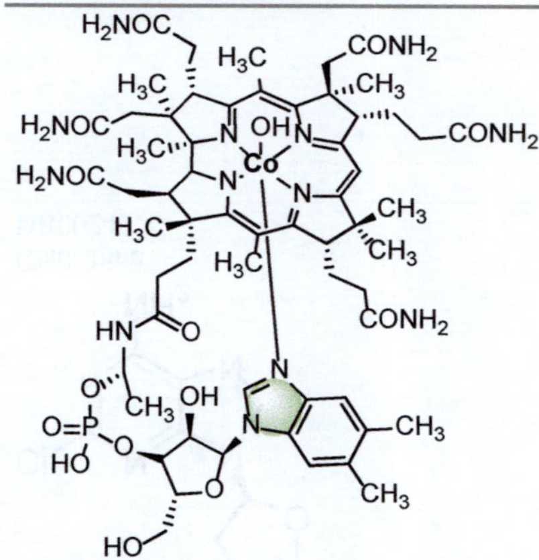 Hydroxocobalamin DB00200	Used in pernicious anaemia therapy, as well as to treat or prevent deficiency of vitamin B_{12}. Also used for treatment of cyanide poisoning, even if suspected or known	Vitamin B_{12} supplement
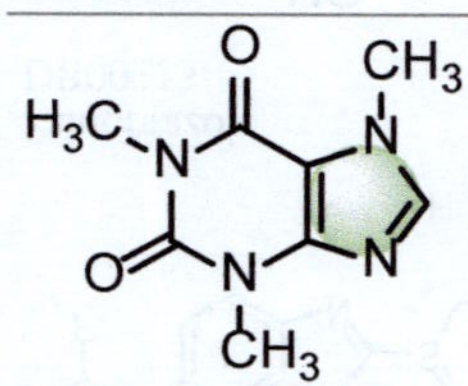 Caffeine DB00201	For the treatment of tiredness, orthostatic hypotension, and apnoea of prematurity in newborns in the short term	In the short term, it is used to treat fatigue, orthostatic hypotension, and prematurity apnoea in infants

(continued)

Table 1 (continued)

Structure of the drug Name of the drug (drug bank ID)	Category/indication	Mechanism of action
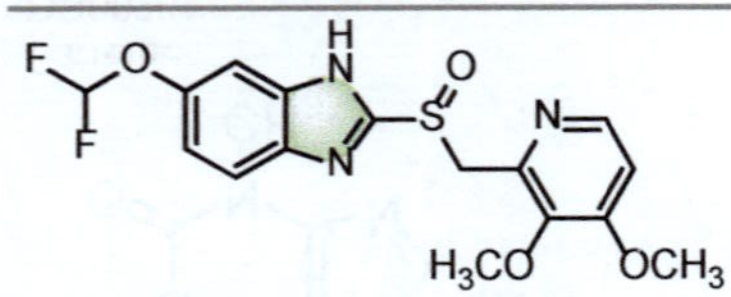 Pantoprazole DB00213	Used for gastric protection in avoiding gastric damages or stomach ulcers resulting from long-term usage of NSAID. Also used to treat gastroesophageal reflux disease (GERD), and Zollinger-Ellison syndrome (ZE), a pathological hyper secretory disorder	Proton pump inhibitor (PPI) of the first generation
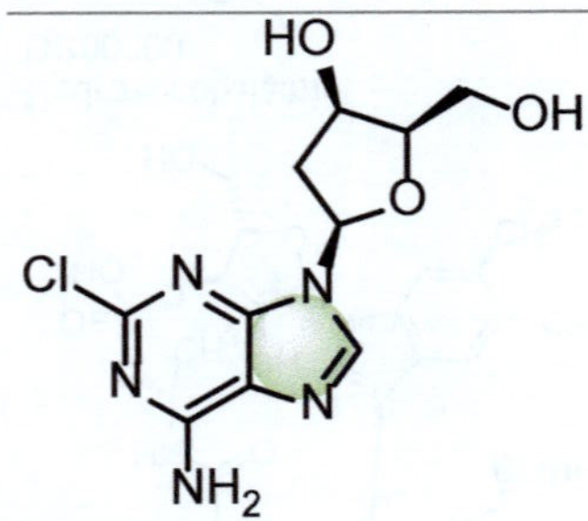 Cladribine DB00242	Hairy cell leukaemia (AHCL) is a kind of cancer that affects the hair follicles (leukaemic reticuloendotheliosis)	DNA strand breaks and DNA synthesis and repair are inhibited. ribonucleotide reductase is inhibited

(continued)

Table 1 (continued)

Structure of the drug Name of the drug (drug bank ID)	Category/indication	Mechanism of action
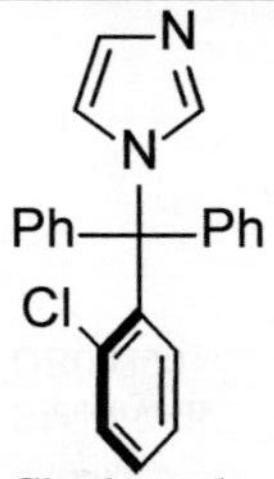 Clotrimazole DB00257	To treat cutaneous infections and oropharyngeal candidiasis	Clotrimazole works by destroying the fungi's cell membrane's permeability barrier. Ergosterol production, an important component of fungal cell membranes, is also inhibited
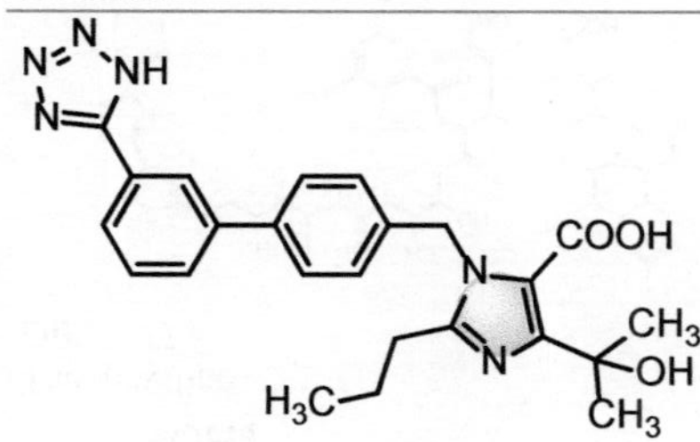 Olmesartan DB00275	It's also used to treat type 2 diabetes-related nephropathy, heart failure, and post-myocardial infarction, especially in those who can't take ACE inhibitors	Angiotensin II receptor blocker (ARB)

(continued)

Table 1 (continued)

Structure of the drug Name of the drug (drug bank ID)	Category/indication	Mechanism of action
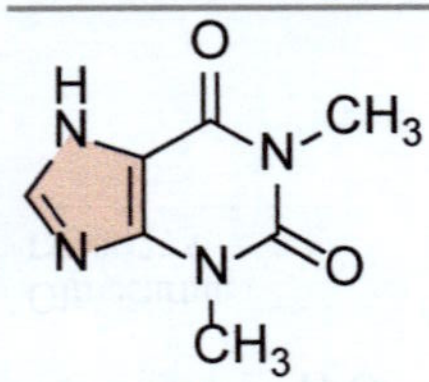 Theophylline DB00277	For the alleviation of persistent asthma symptoms and reversible airflow limitation caused by emphysema and chronic bronchitis (chronic lung diseases)	Phosphodiesterase (PDE) enzyme (type III and type IV), responsible for breakdown of cyclic adenosine monophosphate (AMP) in vascular smooth muscle cells, are competitively inhibited by theophylline, perhaps leading to bronchodilation
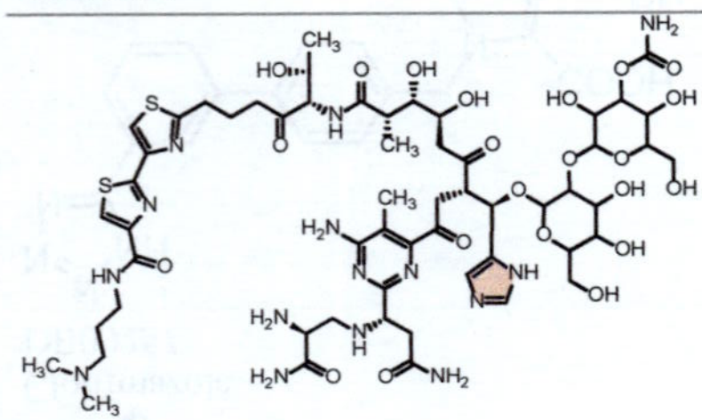 Bleomycin DB00290	Palliative care for lymphoma, squamous cell carcinoma, and malignant neoplasms (trachea, bronchus, lung)	The suppression of DNA synthesis is the primary mechanism of action

(continued)

Table 1 (continued)

Structure of the drug Name of the drug (drug bank ID)	Category/indication	Mechanism of action
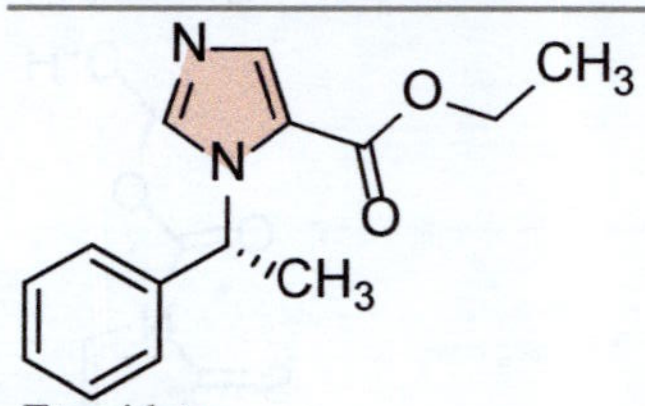 Etomidate DB00292	Used in the induction of general anaesthesia	Etomidate binds to a particular binding site on the GABA receptor that is linked to a Cl-ionophore, allowing the Cl-ionophore to stay open for longer. As a result, GABA's thalamic postsynaptic inhibitory effect is boosted
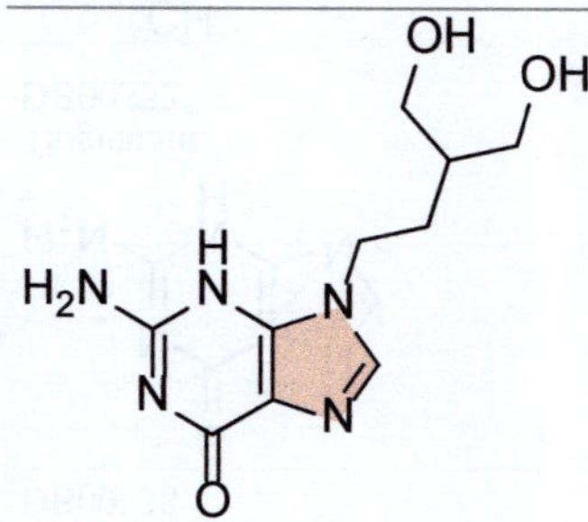 Penciclovir DB00299	Employed in treating herpes virus caused recurrent cold sores appeared on lips and face	By competing with deoxyguanosine triphosphate, inhibits viral DNA polymerase

(continued)

Table 1 (continued)

Structure of the drug Name of the drug (drug bank ID)	Category/indication	Mechanism of action
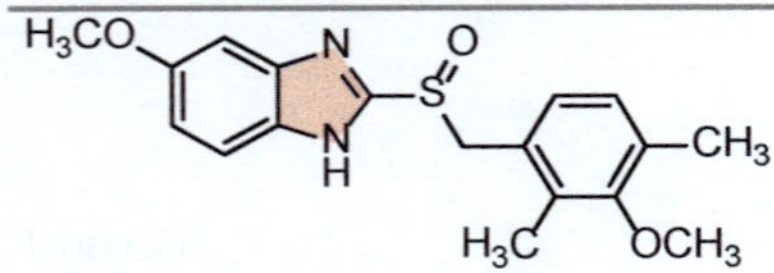 Omeprazole DB00338	Adults with active duodenal ulcers, active benign gastric ulcers, or symptomatic acid reflux disease or gastroesophageal reflux disease (GERD) aged one year and up are treated	Stops gastric acid release by inhibiting the H^+/K^+ ion channels selectively
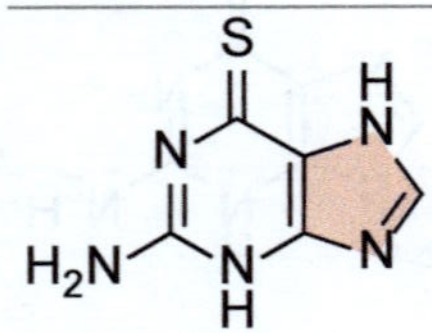 Tioguanine DB00352	For treatment of acute non-lymphocytic leukaemia's	TGMP inhibits guanine nucleotide synthesis by inhibiting purine biosynthesis by pseudo feedback to inhibit glutamine-5-phosphoribosylpyrophosphate amidotransferase, the first enzyme specific to the purine ribonucleotide de novo pathway
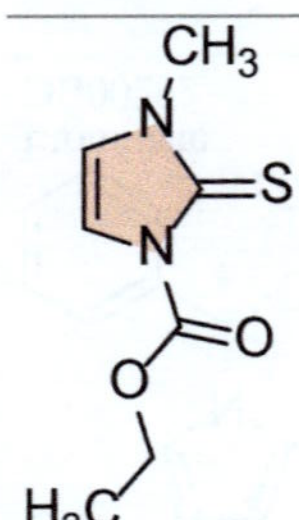 Carbimazole DB00389	Hyperthyroidism and thyrotoxicosis are treated with this medication. It's also used to get people ready for thyroid surgery	Reduces inorganic iodine absorption and concentration in the thyroid, as well as the production of di-iodotyrosine and thyroxine

(continued)

Table 1 (continued)

Structure of the drug Name of the drug (drug bank ID)	Category/indication	Mechanism of action
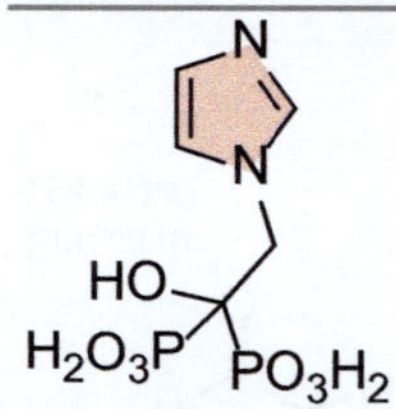 Zoledronic acid DB00399	For the treatment of malignancy-related hyperkalemia as well as individuals with multiple myeloma	Inhibition on farnesyl pyrophosphate synthase (FPP), which is a mevalonate pathway enzyme, by acting as mimics of isoprenoid diphosphate lipids
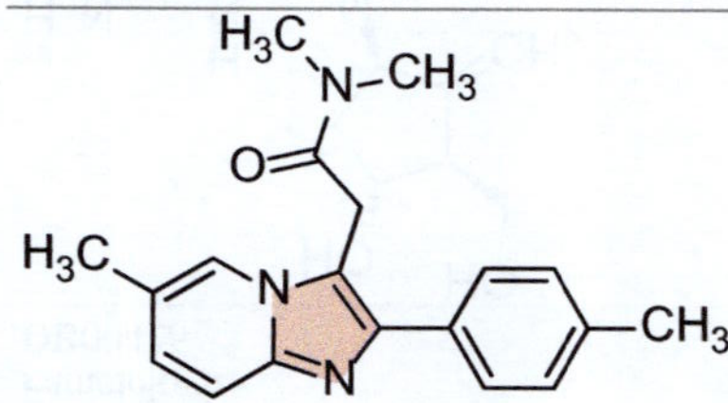 Zolpidem DB00425	This medication is used to assist patients suffering from insomnia for shorter periods in individuals who have trouble falling asleep	It interacts with a complex of GABA-BZ receptors

(continued)

Table 1 (continued)

Structure of the drug Name of the drug (drug bank ID)	Category/indication	Mechanism of action
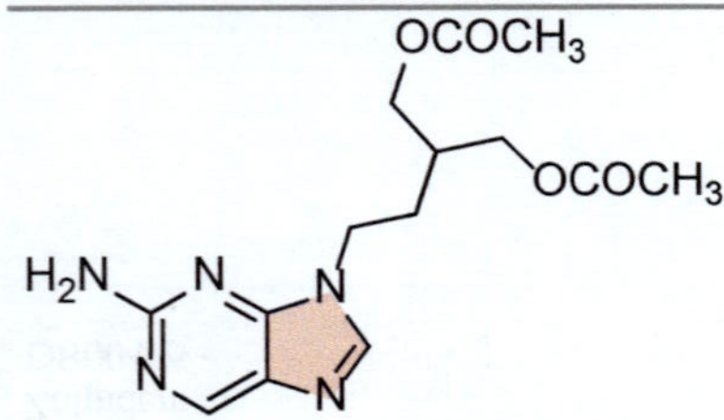 Famciclovir DB00426	Treatment of shingles and recurrent mucocutaneous herpes simplex infections in HIV-positive people as well as to suppress recurrent genital herpes in immunodeficient people	It suppresses herpes simplex virus types 1 and 2 (HSV-1 and HSV-2) as well as varicella-zoster virus, is swiftly biotransformed into famciclovir (VZV)
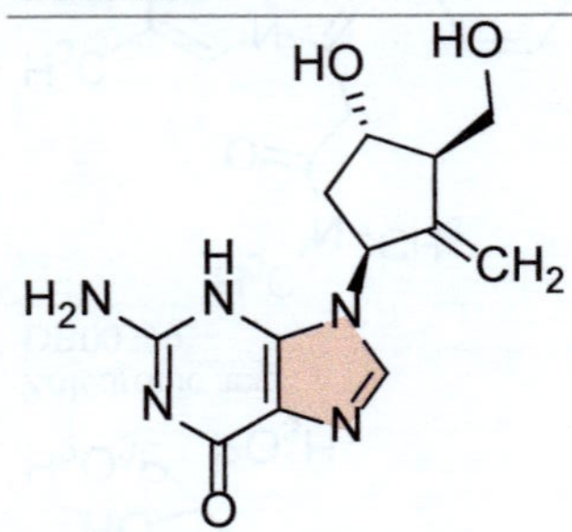 Entecavir DB00442	Persons with chronic hepatitis B virus infection and active viral replication, as well as persistent increases in serum aminotransferases (ALT or AST) or histologically active disease, should take this medication	It blocks the activities of the HBV polymerase (reverse transcriptase, rt) in comparison with naturally occurring deoxyguanosine triphosphate. The HBV DNA production process includes processes such as synthesis of the positive strand of HBV DNA, reverse transcription of the negative strand from pregenomic messenger RNA and base priming

(continued)

Table 1 (continued)

Structure of the drug Name of the drug (drug bank ID)	Category/indication	Mechanism of action
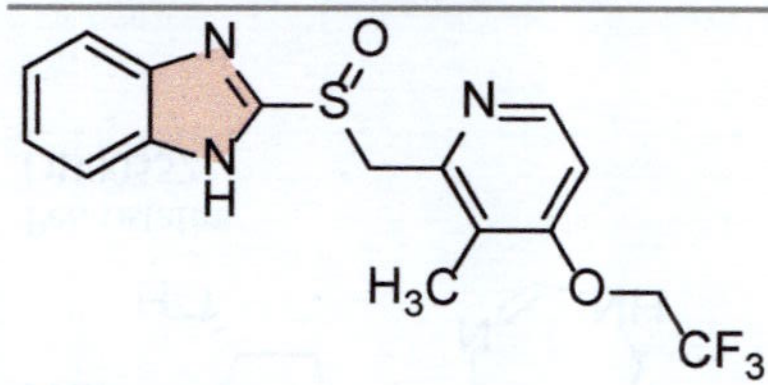 Lansoprazole DB00448	Lansoprazole is a gastric acid suppressor that is useful in treating erosive reflux oesophagitis, active duodenal ulcers, active gastric ulcers, NSAID-induced gastric and duodenal ulcers in the short term and symptomatic gastroesophageal reflux disease	Lansoprazole is a prodrug that must be activated by protonation in an acidic environment. 3 Lansoprazole may react with cysteine residues on parietal H^+/K^+-ATPase, notably Cys813 and Cys321, to produce persistent disulphides once protonated
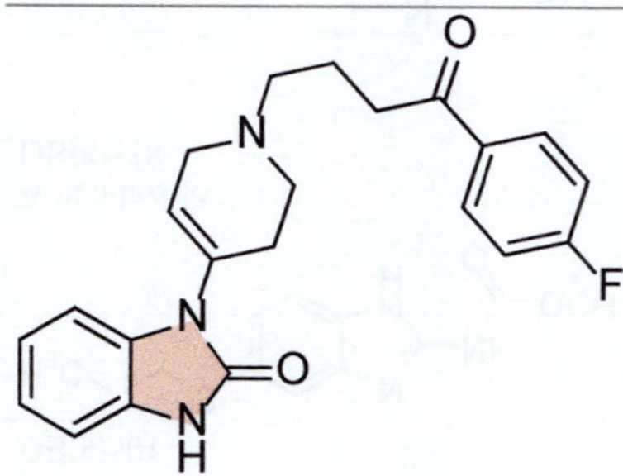 Droperidol DB00450	In surgical and diagnostic operations, droperidol is used to create sedation and to lessen the occurrence of nausea and vomiting	Droperidol induces a depression of the central nervous system that arises in subcortical regions of the midbrain, brain, and brainstem reticular formation, while the specific mechanism of action is uncertain. It may counteract glutamic acid's effects on the extrapyramidal system. It has a high central anti-dopaminergic activity and a modest central anticholinergic activity, and it may block catecholamine receptors and neurotransmitter reuptake

(continued)

Table 1 (continued)

Structure of the drug Name of the drug (drug bank ID)	Category/indication	Mechanism of action
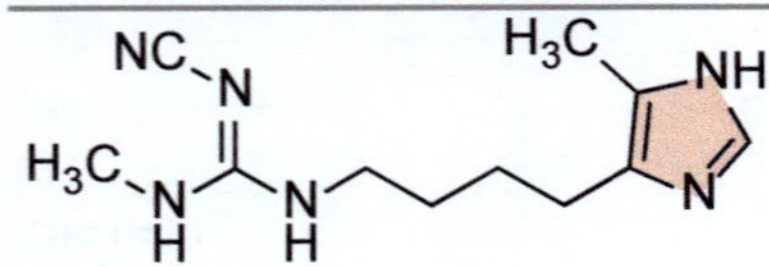 Cimetidine DB00501	This drug is used to treat and control acid reflux diseases (GERD), peptic ulcer disease, heartburn, and acid indigestion	Cimetidine inhibits histamine effects via binding to an H2-receptor on the gastric parietal cell's basolateral membrane. Reduced gastric acid production, as well as gastric volume and acidity, are the outcomes of this competitive inhibition
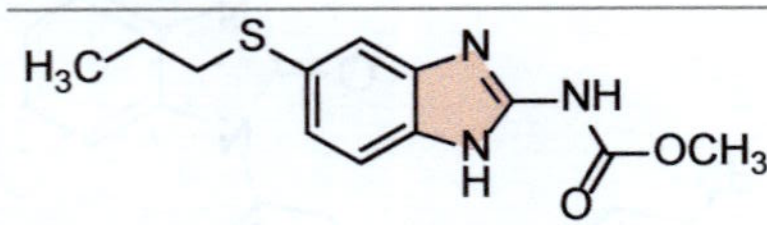 Albendazole DB00518	To treat parenchymal neurocysticercosis resulted from larval *Taenia solium* pig tapeworms, as well as cystic hydatid disease found in lung, liver, and peritoneum due to *Echinococcus granulosus* dog tapeworm larvae	Albendazole produces degenerative changes in the worm's tegument and intestinal cells by reducing its energy generation, resulting in the parasite's immobilisation and death. It operates by attaching to tubulin's colchicine-sensitive region, preventing it from polymerising or assembling into microtubules
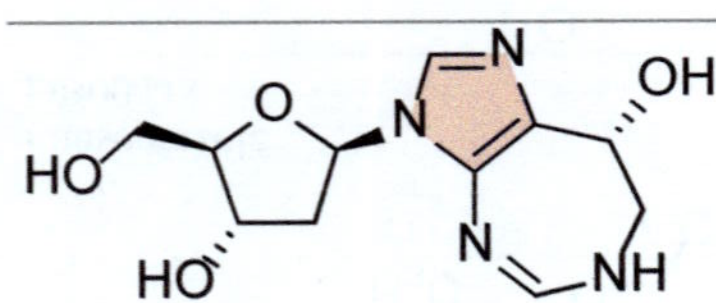 Pentostatin DB00552	Hairy cell leukaemia that has become resistant to alpha-interferon therapy	Pentostatin is a strong adenosine deaminase (ADA) transition-state inhibitor

(continued)

Table 1 (continued)

Structure of the drug Name of the drug (drug bank ID)	Category/indication	Mechanism of action
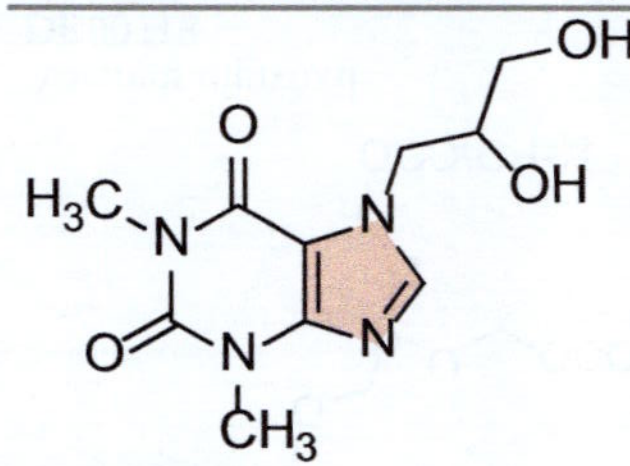 Dyphylline DB00651	For a reversible bronchial spasm correlated with emphysema and chronic bronchitis, as well as acute bronchial asthma	Dyphylline's bronchodilatory effect, probably mediated via competitive phosphodiesterase inhibitor, like other xanthines; resulting in a rise in cAMP and relaxation of bronchial smooth muscle, as well as antagonism of adenosine receptors
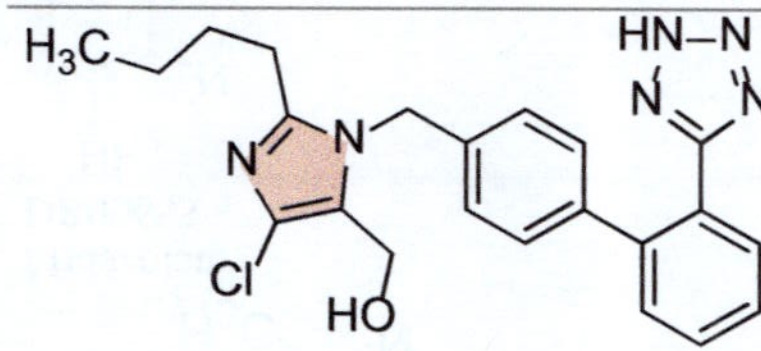 Losartan DB00678	Losartan is used for treating hypertension in people over the age of six, as well as to lower the risk of stroke in people who have left ventricular hypertrophy and hypertension	Losartan inhibits angiotensin II in the adrenal gland and vascular smooth muscle, binding to the AT1 receptor in a reversible and competitive way. Losartan and its active metabolite have 1000 times' higher affinity for AT1 receptor in comparison to the AT2 receptor

(continued)

Table 1 (continued)

Structure of the drug Name of the drug (drug bank ID)	Category/indication	Mechanism of action
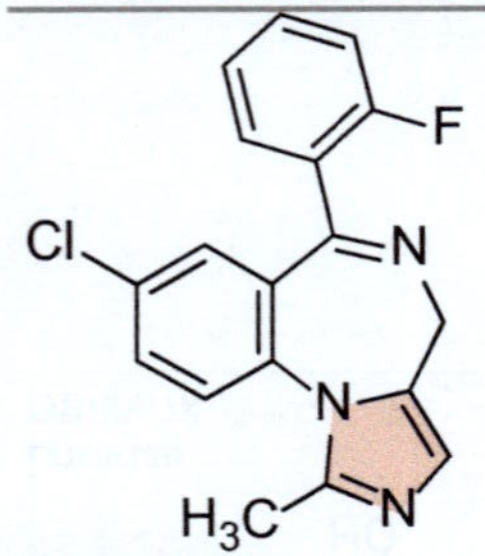 Midazolam DB00683	**Intravenous**: Preoperative sedation, anxiolysis, anaesthesia induction, or forgetfulness are all possible side effects **Intramuscular**: Adults with status epilepticus should be treated with this drug	Benzodiazepines are mediated via the inhibition of inhibitory neurotransmitters in the CNS, gamma-aminobutyric acid (GABA); like midazolam, bind to GABA-A receptors by the benzodiazepine site, causing chloride channels to open more often and therefore potentiating GABA's actions
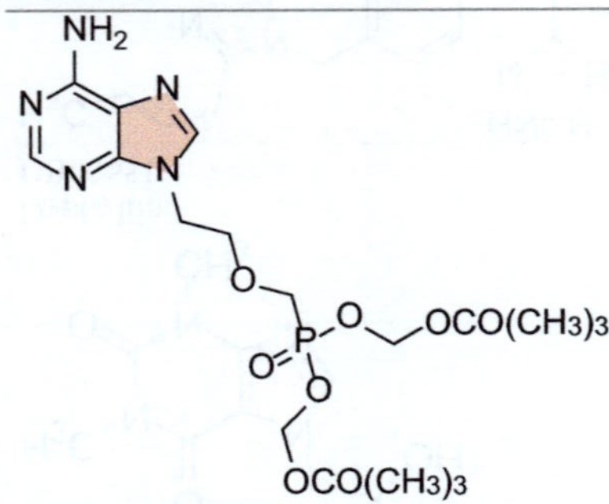 Adefovir dipivoxil DB00718	This medicine should be used to treat adults with chronic hepatitis B with persisting viral replication along with elevations in serum aminotransferases (ALT or AST) or histologically active disease	It is a prodrug of adefovir and also an acyclic nucleotide which is an adenosine monophosphate analogue that is phosphorylated by cellular kinases into the active metabolite adefovir diphosphate. HBV DNA polymerase is inhibited by adefovir diphosphate, which competes with deoxy adenosine triphosphate as a natural substrate and causes DNA chain termination after incorporation into viral DNA (reverse transcriptase)

(continued)

Table 1 (continued)

Structure of the drug Name of the drug (drug bank ID)	Category/indication	Mechanism of action
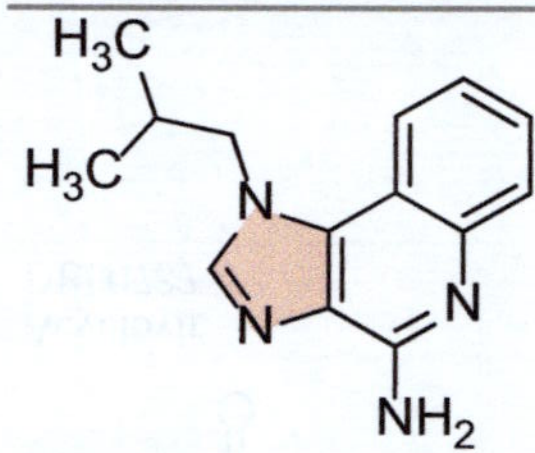 Imiquimod DB00724	Imiquimod is used on the face or scalp of immunodeficient persons as a topical therapy of clinically normal, non-hypertrophic actinic, non-hyperkeratotic keratoses. This is also used as a medication of external genital and perianal parts/condyloma acuminata in persons aged 12 and up	Innate and acquired immune responses are stimulated by imiquimod; induced the inflammatory cell infiltration and disease death in the drug's target area. Imiquimod also upregulates interleukin (IL)-6, IL-8, and tumour necrosis factor-alpha genes. Imiquimod appears to operate as a tissue toll-like receptor-7 agonist for treating basal cell carcinoma
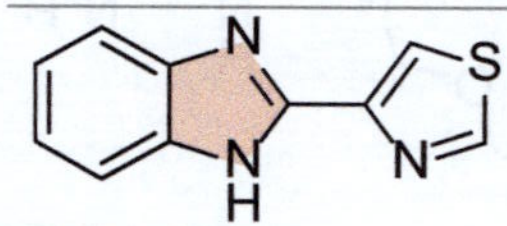 Thiabendazole DB00730	Strongyloidiasis (threadworm), visceral larva migraines, cutaneous larva migraines (creeping eruption) and trichinosis (parasitic disease) are all treated with this drug	Thiabendazole's precise mechanism of action on the parasite is uncertain, however, it is thought to block the helminth-specific enzyme fumarate reductase
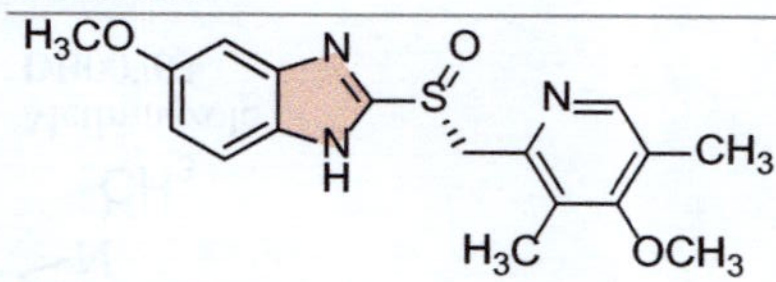 Esomeprazole DB00736	It's used to treat acid reflux disorders like symptomatic gastroesophageal reflux disease (GERD), maintenance of healing of erosive esophagitis, peptic ulcer disease, H. pylori eradication, gastrointestinal bleeding prevention with NSAIDs, and long-term medication of pathological hypersecretory conditions like Zollinger-Ellison (ZE) Syndrome	It blocks the production of gastric acid in the final stage at the secretory surface of gastric parietal cells through covalent bonding with sulfhydryl groups of cysteine's on the (H^+, K^+)-ATPase enzyme. The basal and induced stomach acid secretion was reduced by this effect regardless of the stimuli

(continued)

Table 1 (continued)

Structure of the drug Name of the drug (drug bank ID)	Category/indication	Mechanism of action
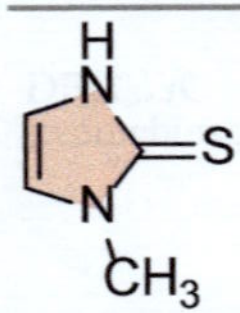 Methimazole DB00763	Methimazole is licenced to treat hyperthyroidism in the United States, to the people with Graves' disease or toxic multi nodular goitre who haven't responded to thyroidectomy or radioactive iodine therapy. It is often used to treat hyperthyroid symptoms before undergoing thyroid surgery or radioactive iodine treatment	Methimazole's main mode of action appears to be interference with thyroid peroxidase (TPO), an early stage in thyroid hormone production. The specific mechanism by which methimazole inhibits this phase is unknown
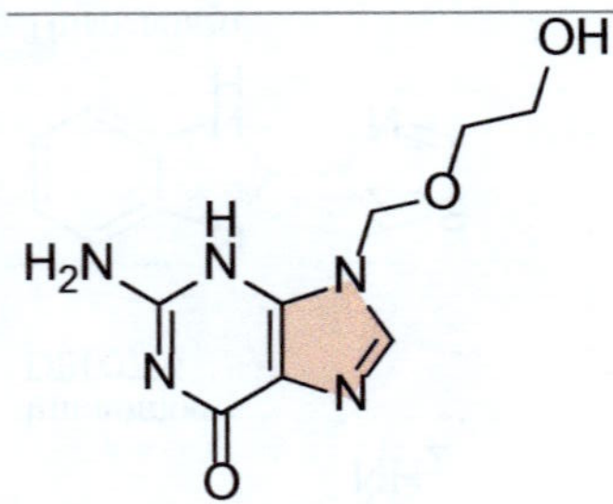 Acyclovir DB00787	In immunocompetent individuals aged 12 and up, an acyclovir topical cream is used to treat recurrent herpes labialis. Treatments for genital herpes, herpes zoster and chickenpox include acyclovir oral pills, capsules, and suspensions	Acyclovir triphosphate binds more firmly to viral DNA polymerase than cellular DNA polymerase and integrates into DNA, ending DNA chains owing to missing 2′ and 3′ carbons. Acyclovir triphosphate fights so aggressively for viral DNA polymerase in some situations that other bases are unable to attach to it, making the enzyme inactive

(continued)

Table 1 (continued)

Structure of the drug Name of the drug (drug bank ID)	Category/indication	Mechanism of action
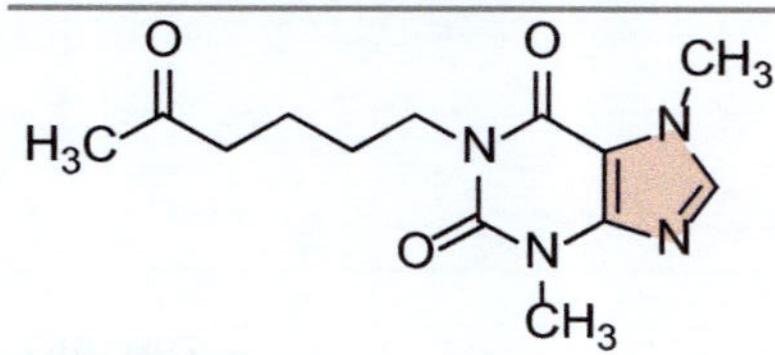 Pentoxifylline DB00806	Chronic occlusive artery disease of the limbs causes intermittent lameness or immobility in patients, which are treated with this medication	Pentoxifylline increases erythrocyte cAMP activity by inhibiting erythrocyte phosphodiesterase. The erythrocyte membrane becomes more resistant to deformation as a result. Blood viscosity is reduced by pentoxifylline through decreasing plasma fibrinogen concentrations and enhancing fibrinolytic activity. In addition to erythrocyte activity, it's an adenosine receptor antagonist that isn't selective
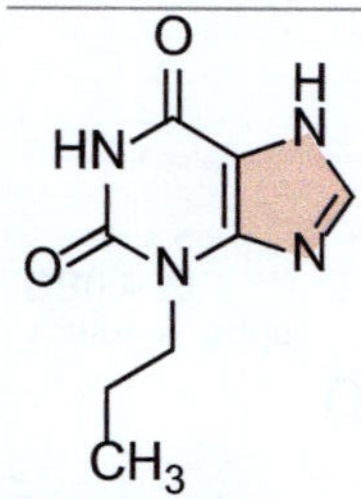 Enprofylline DB00824	Asthma symptoms are managed using this medication. It's also utilised to treat peripheral vascular disease, as well as diabetic neuropathy, sickle cell disease, and cerebrovascular insufficiency	An increase in erythrocyte cAMP activity is caused by enprofylline via inhibiting erythrocyte phosphodiesterase which results in the erythrocyte membrane being more resistant to deformation. Enprofylline reduces blood viscosity via lowering plasma fibrinogen levels and enhancing fibrinolytic activity, in addition to its erythrocyte action

(continued)

Table 1 (continued)

Structure of the drug Name of the drug (drug bank ID)	Category/indication	Mechanism of action
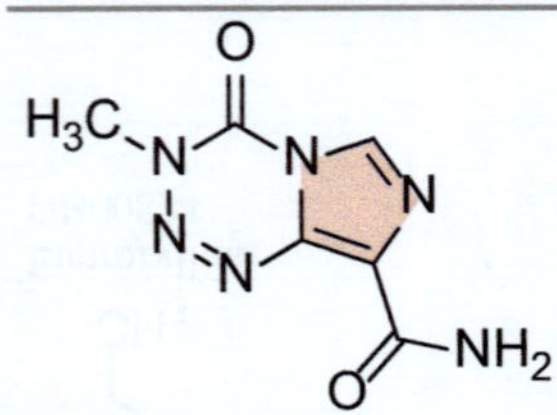 Temozolomide DB00853	To treat newly diagnosed glioblastoma multiforme in conjunction with radiation therapy, as well as for the treatment of anaplastic astrocytoma of adult patients, whose condition has worsened following treatment with procarbazine and nitrosourea. Glioblastoma multiforme is also treated with this drug as a maintenance treatment	Temozolomide is inactive until it is metabolised to MTIC at physiological pH. MTIC is thought to alkylate DNA at the N7 position of guanine, the O3 position of adenosine, and the O6 position of guanosine, with the N7 location being the most prevalent. The methylation of guanine residues resulted in single and double-strand DNA breakage, as well as apoptosis
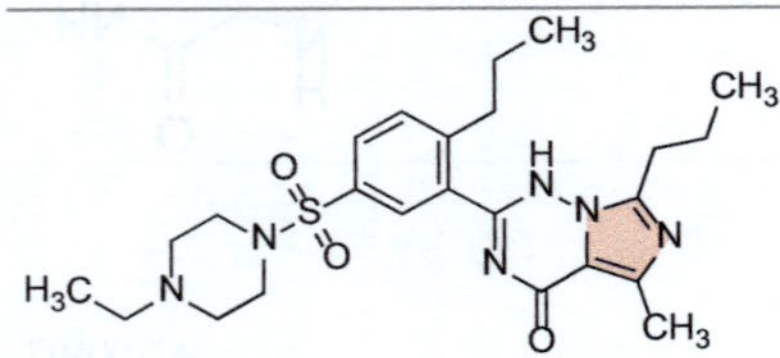 Vardenafil DB00862	It's used to treat erectile dysfunction	Vardenafil inhibits the phosphodiesterase type 5 (PDE5) enzyme which can degrade cGMP in the corpus cavernosum, which is found surrounding the penis. Increase in penile blood flow generates penile erection; produced by the relaxation of corpus cavernosal smooth muscle and penile arteries during sexual stimulation

(continued)

Table 1 (continued)

Structure of the drug Name of the drug (drug bank ID)	Category/indication	Mechanism of action
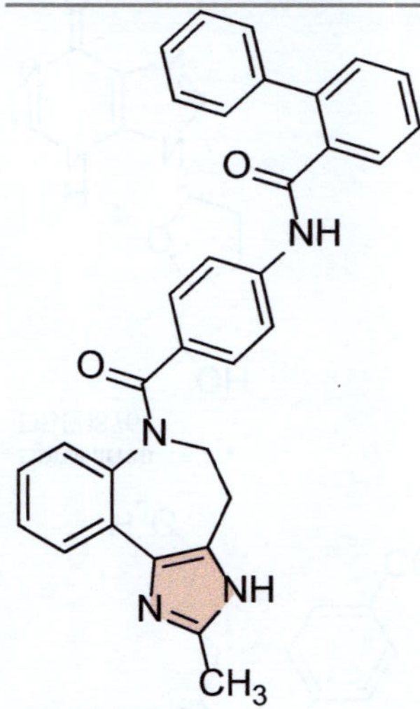 Conivaptan DB00872	To treat hypervolemic hyponatremia or euvolemic hyponatremia in hospitalised patients (for instance, the syndrome of incorrect antidiuretic hormone production, or in the mounting of hypothyroidism, pulmonary problems, adrenal insufficiency, etc.)	Conivaptan is a dual AVP antagonist that binds to the human arginine vasopressin V1A and V2 receptors at nanomolar concentrations in vitro. Aquaresis, or the excretion of free water, is caused by this antagonism in the renal collecting ducts

(continued)

Table 1 (continued)

Structure of the drug Name of the drug (drug bank ID)	Category/indication	Mechanism of action
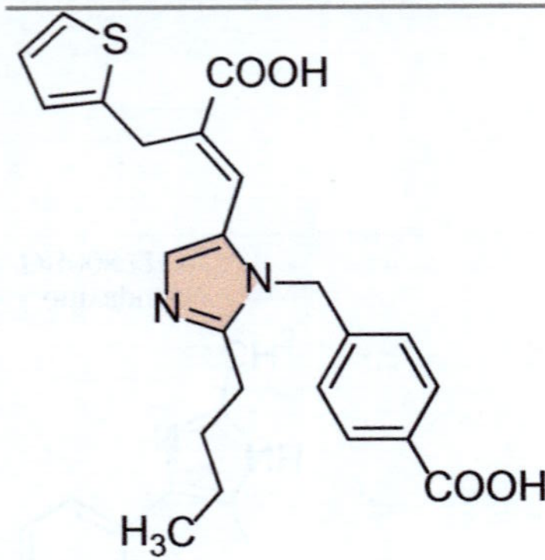 Eprosartan DB00876	Hypertension treatment (used either alone or in conjunction with other antihypertensive medications). Eprosartan Is recognised as a first-line drug to treat diabetic nephropathy and also second-line therapy for treating congestive heart failure (CHF) (only in those intolerant to ACE inhibitors)	Eprosartan prevents the binding of angiotensin II to the AT1 receptor, which is found in many tissues, and so reduces its vasoconstrictor and aldosterone-secreting effects (e.g., vascular smooth muscle, adrenal gland). Although in many tissues, the AT2 receptor is found; it has not been connected to cardiovascular homeostasis
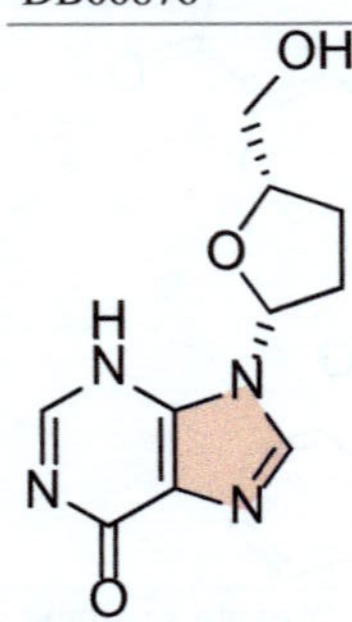 Didanosine DB00900	When taken in concert with other antiretroviral medicines, it is used to treat HIV-1 infection in adults	A sequence of cellular enzymes intracellularly transformed didanosine (ddI) to dideoxyadenosine triphosphate (ddATP), which competes with natural dATP to block the HIV reverse transcriptase enzyme

(continued)

Table 1 (continued)

Structure of the drug Name of the drug (drug bank ID)	Category/indication	Mechanism of action
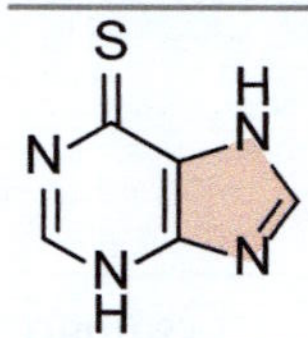 Mercaptopurine DB01033	Acute lymphoblastic leukaemia (ALL) remission induction and maintenance treatment	The Mercaptopurine and hypoxanthine and guanine were in competition towards the enzyme hypoxanthine-guanine phosphoribosyltransferase (HGPRTase) to get converted to thioinosinic acid (TIMP). The conversion of inosinic acid (IMP) to xanthic acid (XMP) or adenylic acid (AMP) via adenylosuccinate (SAMP) is inhibited by TIMP. TIMP is transformed to 6-methylthio inosinate (MTIMP), which inhibits both glutamine-5-phosphoribosylpyrophosphate amidotransferase and glutamine-5-phosphoribosylpyrophosphate amidotransferase. The glutamine-5-phosphoribosylpyrophosphate amidotransferase enzyme resembles the de novo purine ribonucleotide synthesis pathway
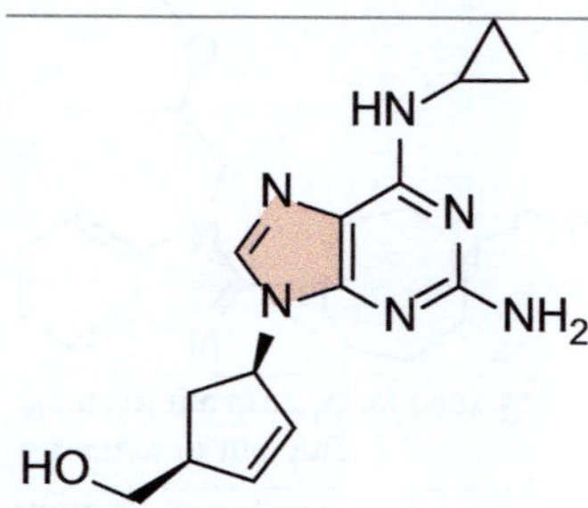 Abacavir DB01048	Medication of HIV-1 infection with combination of other antiretroviral drugs	Abacavir is an antiviral drug and a carbocyclic synthetic nucleoside analogue. Abacavir is transformed through intracellular pathway to carbovir triphosphate which is a deoxyguanosine-5′-triphosphate analogue via cellular enzymes (dGTP). carbovir triphosphate inhibits HIV-1 reverse transcriptase by, competing with the natural substrate dGTP and integrating itself into viral DNA (RT)

(continued)

Table 1 (continued)

Structure of the drug Name of the drug (drug bank ID)	Category/indication	Mechanism of action
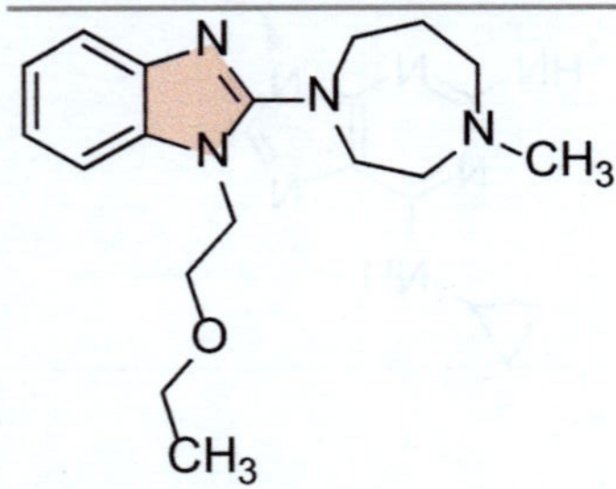 Emedastine DB01084	For the temporary alleviation of allergic conjunctivitis symptoms and indications	Emedastine is a histamine H1 antagonist with a somewhat narrow therapeutic window. In vitro studies of emedastine's affinity for histamine receptors show that it prefers the H1 receptor. Following topical ocular application, in vivo experiments revealed the reduction of histamine-stimulated vascular permeability (concentration-dependent) in the conjunctiva
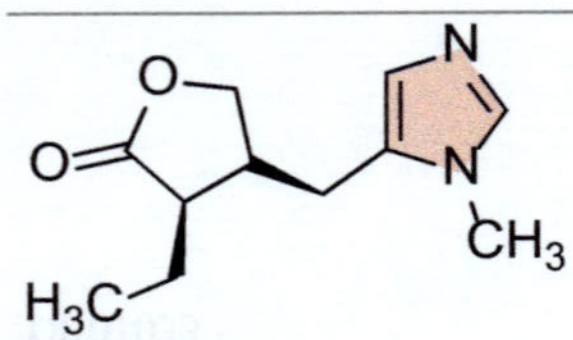 Pilocarpine DB01085	Used to treat radiation-induced dry mouth (xerostomia) and symptoms of dry mouth in patients with Sjögren's syndrome	Pilocarpine is a parasympathomimetic for cholinergic receptors. By primarily activating muscarinic receptors, it increases exocrine gland production and causes contraction of the ciliary muscle and iris sphincter (when applied topically to the eyes)

(continued)

Table 1 (continued)

Structure of the drug Name of the drug (drug bank ID)	Category/indication	Mechanism of action
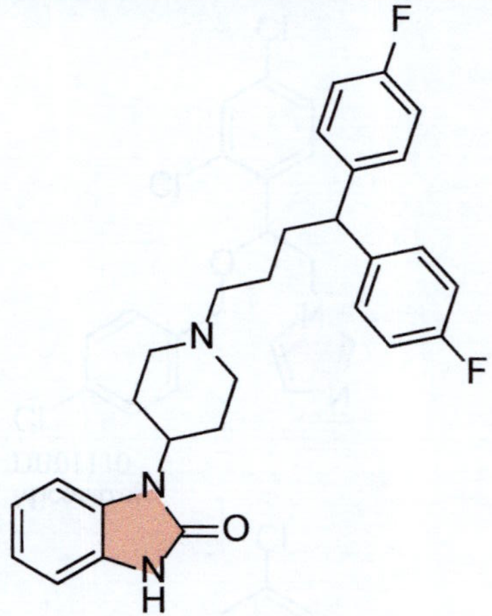 Pimozide DB01100	In individuals with Tourette's disorder who haven't responded well to normal therapy, this medication is used to reduce motor and phonic tics	Pimozide's ability to control motor and phonic tics in Tourette's syndrome is assumed to be due to its dopaminergic inhibiting effect. The dopamine D2 receptor in the CNS bounded by Pimozide and inhibits it

(continued)

Table 1 (continued)

Structure of the drug Name of the drug (drug bank ID)	Category/indication	Mechanism of action
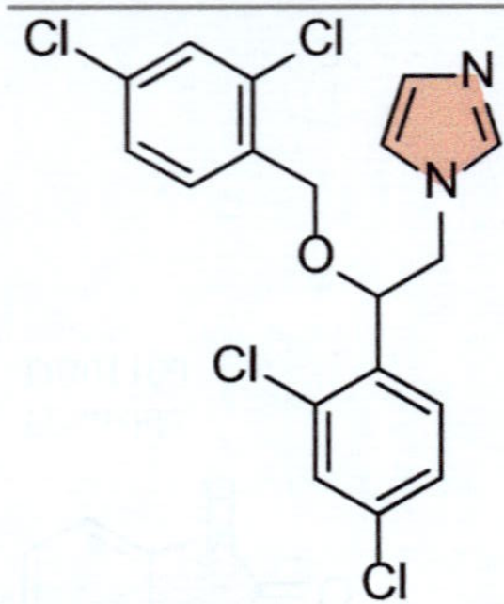 Miconazole DB01110	*Trichophyton rubrum*, *Epidermophyton floccosum*, and *Trichophyton mentagrophytes* cause tinea cruris, tinea pedis (athlete's foot), tinea corporis and For the treatment of cutaneous candidiasis (moniliasis) and tinea versicolor when applied topically	Miconazole binds to cytochrome P-450 enzyme 14-demethylase, which is involved in the conversion of lanosterol to ergosterol. Because ergosterol is a crucial constituent of the fungal cell membrane which decreases the synthesis by increasing cellular permeability, results in cellular contents leaking
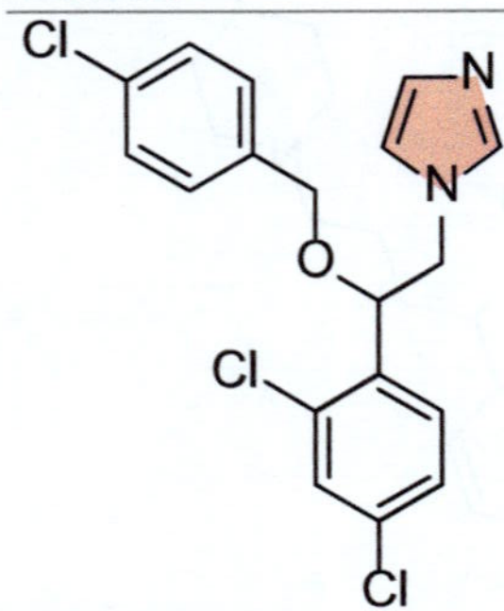 Econazole DB01127	Used for treating tinea corporis, tinea cruris and tinea pedis caused by *Trichophyton rubrum*	Econazole binds to cytochrome P-450 enzyme 14-demethylase, which is involved in the conversion of lanosterol to ergosterol. Because ergosterol is a crucial constituent of the fungal cell membrane which decreases the synthesis by increasing cellular permeability, results in cellular contents leaking

(continued)

Table 1 (continued)

Structure of the drug Name of the drug (drug bank ID)	Category/indication	Mechanism of action
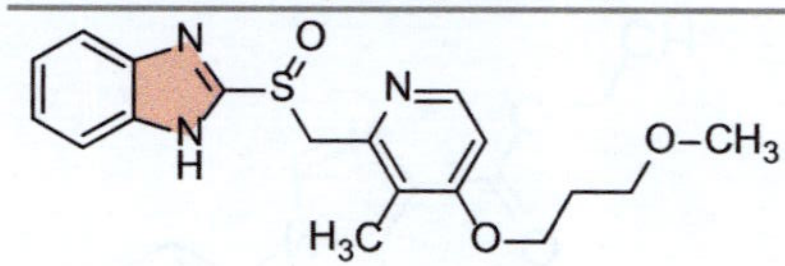 Rabeprazole DB01129	NSAIDs are used to treat gastroesophageal reflux disease (GERD), H. pylori eradication, peptic ulcer disease and gastrointestinal bleed prevention	Rabeprazole is an antisecretory medication (substituted benzimidazole proton pump inhibitor) that inhibits the gastric H^+/K^+-ATP as eat the secretory membrane of the gastric parietal cell, lowering gastric acid production
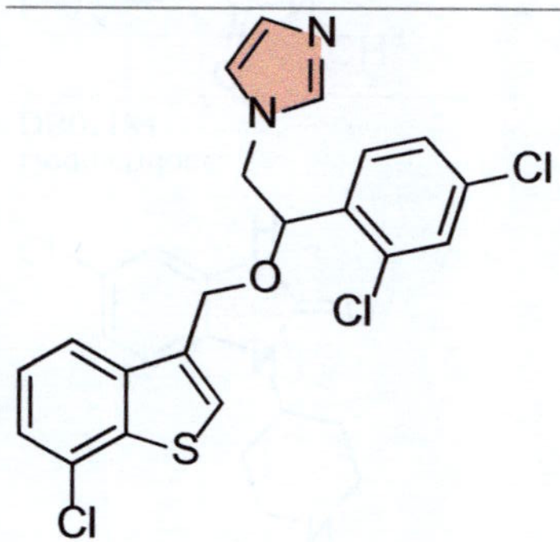 Sertaconazole DB01153	Used for the topical treatment of immuno competent patients aged 12 and above suffering from interdigital tinea pedis	Sertaconazole binds to 14-α demethylase, a cytochrome P-450 enzyme that is responsible for converting lanosterol to ergosterol. Because ergosterol is a crucial constituent of the fungal cell membrane which decreases the synthesis by increasing cellular permeability, results in cellular contents leaking

(continued)

Table 1 (continued)

Structure of the drug Name of the drug (drug bank ID)	Category/indication	Mechanism of action
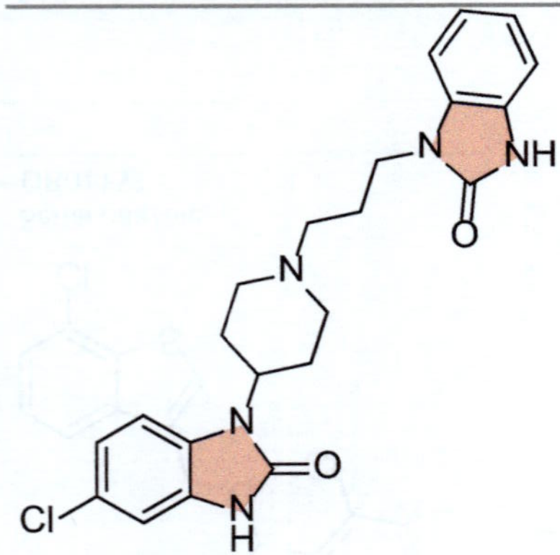 Domperidone DB01184	Dyspepsia, heartburn, epigastric discomfort, nausea, and vomiting can all be treated with this medication	Domperidone is a delayed stomach emptying adjuvant and a peristaltic stimulant. Domperidone's gastroprokinetic effects are due to its ability to inhibit peripheral dopamine receptors. Domperidone improves gastric and esophageal peristalsis and lowers the pressure on esophageal sphincter, which accelerates stomach emptying and reduces small bowel transit time
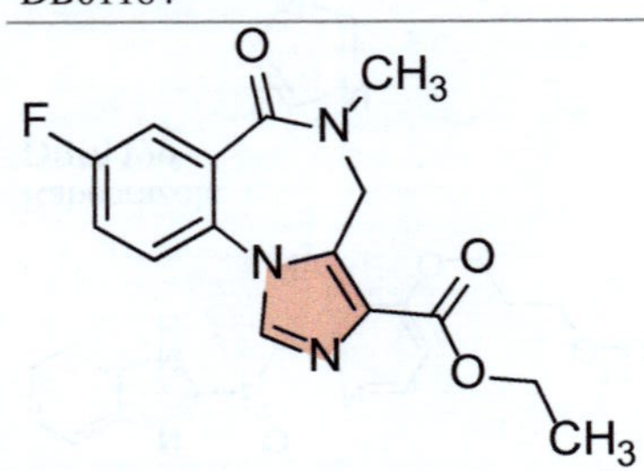 Flumazenil DB01205	Used for the partial or total reversal of sedative effects caused by benzodiazepine in cases where general anaesthesia was established and/or maintained with benzodiazepines, as well as sedation for diagnostic and therapeutic procedures	Flumazenil is a benzodiazepine antagonist that is an imidazo benzodiazepine derivative. The benzodiazepine binding site on the GABA/benzodiazepine receptor combination is competitively inhibited. As a mild partial agonist in various animal activity models, flumazenil did not have potent agonist activity in humans

(continued)

Table 1 (continued)

Structure of the drug Name of the drug (drug bank ID)	Category/indication	Mechanism of action
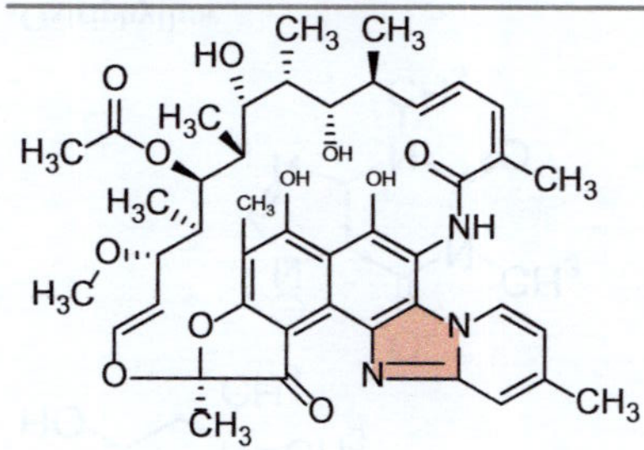 Rifaximin DB01220	The FDA has approved rifaximin for a variety of uses, it is also used to treat patients with diarrhoea (12 years of age) induced by *Escherichia coli* strains. It is administered in patients of 18 years of age to reduce overt hepatic encephalopathy recurrence. Used for the treatment of adult men and women suffering from irritable bowel syndrome with diarrhoea (IBS-D)	The suppression of RNA synthesis in sensitive bacteria by rifaximin through binding to the beta subunit of the DNA-dependent ribonucleic acid (RNA) polymerase enzyme. Translocation is blocked as a result of this binding, and transcription is stopped
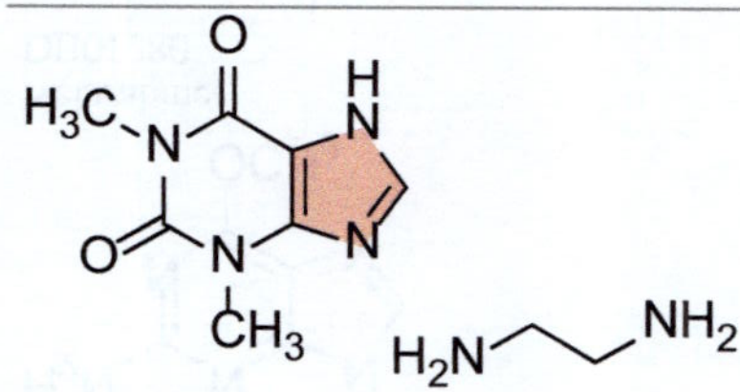 Aminophylline DB01223	Bronchospasm caused by asthma, emphysema, or chronic bronchitis is treated with this medication	The ethylenediamine salt of theophylline is known as aminophylline. Following ingestion, the smooth muscle of the bronchial airways relaxed through theophylline and the pulmonary blood vessels, lowering airway responsiveness to histamine, methacholine, adenosine, and allergen. Phosphodiesterase (PDE) enzyme (type III and type IV), responsible for breakdown of cyclic adenosine monophosphate (AMP) in vascular smooth muscle cells, are competitively inhibited by theophylline, perhaps leading to bronchodilation

(continued)

Table 1 (continued)

Structure of the drug Name of the drug (drug bank ID)	Category/indication	Mechanism of action
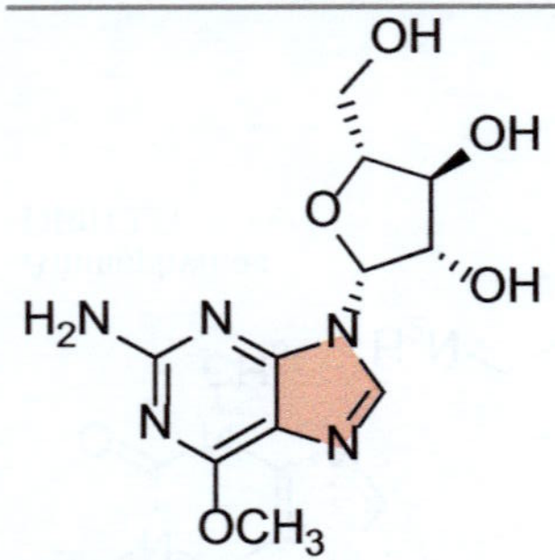 Nelarabine DB01280	Used to treat patients suffering from T-cell lymphoblastic lymphoma and acute T-cell lymphoblastic leukaemia and are not responding or relapsed after undergoing at least two chemotherapy treatments	The metabolite gets accumulated in leukaemic blasts and binds to DNA upon the conversion of nelarabine into ara-GTP. This causes fragmentation and apoptosis. For incorporation into DNA, ara-GTP contend with endogenous deoxyGTP (dGTP). Further DNA elongation is prevented after ara-GTP is integrated at the 3′ end of DNA, signalling apoptosis and resulting in apoptosis
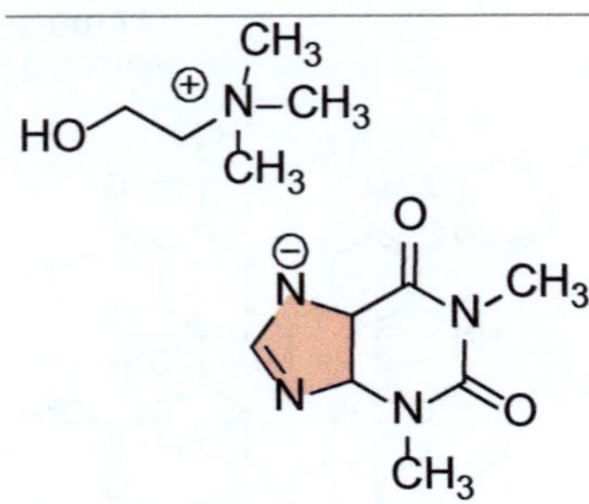 Oxtriphylline DB01303	Asthma, bronchitis, COPD, and emphysema symptoms are treated with this drug	The choline salt of theophylline is oxytriphylline. After ingestion, theophylline relaxes pulmonary blood vessels and bronchial smooth muscle and also decreases airway responsiveness to adenosine, methacholine, histamine, and allergens. Theophylline inhibits the type III and type IV phosphodiesterase (PDE) enzyme which is responsible for breaking down cyclic AMP in smooth muscle cells

(continued)

Table 1 (continued)

Structure of the drug Name of the drug (drug bank ID)	Category/indication	Mechanism of action
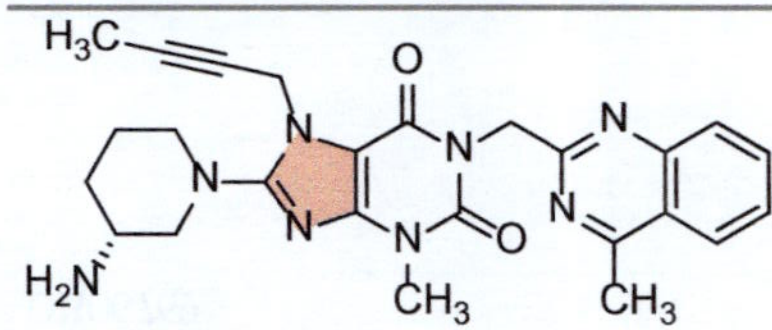 Linagliptin DB08882	In addition to diet and exercise, linagliptin is prescribed for the treatment of type 2 diabetes. It is not advised to treat diabetic ketoacidosis or type 1 diabetes	Linagliptin is a reversible, competitive DPP-4 inhibitor. GLP-1 and glucose-dependent insulinotropic polypeptide breakdown is slowed when this enzyme is inhibited (GIP). Insulin gets released from pancreatic beta cells upon GLP-1 and GIP stimulation while inhibiting glucagon release
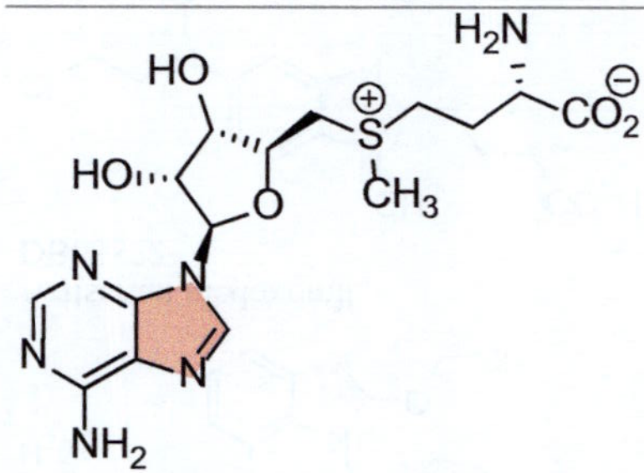 Ademetionine DB00118	In Europe, S-adenosylmethionine (SAMe) is used to treat liver diseases, fibromyalgia, depression and osteoarthritis. It's also been released as a nutritional supplement in the United States	S-Adenosylmethionine (SAMe) is a naturally occurring chemical found in human cells. It's the important amino acid L-immediate methionine's metabolite. By transmethylation, SAMe serves an important metabolic role in the body. The enzyme S-adenosylmethionine synthetase catalyses the reaction between L-methionine and adenosine triphosphate, resulting in SAMe

(continued)

Table 1 (continued)

Structure of the drug Name of the drug (drug bank ID)	Category/indication	Mechanism of action
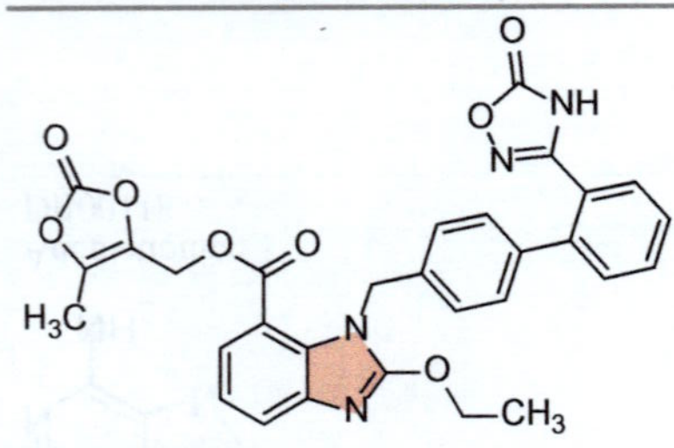 Azilsartan medoxomil DB08822	Hypertension treatment (alone or as an adjunct)	Azilsartan medoxomil prevents angiotensin II from binding and inducing vasoconstriction by blocking the angiotensin II type 1 receptor. One of Azilsartan's most unique characteristics is its capacity to remain securely attached to AT1 receptors for lengthy periods of time following drug washout
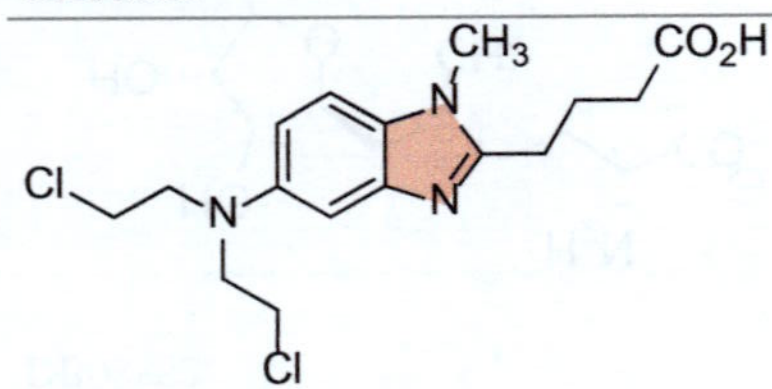 Bendamustine DB06769	Bendamustine is used to treat chronic lymphocytic leukaemia (CLL) and indolent B-cell NHL (non-Hodgkin lymphoma) that has developed after six months of medication with a rituximab-containing regimen or rituximab	It is a bifunctional mechlorethamine derivative that may covalently connect to other compounds by creating electrophilic alkyl groups. Intra- and inter-strand crosslinks between DNA bases caused by this alkylating agent, triggering apoptosis in cells

(continued)

Table 1 (continued)

Structure of the drug Name of the drug (drug bank ID)	Category/indication	Mechanism of action
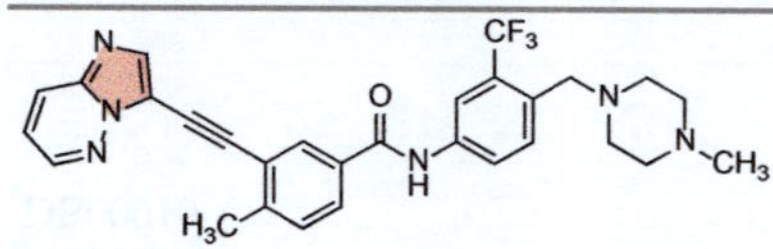 Ponatinib DB08901	Ponatinib exhibits inhibition of tyrosine kinase inhibitor and it is widely recommended for treating adult patients with accelerated, chronic, or blast phase chronic myeloid leukaemia (CML). The drugs work well for patients who have been resistant or intolerant towards prior tyrosine kinase inhibitor therapy, as well as Philadelphia chromosome positive acute lymphoblastic leukaemia (Ph+ALL)	Ponatinib targets the Bcr-Abl tyrosine kinase protein, which is actively participating and promoting CML development. Due to its suppressing activity against tyrosine kinase activity of Abl and T315I mutant kinases, it is beneficial in the treatment of resistant CML. The Philadelphia chromosome, which contains the fused Bcr and Abl genes, produces this protein
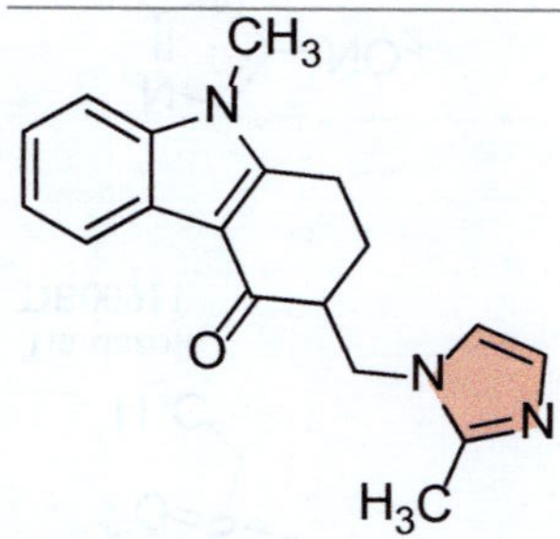 Ondansetron DB00904	(i) Ondansetron tablets which are orally administered and orally disintegrating tablets (ODT) are used to treat vomiting and nausea because of emetogenic cancer chemotherapy, such as high-dose (50 mg/m^2) cis-platin therapy and radiotherapy, as well as to prevent and to treat postoperative nausea and vomiting (ii) Intravenous ondansetron injectable formulations are used to reduce nausea and vomiting caused by emetogenic cancer treatment, especially high-dose chemotherapy	Ondansetron is a selective 5-HT3 serotonin receptor antagonist

(continued)

Table 1 (continued)

Structure of the drug Name of the drug (drug bank ID)	Category/indication	Mechanism of action
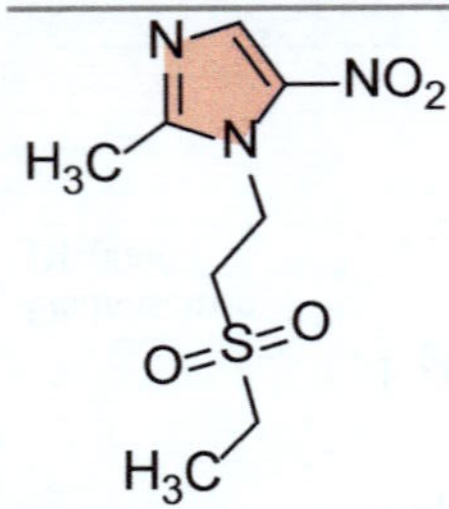 Tinidazole DB00911	In both female and male patients suffering from trichomoniasis caused by T. vaginalis. In addition, it is used to treat giardiasis produced by Giardia duodenalis in adults and children over the age of three, as well as intestinal amoebiasis and amoebic liver abscess produced by *E. histolytica* in adults and children over the age of three	Tinidazole is an antiprotozoal and prodrug. In Trichomonas, a ferredoxin-mediated electron-transport mechanism reduces the nitro group of tinidazole. The antiprotozoal action is thought to be due to the free nitro radical formed by this reduction. The harmful free radicals are thought to attach to DNA covalently, causing DNA damage and cell death. Tinidazole's method of action against Giardia and Entamoeba species is unknown, albeit it is most likely comparable
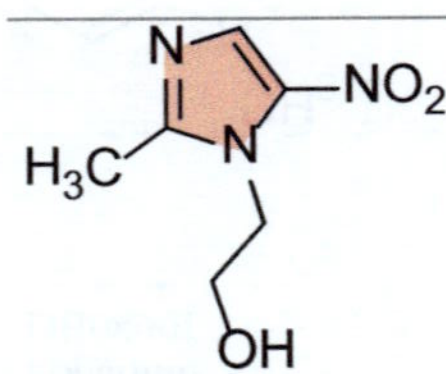 Metronidazole DB00916	Metronidazole is used to treat trichomoniasis caused by Trichomonas vaginalis and the patient's sexual partners (unless during the first trimester of pregnancy), bacterial vaginosis, some kinds of amoebiasis, and different anaerobic diseases. It's also used off-label for Crohn's disease and rosacea, as a post-surgery prophylactic, and for Helicobacter pylori infection therapy	Metronidazole's specific method of action is unknown; however, it's probable that an intermediate formed during the reduction of metronidazole which is exclusively produced by anaerobic bacteria and protozoa. It acts by binding to deoxyribonucleic acid and electron-transport proteins in organisms and hence prevents the production of nucleic acid

(continued)

Table 1 (continued)

Structure of the drug Name of the drug (drug bank ID)	Category/indication	Mechanism of action
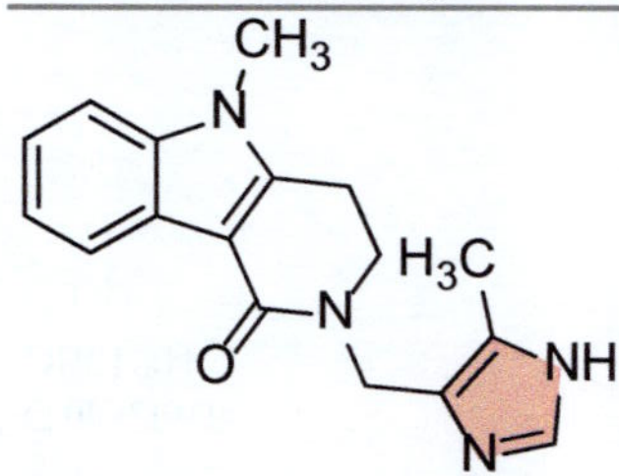 Alosetron DB00969	It acts only in women suffering from severe diarrhoea-predominant irritable bowel syndrome (IBS) symptoms who do not have any abnormalities with anatomic or biochemical GI and are failed to respond to traditional therapies	Alosetron is a 5-HT3 receptor antagonist that is both powerful and selective. 5-HT3 receptors are found throughout the human gastrointestinal system, peripheral and central locations as they are nonselective cation channels. Upon activating these channels with an accompanying neuronal depolarization results in regulation of colonic transit, visceral pain and gastrointestinal secretions, all of which are associated with the aetiology of irritable bowel syndrome (IBS)
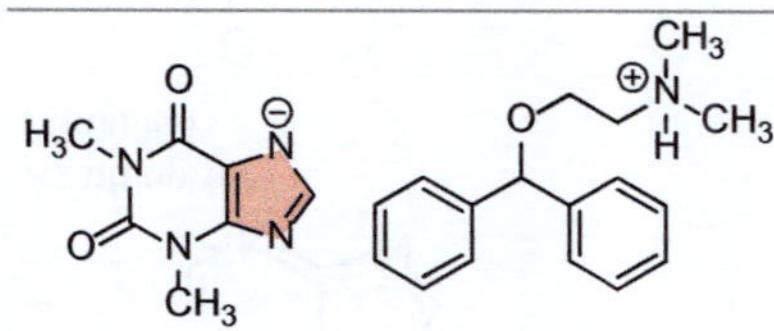 Dimenhydrinate DB00985	Dimenhydrinate is used to prevent and cure motion sickness symptoms such as nausea, vomiting, and vertigo	The exact mechanism of action for certain antihistamines is unknown although they exert their antiemetic, anti-motion sickness, and anti-vertigo properties by their connection to central anticholinergic effects. It results in reduction of vestibular stimulation and labyrinthine function. The antiemetic effect may possibly be due to an impact on the medullary chemoreceptive trigger zone. Dimenhydrinate is a good antagonist in comparison with histamine H1 receptor, found throughout the brain. In the brain's vestibular system, dimenhydrinate shows antiemetic action; which is related to H1 antagonism

(continued)

Table 1 (continued)

Structure of the drug Name of the drug (drug bank ID)	Category/indication	Mechanism of action
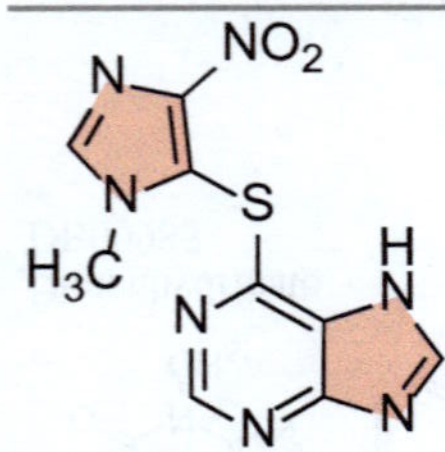 Azathioprine DB00993	For rheumatoid arthritis, kidney transplant rejection prevention, Crohn's disease, and colitis	Azathioprine inhibits DNA, RNA, and protein production via inhibiting purine metabolism. The process of mitosis gets hindered and interfere with cellular metabolism upon intake. The chain termination and cytotoxicity were caused due to the integration of thiopurine analogues into the DNA structure
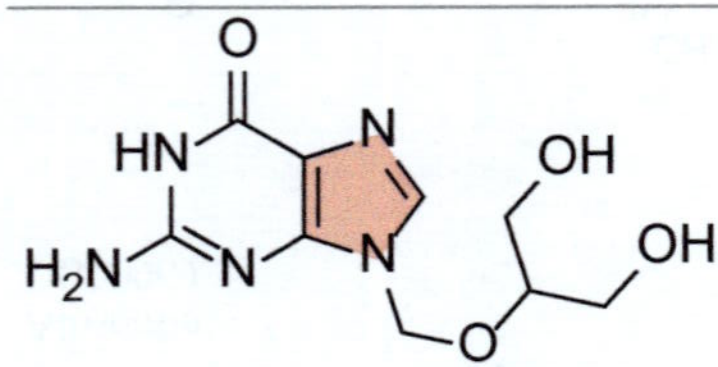 Ganciclovir DB01004	Useful for treating cytomegalovirus (CMV) retinitis in immunocompromised individuals, especially those with acquired immunodeficiency syndrome (AIDS). In immunocompromised individuals, it is also used to treat severe cytomegalovirus (CMV) illness, such as CMV gastrointestinal sickness, CMV pneumonia and disseminated CMV infections	The antiviral action of ganciclovir prevents the virus from reproducing. Because the medication needs to be transformed into an active form through thymidine kinase, a virus-encoded cellular enzyme, this inhibitory activity is very selective (TK). TK catalyses the phosphorylation of ganciclovir to the monophosphate. It is subsequently transformed into diphosphate by cellular guanylate kinase and variety of cellular enzymes transform into a triphosphate The herpes virus DNA is stopped by ganciclovir triphosphate. The competition of ganciclovir triphosphate with dATP act as a substrate for viral DNA polymerase to cause the creation of 'faulty' DNA

(continued)

Table 1 (continued)

Structure of the drug Name of the drug (drug bank ID)	Category/indication	Mechanism of action
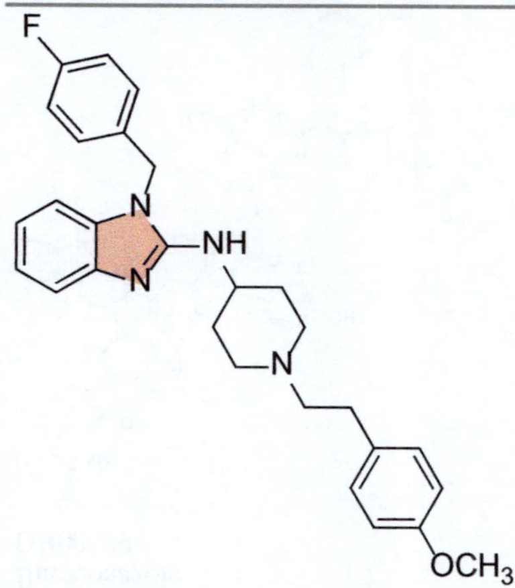 Astemizole DB00637	Astemizole was designed to aid those who suffer from allergies, particularly rhinitis and conjunctivitis. However, due to concerns of arrhythmias, it was withdrawn off the market	Astemizole on competition with histamine to bind with H1-receptor sites in the major blood arteries, uterus, GI tract, and bronchial muscle. Astemizole's reversible binding to H1-receptors prevents edoema, flare, and pruritus from developing as a result of histaminic action
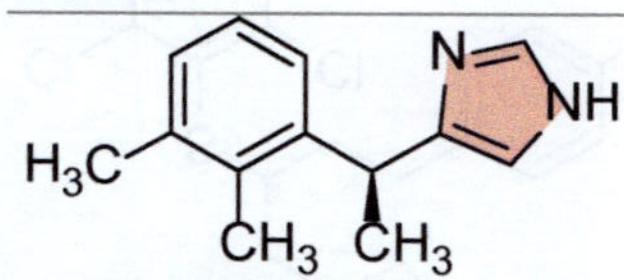 Dexmedetomidine DB00633	It is used to sedate intubated and mechanically ventilated patients In critical care units, as well as pain control, anxiety reduction, and analgesia	Dexmedetomidine is an alpha-2 adrenoceptor agonist that is both specific and selective. It suppresses the norepinephrine release and hence stops the pain signals transmission upon binding with presynaptic alpha-2 adrenoceptors. Activation of the postsynaptic alpha-2 adrenoceptors reduces blood pressure and heart rate by inhibiting sympathetic activity

(continued)

Table 1 (continued)

Structure of the drug Name of the drug (drug bank ID)	Category/indication	Mechanism of action
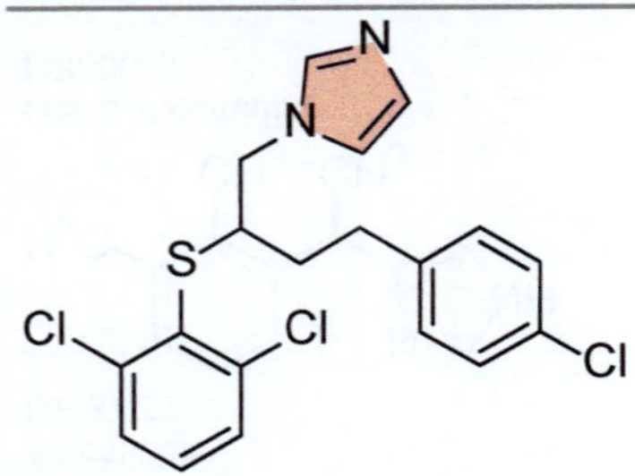 Butoconazole DB00639	For the treatment of vulvovaginal candidiasis on a local basis (infections caused by Candida)	Butoconazole's antifungal action is uncertain, however, it is thought to work in the same way as other imidazole derivatives by inhibiting steroid production. It helps lanosterol to get converted into ergosterol by hindering the enzyme cytochrome P450 14-α-demethylase, which results in lipid composition shift of fungal cell membranes
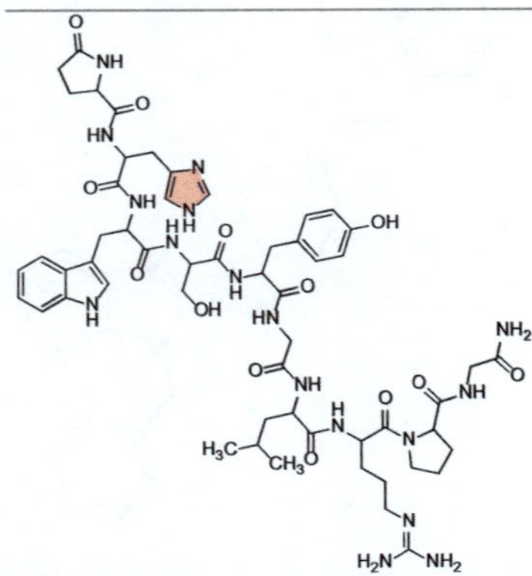 Gonadorelin DB00644	To assess the functional capacity and reaction of the anterior pituitary's gonadotropes, as well as to assess the pituitary's remaining gonadotropic function following surgery and/or irradiation for a pituitary tumour	Gonadotropin-releasing hormone (GnRH) is a naturally occurring hormone that instigates anterior pituitary gland to produce and release luteinising hormone (LH). The synthesis and liberation of folicle-stimulating hormone (FSH) are likewise stimulated by gonadorelin, though to a lower extent

(continued)

Table 1 (continued)

Structure of the drug Name of the drug (drug bank ID)	Category/indication	Mechanism of action
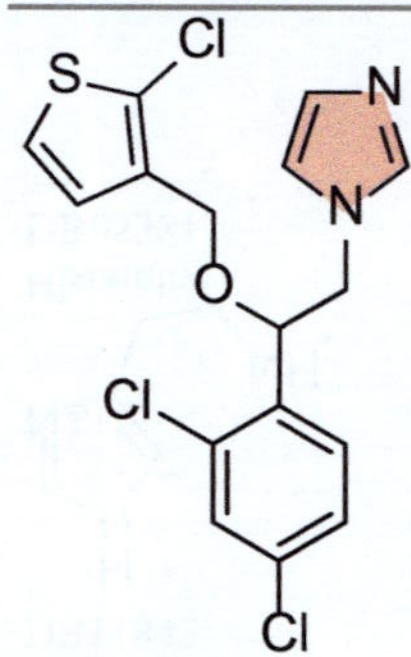 Tioconazole DB01007	Useful for treating vulvovaginal candidiasis (moniliasis) on a local level	Tioconazole binds to 14-α demethylase, a cytochrome P-450 enzyme that transforms lanosterol to ergosterol, an important constituent of yeast membranes. The ergosterol production was suppressed upon intake of Tioconazole which leads to an increase in cellular permeability. It can also assist the following processes, blocking of endogenous respiration, membrane phospholipids interactions, inhibiting the transformation of yeast to mycelial forms and purine uptake, impairs biosynthesis of triglyceride and/or phospholipid, movement of calcium and potassium ions across the cell membrane was inhibited by blocking the Gardos channel
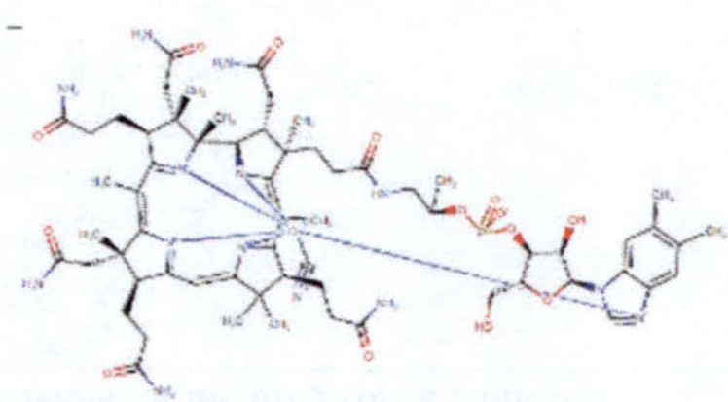 Cyanocobalamin DB00115	Used in Pernicious anaemia, vitamin B12 deficiency	Used as a supplement for Vitamin B12

(continued)

Table 1 (continued)

Structure of the drug Name of the drug (drug bank ID)	Category/indication	Mechanism of action
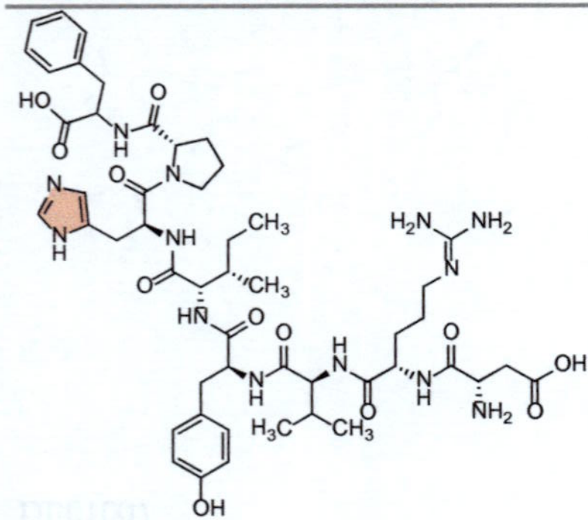 Angiotensin-II DB11842	Vasoconstrictor for increasing blood pressure	On vascular smooth muscle cells, binds to the G-protein-coupled angiotensin II receptor type 1 (AT-1); which facilitates its activity, that induces smooth muscle contraction. and promotes Ca^{2+}/calmodulin-dependent phosphorylation of myosin
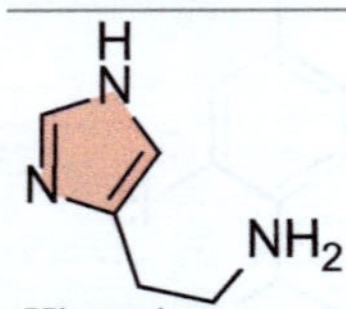 Histamine DB05381	Gastric acid secretory function is evaluated using this diagnostic tool	Capillary dilatation causes an increase in capillary permeability, which results in the outward movement of fluid and plasma protein into the extracellular spaces, a growth in protein content and lymph flow as well the production of edoema

(continued)

Table 1 (continued)

Structure of the drug Name of the drug (drug bank ID)	Category/indication	Mechanism of action
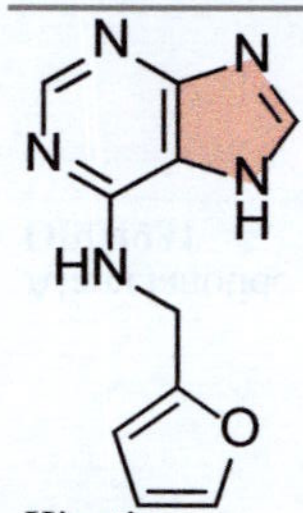 Kinetin DB11336	Not available	Not available
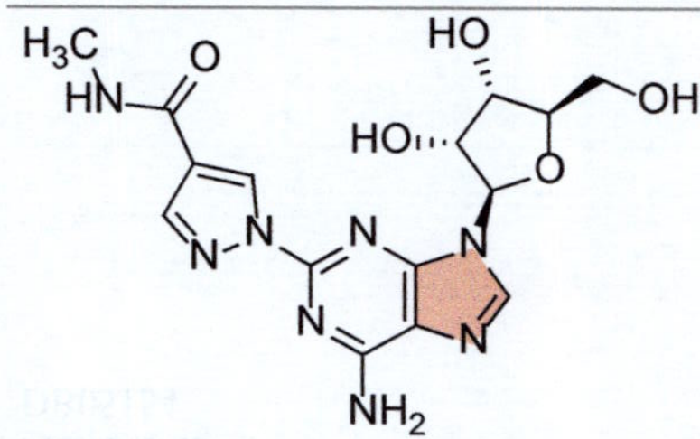 Regadenoson DB06213	Radionuclide myocardial perfusion imaging (MPI) diagnostic agent	Not clear

(continued)

Table 1 (continued)

Structure of the drug Name of the drug (drug bank ID)	Category/indication	Mechanism of action
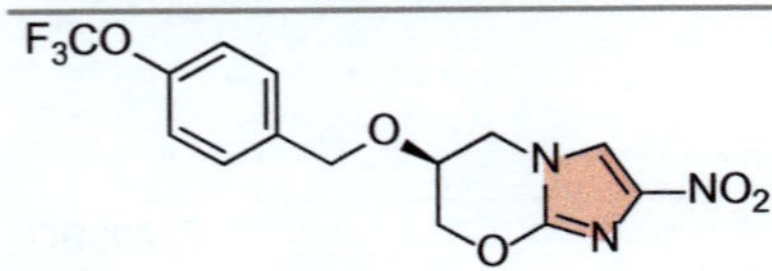 Pretomanid DB05154	It's used to treat extensively drug-resistant (XDR), nonresponsive multidrug-resistant (MDR), and as well treatment-intolerant pulmonary TB in adults when used in conjunction with bedaquiline and linezolid	It's a prodrug that's activated by the Ddn nitro reductase enzyme, which largely causes nitric oxide induction. The pretomanid-activating nitro reductase enzyme is deazaflavin-dependent and relies on the reduced cofactor F420. The enzyme glucose-6-phosphate dehydrogenase reduces cofactor F420. The imidazole ring in pretomanid; reduced at the C-3 position, resulting in the generation of metabolites, including a des-nitro derivative
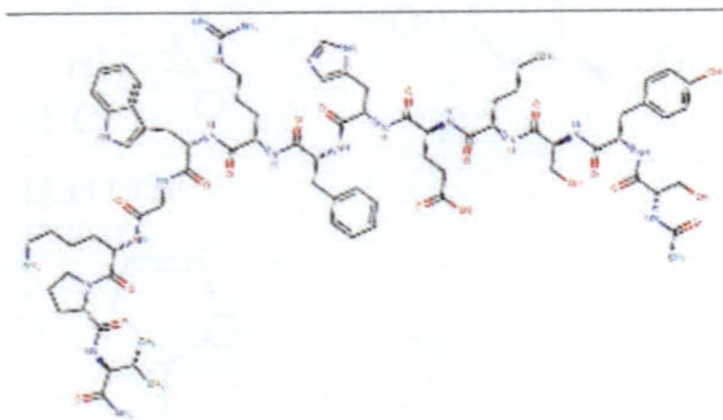 Afamelanotide DB04931	In adult patients with erythropoietic protoporphyria, afamelanotide is used to prevent phototoxicity	Independent of UV exposure, afamelanotide promotes eumelanin production

(continued)

Table 1 (continued)

Structure of the drug Name of the drug (drug bank ID)	Category/indication	Mechanism of action
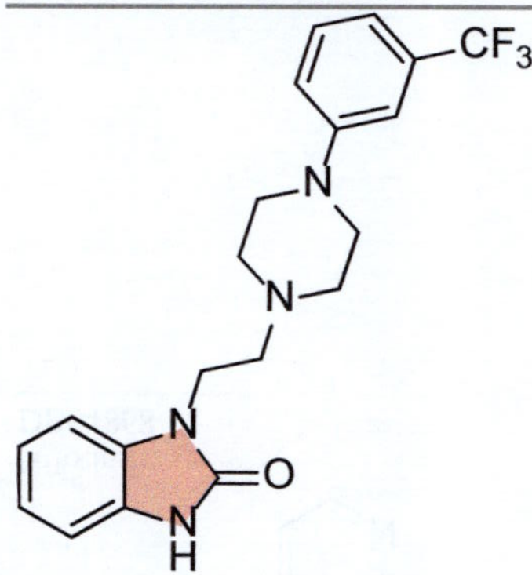 Flibanserin DB04908	Used to treat premenopausal women affected with hypoactive sexual desire disorder (HSDD)	Its impact on neurotransmitter receptors might lead to a reduction in serotonin levels and an enhancement in dopamine and norepinephrine, all of which might play a role in reward processing
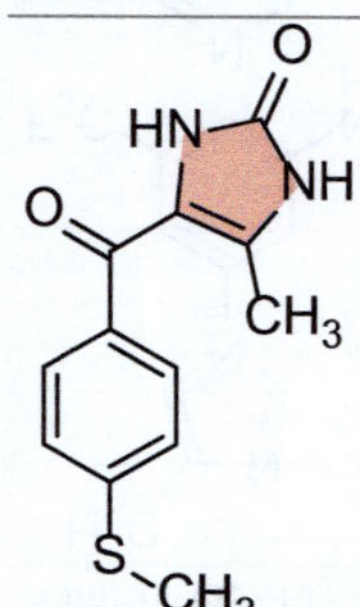 Enoximone DB04880	Congestive heart failure is treated with this drug	PDE3 inhibition prevents cGMP breakdown. This promotes NO production and vascular relaxation

(continued)

Table 1 (continued)

Structure of the drug Name of the drug (drug bank ID)	Category/indication	Mechanism of action
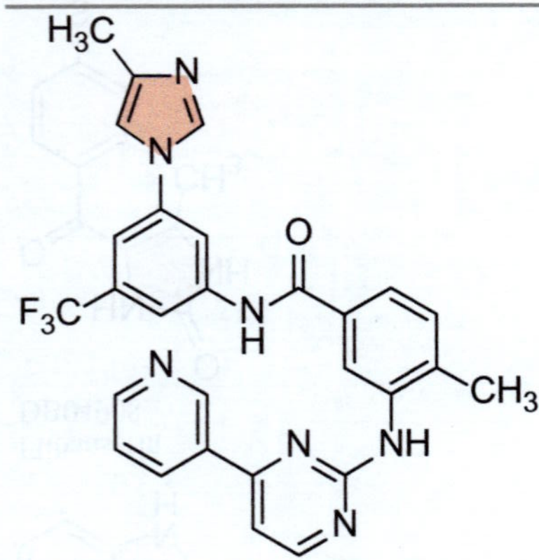 Nilotinib DB04868	To treat a variety of leukaemia's, including chronic myeloid leukaemia (CML)	Nilotinib inhibits the tyrosine kinase activity of the BCR-ABL protein. Nilotinib has a stronger affinity for the ATP-binding site of the BCR-ABL protein than imatinib, allowing it to overcome mutation-induced resistance

(continued)

Table 1 (continued)

Structure of the drug Name of the drug (drug bank ID)	Category/indication	Mechanism of action
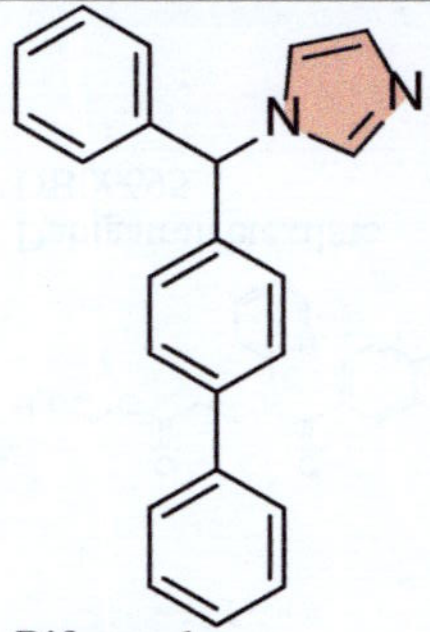 Bifonazole DB04794	It is recommended for treating a variety of fungal diseases on the skin, including sports foot (tinea pedis)	It prevents the synthesis of ergosterol which is a key constituent of fungal cell membranes. It works by destabilising the enzyme cytochrome p450 51 in fungi (lanosterol 14-alpha demethylase). This is essential for the fungus' cell membrance structure. Cells are lysed when it is inhibited
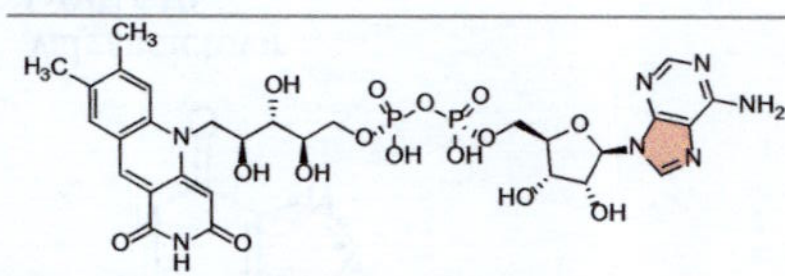 Flavin adenine dinucleotide DB03147	Vitamin B2 deficiency causes keratitis and blepharitis, which can be treated with this medication	Not available

(continued)

Table 1 (continued)

Structure of the drug Name of the drug (drug bank ID)	Category/indication	Mechanism of action
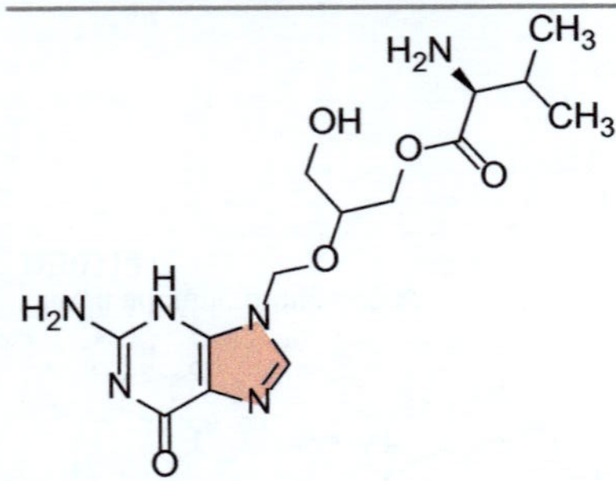 Valganciclovir DB01610	The antiviral drug valganciclovir is used to treat cytomegalovirus infections	The antiviral drug valganciclovir is used to treat cytomegalovirus infections
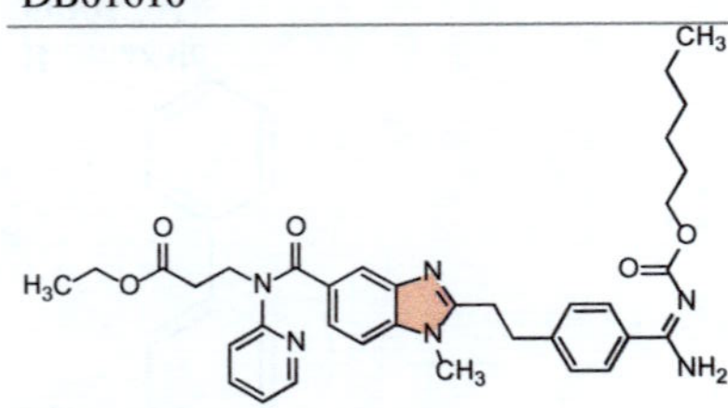 Dabigatran etexilate DB06695	Dabigatran is used to prevent venous thromboembolic events after hip or knee replacement surgery	An esterase that causes hydrolysis hydrolyzes dabigatran to produce active dabigatran. The pharmacological action of dabigatran and its glucuroniated metabolites is the same. The serine protease thrombin is inhibited by dabigatran and its metabolites

(continued)

Table 1 (continued)

Structure of the drug Name of the drug (drug bank ID)	Category/indication	Mechanism of action
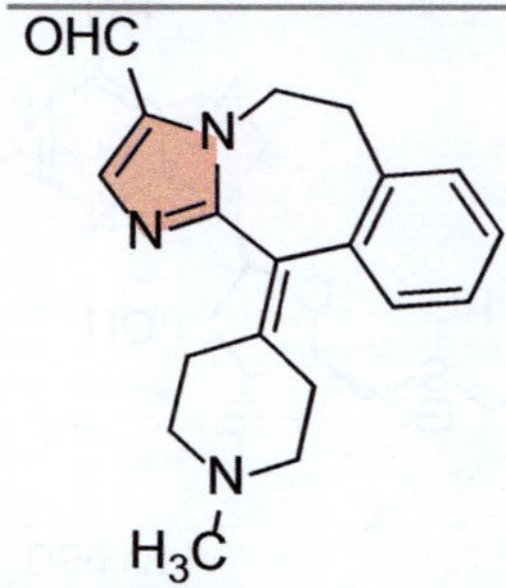 Alcaftadine DB06766	It is used to relieve irritation caused by allergic conjunctivitis	Alcaftadine inhibited the release of histamine from mast cells and an antagonist of the H1 histamine receptor
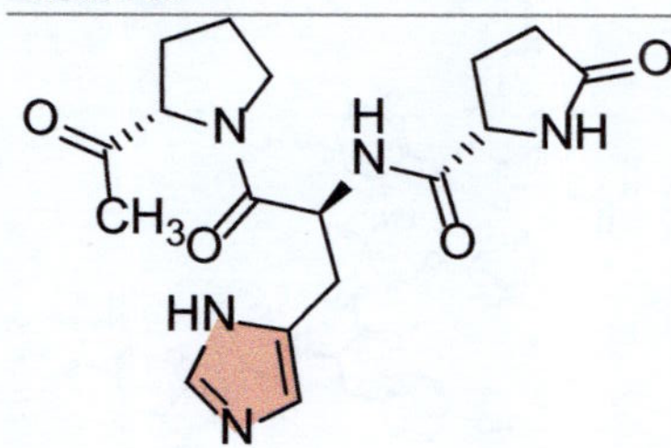 Protirelin DB09421	Not available	Not available

(continued)

Table 1 (continued)

Structure of the drug Name of the drug (drug bank ID)	Category/indication	Mechanism of action
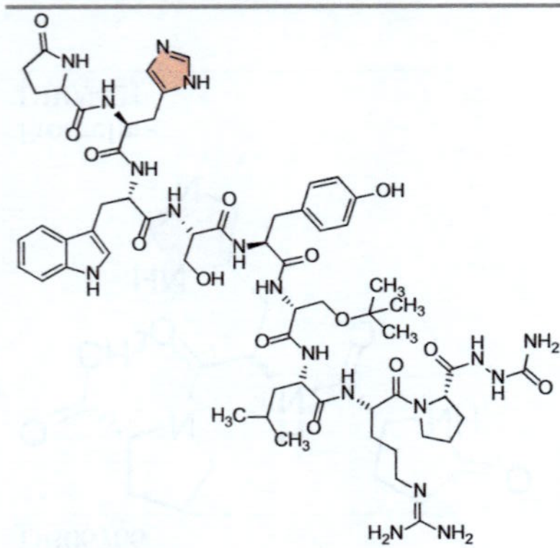 Goserelin DB00014	Use to treat advanced breast cancer in pre- and perimenopause. It acts as an endometrial-thinning agent prior to endometrial ablation for dysfunctional uterine bleeding and also as an endometrial-thinning agent before endometrial ablation for dysfunctional uterine haemorrhage	The inhibition of LH and serum testosterone levels is maintained
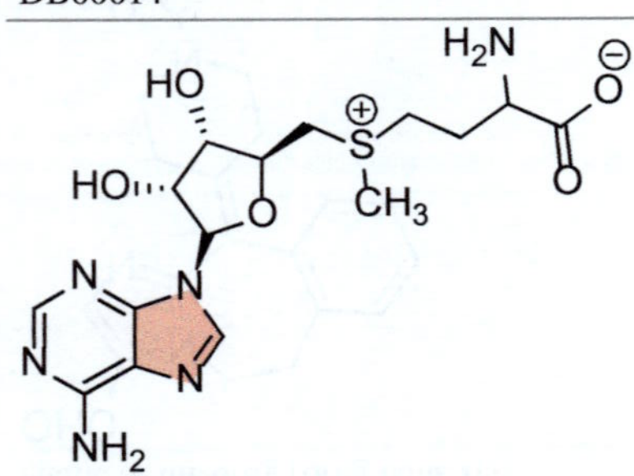 Ademetionine DB00118	Depression, liver diseases, fibromyalgia, osteoarthritis, and nutritional supplements are all treated with it	In a process known as trans methylation, SAMe contributes a one-carbon methyl group. SAMe acts as methyl-group donor in the biosynthesis of proteins, DNA and RNA nucleic acids, melatonin, phospholipids, creatine, epinephrine, and other compounds. It is generated by the reaction of L-adenosine triphosphate and methionine mediated by the enzyme S-adenosylmethionine synthetase

(continued)

Table 1 (continued)

Structure of the drug Name of the drug (drug bank ID)	Category/indication	Mechanism of action
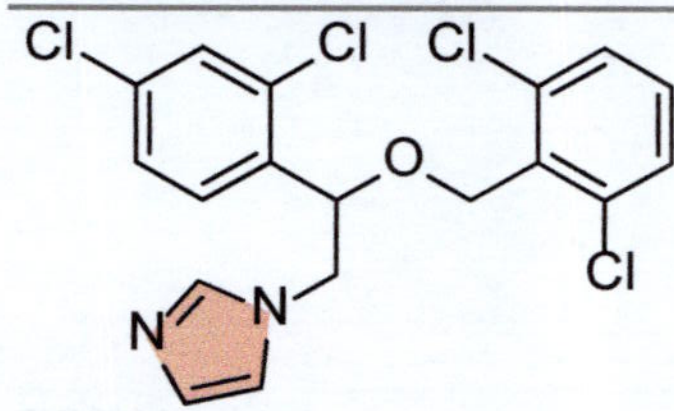 Isoconazole DB08943	Not available	Not available
Cobamamide DB11191	Not available	Not available

(continued)

Table 1 (continued)

Structure of the drug Name of the drug (drug bank ID)	Category/indication	Mechanism of action
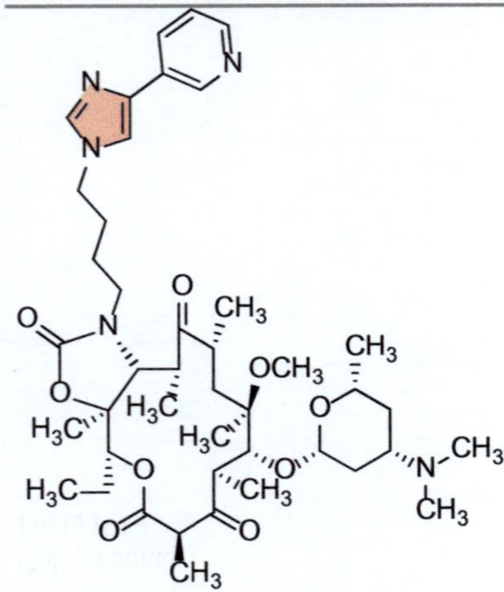 Telithromycin DB00976	Treated as an antibiotic in pneumococcal infection, acute bronchitis and bronchiolitis, acute bacterial tonsillitis, lower respiratory tract infection, acute sinusitis, and lobar (pneumococcal) pneumonia	It operates by binding to domains II and V of 23S rRNA associated with the 50S ribosomal subunit

(continued)

Table 1 (continued)

Structure of the drug Name of the drug (drug bank ID)	Category/indication	Mechanism of action
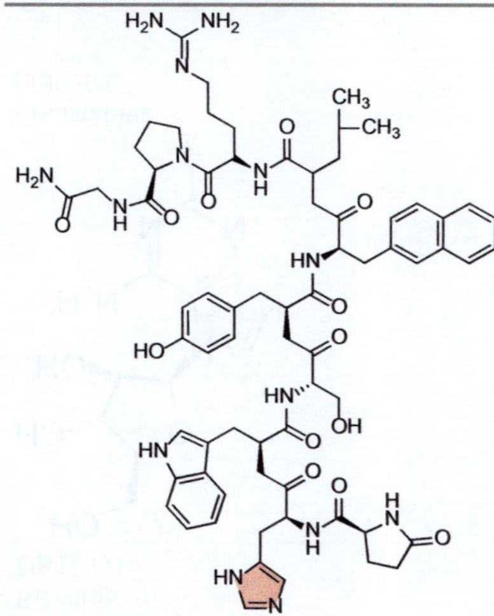 Nafarelin DB00666	For the treatment of real premature puberty, GnRH-dependent precocious precocity, and full iso-sexual precocity in both boys and girls, as well as endometriosis	Nafarelin binds to the GnRH receptor in the pituitary gland, causing a reversible down-regulation of GnRH receptors and desensitisation of pituitary gonadotropes. This results in a large and long-term decrease in LH and FSH production

(continued)

Table 1 (continued)

Structure of the drug Name of the drug (drug bank ID)	Category/indication	Mechanism of action
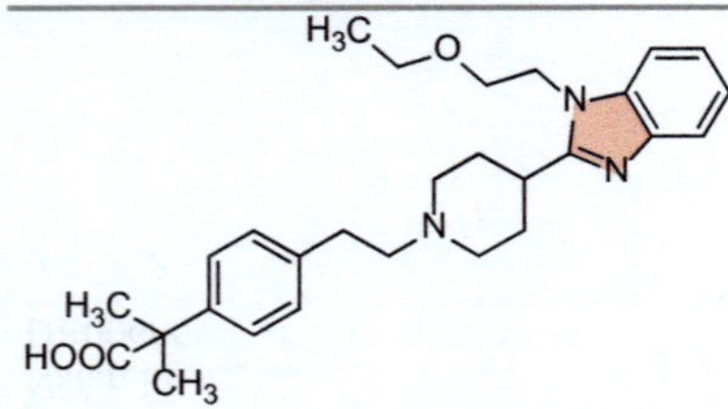 Bilastine DB11591	It is used to treat nasal and non-nasal symptoms of seasonal rhinitis in patients 12 years of age and older, as well as chronic spontaneous urticaria	It works as a selective antagonist of the histamine H1 receptor (Ki = 64 nM). It decreases the development of allergy symptoms caused by the f histamine release from mast cells by binding to and inhibiting the activation of the H1 receptor
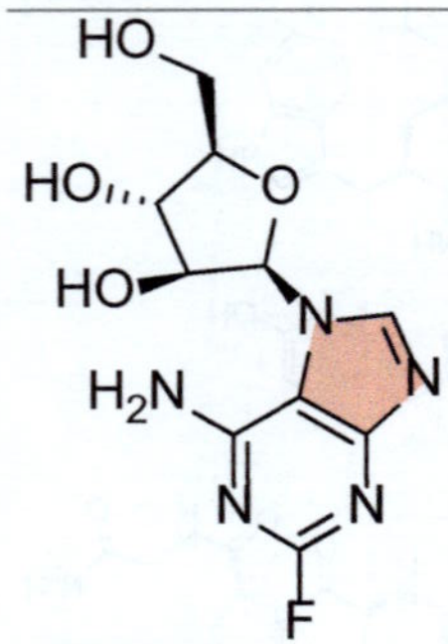 Fludarabine DB01073	B-cell chronic lymphocytic leukaemia (CLL) is treated with this drug	Fludarabine phosphate is quickly dephosphorylated to 2-fluoro-ara-A, which is subsequently phosphorylated in an intracellular manner by deoxycytidine kinase to 2-fluoro-ara-ATP. It stops DNA synthesis by inhibiting ribonucleotide reductase, DNA primase and DNA polymerase alpha

(continued)

Table 1 (continued)

Structure of the drug Name of the drug (drug bank ID)	Category/indication	Mechanism of action
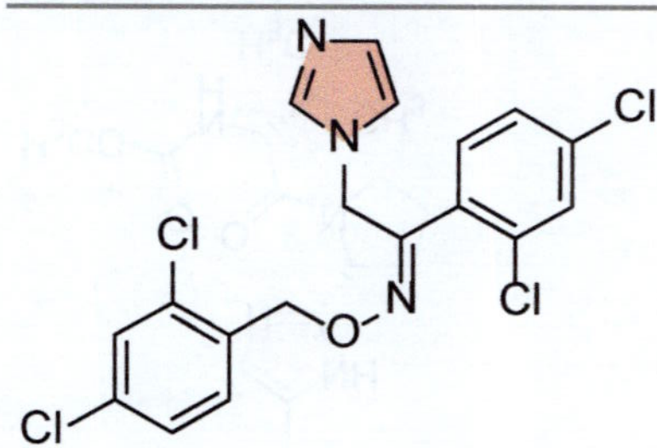 Oxiconazole DB00239	Used to treat dermal fungal infection	Lanosterol 14-alpha demethylase is destabilised by it. Cell lysis occurs when it is inhibited. It also slows DNA synthesis, lowers intracellular ATP levels, and increases zinc membrane permeability
Mipomersen DB05528	As an addition to diet and other lipid-lowering drugs in individuals with homozygous familial hypercholesterolemia	Mipomersen binds to apoB-100 mRNA. This interaction produces double-stranded RNA, which is destroyed by RNase H, preventing the mRNA from being translated into the apo B-100 protein

(continued)

Table 1 (continued)

Structure of the drug Name of the drug (drug bank ID)	Category/indication	Mechanism of action
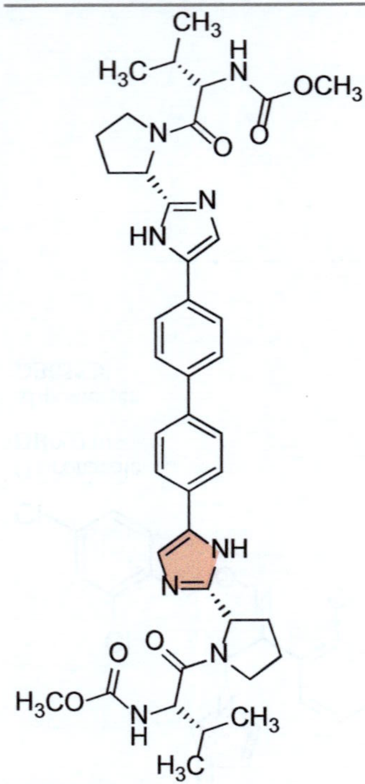 Daclatasvir DB09102	Used to treat Chronic HCV genotype 1a/b or 3 infections	The cis-acting role of basally phosphorylated NS5A maintains the HCV replication complex, whereas the trans-acting role of hyperphosphorylated NS5A modifies HCV assembly and infectious particle production. Daclatasvir inhibits the activity of new HCV replication complexes by disrupting hyperphosphorylated NS5A proteins. In vivo, daclatasvir has been shown to disrupt both intracellular viral RNA production and virion assembly/secretion

(continued)

Table 1 (continued)

Structure of the drug Name of the drug (drug bank ID)	Category/indication	Mechanism of action
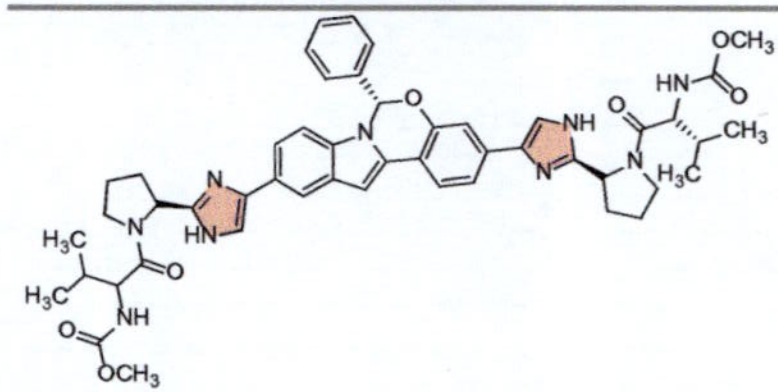 Elbasvir DB11574	Adults with persistent HCV genotypes 1 or 4 infections have been prescribed this drug	Blocking signalling contacts, migration of NS5A to the surface of lipid droplets from the endoplasmic reticulum, and alteration of the HCV replication complex are all possible mechanisms of action for NS5A inhibitors like Elbasvir
Methylcobalamin DB03614	Not available	Not available
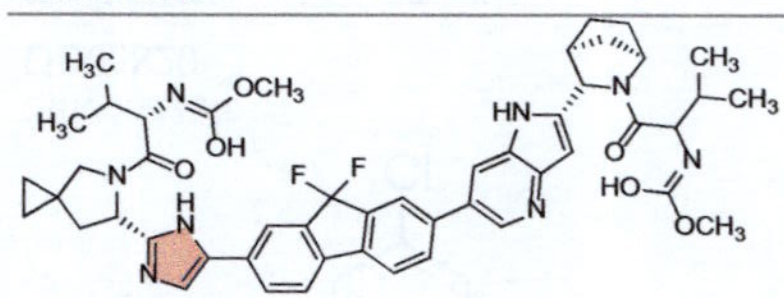 Ledipasvir DB09027	Based on the volume of cirrhosis or liver damage, ledipasvir is used with or without Ribavirin in HCV genotypes 1, 4, 5, and 6. It is also effective in treating HCV in HIV-positive individuals	Ledipasvir is a Hepatitis C Virus (HCV) inhibitor of the NS5A protein, which is required for viral RNA replication and viriogenesis. Although the exact mode of action is unknown, it is believed to prevent NS5A hyperphosphorylation, which is required for viral replication

(continued)

Table 1 (continued)

Structure of the drug Name of the drug (drug bank ID)	Category/indication	Mechanism of action
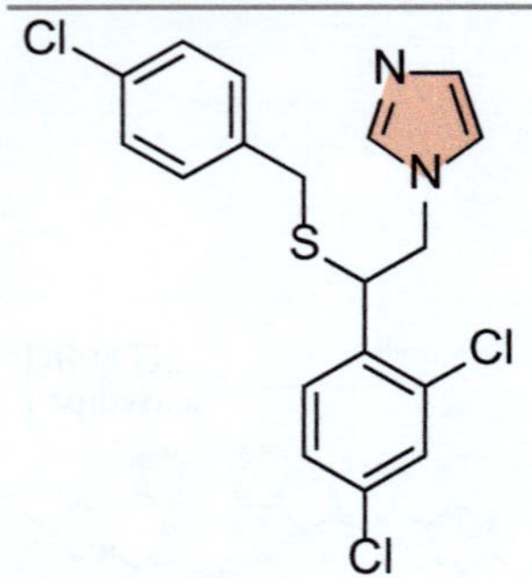 Sulconazole DB06820	It is used with sulconazole solution 1.0% to treat tinea corporis, tinea cruris and tinea versicolor caused by fungus	Not available
Corticorelin ovine triflutate DB09067	In individuals with ACTH-dependent Cushing's disease, corticorelin is used to distinguish between pituitary and ectopic ACTH production	Corticorelin is a powerful activator of anterior pituitary adrenocorticotropic hormone (ACTH) secretion. It is used as a diagnostic tool to assess the condition of the pituitary-adrenal axis in order to distinguish between a pituitary source of excessive ACTH production and an ectopic source

(continued)

Table 1 (continued)

Structure of the drug Name of the drug (drug bank ID)	Category/indication	Mechanism of action
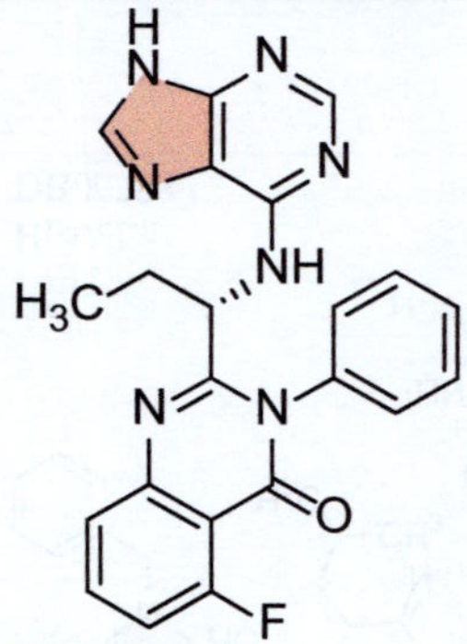 Idelalisib DB09054	Used to treat relapsed small lymphocytic lymphoma (SLL), relapsed or refractory follicular B-cell non-Hodgkin lymphoma (FL) and chronic lymphocytic leukaemia (CLL)	Phosphatidylinositol-4,5-bisphosphate 3-kinase, also known as PI-3K, is inhibited by idelalisib exclusively in the delta isoform (P110δ). Idelalisib inhibits various cell signalling pathways, comprising C-X-C chemokine receptors type 5 and type 4, B-cell receptor (BCR) signalling which is important in B-cell trafficking and homing to lymph nodes and bone marrow, by inhibiting this enzyme

(continued)

Table 1 (continued)

Structure of the drug Name of the drug (drug bank ID)	Category/indication	Mechanism of action
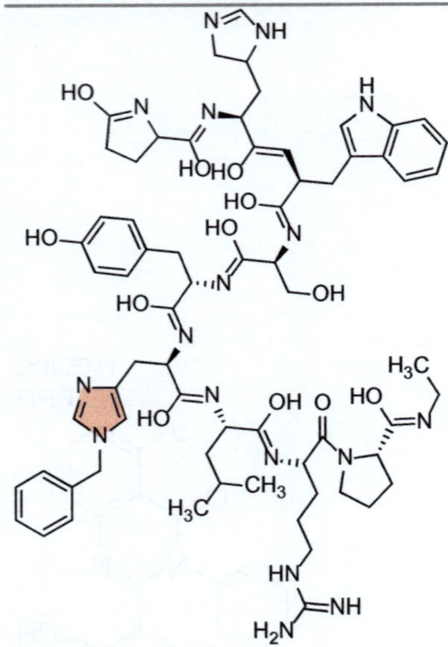 Histrelin DB06788	Histrelin is used as FDA approved drug, Supprelin LA to treat children with central precocious puberty (CPP). Also used for the palliative treatment of advanced prostate cancer under the brand name Vantas	Because of a reversible down-regulation of GnRH receptors in the pituitary gland and desensitisation of the pituitary gonadotropes, continuous administration of histrelin acetate leads to lower levels of LH and FSH

(continued)

Table 1 (continued)

Structure of the drug Name of the drug (drug bank ID)	Category/indication	Mechanism of action
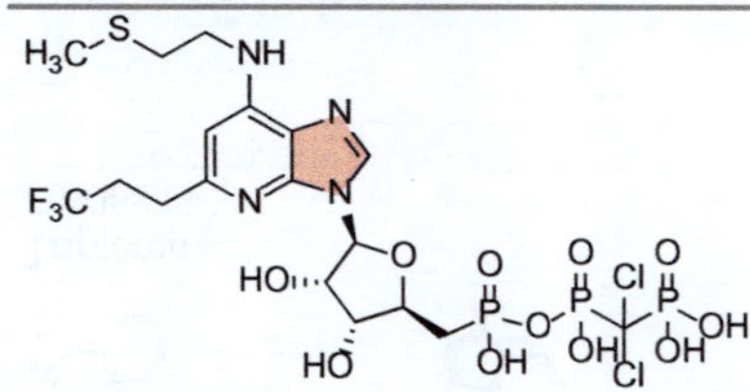 Cangrelor DB06441	It is used as a percutaneous coronary intervention (PCI) to lower the risk factor associated with stent thrombosis (ST), repeat coronary revascularisation and periprocedural myocardial infarction (MI). It is administered for patients who are not being medicated P2Y12 platelet inhibitor or a glycoprotein IIb/IIIa inhibitor	It is a P2Y12 platelet receptor antagonist that suppresses ADP platelet aggregation in a selective, reversible manner. By inhibiting adenylyl cyclase with a Gi protein, ADP binds to P2Y12 to accelerate and complete platelet aggregation, potentiating dense granule production and boosting coagulation activity
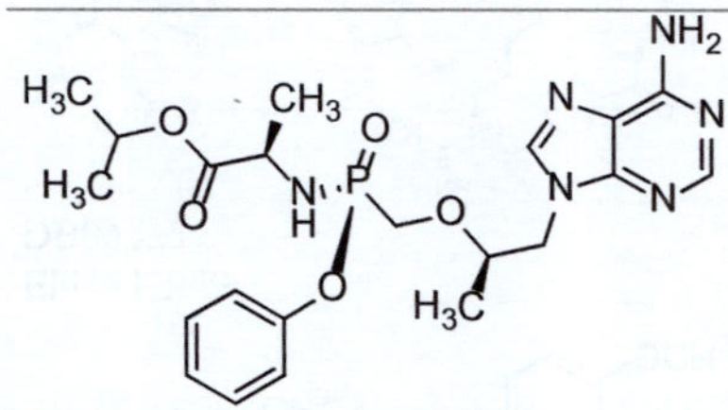 Tenofovir alafenamide DB09299	Adult patients with compensated liver disease are treated for chronic hepatitis B	When compared to red blood cells, prodrug accumulates more in peripheral blood mononuclear cells. Tenofovir works by a variety of mechanisms once activated, including inhibition of viral polymerase, which causes chain termination, and suppression of viral synthesis

(continued)

Table 1 (continued)

Structure of the drug Name of the drug (drug bank ID)	Category/indication	Mechanism of action
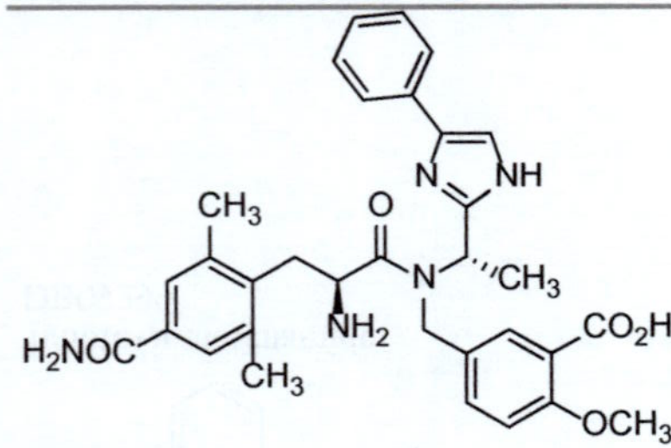 Eluxadoline DB09272	Irritable bowel syndrome with diarrhoea (IBS-D) is treated with this medication	It is an agonist for mu-opioid receptor, as well as a kappa and delta opioid receptor agonist. The constipating impact of mu-opioid receptor agonists alone is reduced by antagonistic action at the delta receptor
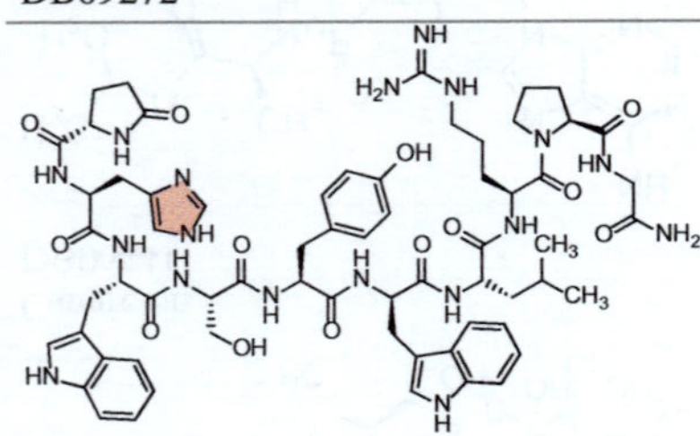 Triptorelin DB06825	Medication for advanced prostate cancer	Triptorelin is an agonist analogue for synthetic gonadotropin-releasing hormone (GnRH). Triptorelin demonstrated a 13-fold higher activity in luteinising hormone releasing and a 21-fold greater activity in follicle-stimulating hormone releasing in animal experiments when compared with native GnRH

(continued)

Table 1 (continued)

Structure of the drug Name of the drug (drug bank ID)	Category/indication	Mechanism of action
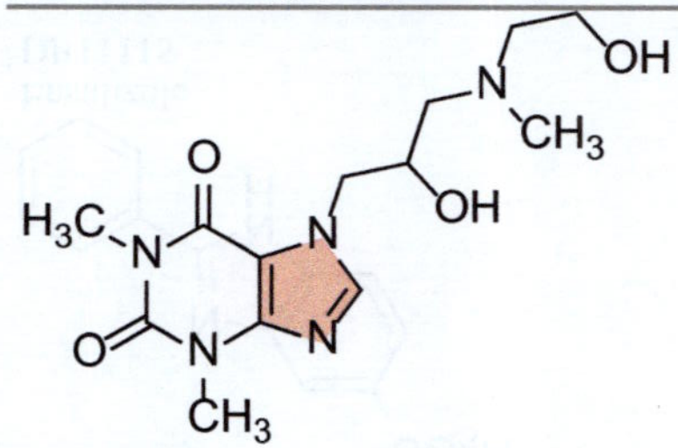 Xanthinol DB09092	Xanthinol is used to treat hyperlipidemia, as well as cerebrovascular and peripheral vascular disorders	The current method of action is thought to be the influence on cell metabolism via the nucleotides NAD and NADP. Nicotinic acid also functions as a cofactor for a number of proteins that are involved in tissue respiration (Embden-Meyerhof and citrate cycle). The activity of xanthinol nicotinate increases glucose metabolism and energy gain
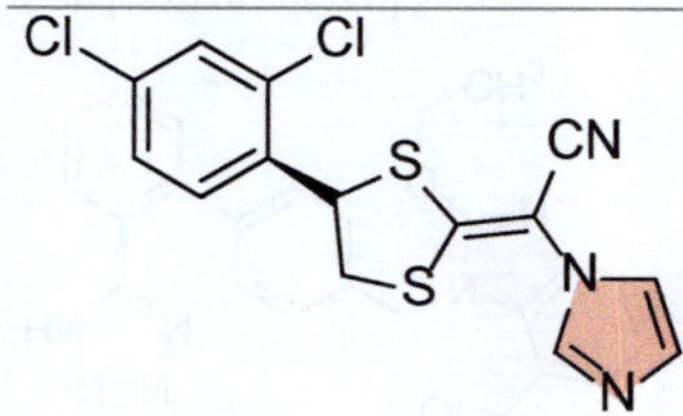 Luliconazole DB08933	Treatment especially for tinea pedis, cruris, and corporis, are fungal diseases caused by Trichophyton rubrum and Epidermophyton floccosum	Although the specific mechanism of action for luliconazole's antifungal effect is unknown, it is assumed that it inhibits the enzyme lanosterol demethylase. Ergosterol is a key constituent of fungal cell membranes, and lanosterol demethylase is required for its formation

(continued)

Table 1 (continued)

Structure of the drug Name of the drug (drug bank ID)	Category/indication	Mechanism of action
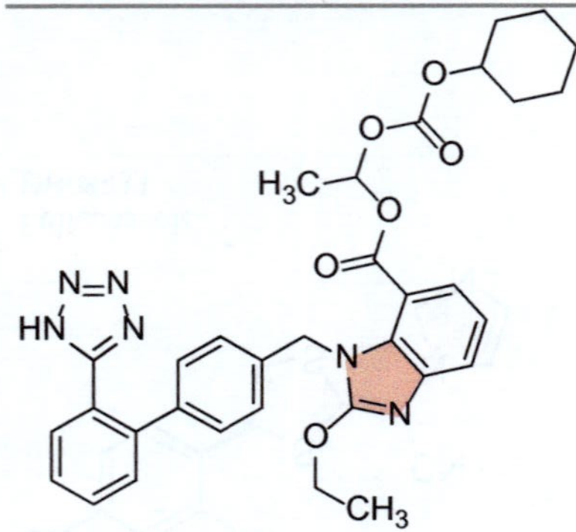 Candesartan cilexetil DB00796	As a first-line medication, it can be utilised in treating left ventricular hypertrophy, isolated systolic hypertension and uncomplicated hypertension. It might be used to treat diabetic nephropathy as first-line therapy. Candesartan is a second-line treatment for myocardial infarction, congestive heart failure, systolic dysfunction, and coronary artery disease in patients intolerant to ACE inhibitors	Angiotensin II binding to AT1 is inhibited by Candesartan in a range of tissues, including adrenal glands and the vascular smooth muscle. AT1 inhibits the vasoconstrictive and aldosterone-secreting actions of angiotensin II, lowering blood pressure. AT1 selectivity is 10,000 times higher in Candesartan than in AT2. Aldosterone secretion inhibition may decrease potassium excretion while increasing salt and water excretion
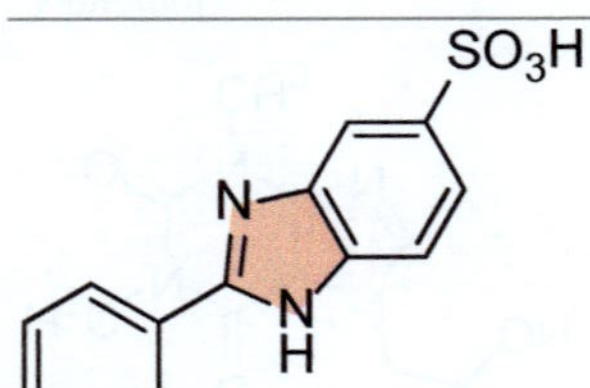 Ensulizole DB11115	Indicated for use in sunscreen compositions as a UV-B-absorbing molecule	Following UV-A or UV-B irradiation, ensulizole damages DNA produces DNA strand breaks and photosensitisers the production of oxidised guanines via type I and II photosensitisation processes. Ensulizole can produce reactive oxygen species, particularly singlet oxygen when exposed to light

(continued)

Table 1 (continued)

Structure of the drug Name of the drug (drug bank ID)	Category/indication	Mechanism of action
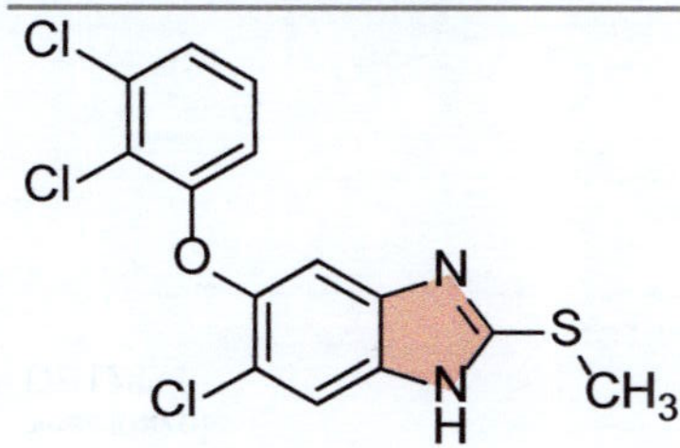 Triclabendazole DB12245	It's used to treat fascioliasis	It inhibits the function of tubulin, as well as protein and enzyme production. Motility is inhibited, the worm's outer surface is disrupted, and spermatogenesis and egg/embryonic cells are inhibited as a result of these metabolic disruptions
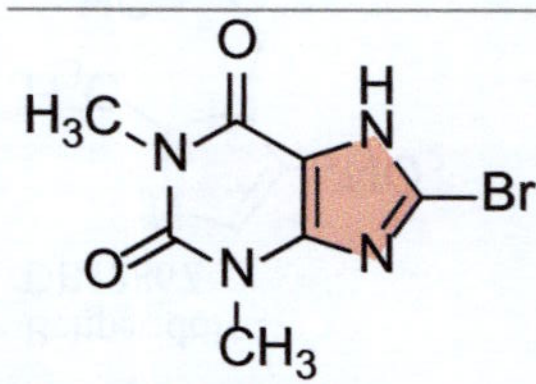 Bromotheophylline DB14018	It is a diuretic that is also used in conjunction with acetaminophen to relieve transitory water weight gain, bloating, edoema, and a full sensation related to menstrual cycles	Increases the permeability of the renal tubule, increases glomerular filtration rate and inhibits the sodium reabsorption in the proximal tubule

(continued)

Table 1 (continued)

Structure of the drug Name of the drug (drug bank ID)	Category/indication	Mechanism of action
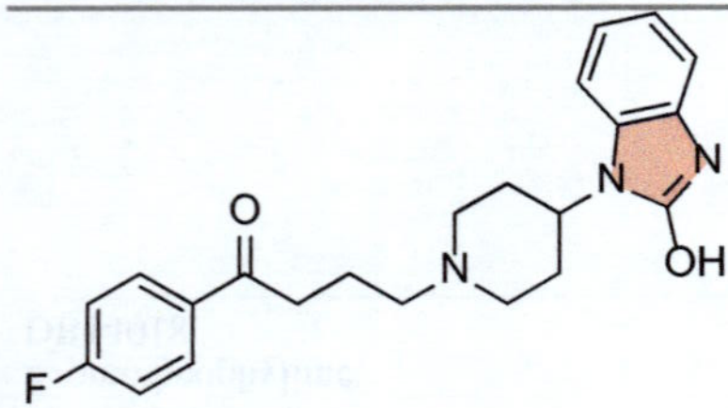 Benperidol DB12867	Not available	Not available
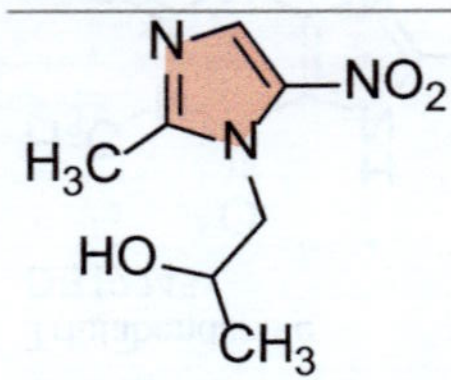 Secnidazole DB12834	Treatment of bacterial vaginosis	Secnidazole is a prodrug that penetrates the bacterial cell but has no antibacterial action. Bacterial enzymes reduce nitro groups to radical anions, which converts the medication into an active form. Susceptible isolates' bacterial DNA synthesis is hypothesised to be hampered by the radical anions

(continued)

Table 1 (continued)

Structure of the drug Name of the drug (drug bank ID)	Category/indication	Mechanism of action
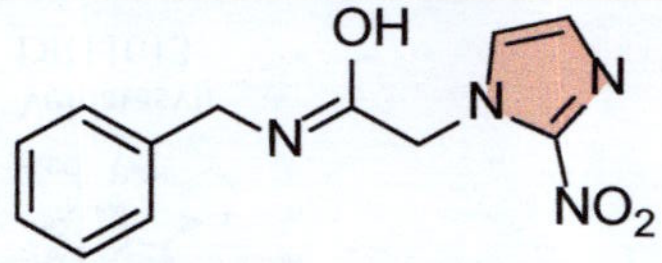 Benznidazole DB11989	Treatment of Chagas disease	Interferon, which is presumably present in greater proportions due to inflammation induced by macromolecule damage, has been reported to promote trypanosomal death by benzinidazole. DNA in benznidazole-affected parasites has been observed to undergo significant unpacking, with upregulation of DNA repair proteins, supporting the concept that DNA damage plays a role in the drug's action
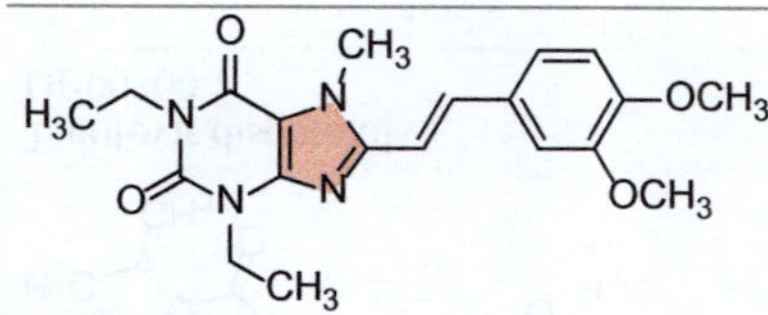 Istradefylline DB11757	For Parkinson's disease medication, istradefylline is used in conjunction with levodopa and carbidopa	Istradefylline is an adenosine A2A receptor inhibitor with a narrow therapeutic window. These receptors are situated in the basal ganglia, a brain area that is heavily engaged in motor control and suffers from degradation in Parkinson's disease. Within the indirect striatopallidal circuit, A2A receptors are also found on GABAergic medium spiny neurons. As a result, this pathway's GABAergic function is diminished. The affinity of istradefylline for A2A receptors is 56 times greater than that of A1 receptors

(continued)

Table 1 (continued)

Structure of the drug Name of the drug (drug bank ID)	Category/indication	Mechanism of action
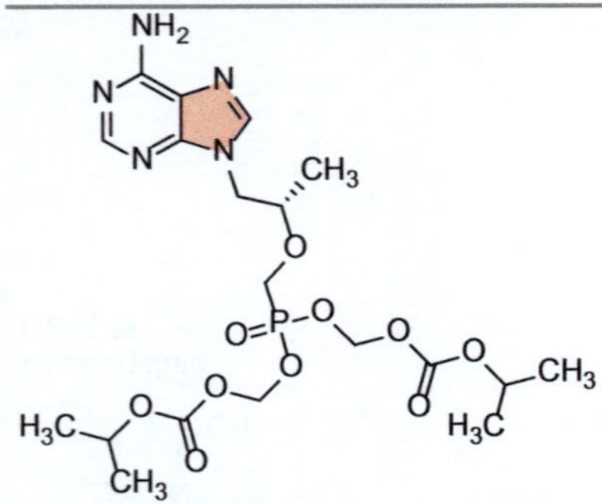 Tenofovir disoproxil DB00300	HIV infection and chronic hepatitis B infection are treated with this drug	HIV-1 reverse transcriptase and Hepatitis B polymerase inhibited by tenofovir diphosphate compared to the natural deoxyribonucleotide substrate (deoxyadenosine 5′-triphosphate), which results in viral DNA chain termination following integration into DNA
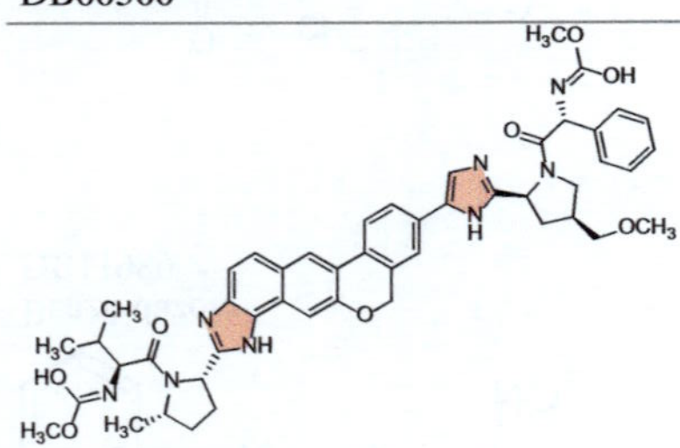 Velpatasvir DB11613	Hepatitis C virus infection (HCV) causes chronic hepatitis C which is liver disease and is treated with a combination of direct-acting antiviral (DAA) medicines. Depending on the severity of the decompensating or cirrhosis, combination therapy may additionally contain Ribavirin	Other selective NS5A inhibitors bind to domain I of NS5A, which contains amino acids 33–202. At this location, NS5A inhibitors compete with RNA for binding. NS5A inhibitors are also thought to engage the target during replication when the binding site is accessible. The protein is also known to redistribute to lipid droplets when NS5A is inhibited. Although it is recognised to be a crucial component in RNA replication, the specific role of NS5A remains unknown

(continued)

Table 1 (continued)

Structure of the drug Name of the drug (drug bank ID)	Category/indication	Mechanism of action
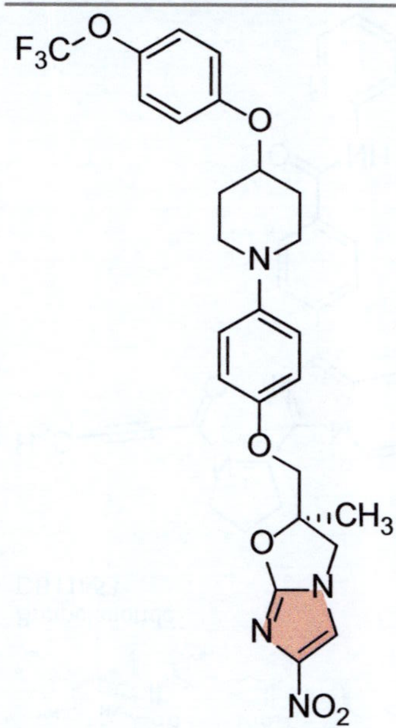 Delamanid DB11637	In a combination regimen, it is used to treat tuberculosis (TB) which is multidrug-resistant and extensively drug-resistant	Inhibits the production of mycolic acid in the bacterial cell wall

(continued)

Table 1 (continued)

Structure of the drug Name of the drug (drug bank ID)	Category/indication	Mechanism of action
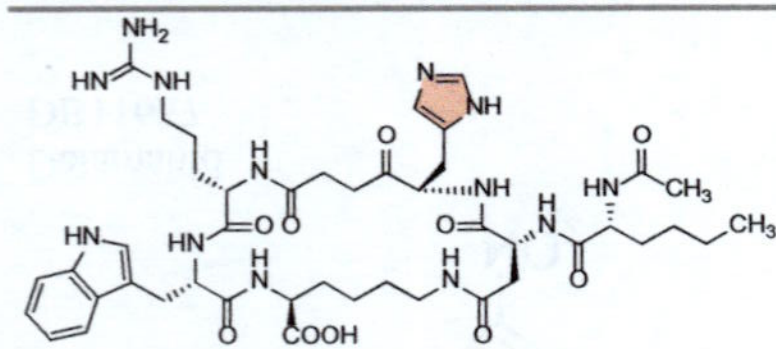 Bremelanotide DB11653	Used to treat premenopausal women with hypoactive sexual drive condition	Many melanocortin receptors are agonists, including the MC1R, MC4R, MC3R, MC5R, and MC2R, in order of potency. There is no clear mechanism. Bremelanotide, according to one theory, activates dopamine in the medial preoptic region, which is implicated in sexual behaviour in a variety of creatures
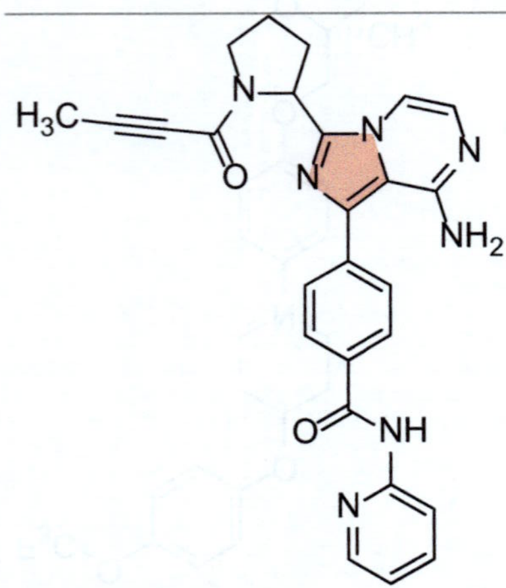 Acalabrutinib DB11703	Recommended for adult patients who are at least once undergone Mantle Cell Lymphoma (MCL) therapy	The signalling molecule Bruton Tyrosine Kinase (BTK) is used in both the B-cell antigen receptor and the cytokine receptor pathways. B-cell proliferation, chemotaxis, trafficking, and adhesion pathways are getting activated upon BTK signalling. Acalabrutinib, a BTK inhibitor, is a very small molecule. BTK enzymatic activity was inhibited by creating a covalent bond between cysteine residue (Cys481) in the active site and acalabrutinib along with its active metabolite (ACP-5862). Due to this effect acalabrutinib prevents BTK from activating downstream signalling proteins CD86 and CD69, halting the proliferation and survival of malignant B cells

(continued)

Table 1 (continued)

Structure of the drug Name of the drug (drug bank ID)	Category/indication	Mechanism of action
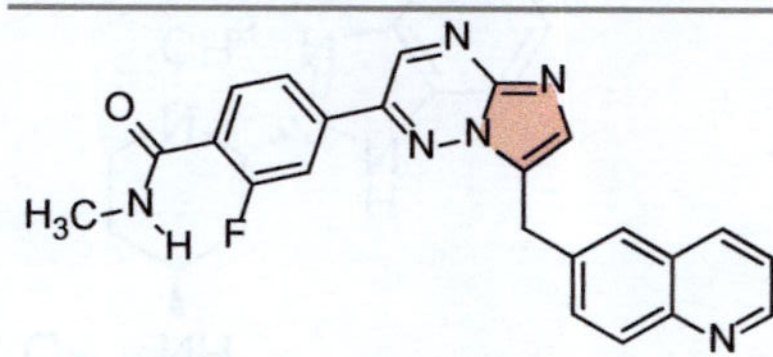 Capmatinib DB11791	Melanoma, gliosarcoma, solid tumours, colorectal cancer, and hepatic impairment are among the conditions studied in clinical trials	No information available
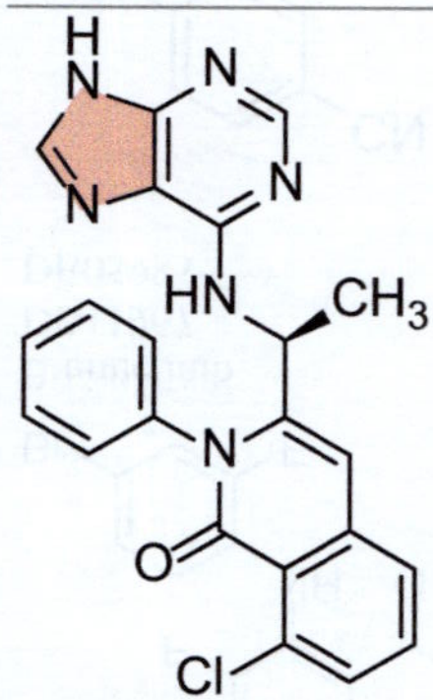 Duvelisib DB11952	Patients with small lymphocytic lymphoma (SLL) relapsed or refractory chronic lymphocytic leukaemia (CLL) after at least two previous therapies should use this medication	Strong reversible inhibitor of the phosphoinositide3-kinase (PI3K) isoforms gamma and delta. PI3K plays a significant role in innate and adaptive immunity, and inhibition of the isoforms delta and gamma has proved crucial for immune suppression

(continued)

Table 1 (continued)

Structure of the drug Name of the drug (drug bank ID)	Category/indication	Mechanism of action
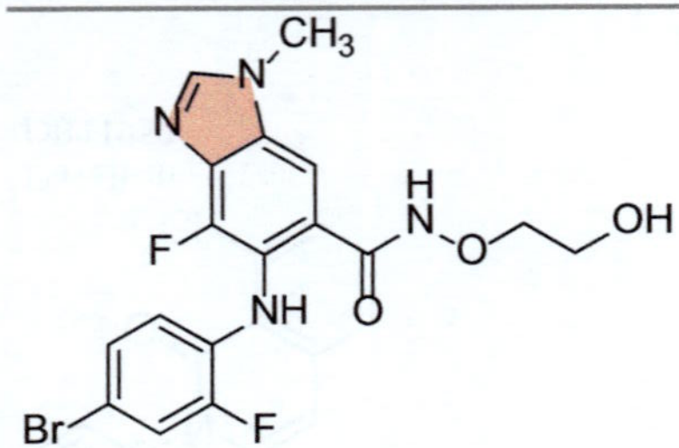 Binimetinib DB11967 DB05985	Adult patients with metastatic or unrespectable melanoma who have a BRAF V600E or V600K mutation, as identified by a US FDA, can take encorafenib as well binimetinib in combination	Binimetinib binds to MEK1/2 and suppresses its activity in a non-competitive manner with ATP. MEK1/2 inhibition inhibits MEK1/2-dependent effector proteins and transcription factors from being activated. Growth factor-mediated cell signalling can be inhibited as a result of this mechanism. This may prevent tumour cell growth as well as the generation of inflammatory cytokines such as interleukin-1, -6, and tumour necrosis factor
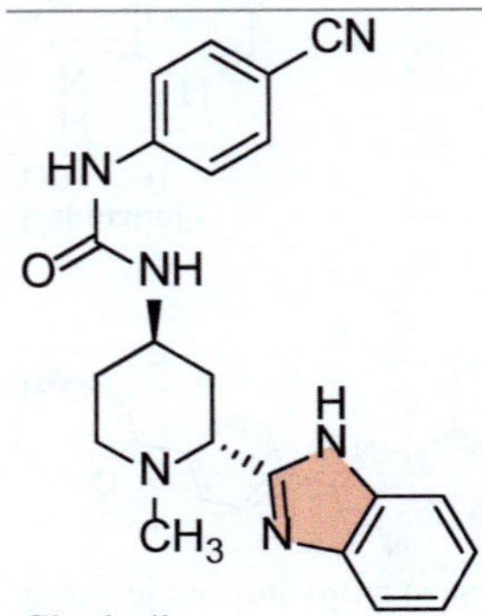 Glasdegib DB11978	Adults over the age of 75, it is approved to treat newly diagnosed acute myeloid leukaemia, glasdegib in combination with cytarabine can be used	Binds to the smoothened (SMO) receptor and is a strong and specific inhibitor of the hedgehog signalling pathway

(continued)

Table 1 (continued)

Structure of the drug Name of the drug (drug bank ID)	Category/indication	Mechanism of action
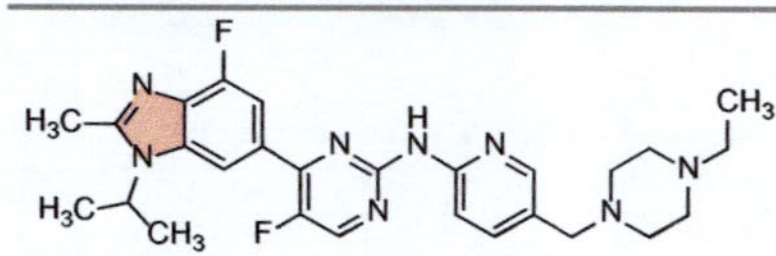 Abemaciclib DB12001	In the metastatic context, it is indicated as immunotherapy to treat adult patients with advanced HER2-negative or metastatic breast cancer, HR-positive who have improved post endocrine therapy and previous chemotherapy	Inhibits the uncontrolled activity of cyclin-dependent kinases 4 (CDK4) and 6 (CDK6), which are involved in the cell cycle and promote cancer cell proliferation
Inosine pranobex DB13156	It's an antiviral medication made up of inosine, acetamidobenzoic acid, and dimethylaminoisopropanol. As an adjuvant therapy to podophyllin or carbon dioxide laser, it's also used to treat herpes simplex virus (types 1 and 2) mucocutaneous infections and genital warts	Immune responses mediated by cells are stimulated in response to viral infections
Lixisenatide DB09265	Medication of diabetes mellitus type 2 as an anti-hyperglycaemic agent	The activation of adenylyl cyclase occurs when the GLP-1 receptor is activated by lixisenatide. Protein kinase A (PKA), as well as Epac1 and Epac2, are activated when the quantity of cyclic adenosine monophosphate in the cell rises. When the triggering route is triggered, PKA, Epac1, and Epac2 are engaged in the release of Ca^{2+} from the endoplasmic reticulum, which is known as the "amplification" route, which enhances insulin secretion

(continued)

Table 1 (continued)

Structure of the drug Name of the drug (drug bank ID)	Category/indication	Mechanism of action
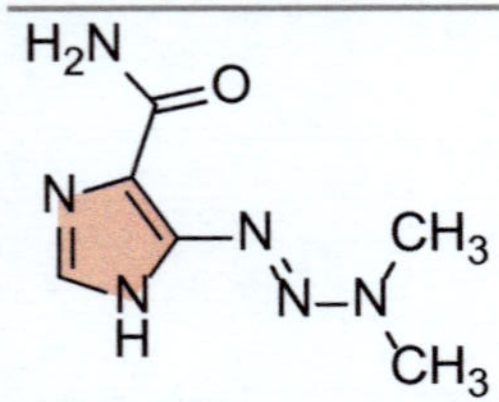 Dacarbazine DB00851	To treat malignant melanoma that has dispersed to other body parts. Furthermore, dacarbazine is approved as a secondary-line therapy for Hodgkin's disease when combined with other anticancer drugs	By acting as an alkylating agent, it appears to have cytotoxic properties. DNA synthesis inhibition as a purine analogue and association with SH groups are two further hypotheses. Dacarbazine doesn't have a defined cell cycle phase
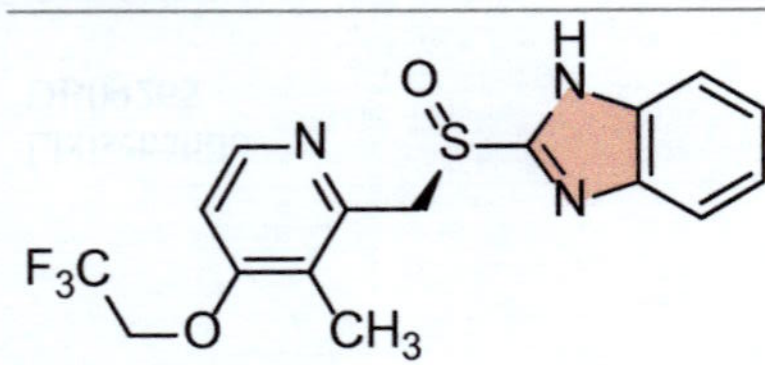 Dexlansoprazole DB05351	It's used to treat heartburn affiliated with symptomatic non-erosive gastroesophageal reflux disease, as well as to heal all categories of erosive esophagitis (EE), maintain and heal EE, and relieve heartburn (GERD)	The H^+/K^+-ATPase enzyme is involved in hydrolyzing ATP, secretion of hydrochloric acid and exchanging H^+ ions for K^+ ions from the cytoplasm in the secretory canaliculus. It results in secretion of HCl into the stomach lumen

(continued)

Table 1 (continued)

Structure of the drug Name of the drug (drug bank ID)	Category/indication	Mechanism of action
Tetracosactide DB01284	For use as a screening tool in patients suspected of having adrenocortical insufficiency	tetracosactide (also known as cosyntropin) is a synthetic peptide that is similar to the 24-amino-acid section at the *N*-terminus of adrenocorticotropic hormone (sequence: SYSMEHFRWGKPVGKKRRPVKVYP). The biological activity that promotes corticosteroid synthesis in the adrenal cortex is included in ACTH (1–24), a section that is identical in all animals. In terms of all biological functions, tetracosactide has the same activity as natural ACTH
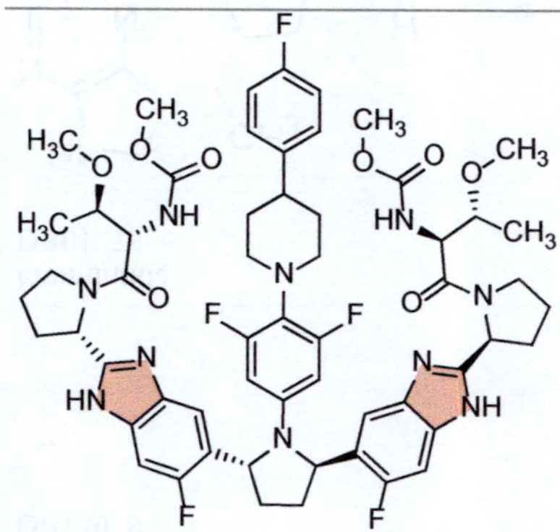 Pibrentasvir DB13878	This medication (Child–Pugh A) is advised for adult patients with chronic hepatitis C virus (HCV) genotypes 1, 2, 3, 4, 5, or 6 infections without compensated cirrhosis or cirrhosis	Hepatitis C virus (HCV) NS5A inhibitor and direct-acting antiviral drug that inhibits viral RNA replication and virion assembly

(continued)

Table 1 (continued)

Structure of the drug Name of the drug (drug bank ID)	Category/indication	Mechanism of action
Semaglutide DB13928	Used as a supplement to enhance glycaemic levels in diabetes mellitus type 2 patients	Glycaemic control mechanism GLP-1 is a physiological hormone that supports glycaemic control through a variety of processes, including insulin secretion, stomach emptying, and postprandial glucagon secretion reduction. Reduces atherosclerosis development by lowering intestinal permeability and reducing inflammation
Pramlintide DB01278	To treat type 1 and type 2 diabetes mellitus in individuals who do not have satisfactory glycaemic control with pre-prandial insulin therapy	As compared to the naturally occurring pancreatic hormone amylin, an amylinomimetic
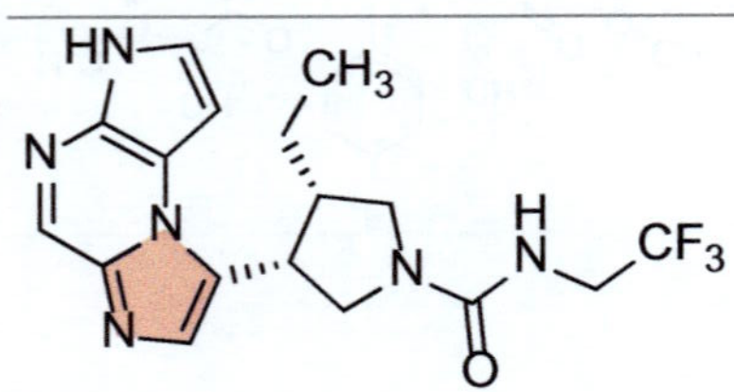 Upadacitinib DB15091	Upadacitinib is used to treat adults with rheumatoid arthritis in moderate to severe levels, who have an unsatisfactory response or sensitivity to methotrexate	Not available

(Fig. 11). In a nutshell, these compounds exhibit remarkable anticancer properties. The intriguing anticancer capabilities of imidazoles prompted the creation of a slew of derivatives in the hopes of boosting efficacy while minimising adverse effects. According to experimental proof, changes in the structure and operation of VEGF (vascular endothelial growth factor), topoisomerases I and II, DNA (deoxyribonucleic acid), histone deacetylases, mitotic spindle microtubules, receptor tyrosine kinases, and the CYP26A1 enzyme causes cancer. DNA is a superior target for anticancer compounds among the several aims. As a result, scientists have concentrated their efforts on detecting and documenting the association of small molecules with DNA in order to help in the creation of innovative and effective therapeutic medications that modulate gene interpretation [41–43]. Imidazole inhibits cell proliferation by interacting with DNA through covalent or non-covalent interactions (electrostatic interaction and groove binding) [44]. There are around 1180 research articles on imidazoles with noteworthy anticancer discoveries, according to the literature survey. An examination of the number of publications on imidazole anticancer activity from 2002 to 2016 shows a steady increase in interest in the field (Fig. 12) [45–48].

Imidazoles have anticancer drug properties, according to several study articles. The development of imidazoles as antineoplastic drugs began with dacarbazine,

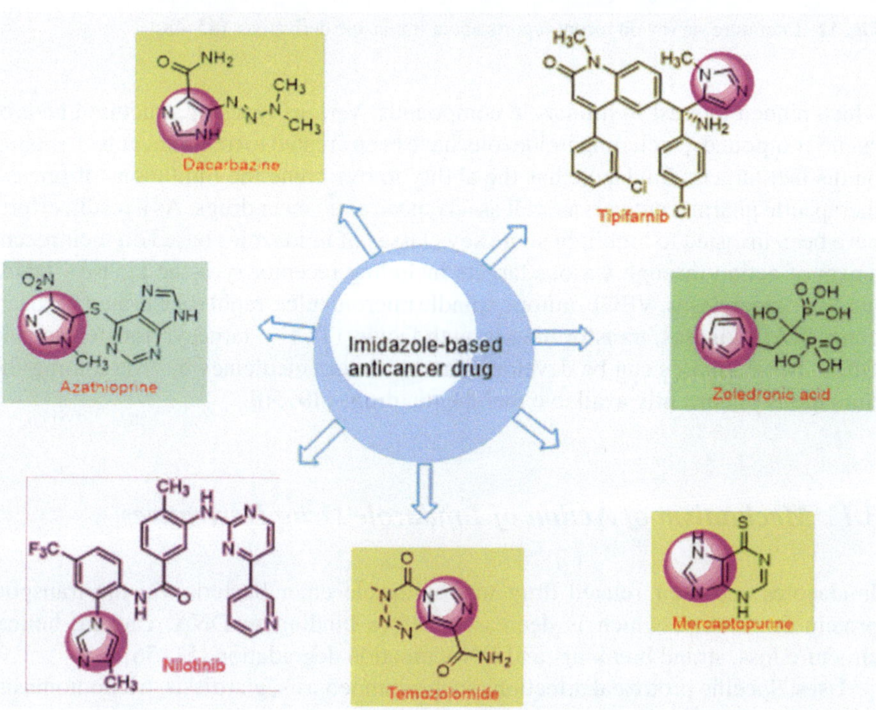

Fig. 11 Imidazole-based anticancer drugs' chemical structures [23–34]

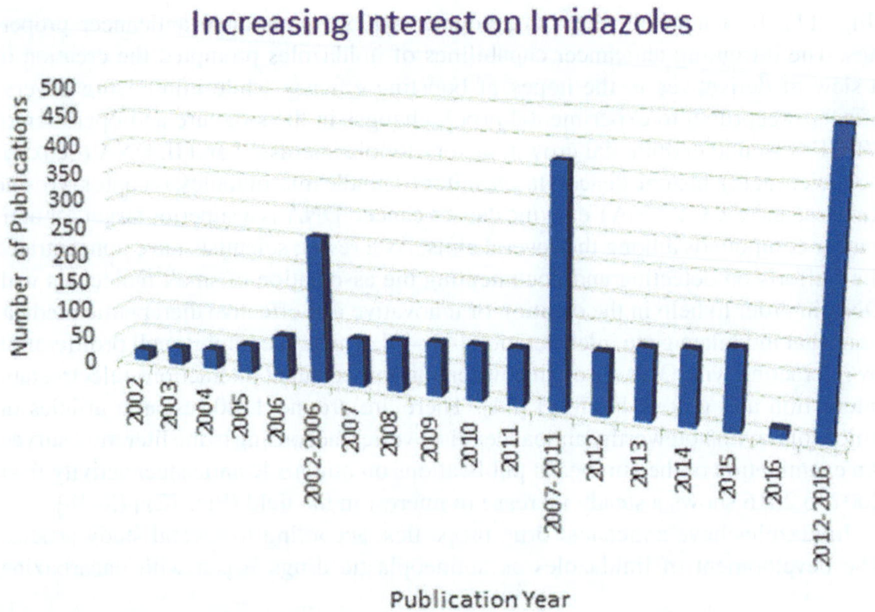

Fig. 12 Literature survey on recent reports about imidazole derivatives [45–48]

which piqued interest in imidazole compounds. Various types of structured heterocyclic compounds, including imidazole, have been created to treat cancer by focusing on distinct targets. Imidazole has the ability to overcome the limitations of present therapeutic pharmaceuticals as well as advanced anticancer drugs. As a result, efforts have been invested to highlight some key classes of imidazoles based on their mechanism of action through various targets including receptor tyrosine kinases, DNA, histone deacetylases, VEGF, mitotic spindle microtubules, rapid accelerated fibrosarcoma (RAF) kinases, transforming growth factor- (TGF-), farnesyl transferase, and DNA. These kinases can be developed as anticancer medicines by overcoming the limitations of currently available therapeutic drugs [49, 50].

3.1 Mechanism of Action of Imidazole Drug Derivatives

Imidazoles and their related drug metronidazole enter bacteria via the transport protein ferredoxin, which is decreased before binding to DNA, causing helical structure loss, strand breaking, and DNA function degradation [51–56].

Uses: Specific protozoal infections such as amoebiasis, giardiasis, trichomoniasis, and balantidiasis are treated.

Adverse Effect: Diarrhoea, nausea, insomnia, ataxia, vomiting.

Metabolism: The cytochrome P450 system is responsible for its metabolism. As both are active metabolites, the methyl group is oxidised to hydroxy methyl and the ethanol side chain is oxidised to an acid group.

3.2 Molecular Mechanism of Imidazole Drugs as Anticancer Agents

Imidazoles work as anticancer drugs through a variety of pathways including VEGF, topoisomerases, DNA, histone deacetylases, mitotic spindle microtubules, receptor tyrosine kinases, CYP26A1 enzyme and rapid accelerated fibrosarcoma (RAF) kinases.

Imidazole ring-containing drugs such as zoledronic acid, nilotinib, Dacarbazine, temozolomide, tipifarnib, mercaptopurine and others are used to treat different malignancies [57].

4 Imidazoles as Anti-Angiogenic Agents by Targeting VEGF

Angiogenesis is the emergence of new blood vessels and capillaries. In female embryogenesis, tissue healing and the reproductive cycle, it is a critical physiological function. It has a function in psoriasis, diabetic retinopathy, rheumatoid arthritis and tumour growth, among other pathological conditions.

The expression of VEGF is elevated under hypoxia, notably in cancer cells. H1F1α interacts with H1F1β in hypoxic circumstances, resulting in dimer formation, as seen in the diagram above. It passes through the gene transcription process and produces VEGF. VEGF-A, VEGF-B, VEGF-C, VEGF-D and placenta growth factors are all members of the VEGF family in humans (PLGF). VEGF-A is the most important factor in angiogenesis (Fig. 13).

VEGF ligands bind to and activate three tyrosine kinase receptors: VEGFR1, VEGFR2, and VEGFR3. In the tumour vasculature, VEGFR2 levels are increased. Both VEGFR-1 and VEGFR2 are activated by VEGF-A binding, however, VEGFR2 is more active in cancer cells because it is significantly increased in cancer cells [58].

The structure of VEGFR1 and VEGFR2 contains seven immunoglobulin (Ig-like) domains. VEGF-A binds to the second domain of VEGFR1 and the second or third domain of VEGFR2. Other Ig domains, such as the fourth through seventh, are important in receptor activation and dimerization (Fig. 14).

The binding of VEGF ligand in dimeric form to VEGFR2 improves the binding capacity of the second receptor monomer to the tethered ligand, according to electron microscope observations of the VEGF/VEGFR2 complex. Both receptors will be

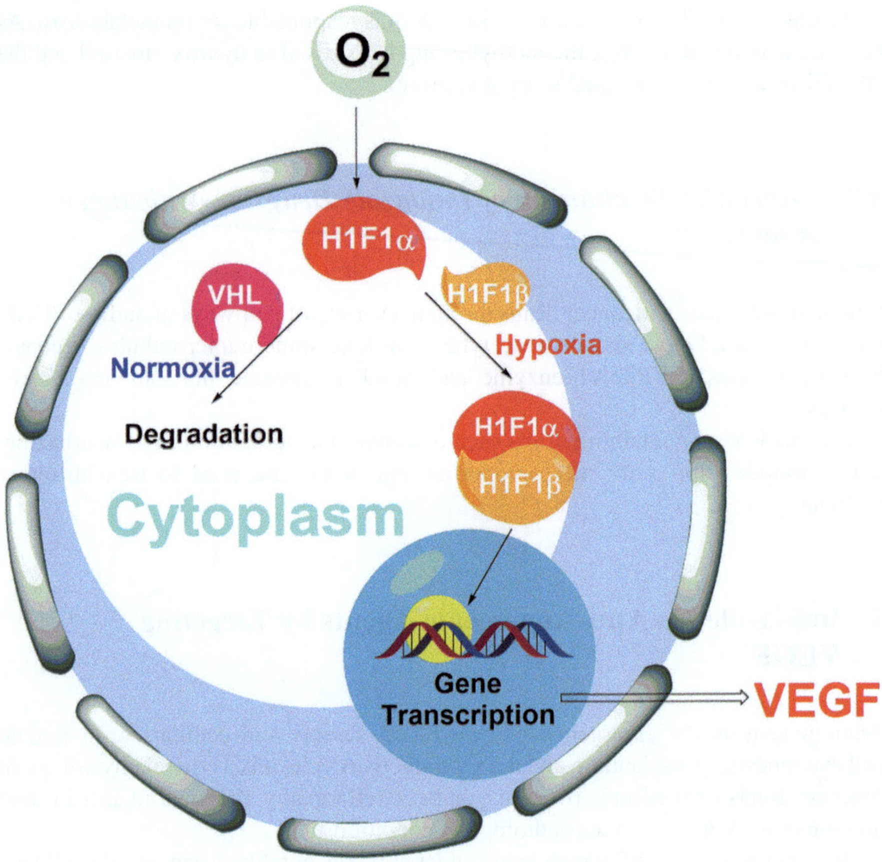

Fig. 13 Role of H1F1 in different levels of oxygen [58]

cross-linked, and the low-affinity connections between the two domains will stabilise, resulting in receptor dimerization.

Imidazoles reduce vasculogenesis and angiogenesis by inhibiting the activity of VEGFR-2. The hydrogen bonding between the amino acids in VEGFR-2's active site and the Imidazole moiety is what causes the inhibition (Fig. 15) [59–63].

5 Imidazoles Targeting Mitotic Spindle Microtubules

Microtubules are an important part of the cytoskeleton. It is required for crucial cellular functions such as cell shape preservation, cell migration, cell division and intracellular transport. During cell division, microtubules shape the mitotic spindle, which is accountable for aligning replicated chromosomes to the equatorial plane

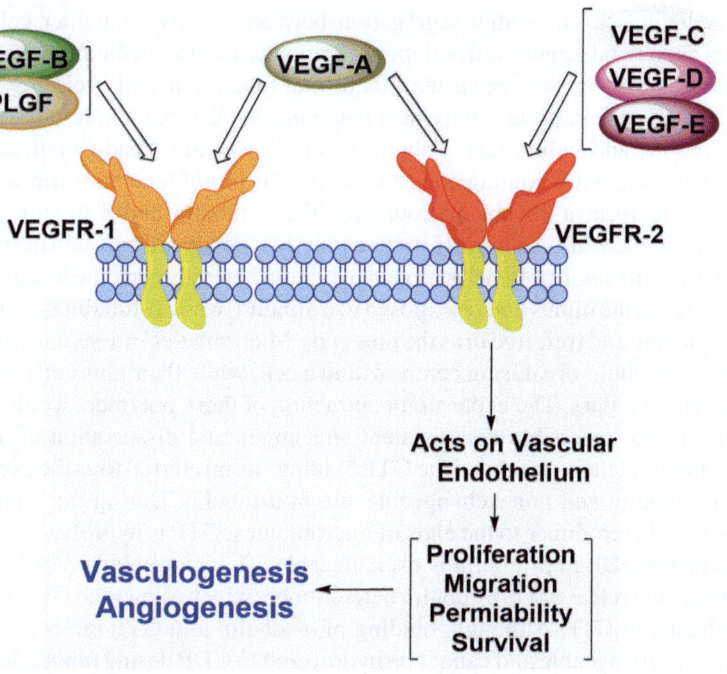

Fig. 14 Activation VEGFR-2 by VEGF-A [59–63]

Fig. 15 Imidazoles inhibit VEGFR-2 activity [59–63]

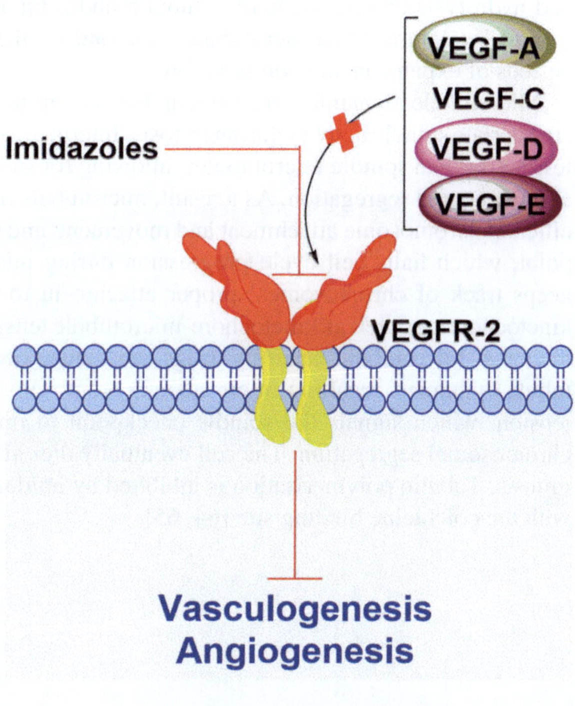

and mediating chromosomal segregation between the two daughter cells. Microtubules are a good target for developing chemotherapeutic medicines against rapidly proliferating cancer cells because of their vital function in cell cycle.

Microtubules are protein polymers made up of α/β-tubulin heterodimers that are long, hollow, and cylindrical. Tubulin heterodimers link head-to-tail, resulting in protofilaments. After joining longitudinally, the protofilaments form a sheet that closes up to form a 25 nm microtubule. The α-tubulin and β-tubulin are almost identical amino acids. An intrinsic heterogeneity is generated between the two ends of the microtubule upon polymerisation of α-tubulin heterodimers; the less dynamic end (referred to as the minus end) is exposed to α-tubulin, while β-tubulin is exposed at the more dynamic end (referred to as the plus end). Microtubules' minus ends are attached to the microtubule-organising centre within a cell, while their plus ends are disposed to the cell boundary. The expansion/contraction of these polymers would take place because of the reversible non-covalent attachment and dissociation of α/β-tubulin heterodimers at their two ends. The GTP binding site is referred to as the exchangeable site in β-tubulin and nonexchangeable site in α-tubulin. During the binding of the α/β-tubulin heterodimer to the ends of microtubules, GTP is hydrolyzed to GDP, and the resultant GDP in β-tubulin is exchangeable. The microtubule depolymerization resulted in the release of α/β-tubulin heterodimers which allows the GDP in β-tubulin to exchange to GTP. Although, binding of α-tubulin to a GTP molecule makes the GTP nonexchangeable and cannot be hydrolyzed to GDP during tubulin heterodimer addition to the ends of microtubules. Microtubules have two uncommon dynamic properties: dynamic instability and treadmilling, due to the unique GTP binding and hydrolysis characteristic of α-tubulin and β-tubulin. Tread milling defines the overall development of a microtubule at one end and the stochastic switching between periods of expansion and compression.

Microtubule dynamics are critical for numerous cellular activities, including appropriate spindle activity during mitosis. Interphase microtubules are 10–100 times less active than spindle microtubules, allowing for successful chromosomal capture, alignment, and segregation. As a result, microtubule dynamics suppression prevents efficient chromosome attachment and movement, and so activates the spindle checkpoint, which halts cell cycle progression during mitosis. The spindle checkpoint keeps track of chromosomes' proper attachment to spindle microtubules at their kinetochores, as well as kinetochore-microtubule tension exerted across paired kinetochores. Microtubule-targeting drugs can induce even slight alterations in microtubule dynamics, resulting in erroneous chromosomal attachment and kinetochore tension, which signals the spindle checkpoint to impede anaphase initiation and chromosomal segregation. The cell eventually dies after an aberrant departure from mitosis. Tubulin polymerisation is inhibited by imidazole analogues, which interact with the colchicine binding site [64, 65].

6 Imidazole as Histone Deacetylase Inhibitors

Carcinogenesis cannot be explained solely by genetic changes; epigenetic mechanisms are also involved (histone modifications, DNA methylation, and non-coding RNA deregulation). Chromatin decondensation is caused by histone changes such lysine deacetylation on H3 and H4. These variants play a crucial role in regulating gene expressions by altering the convenience of transcriptional factors to DNA through conformational modifications in the nucleosome structure. The histone acetyltransferase (HAT) and histone deacetylase (HDAC) activity were controlled by the balancing of histone acetyltransferase (HAT) and histone deacetylase (HDAC) [66]. Inhibition of HDACs causes the production of genes that are responsible for growth arrest, terminal differentiation, and/or death in a variety of cancer cells, and this imbalance has been related to cancer [37, 38].

The HDAC active site consists of a narrow, hydrophobic tunnel arising at the interface of protein–solvent and leads a Zn^{2+} containing cavity of 8 Å. HDAC inhibitor comprises of bidentate chelation tethered to binding constituent that interrelates with the mouth of the tunnel at the interface of protein–solvent.

The characteristic feature of HDAC inhibition is the p21waf activation, which is hypothesised to underlie the antiproliferative effects found with intracellular HDAC inhibition. Human histone deacetylases are inhibited by benzimidazole and imidazole compounds at nanomolar concentrations [38]. HDAC inhibitors cause cell cycle arrest by a number of mechanisms as the most important one is appeared to be increased expression caused by cell cycle genes such as CDKN1A (cyclin-dependent kinase inhibitor p21), which has been observed in a number of cancer cells. Its product prevents cyclins and cyclin-dependent kinases (CDKs) from forming dimers, which cause cell cycle arrest and differentiation inhibition (Fig. 16) [37].

7 Summary/Conclusion

In-depth research and continuous efforts towards imidazole-based pharmaceuticals have been dedicated owing to their potency, less toxicity and better pharmacokinetic aspects. The evolution of imidazole-based drugs in other disciplines such as molecular biology, pharmacology, cell biology, materials science and organic chemistry marks them as effective diagnostic agents. This chapter would serve as a comprehensive basis and reference source for allowing researchers to develop new perspectives on imidazole-based pharmaceuticals pursuing diverse pharmacological effects.

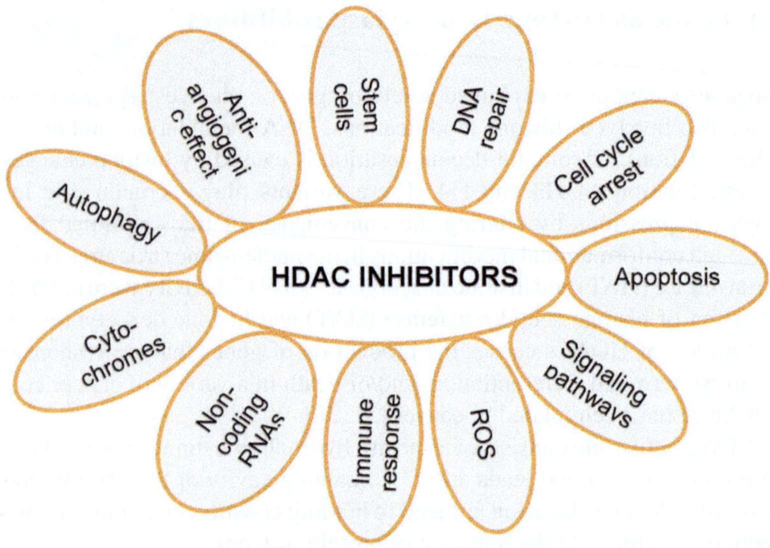

Fig. 16 HDAC inhibitors have a diversity of anticancer mechanisms [37]

References

1. Romero DH, Heredia VET, García-Barradas O, López MEM, Pavón ES (2014) Synthesis of imidazole derivatives and their biological activities. J Chem Biochem 2:45–83
2. Weinreb SM (2007) Some recent advances in the synthesis of polycyclic imidazole-containing marine natural products. Nat Prod Rep 24:931–948
3. Jin Z (2006) Imidazole, oxazole and thiazole alkaloids. Nat Prod Rep 23:464–496
4. Venkatakrishnan K, Von Moltke LL, Greenblatt DJ (2000) Effects of the antifungal agents on oxidative drug metabolism. Clin Pharmacokinet 38:111–180
5. Niwa T, Shiraga T, Takagi A (2005) Effect of antifungal drugs on cytochrome P450 (CYP) 2C9, CYP2C19, and CYP3A4 activities in human liver microsomes. Biol Pharm Bull 28:1805–1808
6. Bellmann R (2007) Clinical pharmacokinetics of systemically administered antimycotics. Curr Clin Pharmacol 2:37–58
7. Steinman RA, Brufsky AM, Oesterreich S (2012) Zoledronic acid effectiveness against breast cancer metastases—a role for estrogen in the microenvironment? Breast Cancer Res 14:213–221
8. Ashley ESD (2010) Pharmacology of azole antifungal agents. In: Ghannoum MA, Perfect JR (eds) Antifungal therapy. Taylor and Francis Group, New York, pp 199–218
9. Mishra R, Ganguly S (2012) Imidazole as an anti-epileptic: an overview. Med Chem Res 21:3929–3939
10. Carrillo-Muñoz AJ, Giusiano G, Ezkurra PA, Quindós G (2005) Sertaconazole: updated review of a topical antifungal agent. Expert Rev Anti Infect Ther 3:333–342
11. Fothergill AW (2006) Miconazole: a historical perspective. Expert Rev Anti Infect Ther 4:171–175
12. Natarajan SR, Doherty JB (2005) P38MAP kinase inhibitors: evolution of imidazole-based and pyridopyrimidin-2-one lead classes. Curr Top Med Chem 5:987–1003
13. De Luca L (2006) Naturally occurring and synthetic imidazoles: their chemistry and their biological activities. Curr Med Chem 13:1–23

14. Kumar JR (2010) Review of imidazole heterocyclic ring containing compounds with their biological activity. Pharmacophore 1:167–177
15. Zhang L, Peng X-M, Damu GLV, Geng R-X, Zhou C-H (2014) Comprehensive review in current developments of imidazole-based medicinal chemistry. Med Res Rev 34:340–437
16. Xiong F, Chen X, Chen F (2010) An improved asymmetric total synthesis of (+)-biotin via the enantioselective desymmetrization of a meso-cyclic anhydride mediated by cinchona alkaloid-based sulphonamide. Tetrahedron Asymmetry 21:665–669
17. Roue M, Domart-Coulon I, Ereskovsky A, Djediat C, Perez T, Bourguet-Kondracki M-L (2010) Cellular localization of clathridimine, an antimicrobial 2-aminoimidazole alkaloid produced by the mediterranean calcareous sponge *Clathrina clathrus*. J Nat Prod 73:1277–1282
18. Liao S-M, Du Q-S, Meng J-Z, Pang Z-W, Huang R-B (2013) The multiple roles of histidine in protein interactions. Chem Cent J 7:44–56
19. Rigden DJ (2008) The histidine phosphatase superfamily: structure and function. Biochem J 409:333–348
20. Shahid M, Tripathi T, Sobia F, Moin S, Siddiqui M, Khan RA (2009) Histamine, histamine receptors, and their role in immunomodulation: an updated systematic review. Open Immunol J 2:9–41
21. Daniels KG, Beckett D (2010) Biochemical properties and biological function of a monofunctional microbial biotin protein ligase. Biochemistry 49:5358–5365
22. Wishart DS, Feunang YD, Guo AC, Lo EJ, Marcu A, Grant JR, Sajed T, Johnson D, Li C, Sayeeda Z, Assempour N, Iynkkaran I, Liu Y, Maciejewski A, Gale N, Wilson A, Chin L, Cummings R, Le D, Pon A, Knox C, Wilson M (2018) DrugBank 5.0: a major update to the DrugBank database for 2018. Nucleic Acids Res 46:D1074–D1082
23. Reichert JM, Wenger JB (2008) Development trends for new cancer therapeutics and vaccines. Drug Discovery Today 13:30–37
24. SathiyaPriya E, Mary venila S (2017) A study on classification algorithms and performance analysis of data mining using cancer data to predict lung cancer disease. Int J New Technol Res 3:88–93
25. Ramana Murty NV, Prasad Babu MS (2017) A critical study of classification algorithms for lung cancer disease detection and diagnosis. Int J Comput 13:1041–1048
26. American Cancer Society (2008) Cancer facts and figures 2008. American Cancer Society, Atlanta
27. Molina JR, Yang P, Cassivi SD, Schild SE, Adjei AA (2008) Non-small cell lung cancer: epidemiology, risk factors, treatment, and survivorship. Mayo Clin Proc 83:584–594
28. Navada S, Lai P, Schwartz AG, Kalemkerian GP (2016) Temporal trends in small cell lung cancer: analysis of the national surveillance, epidemiology, and end-results (SEER) database. J Clin Oncol 24:7082–7082
29. Grasso CS, Wu Y-M, Robinson DR, Cao X, Dhanasekaran SM, Khan AP, Quist MJ, Jing X, Lonigro RJ, Brenner JC, Asangani IA, Ateeq B, Chun SY, Siddiqui J, Sam L, Anstett M, Mehra R, Prensner JR, Palanisamy N, Rysilk GA, Vandin F, Raphael BJ, Kunju LP, Rhodes DR, Pienta KJ, Chinnaiyan AM, Tomlins SA (2012) The mutational landscape of lethal castration-resistant prostate cancer. Nature 487:239–243
30. McGuire S (2016) World cancer report 2014. Geneva, Switzerland: World Health Organization, International Agency for Research on Cancer, WHO Press, 2015. Adv Nutr 7:418–419
31. Bray F, Møller B (2006) Predicting the future burden of cancer. Nat Rev Cancer 6:63–74
32. Ali I, Wani WA, Saleem K (2011) Cancer scenario in India with future perspectives. Cancer Ther 8:56–70
33. Pereira C, Leão M, Soares J, Bessa C, Saraiva L (2012) New therapeutic strategies for cancer and neurodegeneration emerging from yeast cell-based systems. Curr Pharm Des 18:4223–4235
34. Ali I, Lone MN, Al-Othman ZA, Al-Warthan A, Sanagi MM (2015) Heterocyclic scaffolds: centrality in anticancer drug development. Curr Drug Targets 16:711–734
35. Ali I, Lone MN, Aboul-Enein HY (2017) Imidazoles as potential anticancer agents. Med Chem Commun 12:3–35
36. Hill RA (2009) Marine natural products. Annu Rep Prog Chem Sect B Org Chem 105:150–166

37. Forte B, Malgesini B, Piutti C, Quartieri F, Scolaro A, Papeo G (2009) A submarine journey: the pyrroleimidazole alkaloids. Mar Drugs 7:705–753
38. Jin Z (2011) Muscarine, imidazole, oxazole, and thiazole alkaloids. Nat Prod Rep 28:1143–1191
39. Gao G, Xiao R, Yuan Y, Zhou CH, You JS, Xie RG (2002) Efficient imidazolium catalysts for the benzoin condensation. J Chem Res 2002:262–263
40. Jiang HY, Zhou CH, Luo K, Chen H, Lan JB, Xie RG (2006) Chiral imidazole metalloenzyme models: synthesis and enantioselective hydrolysis for α-amino acid esters. J Mol Catal A Chem 260:288–294
41. Li LX, Hu JY, Wang H, Yu QB, Yue LH (2012) Molecular spectroscopy evidence of berberine binding to DNA: comparative binding and thermodynamic profile of intercalation. Biomacromolecules 13:873–880
42. Alniss HY, Anthony NG, Khalaf AI, Mackay SP, Suckling CJ, Waigh RD, Wheate NJ, Parkinson JA (2012) Rationalising sequence selection by ligand assemblies in the DNA minor groove: the case for thiazotropsin A. Chem Sci 3:711–722
43. Ketron AC, Denny WA, Graves DE, Osheroff N (2012) Amsacrine as a topoisomerase II poison: importance of drug-DNA interactions. Biochemistry 51:1730–1739
44. Wang X, Li Y, Gong S, Fu D (2002) A spectroscopic study on the DNA binding behavior of the anticancer drug dacarbazine. Spectrosc Lett 35:751–756
45. Baroniya S, Anwer Z, Sharma PK, Dudhe R, Kumar N (2010) Recent advancement in imidazole as anti-cancer agents: a review. Der Pharmacia Sinica 1:172–182
46. Iradyan MA, Iradyan NS, Stepanyan GM, Arsenyan FG, Garibdzhanyan BT (2010) Antitumor activity of imidazole derivatives: dacarbazine and the new alkylating agent imidazene (review). Pharm Chem J 44:175–182
47. Gaba M, Mohan C (2016) Development of drugs based on imidazole and benzimidazole bioactive heterocycles: recent advances and future directions. Med Chem Res 25:173–210
48. Verma A, Joshi S, Singh D (2013) Imidazole: having versatile biological activities. J Chem 2013:329412, 12 pages
49. Cai JL, Li S, Zhou CH, Wu J (2009) Advance in research of imidazoles as anti-tumor agents. Chin J New Drugs 18:598–608
50. Zhou C, Hassner A (2001) Synthesis and anticancer activity of novel chiral glucose derived bis-imidazoles and their analogs. Carbohydr Res 333:313–326
51. Jone EB (2011) Antimicrobial agents: antifungal agents, in "Goodman and Gilman's". In: Laurence LB, Chabner BA, Knollmann BC (eds) The pharmacological basis of therapeutics, 11th edn. McGraw-Hill Inc Health Professions Division, New York
52. Gubbins P, Anaissie E (2006) Overview of antifungal agents. Pharmacy Practice News
53. Sheppard D, Lampiris HW (2004) Antifungal agents. In: Katzung BG (ed) Basic and clinical pharmacology, 9th edn. Appleton and Lange. Norwalk, Connecticut
54. Fromtling RA (1988) Overview of medically important antifungal azole derivatives. Clin Microbiol Rev 1:187–217
55. Ralph HR, Roy LH (1999) Antifungal agents. In: Munson PL (ed) Principles of pharmacology, basic concepts and clinical applications. Chapman and Hall, New York
56. Hafeez A, Kazmi I (2017) Dacarbazine nanoparticle topical delivery system for the treatment of melanoma. Sci Rep 7:16517. https://doi.org/10.1038/s41598-017-16878-1
57. Friis T, Engel AE, Bendiksen CD, Larsen LS, Houen G (2013) Influence of levamisole and other angiogenesis inhibitors on angiogenesis and endothelial cell morphology in vitro. Cancers 5:762–785
58. Ahmadvand D, Rahbarizadeh F, Iri-Sofla FJ, Namazi G, Khaleghi S, Geramizadeh B, Pasalar P, Karimi H, Bakhtiari SHA (2010) Inhibition of angiogenesis by recombinant VEGF receptor fragments. Science 41:417–422
59. Holmes K, Roberts OL, Thomas AM, Cross MJ (2007) Vascular endothelial growth factor receptor-2: structure, function, intracellular signalling and therapeutic inhibition. Cell Signal 19:2003–2012

60. Ruch C, Skiniotis G, Steinmetz MO, Walz T, Ballmer-Hofer K (2007) Structure of a VEGF-VEGF receptor complex determined by electron microscopy. Nat Struct Mol Biol 14:249–250
61. Kiselyov AS, Semenova M, Semenov VV (2006) Hetaryl imidazoles: a novel dual inhibitors of VEGF receptors I and II. Bioorg Med Chem Lett 16:1440–1444
62. Abdullaziz MA, Abdel-Mohsen HT, Kerdawy AME, Ragab FAF, Ali MA, Abu-bakr SM, Girgis AS, Diwani HIE (2017) Design, synthesis, molecular docking and cytotoxic evaluation of novel 2-furybenzimidazoles as VEGFR-2 inhibitors. Eur J Med Chem 136:315–329
63. Zhou J, Giannakakou P (2005) Targeting microtubules for cancer chemotherapy. Curr Med Chem Anti Cancer Agents 5:65–71
64. Dong M, Liu F, Zhou H, Zhai S, Yan B (2016) Review on novel natural product- and privileged scaffold-based tubulin inhibitors targeting the colchicine binding site. Molecules 21:1375
65. Eckschlager T, Plch J, Stiborova M, Hrabeta J (2017) Review on histone deacetylase inhibitors as anticancer drugs. Int J Mol Sci 18:1414
66. Bressi JC, de Jong R, Wu Y, Jennings AJ, Brown JW, O'Connell S, Tari LW, Skene RJ, Vu P, Navre M, Cao X, Gangloff AR (2010) Benzimidazole and imidazole inhibitors of histone deacetylases: synthesis and biological activity. Bioorg Med Chem Lett 20:3138–3141

Chapter 7
Overview on Biological Activities of Pyrazole Derivatives

Arup K. Kabi, Sattu Sravani, Raghuram Gujjarappa, Aakriti Garg, Nagaraju Vodnala, Ujjawal Tyagi, Dhananjaya Kaldhi, Virender Singh, Sreya Gupta, and Chandi C. Malakar

Abbreviations

US FDA	United States Food and Drug Administration
ZE	Zollinger–Ellison syndrome
ARB	Angiotensin II receptor blocker
DNA	Deoxyribonucleic acid
CNS	Central nervous system
COX	Cyclooxygenase
GABA	Gamma-aminobutyric acid
NSAID	Non-steroidal anti-inflammatory drugs
ADP	Adenosine diphosphate
JAK	Janus kinase
PARP	Poly(ADP-ribose)polymerase
RNA	Ribonucleic acid
HCV	Hepatitis C virus
FGFR	Fibroblast growth factor receptor
NRS	Nuclear hormone receptors
MRSA	Methicillin-resistant staphylococcus aureus
PRSA	Penicillin-resistant S aureus

A. K. Kabi · R. Gujjarappa · N. Vodnala · U. Tyagi · D. Kaldhi · C. C. Malakar (✉)
Department of Chemistry, National Institute of Technology Manipur, Langol, Imphal, Manipur 795004, India
e-mail: cmalakar@nitmanipur.ac.in

S. Sravani · A. Garg · S. Gupta (✉)
Department of Medicinal Chemistry, National Institute of Pharmaceutical Education and Research (NIPER), Chunilal Bhawan, 168, Maniktala Main Road, Kolkata 700054, India

V. Singh
Department of Chemistry, Department of Chemistry, Central University of Punjab, Bathinda, Punjab 151001, India

B. P. Swain (ed.), *Nanostructured Biomaterials*, Materials Horizons: From Nature to Nanomaterials, https://doi.org/10.1007/978-981-16-8399-2_7

VRE	Vancomycin-resistant enterococcus
HepG2	Human liver cancer cell line
GIP	Gastric inhibitory polypeptide
BGC823	Human gastric carcinoma cell line
BT474	Human breast tumour cell line
VEGF	Vascular endothelial growth factor
RT-PCR	Reverse transcription-polymerase chain reaction

1 Introduction

Pyrazole moieties are *N*-containing five-membered ring size heterocycles which constitute a class of molecules especially useful in drug synthesis. Among the whole azole family, pyrazoles are one of the most investigated classes of compounds because of their broad range of pharmacological properties [1, 2]. Over the decades, several protocols for synthesis as well synthetic analogues and pharmaceuticals are well known [3]. The existence of pyrazole units in various structures has led to the expansion of applications in various fields such as medicine, engineering and agriculture. Pyrazoles are designated as anti-tuberculosis, anticancer, antifungal, anti-inflammatory, antidepressant, antibacterial, antioxidant antiviral agents as well as inhibitors of protein glycation [4, 5]. To date, a large number of compounds containing pyrazole have been effectively commercialized, such as Celecoxib, Rimonabant, Sulfaphenazole and Penthiopyrad (Fig. 1) [6–13]. This chapter presents descriptions and discussions on the pharmacological values of pyrazole-containing heterocyclic scaffolds.

Phenylbutazone is a non-steroidal anti-inflammatory drug (NSAID) active in treating fever, pain and inflammation in the body. As a group, NSAIDs are known as non-narcotic relievers of minor to reasonable pain related to musculoskeletal conditions, menstrual cramps, injuries and arthritis (Fig. 2).

The list of FDA-approved pyrazole-containing drugs has been represented in Table 1 [15].

2 Mechanism of Action of Pyrazole NSAIDs

A series of new pyrazole analogues were prepared, and their activity was screened for NSAID activity; their results are compared with the standard drug, Ibuprofen (Fig. 3). The new compounds consist of pyrazole moiety, a sulphonamide and a linker. The strategy for designing the compound in this way is pyrazole which is known for its anti-inflammatory activity, analgesic and COX inhibition, and it is proven by the various marketed drugs like Celecoxib, Deracoxib, Metamizole, Antipyrine and Lonazolac. The importance of sulphonamide is for its selectivity towards the COX-II

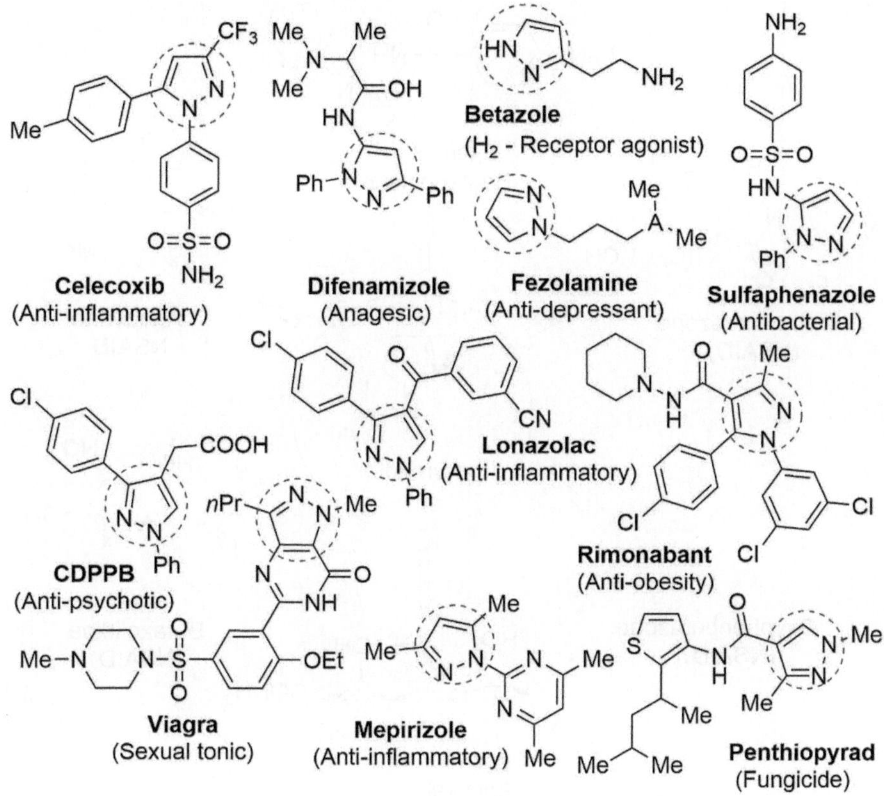

Fig. 1 Pharmaceutically important compounds containing pyrazole unit [6–13]

receptor. *N*-phenyl/phenyl sulfonamide moiety with various substitutions on it was synthesized and evaluated. Two compounds exhibited excellent anti-inflammatory activity, and their percentage of inhibition was found to be 80.87–76.56% and 80.63–78.09%, whereas Ibuprofen exhibited 81.32–79.23% after 3 and 4 h. The analgesic activity of the above two compounds was found to be 73.56 and 73.22%, whereas the analgesic activity of the Ibuprofen was found to be 74.12%. These compounds were also tested for their selective index, and they have exhibited high selectivity of 75 and 73%. This selectivity makes these compounds the least ulcerogenic. The ulcerogenic activity of these compounds was studied, and no induction of gastric ulceration or rupture of the gastric mucosal layer at a drug conc. of 60 mg/kg was found [14].

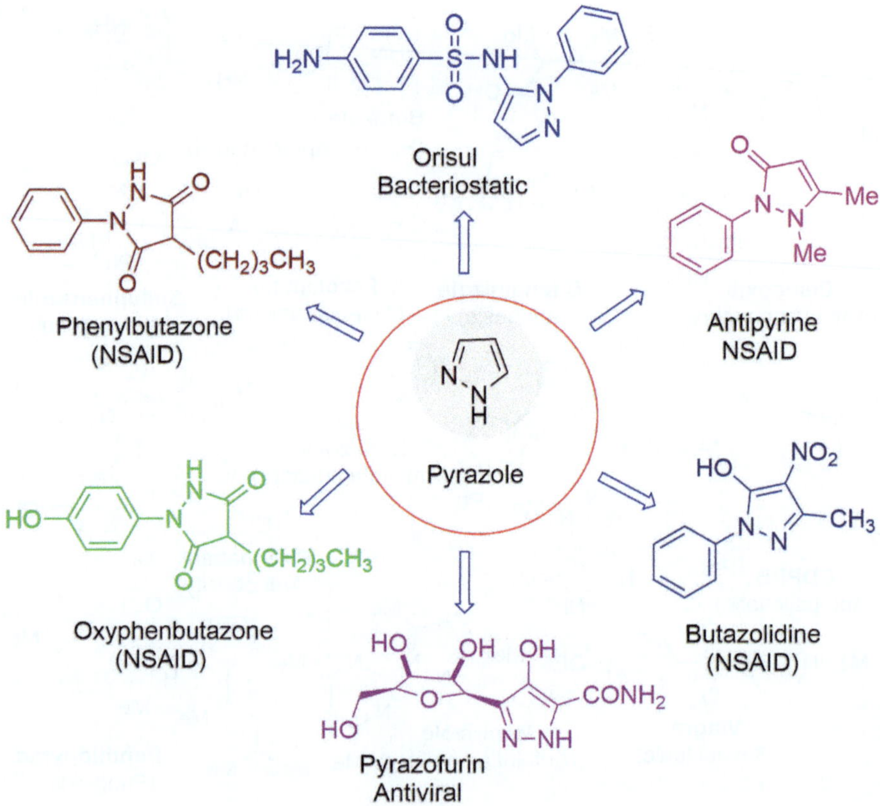

Fig. 2 Biologically important pyrazole derivatives [14]

3 Antibacterial and Antifungal Activity of Pyrazole Derivatives

Over the past 20 years, microbial diseases have been found as a main source of disease and often as a suppressor of immune power. Microbes are liable for a broad range of harmful syndromes and predominant widespread diseases in human civilizations. Microbial diseases such as typhoid, plague, diphtheria, cholera, tuberculosis and pneumonia have taken the highest position in recent pasts. Recently, some new derivatives of 1*H*-pyrazole-3-carboxylic acid were investigated for their antibacterial performances against *Staphylococcus aureus*, *Bacillus cereus*, *Pseudomonas putida and Escherichia coli*. The results indicated that the molecule **1** shows antibacterial activity against both gram-negative and gram-positive bacteria. The antibacterial activity of pyrazole derivatives embedded in a quinolinyl chalcone group was examined. Molecule **2** was found to be very effective against fungal and bacterial strains. Pyrazole-3,4-dicarboxylic acid and pyrazole-3-carboxylic acid derivatives are also distinguished for their antifungal and antibacterial activities against five fungal and

Table 1 FDA-approved pyrazole-containing drugs [15]

Structure of the drug Name of the drug (Drug Bank ID)	Category/Indication	Mechanism of action
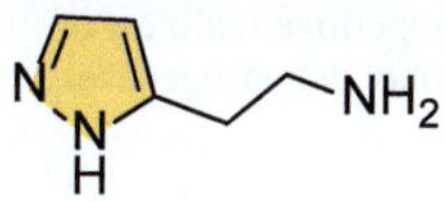 Betazole DB00272	Used clinically to test gastric secretory function	Histamine analogue, binds the H_2-receptor which is responsible for secretion of gastric acid
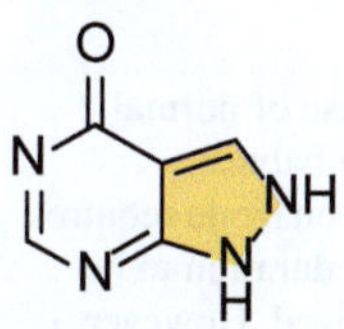 Allopurinol DB00437	Management of symptoms of secondary and primary gout, the control of patients with recurrent calcium oxalate calculi	Structural analogue of the purine bases, referred as xanthine oxidase enzyme inhibitors
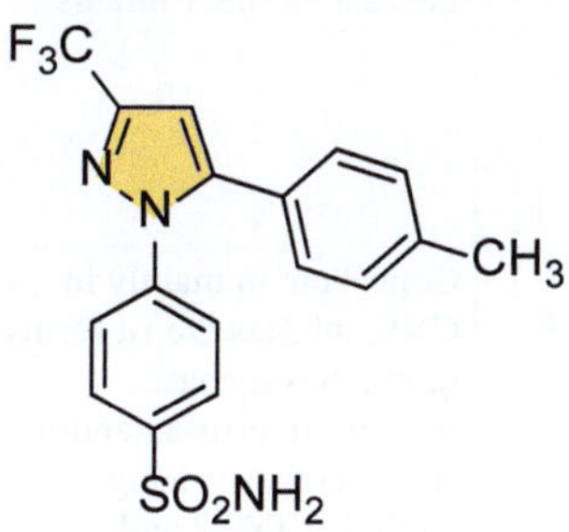 Celecoxib DB00482	Referred for the treatment of symptomatic cases of adult rheumatoid arthritis (RA) and adult osteoarthritis	cyclooxygenase-2 (COX-2) enzyme inhibitors in selective and non-competitive manner
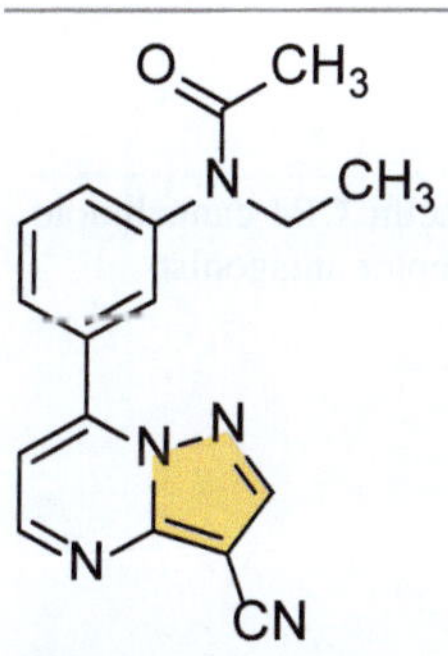 Zaleplon DB00962	Short-term treatment of insomnia in adults	Acts using $GABA_BZ$ chloride channel receptor subunit, binds effectively to the omega-1 receptor of the brain. This occurs on the GABA-A/chloride (alpha subunit) ion channel receptor complex potentiates t-butyl bicycle-phosphorothionate (TBPS)

(continued)

Table 1 (continued)

Structure of the drug Name of the drug (Drug Bank ID)	Category/Indication	Mechanism of action
 Fomepizole DB01213	Antidote for ethylene glycol or methanol poisoning	Competitive inhibitor of alcohol dehydrogenase
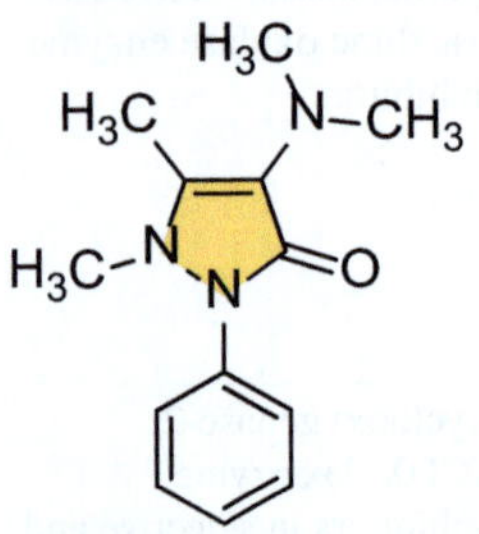 Aminophenazone DB01424	Employed as an efficient analgesic, antipyretic in common colds, neuritis and rheumatism. Also, it is important to determine total body water	In the case of normal newborn babies, aminophenazone requires a longer duration to be metabolized. However, higher quantities of exhaled $^{13}CO_2$are found in the case of older infants
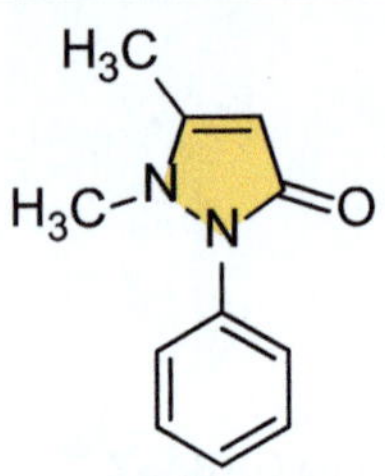 Antipyrine DB01435	To verify the consequences of the different drug molecules on the enzymes of the liver. The antipyrine is referred to as the symptomatic relaxation of acute otitis media which can be appearing from several etiologies	Contribute in mainly in CNS, inhibits the isoforms of cyclooxygenase, requires in prostaglandin (PG) synthesis using COX-1, COX-2 and COX-3 enzymes
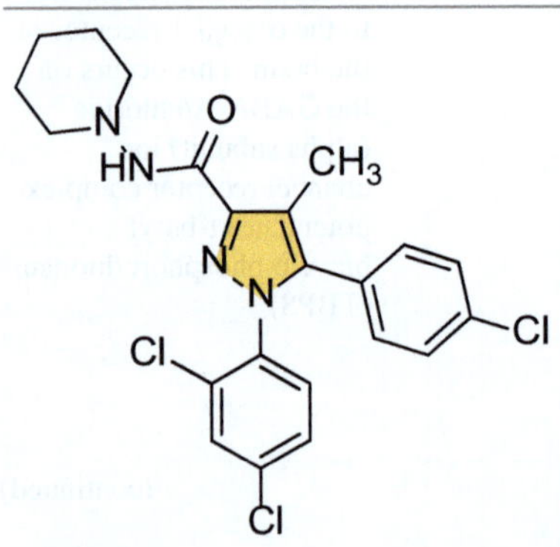 Rimonabant DB06155	Utilized in concurrence with exercise and diet in case patients have body mass index more than 30 kg/m^2, or having BMI higher than 27 kg/m^2 and affiliated risk factors, like dyslipidaemia and type 2 diabetes	Specific CB1 cannabinoid receptor antagonist

(continued)

Table 1 (continued)

Structure of the drug Name of the drug (Drug Bank ID)	Category/Indication	Mechanism of action
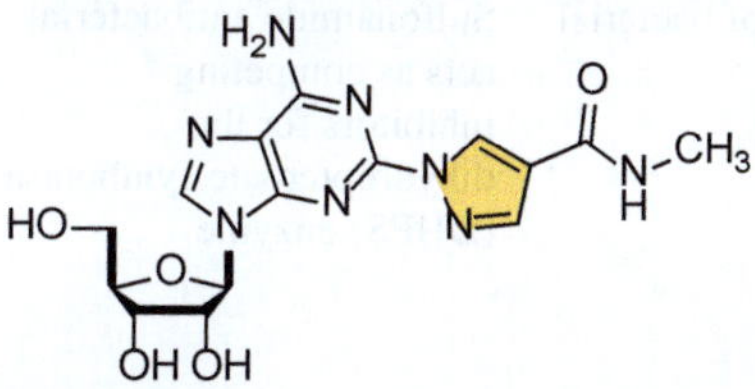 Regadenoson DB06213	Used as Radionuclide Myocardial Perfusion Imaging (MPI) Diagnostic agent	A low-affinity A2A receptor agonist is responsible for coronary vasodilatation that increases myocardial blood flow and the process occurs via mimicking the effects of adenosine
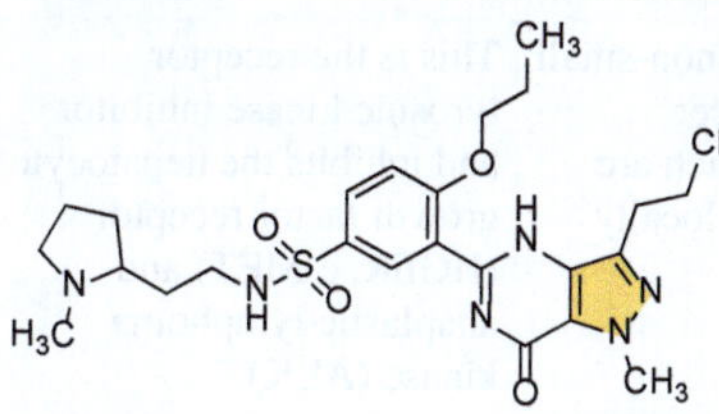 Udenafil DB06267	Used/treatment in erectile dysfunction and hypertension	Inhibits phosphodiesterase type 5 (PDE5) which is cGMP specific. This process is important for the degradation of cGMP in the corpus cavernosum which is located around the penis
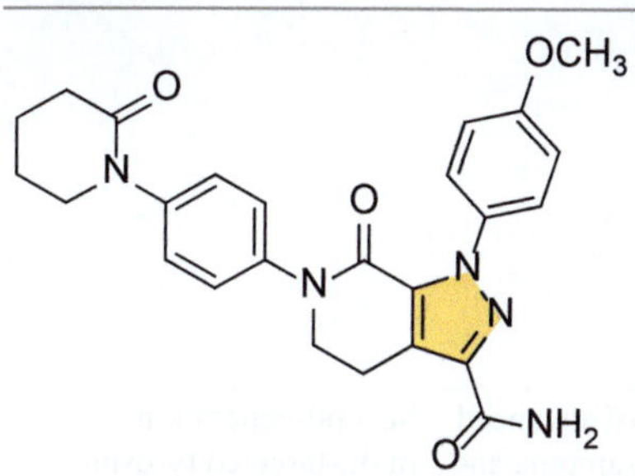 Apixaban DB06605 (DB07828)	Decrease the associated risk of systemic embolism and stroke in patients having prophylaxis of deep vein thrombosis (DVT), non-valvular atrial fibrillation. This results in pulmonary embolism (PE) after a knee or hip alteration surgery. Treatment of PE and DVT that suppresses the danger of reappearance	This selectively inhibits the prothrombinase, and factor Xa in its bound and free forms
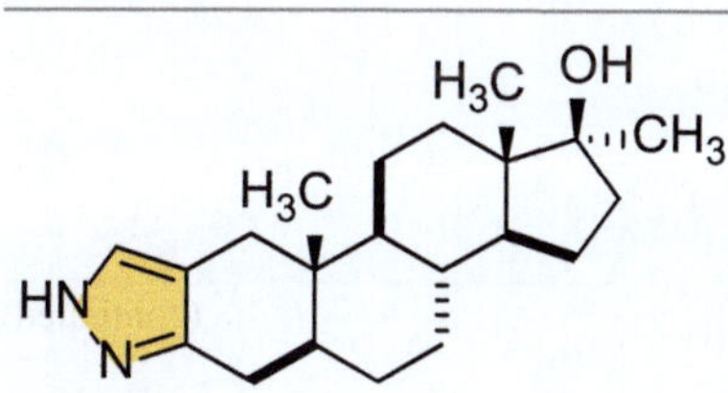 Stanozolol DB06718	Synthetic anabolic steroid treats C1-inhibitor deficient hereditary angioedema	This appends to androgen receptors, like stanozolol-binding protein (STBP) and membrane bound receptor proteins LAGS

(continued)

Table 1 (continued)

Structure of the drug Name of the drug (Drug Bank ID)	Category/Indication	Mechanism of action
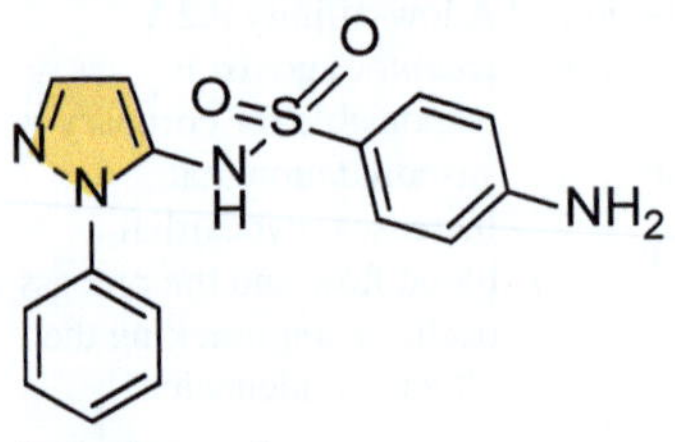 Sulfaphenazole DB06729	Treatment of bacterial infections	Sulfonamide antibacterial acts as competing inhibitors for the dihydropteroate synthetase (DHPS) enzyme
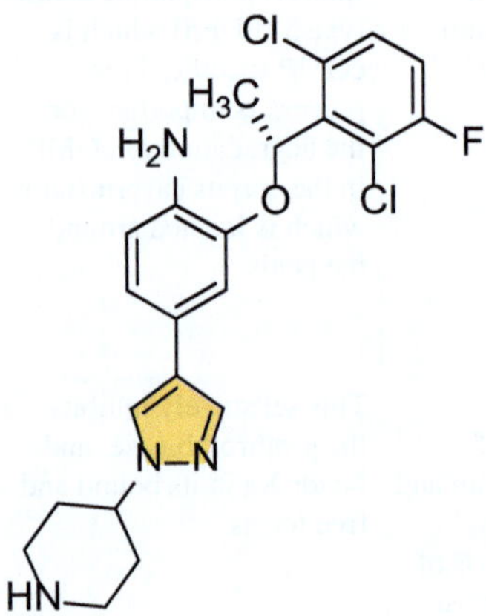 Crizotinib DB08865 (DB08700)	Treatment of non-small cell lung cancer (NSCLC) which are metastatic or locally advanced	This is the receptor tyrosine kinase inhibitor and inhibits the hepatocyte growth factor receptor (HGFR, c-MET) and anaplastic lymphoma kinase (ALK)
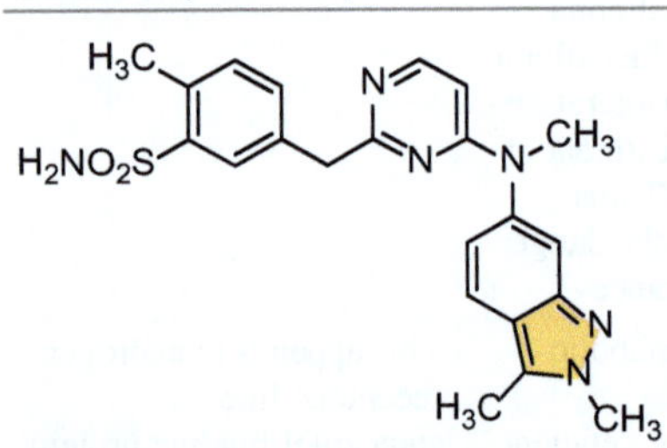 Pazopanib DB06589	Treatment of advanced soft tissue sarcoma and renal cell cancer	Second-generation multi-targeted tyrosine kinase inhibitor against PDGR-R alpha and beta, VEGFR-1, VEGFR-2 and VEGFR-3

(continued)

Table 1 (continued)

Structure of the drug Name of the drug (Drug Bank ID)	Category/Indication	Mechanism of action
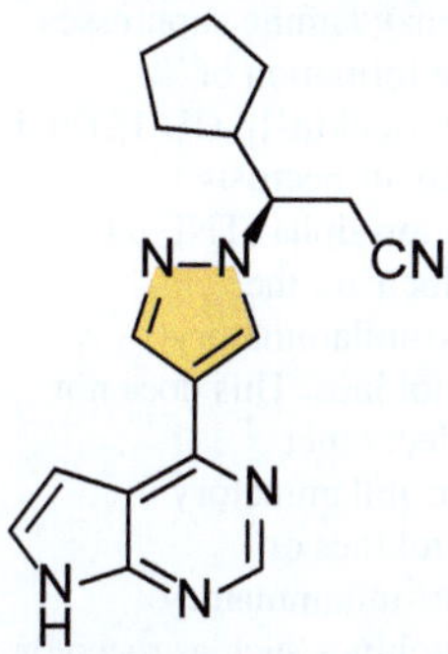 Ruxolitinib DB08877 (DB06164)	Treatment of high-risk or moderate myelofibrosis, includes post-polycythemia vera (post-PV) myelofibrosis, primary myelofibrosis and post-essential thrombocythemia (post-ET) myelofibrosis	Kinase inhibitor that is selective for the Janus Associated Kinases (JAK) 1 and 2
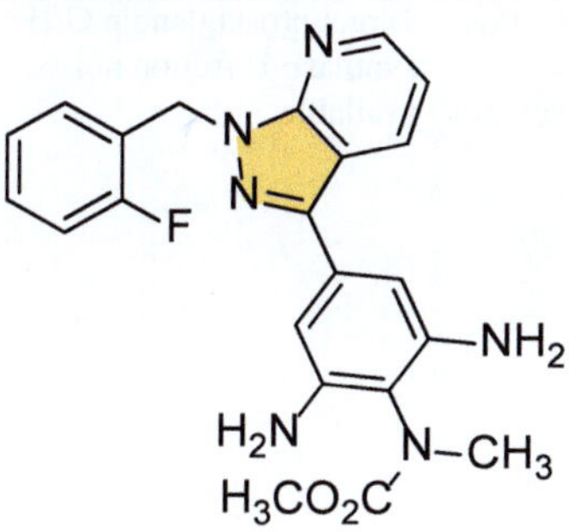 Riociguat DB08931	Referred for the post-surgical treatment of patients with recurrent/persistent chronic thromboembolic pulmonary hypertension (CTEPH)	Stimulator of an enzyme (soluble guanylate cyclase (sGC)) in the cardiopulmonary system and the nitric oxide (NO) receptor. Riociguat sensitizes sGC to intimate NO through stabilization of NO-sGC binding. Riociguat could also stimulate directly the sGC without the assistance of NO using other binding sites. Riociguat also accelerates the formation of cGMP via NO-sGC-cGMP pathways
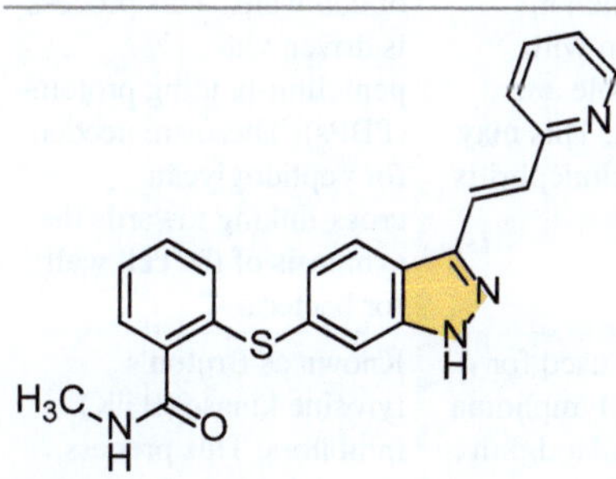 Axitinib DB06626	Treatment of cancer in kidney cells, thyroid and pancreas	Blocks the receptors of the tyrosine kinase VEGFR-1, VEGFR-2 and VEGFR-3

(continued)

Table 1 (continued)

Structure of the drug Name of the drug (Drug Bank ID)	Category/Indication	Mechanism of action
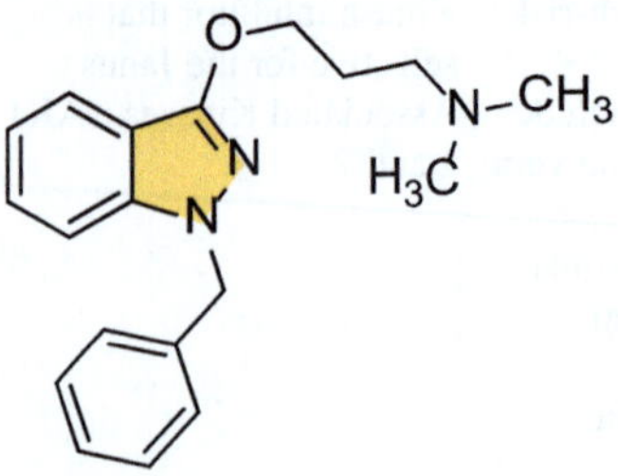 Benzydamine DB09084	Employed as an analgesic, topical cream, oromucosal spray, liquid mouthwash, anti-inflammatory	Benzydamine suppresses the formation of interleukin-1β (IL-1β) and tumour necrosis factor-alpha (TNF-α) which are the pro-inflammatory cytokines. This does not affect other pro-inflammatory cytokines or anti-inflammatory cytokines such as receptor antagonists IL-6, IL-8, IL-1 and IL-10
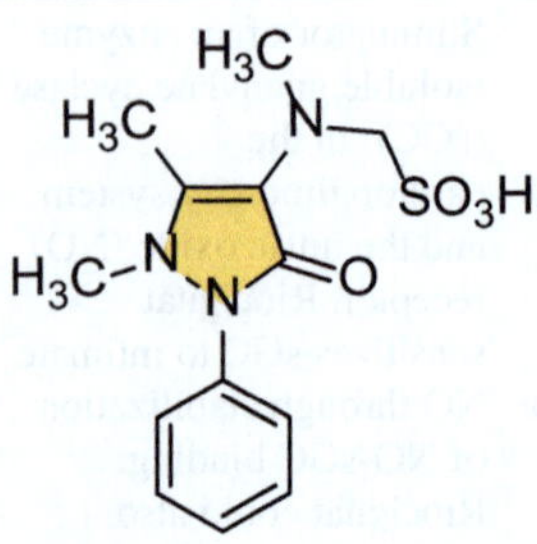 Metamizole DB04817	In earlier decades, this was employed as suppresser of fever and painkillers	Target prostaglandin G/H synthase 1. Action not available
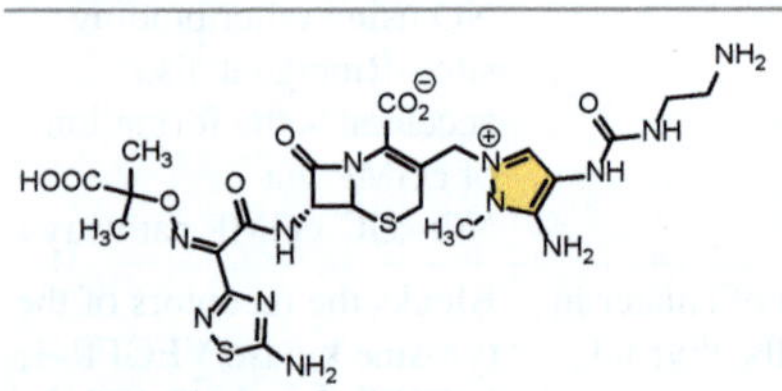 Ceftolozane DB09050	Ceftolozane is recommended for the treatment of infections in the urinary tract and intra-abdomen in combination with metronidazole and tazobactam. This may include pyelonephritis	Ceftolozane is the antibacterial drug under the cephalosporin class that inhibits the generation of cell walls. This process is driven via penicillin-binding proteins (PBPs). These are needed for peptidoglycan cross-linking towards the synthesis of the cell wall for bacteria
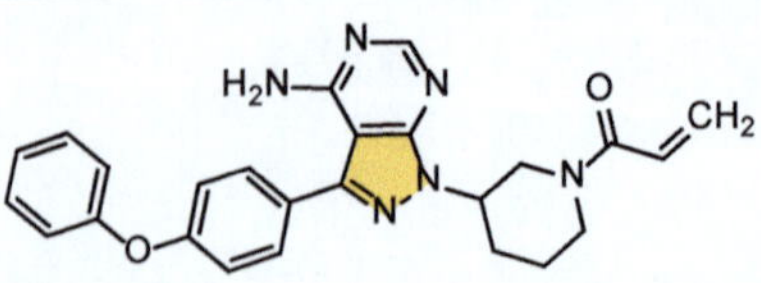 Ibrutinib DB09053	Ibrutinib is used for mantle cell lymphoma (B-cell non-Hodgkin lymphoma)	Known as Bruton's tyrosine kinase (BTK) inhibitors. This process proceeds via covalent bond generation with a cysteine residue which is present in the active site of BTK (Cys481)

(continued)

Table 1 (continued)

Structure of the drug Name of the drug (Drug Bank ID)	Category/Indication	Mechanism of action
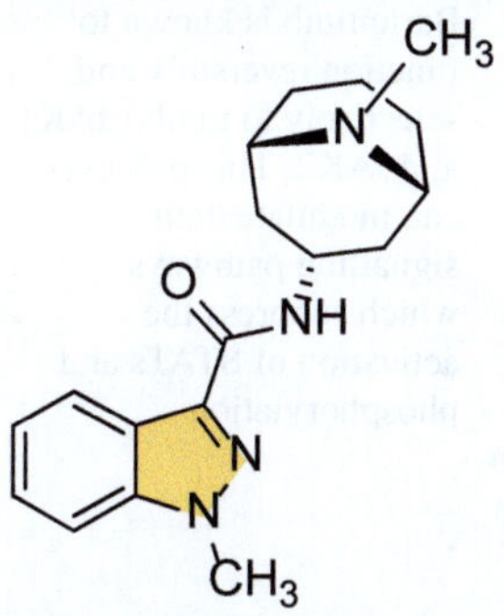 Granisetron DB00889 (APRD01002)	Recommended for vomiting and nausea related to the courses of emetogenic cancer therapy. This is also used for the same symptoms during radiation and postoperative	5-HT3receptor antagonist which functions selectively and effectively. The process for the antiemetic activity of the drug is realized via the prevention of 5-HT3 receptors present centrally and peripherally
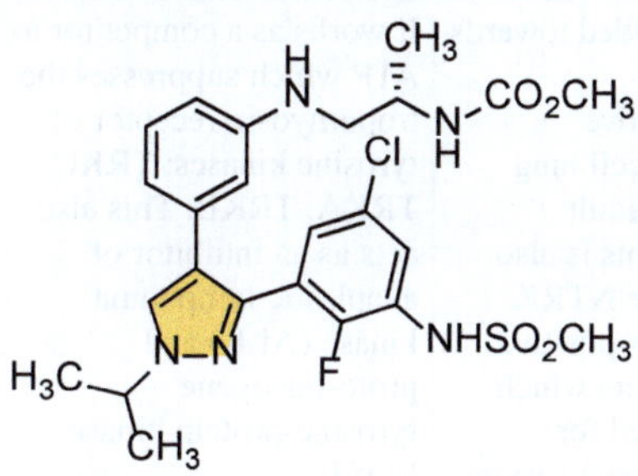 Encorafenib DB11718	Recommended for metastatic melanoma with a V600K or BRAF V600E mutation in addition with Binimetinib	BRAF V600E, wild-type BRAF and CRAF are targeted specifically by this kinase inhibitor. This target during in vitro cell-free assays with IC50 values of 0.47, 0.35 and 0.3 nM, respectively
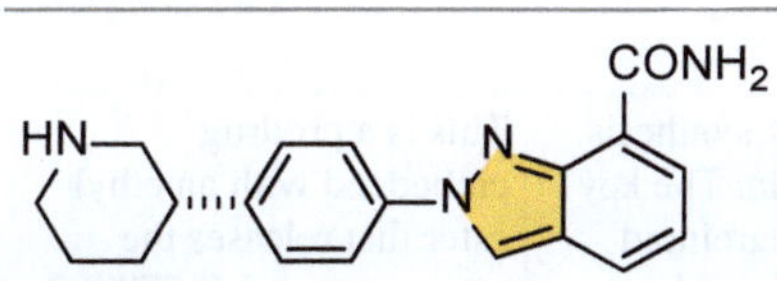 Niraparib DB11793	Recommended for the adult patients respond with platinum-based chemotherapy and suffering from the fallopian tube, recurrent epithelial ovarian, or primary peritoneal cancer	The enzymes poly (ADP-ribose) polymerase (PARP) enzymes, PARP-2 and PARP-1, which assist in DNA repair are inhibited by Niraparib

(continued)

Table 1 (continued)

Structure of the drug Name of the drug (Drug Bank ID)	Category/Indication	Mechanism of action
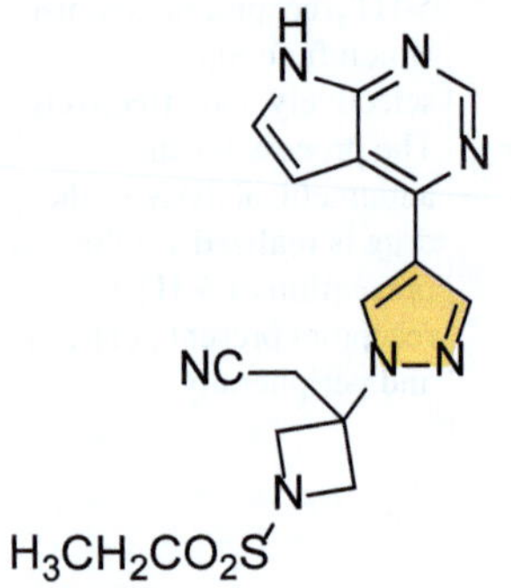 Baricitinib DB11817	Used for adult patients towards the treatment of active rheumatoid arthritis. It is also recommended for intolerant patients to one or more disease-modifying anti-rheumatic. This can function in addition with methotrexate or as monotherapy	Baricitinib is known to function reversibly and selectively to inhibit JAK1 and JAK2. This process can modulate their signalling pathways, which suppress the activation of STATs and phosphorylation
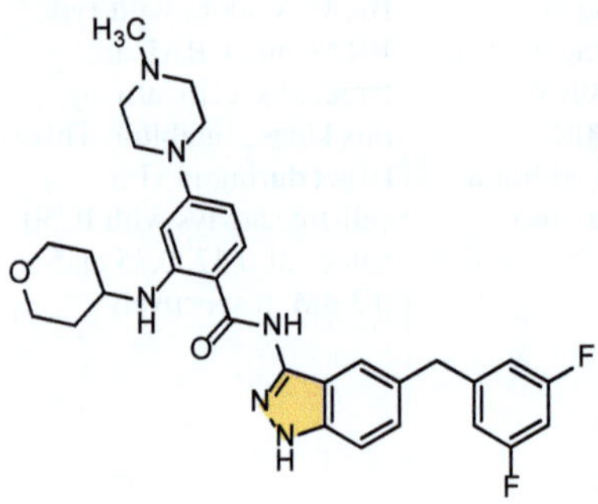 Entrectinib DB11986	Recommended towards metastatic ROS1-positive non-small cell lung cancer for adult patients. This is also used for the NTRK gene fusion-positive solid tumours which are indicated for children over 12 years old and adults	It works as a competitor to ATP which suppresses the tropomyosin receptor of tyrosine kinases: TRKC, TRKA, TRKB. This also acts as an inhibitor of anaplastic lymphoma kinase (ALK) and proto-oncogene tyrosine-protein kinase ROS1
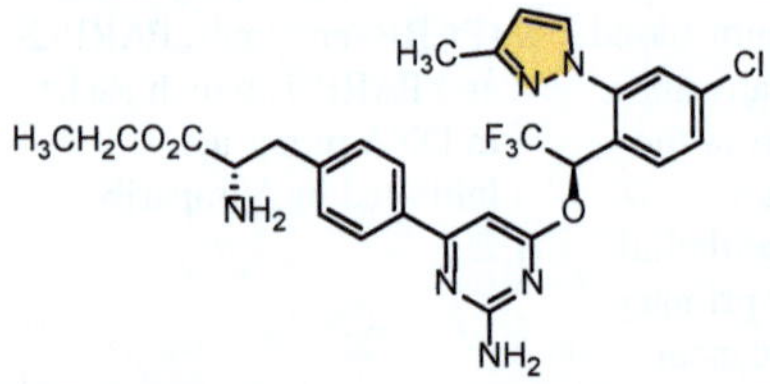 Telotristatethyl DB12095	Inhibitor of synthesis of Serotonin. The key factors of carcinoid syndrome have been taken care of and this is efficient for reducing serotonin levels	This is a prodrug embedded with an ethyl ester that releases the active moiety LP-778902 via hydrolysis both in vivo and in vitro

(continued)

Table 1 (continued)

Structure of the drug Name of the drug (Drug Bank ID)	Category/Indication	Mechanism of action
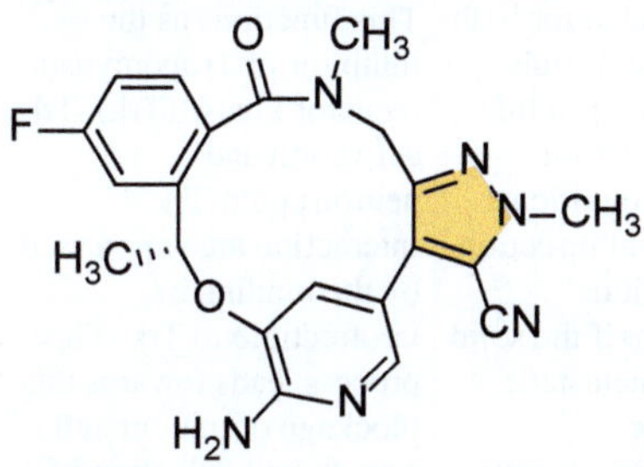 Lorlatinib DB12130	Used for the treatment of anaplastic lymphoma kinase (ALK)-positive NSCLC and ROS1-positive NSCLC	Acts as an inhibitor of ALK tyrosine kinase for the patients having ALK-positive metastatic NSCLC
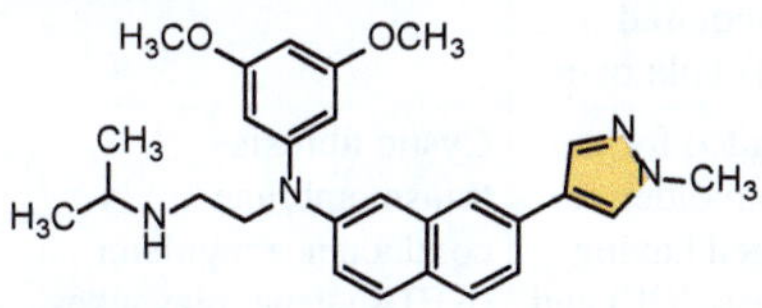 Erdafitinib DB12147	The tyrosine kinase inhibitor of FGFR	Recommended for the metastatic or locally advanced urothelial carcinoma having vulnerable FGFR2 or FGFR3 genetic alterations, which advances following or during platinum-containing chemotherapy
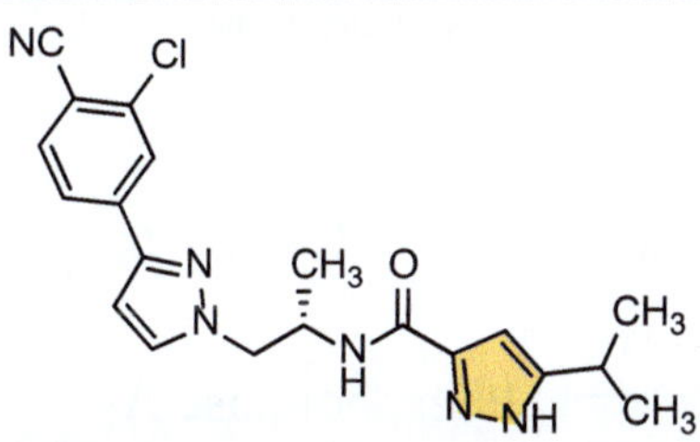 ODM-201 DB12941	Investigated in trials towards understanding the effects in female Breast cancer, Prostate Cancer, Drug Interaction, Castration-Resistant and Pharmacokinetics	No information available
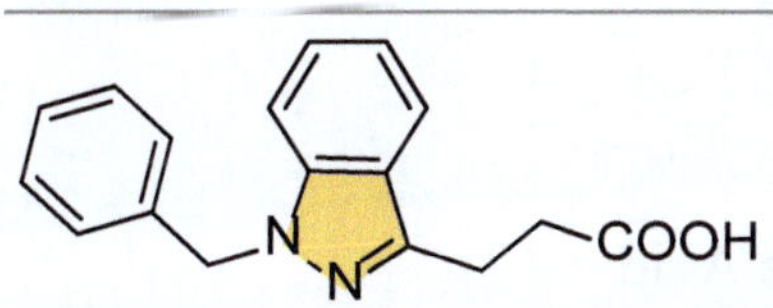 Bendazac DB13501	Used for medical conditions such as inflammation, local pain, eczema, dermatitis, hives, pruritus, burns, insect bites, erythema and several other symptoms	This suppresses the lens proteins glycosylation by sugars such as glucose-6-phosphate and galactose which dependents upon specific doses

(continued)

Table 1 (continued)

Structure of the drug Name of the drug (Drug Bank ID)	Category/Indication	Mechanism of action
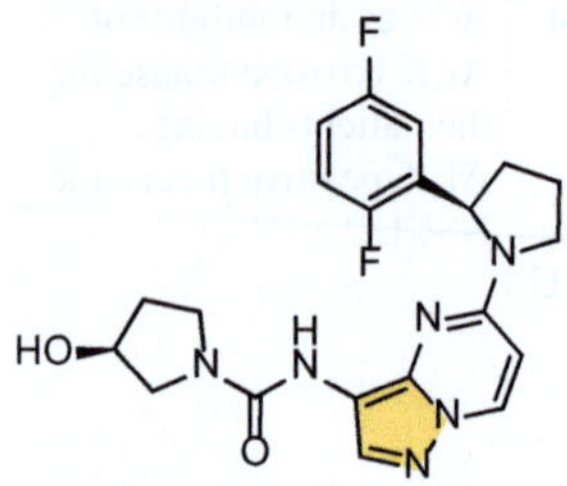 Larotrectinib DB14723	Recommended for both paediatric and adult patients having solid tumours under the following conditions: (a) if surgical resection leads to serious morbidity or if the solid tumour is metastatic, (b) in case of neurotrophic receptor tyrosine kinase (NTRK) gene fusion and if the acquired resistance is unknown	This functions as the inhibitor of Tropomyosin receptor kinase (Trk). Trk activation and neurotrophin-Trk interaction are suppressed by the binding of larotrectinib to Trk. This process leads towards the blockage of tumour cell growth and induction of cellular apoptosis
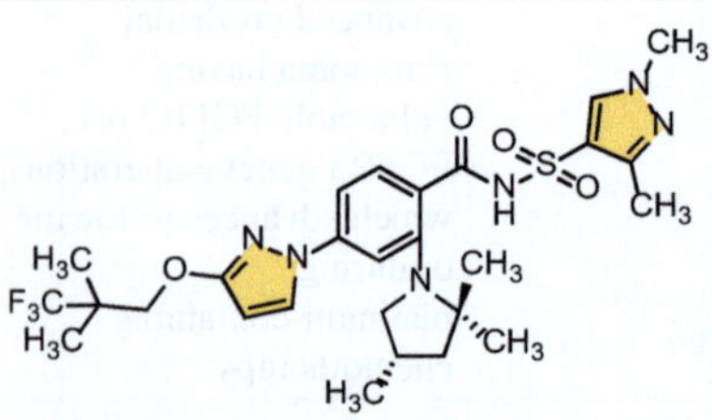 Elexacaftor DB15444	Recommended for the patients both older or 12 years aged having cystic fibrosis (CF) and in the CFTR gene has at least one *F508del* mutation	Cystic fibrosis transmembrane conductance regulator (CFTR) gene, play a key role with CFTR proteins which promotes the trafficking of cell surface for incorporation into the cell membrane

Fig. 3 Mechanism of action of pyrazole NSAIDs [14]

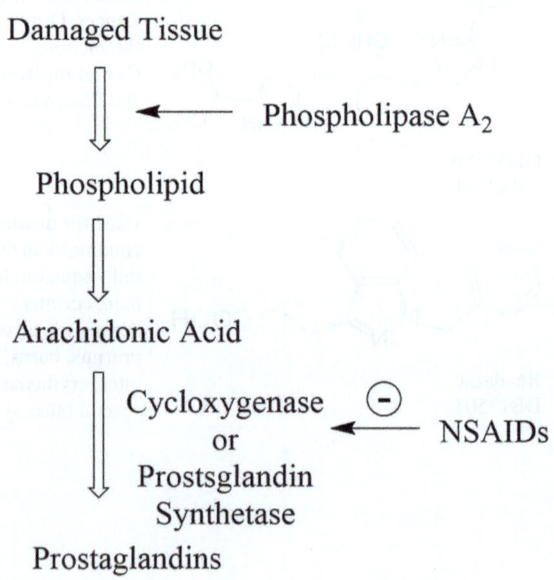

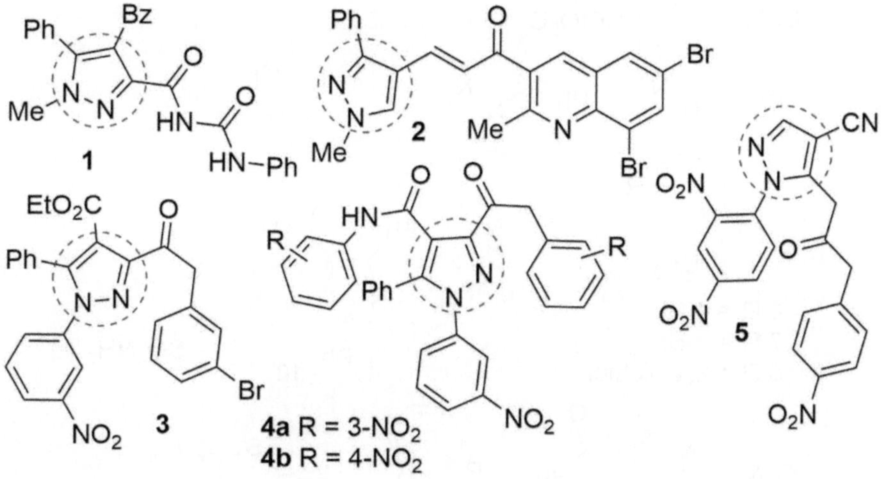

Fig. 4 Structures of some antimicrobial compounds of pyrazole derivatives [16–18]

five bacterial pathogens. Compounds **3**, **4a** and **4b** were identified by some inhibitory effects on the strains of *C. tropicalis*, *C. parapsilosis* and *C. glabrata*. Additionally, 5-amido-1-(2,4-dinitrophenyl)- 1*H*-pyrazole-4-carbonitriles **5** also shows potent antimicrobial activities (Fig. 4) [16–18].

The antibacterial properties of several pyrazole molecules against *Bacillus subtilis*, *Escherichia coli*, *Staphylococcus aureus* and *Pseudomonas aeruginosa* were tested. Of the verified compounds, molecules **6**, **7**, **8** and **9** (Fig. 5) were shown to have potent antibacterial activity against all bacterial strains studied, and the data were compared with the standard drug ceftriaxone [19].

Pyrazolylpyrazolines **10** was identified for its antimicrobial activity against two gram-negative bacteria and two gram-positive bacteria. Compounds **11** and **12** exhibited good antibacterial performances. Compound **13** was realized as an efficient molecule against *E. coli*, while molecule **14** was understood as outstanding against *S. pyogenes and S. aureus* and has very good activity against *C. albicans*.

1,3,4,5 tetra-substituted pyrazole molecules were distinguished by their antimicrobial activity against *E. coli*, *S. aureus*, *C. albicans* and *Aspergillus flavus*. The molecule **15** exhibited very good antifungal and antibacterial activity. The molecule bound with sulfone derivative **16** showed promising antimicrobial activity. Antimicrobial activity against *S. aureus*, *E. coli*, *P. aeruginosa* and *K. pneumonia* was investigated, as well as antifungal activity against *A. fumigatus*, *A. flavus*, *T. mentagrophytes* and *P. marneffei*, using derivatives of the 1,5-diaryl pyrazole molecule. The derivative **17** showed promising antifungal and antibacterial activity. 1,3-Diaryl pyrazole molecules were examined for their antibacterial action contrary to *B. subtilis*, *S. aureus*, *P. aeruginosa*, *E. coli*, and their antifungal movement in contrast to *A.*

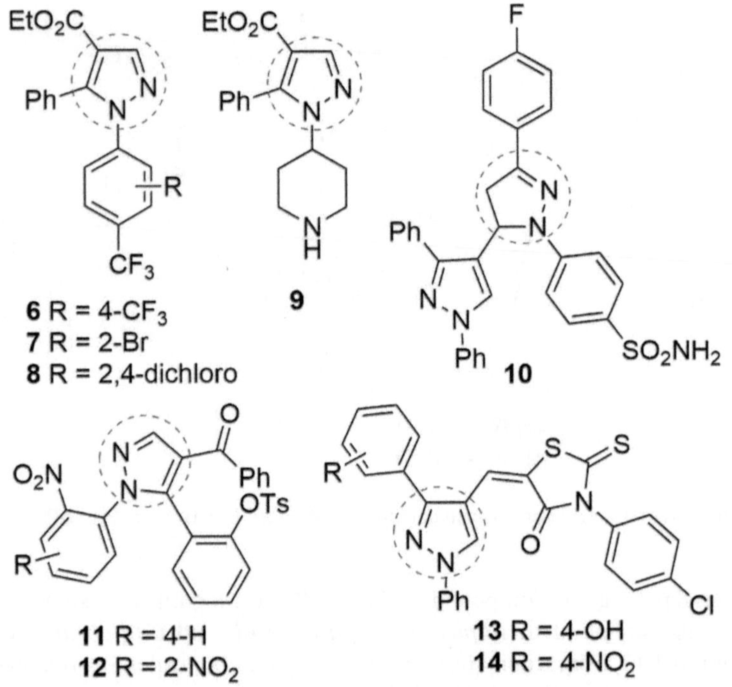

Fig. 5 Structures of some antibacterial compounds of pyrazole derivatives [19–22]

flavus and *A. niger*. Molecule **18** showed moderate antifungal and antibacterial activities against the fungi and bacteria studied. The antifungal activity of several 3-(4-chlorophenyl)-4-substituted pyrazoles was screened against the antibacterial activity and the pathogenic fungal strain for gram-negative and gram-positive organisms. Among the investigated derivatives, compound **19** exhibited outstanding activities against all the investigated pathogenic fungi and bacteria (Fig. 6).

A range of imidazole derivatives embedded with functionalized pyrazole moiety was examined for antibacterial and antifungal activities. Among the prepared compounds, molecule **20** was found to be an effective antimicrobial agent. The antifungal activity of a large number of acylthiourea pyrazole derivatives was verified against *F. oxysporum*, *G. zeae* and *C. mandshurica*. Molecule **21** showed good antifungal activities against the fungi examined. In addition, the *N*- (substituted-pyridyl)-1-methyl(phenyl)-3-trifluoromethyl-1*H*-pyrazole-4 carboxamide molecule targets three types of plant pathogenic fungi (*Fusarium oxysporum*, *G. zeae* and *C. mandshurica*) were tested. The results obtained showed that molecule **22** had more than 73% inhibitory activity against *G. Zeae*.

The antimicrobial compound **23** was discovered to be effective against methicillin-resistant Staphylococcus aureus (MRSA). The derivative **24** revealed efficient antibacterial activity against *E. coli, S. aureus* and Pyrido[1,2-*a*]benzimidazole

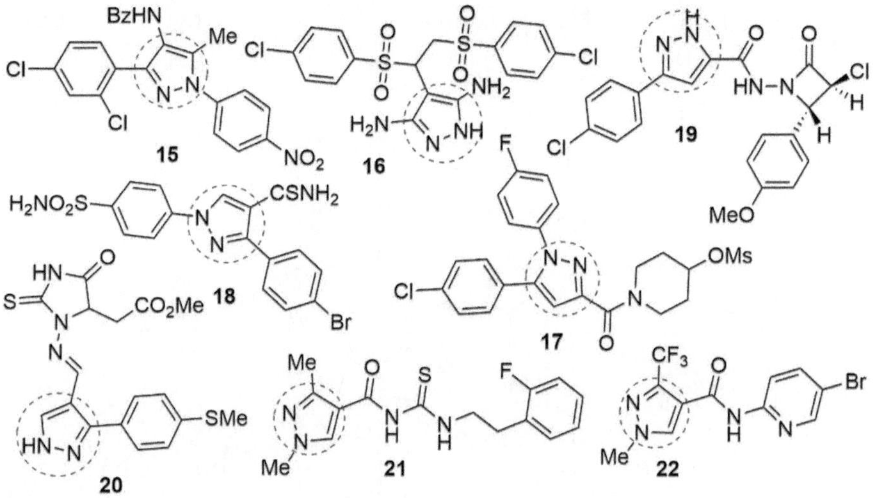

Fig. 6 Antimicrobial activity of pyrazole derivatives [23–30]

derivatives containing the aryloxy pyrazole nucleus have been examined for antimicrobial activity. The molecule **25** was identified as active against realized pathogens. In addition, new Schiff bases embedded pyrazole rings were examined for their antibacterial (*B. subtilis, S. aureus, P. aeruginosa and E. coli*) activity. The outcome furnished that, the derivative **26** exhibited excellent antibacterial activity against the employed microorganisms. A broad range of formyl-pyrazoles molecules were examined for their antifungal and antibacterial activities. The molecule **27** showed interesting antibacterial and antifungal activity. The derivative **28** revealed satisfactory inhibitory activity against examined quinolone-resistant and methicillin-resistant *S. aureus* (QRSA, MRSA). The molecule **29** exhibited generous antimicrobial action in contrast to the verified microorganisms. Nevertheless, the antibacterial activity of the new 5-imidazole-pyrazole derivatives was investigated against a number of pathogenic fungal and bacterial strains. Derivative **30** showed excellent antimicrobial activity when compared to first-line drugs (Fig. 7).

Antimicrobial activity of pyrimidine pyrazole derivatives was assessed against fungi and bacteria to notify compound **31** to be most active against *B. cereus* and *S. aureus*. The fungicidal activities of pyrazole derivatives were testified against *R. solani, T. cucumeris, F. oxysporum, F. graminearum* and *B. cinerea*. The outcomes designated that compound **32** had superior action counter to *T. cucumeris, F. oxysporum, B. cinerea, F. graminearum* and *R. solani*. In addition, 2,5-disubstituted-1,3,4-oxadiazoles embedded with pyrazole motifs were produced and estimated for their antibacterial movement in contrast to *P. aeruginosa, S. aureus* and *E. coli* along with antifungal movement contrary to *A. flavus, C. albicans* and *C. keratinophilum* and the compound **33** showed excellent activity. Compound **34** was recognized for its antimicrobial and antifungal activities among the quinazolin-4(3*H*)-one derivatives which contain 1,3-diphenyl-1H-pyrazol-4-yl nuclei. In addition, compound **35** showed good

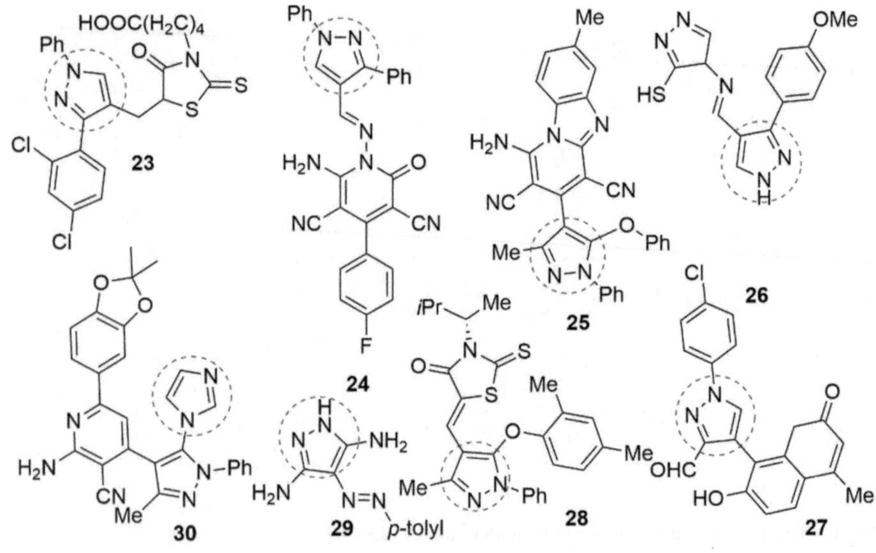

Fig. 7 Antimicrobial activity of pyrazole derivatives [31–38]

to moderate activity against bacteria and fungi among the investigated pyrazole derivatives having 1,3,4-oxadiazoles motifs. Upon verifying antibacterial and antifungal activities of pyrazole derivatives linked thiazole and imidazoles, compounds **36** and **37** showed excellent activities. Additionally, for 3-(difluoromethyl)-1-methyl-1*H*-pyrazol-4- carboxylic acid amides activities against seven phytopathogenic fungi were observed. *N*-(2-(5-bromo-1*H*-indazol-1-yl)-phenyl)-3-(difluoromethyl)-1-methyl-1*H*-pyrazol-4-carboxamide **38** showed a superior activity against the seven phytopathogenic fungi compared to the phytopathogenic fungi boscalid (Fig. 8).

The antifungal and antibacterial activity of a new 2-chloroquinoline-based pyrazole derivative was evaluated. Among those, compound **39** showed moderate activity counter to *A. fumigatus*, *P. notatum*, *B. subtilis* and *E. coli* and compound **40** had

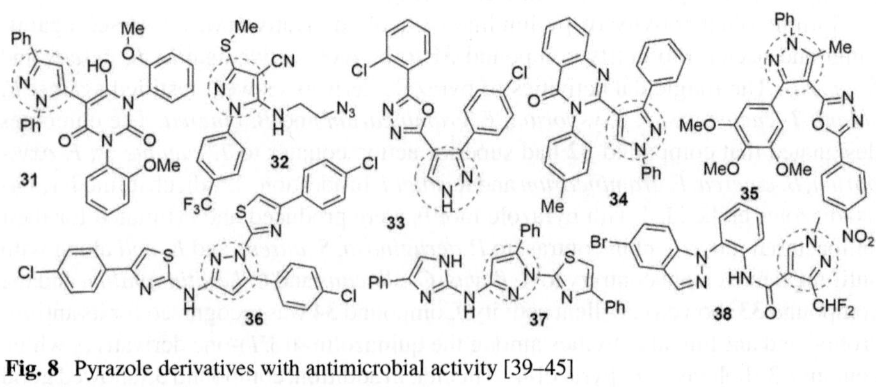

Fig. 8 Pyrazole derivatives with antimicrobial activity [39–45]

extreme potency towards *Fusarium oxysporum f. sp albedinis FAO*. Some pyrazole derivatives screened for antimicrobial action and compound **41** displayed maximum potency in contrast to tested organisms. Isoxazolol pyrazole carboxylate derivatives have been investigated against phytopathogenic fungi such as *A. porri*, *M. coronaria*, *C. petroselini* and *R. solani* and compound **42** displayed noteworthy antifungal action counter to *R. solani*. Moreover, the pyrazole ring with diterpene derivatives was studied in contrast to the *S. aureus* Newman strain and the multi-resistant strains (NRS-1, NRS-70, NRS-100, NRS-108 and NRS-271), and among the analogues evaluated, compound **43** showed good potential compared to five multi-resistant *S. aureus*. The antimicrobial activity of a novel sequence of pyrazole-thiobarbituric acid analogues was examined, and compound **44** was most active against *C. albicans*, *S. aureus*, *B. subtilis* and *E. faecalis*. In addition, an *N*-triazole scaffold containing pyrazole-5-carboxylate derivatives was investigated for their antimicrobial action contrary to three fungi as well three gram-positive and gram-negative bacteria. As a result, compound **45** exhibited significant potential against all bacterial strains as well displayed potential antifungal actions counter to *A. niger* and *C. albicans*. For antibacterial and antifungal activities, pyrazole derivatives including a thiophene moiety were screened and compound **46** showcased potent antibacterial action against *Pseudomonas aeruginosa* as well inhibitory action against *Escherichia coli* (Fig. 9).

A number of distinct pyrazolamide derivatives have been characterized as having antifungal action in contrast to *Pythium ultimum Trow*, *Phytophthora infestans (Mont.) De Bary*, *Corynespora cassiicola*, *Botrytis cinerea* and *Rhizoctonia solani*. The consequences of antifungal activity showed that compound **47** had sensible efficiency (77.78%) in distinction to *P. ultimum*. In addition, the hybrids of coumarin pyrazole were assessed for antimicrobial action, and among those, compound **48** exhibited potent antimicrobial activity against several strains of bacteria and fungi. In another arrangement of pyrazole derivatives, N,N,N′,N′-tetradentate pyrazole analogues, evaluated for their antimicrobial action and remarkably the compound **49** showed potent antifungal activity contrary to budding yeast (*Saccharomyces cerevisiae*). In addition, quinoline-based pyrazole derivatives were screened for their antifungal and antibacterial activities and the compound **50** exhibited good

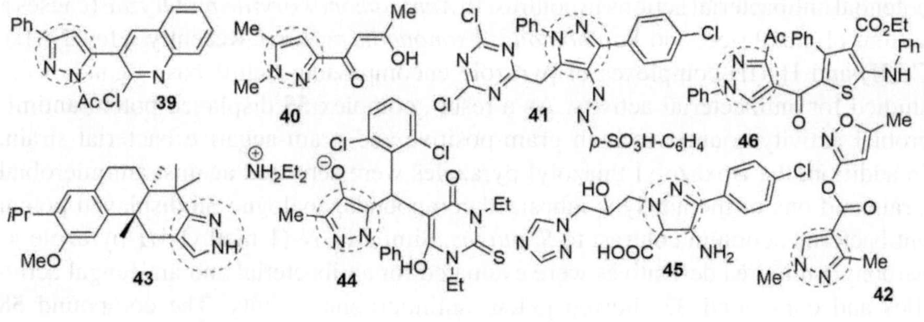

Fig. 9 Pyrazole analogues with antimicrobial activity [46–53]

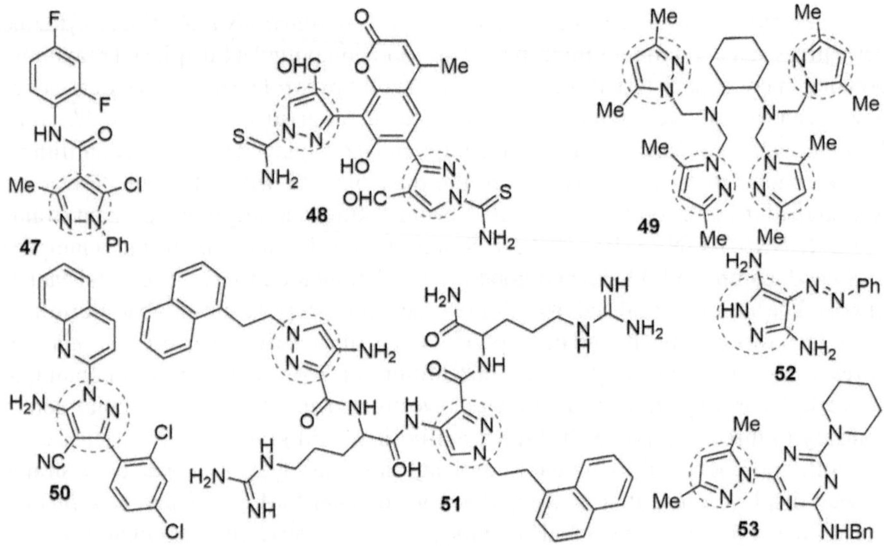

Fig. 10 Pyrazole analogues with antifungal activity [54–60]

potential in contrast to the *Shigella flexneri*, *Aspergillus clavatus*, *Candida albicans*, *Proteus. vulgaris*, *Staphylococcus. epidermidis* and *Aspergillus. fumigatus* (human pathogens strains). Pyrazole-based amino acids and peptidomimetics were also tested for antimicrobial activities, and compound **51** showed a good potency counter to *E. coli*, *P. aeruginosa*, *S. epidermidis* and *S. aureus*. Subsequently, compound **52** showed potential only at 0.075 mg/ml in contrast to the evaluated microorganisms. The pyrazole-containing s-triazine derivatives were screened against antimicrobial and antifungal strain, counter to the growth of several microorganisms, and the compound **53** showed antibacterial action towards bacterial strains: *M. luteus, P. aeruginosa,* and methicillin-resistant *S. aureus* (Fig. 10).

In addition, pyridinium-tailored 5-trifluoromethyl-pyrazoles bearing 1,3,4-oxadiazole scaffolds were investigated against 3 varieties of pathogenic bacteria and 6 fungal strains for antimicrobial activities. Among those, compound **54** exhibited potential antibacterial actions in contrast to *Xanthomonas oryzae pv. Oryzae* (causes a serious blight of rice) and *Xanthomonas axonopodis pv. Citri*. Recently, a few Ni(II), Cd(II) and Hg(II) complexes of pyrazole encompassing Schiff base ligands were studied for antibacterial activity. As a result, complex **55** displayed potent antimicrobial activity counter to both gram-positive and gram-negative bacterial strain. In addition, the isoxazolyl thiazolyl pyrazoles were screened against antimicrobial strain and one of the newly synthesized compounds, analogue **56** displayed potent antibacterial action in contrast to *S. aureus*. Similarly, *N*-(1-methyl-1*H*-pyrazole-4-carbonyl)-thiourea derivatives were evaluated for antibacterial and antifungal activities and compound **57** showed potent antimicrobial activity. The compound **58** exhibited good potential against bacterial strain and selective inhibition counter to

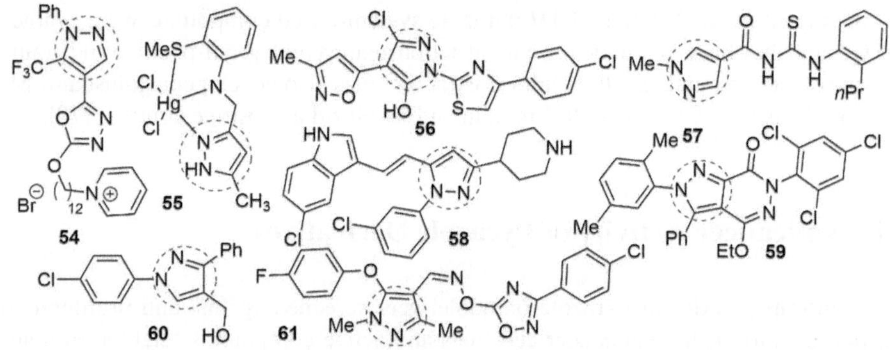

Fig. 11 Pyrazole analogues with antimicrobial activity [61–68]

bacterial topoisomerases (MRSA, PRSA, and VRE). In addition, 4-Acyl-pyrazole-3-carboxylic acids were screened for antibacterial activity and compound **59** showed potent antibacterial action in contrast to *B. subtilis, S. aureus, E. coli, P. aeruginosa* and *K. pneumonia.* Moreover, the antibacterial activity of pyrazolyl alcohols was screened and the compound **60** showed good potential of antibacterial action contrary to *Micrococcus luteus.* However, for fungicidal activities; pyrazole oxime derivatives were screened towards *Pseudoperonospora cubensis* and among tested compounds, analogue **61** exhibited extreme potential of fungicidal action contrary to *P. cubensis* in comparison with the control pyraclostrobin (Fig. 11).

Pyrazole dimedone novel molecules were prepared, and they were screened for antibacterial and antifungal activities against *S. aureus* ATCC 29,213, *E. faecalis* ATCC29212, *B. subtilis* ATCC 10,400 and *C. albicans* ATCC 2091 using agar Cup plate method. For antibacterial and antifungal activity, pyrazoles and dimedones are well known. So, these two compounds were incorporated as a single molecule in order to gain the synergistic effect. A series of these compounds were synthesized, and their activity was evaluated.

One compound exhibited the wonderful activity with MIC = 8 μg/L. against *B. subtilis* and one compound exhibited excellent activity counter to *C. albicans* with MIC = 4 μg/L. one among the compound has shown finest movement in contrast to *B. subtilis* and *E. faecalis* with MIC = 16 μg/L. and also the alternative two compounds exhibited the most effective activity against S. aureus with the MIC = 16 μg/L.

The newly synthesized compounds were evaluated for their biological activity computationally in terms of docking score. The compound which has exhibited the more potent action as antibacterial among these series of the compounds has displayed the docking score of -6.86 kcal/mol which is comparable with the docking score of Ciprofloxacin, the standard drug with the score of -6.9 kcal/mole [69].

Quinazoline-1,33-diphenyl-1*H*-pyrazole system-based compounds were synthesized, and they have exhibited prominent action against both gram-positive and gram-negative bacteria and excellent inhibitory action was also developed against anti-M. tuberculosis with 98% using Rifampicin and Isoniazid as positive controls [70].

4 Anticancer Activity of Pyrazole Derivatives

An innovative series of pyrazole-oxindole were screened against anti-proliferative action on various human cancer cell lines and active compounds established noteworthy cytotoxicity and inhibition of tubulin. The compound **62** as hPKM2 activator works as the most active anticancer agent contrary to A549 cell lines (lung carcinoma) and human non-small cell lung carcinoma (NCI-H1299) cell lines. Pyrazole derivatives were noted for their anti-proliferative activity, as compounds **63** and **64** exhibit potent cytostatic properties. In addition, the inhibitory effect of compound **65** pyrazolo[1,5-a]-pyrazin-4-(5*H*)-ones on the growth of A549 and H322 cancer cells was tested and showed good activity against A549 and H322 cell lines. In another study, pyrazole derivatives were found to have anticancer action in A2780 human ovarian adenocarcinoma cells, A549 human lung carcinoma cells and P388 mouse leukaemia cells, with compound **66** showing the most anti-proliferative activity. Additionally, pyrazole compounds were evaluated towards antineoplastic activity on a panel of 60 cellular lines, and as a result, compound **67** exhibited noteworthy growth inhibition on the evaluated cancer cells. Moreover, derivatives of 3,5-diarylpyrazole were tested counter to 5 cancer cell lines (lung cancer, breast cancer, prostate cancer, promyelocytic leukaemia, colon cancer) and analogue **68** showed excellent anticancer activity in contrast to all selected cell lines (Fig. 12).

A series of pyrazoles derivatives were investigated for their effects on the growth of A549 cell, and as a result, the compounds **69–77** consumed the potent growth inhibition as well encouraged apoptosis of A549 lung cancer cells (Fig. 13).

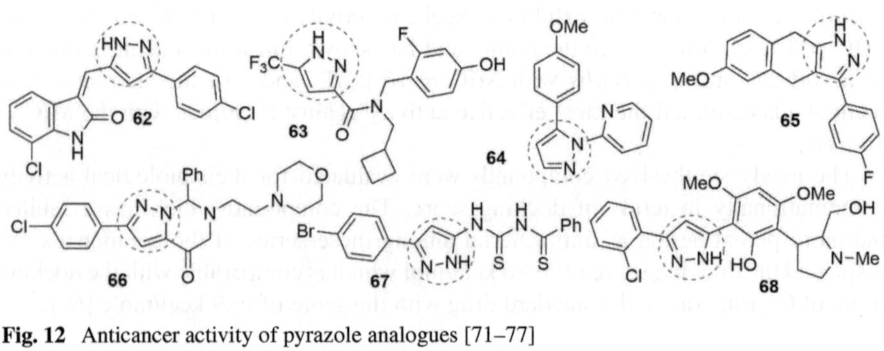

Fig. 12 Anticancer activity of pyrazole analogues [71–77]

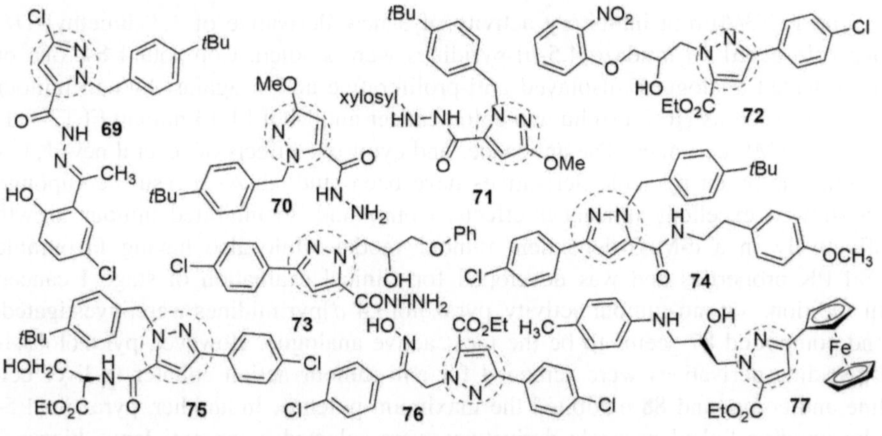

Fig. 13 Pyrazole analogues with anticancer activity [78–87]

Compound **78** exhibited the extremely good anticancer activity counter to Ehrlich ascites carcinoma tumour cells. In continuation, numerous [1,2,4]-triazolo[1,5-*a*]pyridine-based pyrazole scaffolds examined for their ALK5 inhibitory activity and compound **79** exhibited maximal anticancer activity. In addition, compound **80** shows excellent enzymatic activity (B-Raf inhibitor) and various residues of 3-(1*H*-indole-3-yl)-1*H*-pyrazole-5-carbohydrazide were tested for cytotoxicity against A549, HepG2, BGC823 and BT474 cell lines. Similarly, the compound **81** showed good anti-proliferative action. In addition, phenothiazine contains scaffold pyrazole, a new class of protein farnesyltransferase inhibitors; examined for their antineoplastic effect in a panel of the cancer cell line NCI-60. Indenopyrazole **82** showed potential cytostatic action with inhibitory effect on the growth of HCT-116, LOX IMVI and SK-MEL-5 cell lines. Moreover, compared with the HCT-116 cell line (human colon cancer), the anti-proliferative effect of 1,4,5,6-tetrahydropyrrolo[3,4-*c*]pyrazole analogues was mainly tested. Compound **83** also showed significant anticancer activity (Fig. 14).

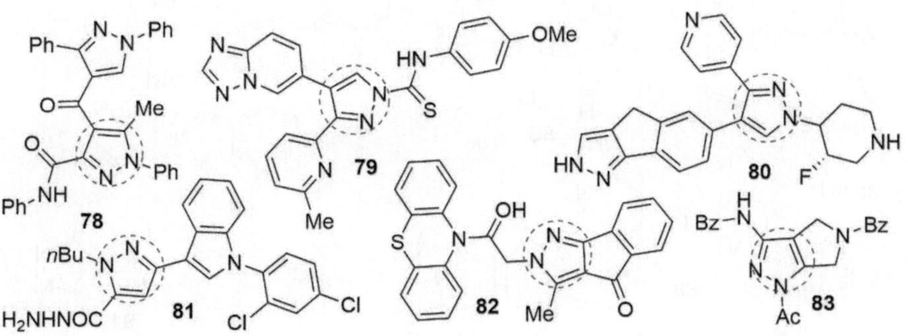

Fig. 14 Anticancer activity of pyrazole derivatives [88–93]

The FLT3/Aurora inhibitory activity of a new derivative of 1,3-dimethyl-1*H*-pyrazole based on imidazo[4,5-*b*]-pyridines were studied. Compound **84**, one of the selected analogues, displayed anti-proliferative action against human tumour cell lines such as HCT-116 human colon cancer and MOLM-13 human FLT3-ITD positive AML cell lines. The anticancer and cytotoxic effects of several new 1,3,4-oxadiazole-based pyrazole derivatives have been studied. As a result, compound **85** showed excellent anticancer effects. Compound 86 inhibited tumour growth effectively in a c-MET-dependent tumour model while also having favourable oral PK properties and was developed for clinical evaluation of stage I cancer. In addition, for anti-tumour activity, pyrazolo[3,4-*d*]pyrimidines were investigated, and compound **87** seems to be the most active analogue. However, pyrazolo[3,4-*b*]pyridine derivatives were screened for anti-tumour action counter to liver cell line and compound **88** exhibited the maximum potency. In another, pyrazolo[1,5-*a*]pyrimidine-linked pyrazole derivatives were selected as an anti-Janus kinase-2 inhibitor. Compound **89** showed a strong inhibitory effect, and the time-dependent decrease of pSTAT5 was found. Further, pyrazolo[3,4-*d*]pyrimidines derivatives have been tested for anti-tumour activity and among those analogues, compound **90** exhibited good potential against NPC-TW01 and NCI-H226 cancer cells. However, the 1*H*-pyrazole-4-carboxamide analogues were selected to have inhibitory activity and anti-proliferative effect on Aurora-A kinase. Compound **91** has excellent biological activity in contrast to MCF-7 and HCT-116 cell lines and exhibited noteworthy Aurora-A kinase inhibitory activity (Fig. 15).

In addition, compared with 4 diverse cell lines (HepG2, WI 38, VERO, and MCF-7), compound **92** showed promising anti-tumour activity. Pyrazole analogues comprising pregnenolone scaffold classified for their cytotoxic activity in contrast to three human tumour cell lines and the compound **93** displayed potent inhibitory effects against non-small cell lung cancer (NCI-H460), CNS cancer (SF-268) and

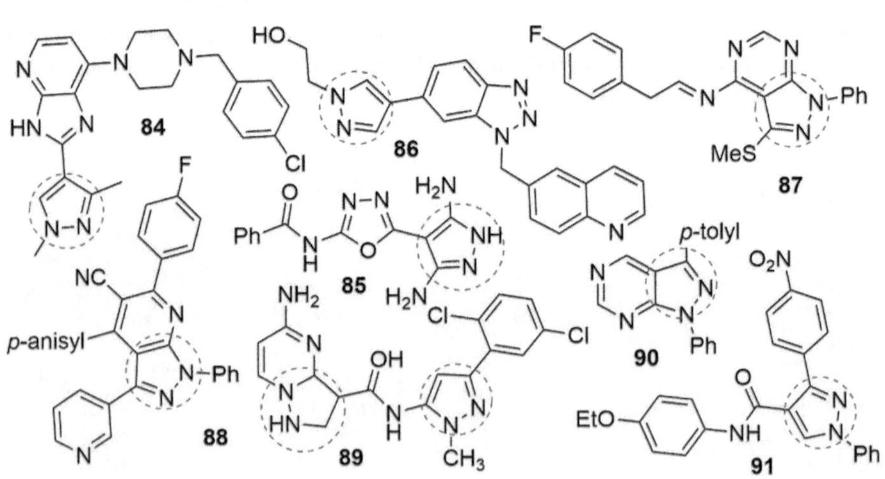

Fig. 15 Pyrazole analogues with anticancer activity [94–101]

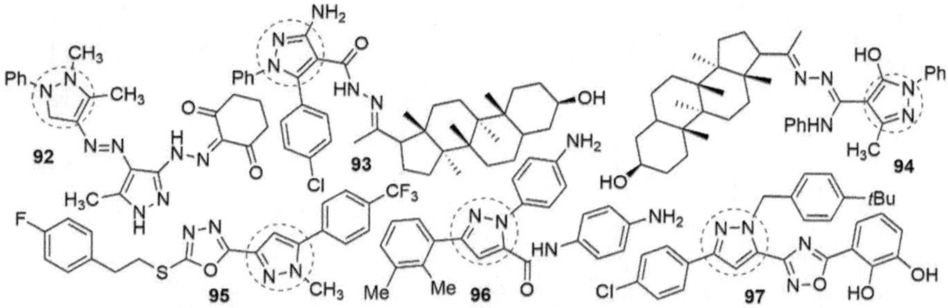

Fig. 16 Pyrazole analogues with anticancer activity [102–107]

adenocarcinoma (MCF-7). Additionally, for cytotoxicity activity, pyrazolyl semicarbazide androstane derivatives were screened; as a result, compound **94** exhibited potent cytotoxicity against NCI-H460, SF-268 and MCF-7 cells. The anticancer activity of pyrazole derivatives containing 5-substituted 1,3,4-oxadiazole residues was evaluated. Among all the verified analogues, the compound **95** exhibited decent anticancer activity against MCF-7 (adenocarcinoma) cells, related to the reference drug, doxorubicin. However, on hepatocellular carcinoma (HCC)-derived cell lines, small pyrazole analogues were assessed for anti-tumour action and the compound **96** showed potency against hepatocellular carcinoma (HCC) as well-exhibited potential inhibition on the growth of SNU449 cell line. A novel pyrazole derivative containing 1,2,4-oxadiazole moiety was tested for its anticancer activity, and compound **97** proved a more powerful and safer anticancer therapy (Fig. 16).

In addition, 4-arylmethyl-1-phenylpyrazole derivatives were investigated for their potential as new-generation androgen receptor (AR) antagonists and they are therapeutically effective against castration-resistant prostate cancer (CRPC). In this context, in the mouse xenograft model, compound **98** showed excellent anticancer activity against the CRPC model of the LNCaP-hr cell line. In continuation, several new pyrazolo[1,5-*a*]-pyrimidines have been tested for their cytotoxicity to Vero cells, and compound **99** has shown promising activity. Additionally, the action of imidazo-[4,5-b]pyridines-based 1,3-dimethyl-1H-pyrazole derivatives on Aurora-A kinase was investigated. Compound **100**, among the analogues examined, had anti-proliferative effects on Aurora-A (GI_{50} = 0.067 μM) and Aurora-A (IC_{50} = 12.71 μM) human tumour cell lines, as well as HCT-116 human colon cancer cells. A pyrazole-based Abl kinase inhibitor was also examined, and compound **101** was identified as a new template and hinge-binding motif, which contrasted with the clinically significant mutant and wild-type Abl kinase. In another, for anticancer activity, pyrazoles derivatives bearing 4β-amido podophyllotoxin rings were screened counter to five human cancer cell lines and among screened compounds, compound **102**, displayed potent antineoplastic action in A549 (lung cancer) cell line. In addition, pyrazole rings linked with acyl thiourea derivatives were assessed for anti-proliferative action and the compound **103** showed good potency. In addition, 1*H*-Pyrazole-5-Carboxamide derivatives have been assessed for their anticancer

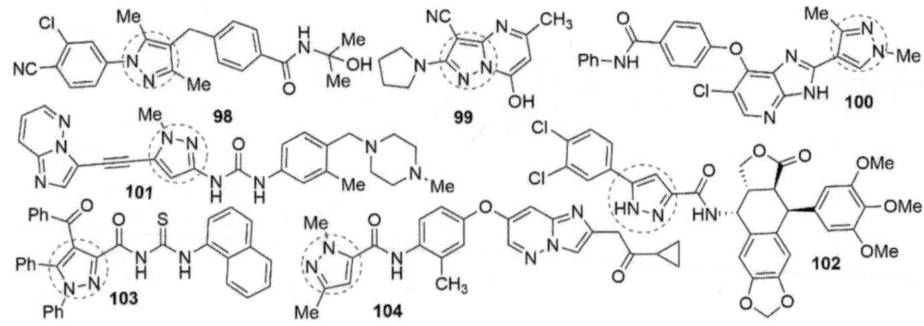

Fig. 17 Pyrazole analogues with anticancer activity [108–114]

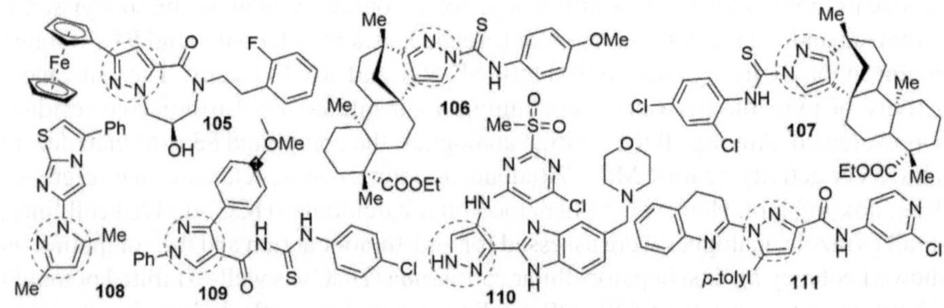

Fig. 18 Pyrazole analogues with anticancer activity [115–121]

activity and compound **104** has demonstrated a strong inhibitory effect on the receptor VEGF 2 kinase with a value of IC_{50} of 0.95 nM (Fig. 17).

In addition, the anticancer activity of the ferrocene framework related to pyrazolo[1,5-*a*][1,4]dia-azepin-4-one derivatives was tested against lung cancer cells A549, H322 and H1299 (Fig. 18). Of the compounds tested, analogue **105** showed cytotoxic activity and selectivity for H322 and A549 lung cancer cells over H1299. Additionally, carbothioamide-pyrazole derivatives were screened for cytotoxicity in contrast to four human tumour cell lines, and as a result, compound **106** showed effective cytotoxic activity against Raji cell. In another, isosteviol-fused pyrazole derivatives were assessed for antineoplastic actions on four human malignant cell lines and as a result, analogue **107** exhibited better cytotoxicity against SGC 7901, A549, Raji and HeLa cells, and their IC_{50} values are comparable to cisplatin. In addition, imidazo[2,1-*b*]thiazoles-bearing pyrazole derivatives were examined for their anticancer activity, and as a result, compound **108** showed strong anticancer activity in contrast to the CNS renal cancer cell lines SNB-75 and UO31. In addition, the anti-proliferative activity of various 1*H*-pyrazole-4-carboxamide derivatives as CDK inhibitors was tested. For example, compound **109** inhibited CDK2 to a better extent. In addition, the anticancer activity and Aurora-A/B kinase inhibitory activity of pyrazole-benzimidazole derivatives on cancer cell lines U937, K562,

A549, LoVo and HT29 were also tested, and compound **110** was found to have significant effects on cancer cell lines and Aurora Kinase A/B inhibition. In another, for antineoplastic actions, novel 5-(*p*-tolyl)-1-(quinolin-2-yl)pyrazole-3-carboxylic acid analogues were screened in contrast to three human cancer cell lines (Huh7, human liver; MCF-7, breast and HCT-116, colon carcinoma cell lines) and the analogue **111** showed potent cytotoxicity against all cell lines (Fig. 18).

In addition, the Cd (II) complexes of the tridentate nitrogen donor ligand 2,6-bis(3,4,5-trimethylpyrazolyl)pyridines have an effect on the human cancer cell lines MCF-7, Hep3B, PC3 and Saos-2, showed that complex **112** is a highly cytotoxic complex of PC3 (prostate adenocarcinoma). In another, compound **113** was examined to inhibit ALK5 phosphorylation among the screened pyrazole derivatives. Furthermore, compound **114** was identified to be an improved cytotoxic agent against MCF-7 cells line (breast adenocarcinoma) among the testified 4-pyrazolyl-1,8-naphthalimide derivatives. Additionally, trisubstituted pyrazole-based compounds were tested for ROS1 inhibitor activity, and among these compounds, analogue **115** has revealed higher degree of selectivity and fivefold potency than crizotinib. The anticancer activity of 1*H*-pyrazolo[4,3-*d*]pyrimidin-7(6*H*)-ones was examined in contrast to human cancer cell lines A549, CAKII, HeLa, PC3, and MiaPaca2 to confirm that compound **116** is an excellent anticancer agent for the apoptosis mechanism and also prevents mTOR with nanomelia potency. The compound **117** showed the most potent inhibiting property for BRAFV600E and against WM266.4 and A375 among the tested 5-phenyl-1*H*-pyrazole derivatives. However, compound **118** has anticancer activities against various cancer cell lines and considerable inhibitory activity against Class I and IIb HDACs (Fig. 19).

The anticancer properties of pyrazolyl hydroxamic acid analogues were investigated counter to the A549 (human lung cancer cell line), with compound **119** emerging as the most effective anti-proliferative agent. In another study, benzofuran-pyrazole also examined its anti-proliferative activity and compound **120** showed a remarkable growth inhibitory pattern of activity in colon cancer HCC-2998, CNS cancer SNB-75 (brain tumour), CCRF-CEM leukaemia, MOLT-4, HOP-92 lung cancer, SK-MEL-2 melanoma, breast cancer HS 578 T, IGROV1 cancer, 786–0

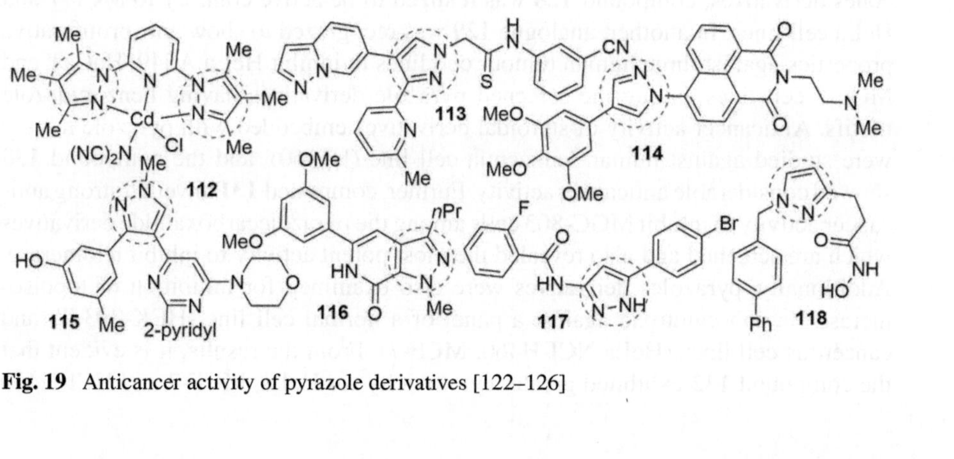

Fig. 19 Anticancer activity of pyrazole derivatives [122–126]

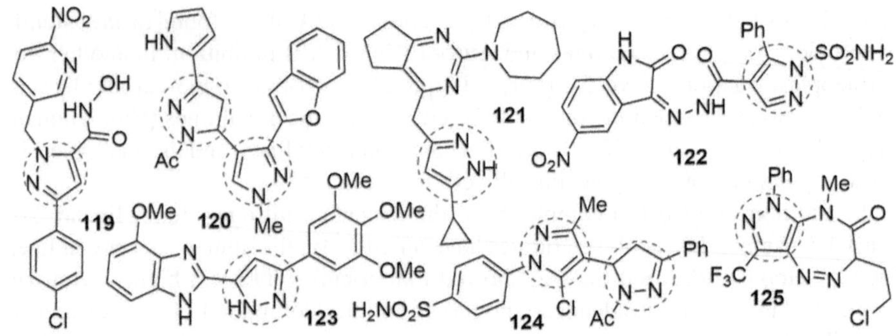

Fig. 20 Pyrazole analogues with anticancer activity [127–135]

kidney cancer, RXF 393 and T-47D. In addition, 4-pyrazole cyclopentyl pyrimidines were selected as the IGF1R tyrosine kinase inhibitor, and compound **121** was found to have high activity. The anti-tumour activity of pyrazole-containing benzenesulfonamide derivatives was evaluated against human hCA isomers I, II, IX and XII and metalloenzyme carbonic anhydrase to notice compound **122** as the most active analogue. The capacity of the arylpyrazole-linked benzimidazole derivatives to suppress the development of 60 cancer cell lines was tested. Compound **123** showed significant inhibition of the growth of most cell lines in the range of 0.3–3 μM. Then, carbonic anhydrase inhibiting activity of pyrazolyl pyrazolines was screened against cytosolic human isozymes to perceive compound **124** as the better inhibitor of hCA II. In another, 3-(2-chloroethyl)-5-methyl-6-phenyl-8-(trifluoromethyl)-5,6-[3,4-*f*][1,2,3,5]tetrazepin-4-(3*H*)-one **125** shows the greatest resistance proliferative activity (Fig. 20).

Detect the apoptosis effect of the new derivative of pyrazole thiourea in human cancer cells. It is observed that compound **126** has a stronger apoptosis-inducing effect and the compound **127** induces significant cell expansion in the G2/M phase, combined with improved verbalization of cyclin A and cyclin B, making it an auspicious anticancer medication. Upon assessment of anticancer activity of pyrazole chalcones derivatives, compound **128** was realized to be active contrary to MCF-7 and HeLa cell lines. In another, analogue **129** was recognized to show anti-proliferative properties against three human tumour cell lines to inhibit HeLa A549, HaCaT and MCF-7 cell lines among the screened pyrazole derivatives having benzimidazole motifs. Anticancer activity of steroidal derivatives embedded with pyrazole moiety were studied against human leukaemia cell line (HL-60), and the compound **130** showed considerable anticancer activity. Further, compound **131** unveiled strong anticancer activity to inhibit MGC-803 cells among the pyrazolecarboxamide derivatives which are screened and also revealed the most potent activity to inhibit telomerase. Additionally, pyrazoles derivatives were also examined for inhibition of topoisomerase IIa and cytotoxic against a panel of a normal cell line (HEK-293 T) and cancerous cell lines (HeLa NCI-H460, MCF-7). From the results, it is evident that the compound **132** exhibited greater cytotoxicity for HeLa, MCF-7 and NCI-H460

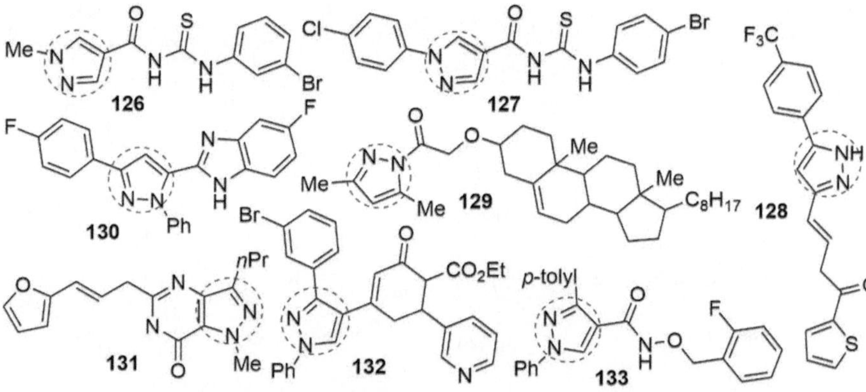

Fig. 21 Pyrazole analogues with anticancer activity [136–143]

cancer cell lines. In addition, anticancer activity of *N*-(Benzyloxy)-1,3-diphenyl-1*H*-pyrazole-4-carboxamide derivatives were assessed as MEK inhibitors and the compound **133** exhibited the most potency to inhibit MEK1 and A549 cells among those compounds (Fig. 21).

Alongside, the anti-tumour activity of 1,4,6,7-tetrahydropyranol[4,3-*c*]pyrazole derivatives on 4 human carcinoma cell lines (HGC-27, EC-109, MCF-7 and PC-3) has been confirmed, and compound **134** displayed the greatest inhibition against HGC-27 and PC-3. Similarly, The HDAC inhibitory properties of *N*-(6-mercaptohexyl)-3-substituted-1*H*-pyrazole-5-carboxamide were examined to confirm that the disulphide compound **135** is an effective cytotoxicity in contrast to a group of 7 cancer cells. The compound is responsible for the cellular hyperacetylation of non-histone and histone proteins and demonstrated significant anti-tumour activity in the HCT-116 xenograft model. In another, pyrazole-benzimidazole derivatives were investigated against novel potent active Chk2 inhibitor activity and compound **136** was identified to have maximum potency. In addition to the enhanced cytotoxicity of doxorubicin and cisplatin, the cytotoxic effect of **136** is the only drug with significant anti-tumour activity in animals with breast cancer. Furthermore, in order to obtain compound **137** with a lower IC_{50} value for HepG2, the anti-tumour activity of 1*H*-pyrazole-3-carboxylate derivatives on tumour cell lines HepG2, NCIH-460, A549, T-24 and BEL-7404 was tested. The cytotoxic activity of steroidal pyrazole derivatives on 293 T cell line and three cancer cell lines (such as HeLa, A549 and MCF-7) was examined. The compound **138** presented the maximum potency for HeLa and 293 T cell lines. The anticancer activity of the scopolamine-pyrazole hybrid against three human cancer cell lines Hun7, HCT-116 and SW620 was selected, and the results showed that compound **139** showed effective cytotoxic activity. The pyrazole derivatives bearing Sorafenib scaffold were assessed for the cytotoxic activity against MCF-7, HepG2, A549 and PC-3 cancer cell lines. Additionally, few of those compounds were advanced for their activity against BRAF, c-Met, CRAF, EGFR, Flt-3 and VEGFR-2/KDR kinases. Compound **140** showed moderate to good activity

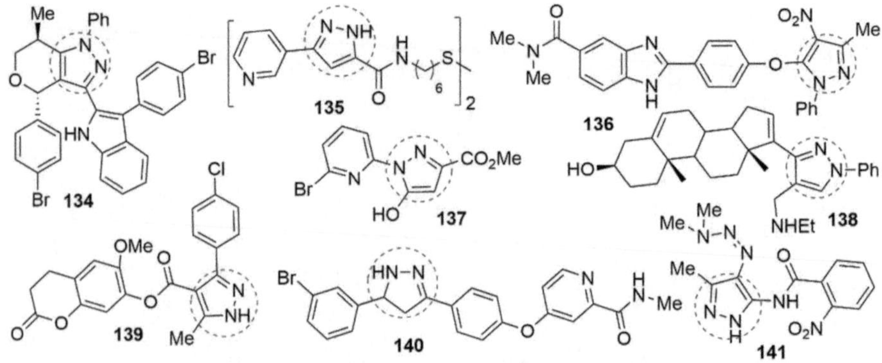

Fig. 22 Pyrazole analogues with anticancer activity [144–151]

against c-Met, moderate to inactive activity against EGFR, CRAF, and Flt3 kinase, and showed higher anti-tumour activity against HepG2, A549 and MCF-7 cell lines. However, The anticancer activity of 4-(3,3-dimethyltriazeno)-5-benzamidopyrazole derivatives on Raji and K562 cell lines was measured, and compound **141** was observed as an anticancer effect on K562 (97.87%) and Raji (99.49%) the most vigorous growth inhibitor (Fig. 22).

Pyrazole derivatives exhibit various pharmacological activities, and one of them is the anticancer activity. The anticancer activity is exhibited by various mechanisms of action because of their high degree of diversity upon change in the structure. It acts as an inhibitor of cyclin-dependent kinase (CDK), aurora kinase inhibitor, heat shock protein inhibitors, break point cluster region-Abelson tyrosine kinase inhibitors, cyclooxygenase and lipo-oxygenase inhibitors, polo-like kinase inhibitors, reticular activating system-neuro endocrine tumour ETS-like transcription factor, epithelial growth factor receptor inhibitors and DNA binding agent [152].

Cyclin-dependent kinases are cell cycle regulating proteins. These proteins are overexpressed in the cancer cells [153]. So these proteins are one of the targets to treat cancer. CDK inhibitory property of pyrazolo[4,3-*h*] quinazoline-3-carboxamides was elucidated [152].

Aurora kinase is involved in various steps of the cell cycle like chromosome segregation, spindle-checkpoint and cytokinesis. Any alteration in the above-mentioned steps would induce aneuploidy, a symptom of cancer cells. Aurora B is highly expressed in various cancer like colon, oral cancer, testicular germ cell tumours, hepatocellular carcinoma, mesothelioma, malignant endometrium, non-small cell lung carcinoma, thyroid, glioblastoma, prostate and ovarian. Aurora B seems to be a good target for anticancer activity as its inhibition rapidly leads to catastrophic mitosis, heading to cell death [153]. The 1,4,5,6-Tetrahydropyrrolo[3,4-*c*]pyrazole bicycle derivative have been prepared to target the active site of these protein kinases. SAR optimization is required for these compounds to increase the potential of inhibition [152].

The BCR-ABR, a chimeric protein, plays a vital role in pathogenesis of Philadelphia chromosome-positive leukaemia, i.e. chronic myeloid leukaemia (CML). BCR-ABL is a fusion protein which is present only in malignant cells and absent in non-malignant cells. So, this makes it specific target for treating CML [154]. 1*H*-pyrazolo[3,4-*b*]pyridine derivative GZD824 has shown excellent anti-proliferative activity for BCR-ACL-positive K562 and Ku812 human CML cells in nanomolar range. It also exhibits good bioavailability, half-life and potent anti-tumour efficacy [152].

Many proteins interact with heat shock protein 90 (HSP90) which is a molecular chaperone and are involved in various cell signalling pathways, proteins related to cell cycle, cell death and cell proliferation [155]. So, these HSP 90 can be the target for cancer cells and its inhibition may accompany the decrease/complete inhibition of cell proliferation. Pyrazolo[3,4-*d*]pyrimidines derivatives have been synthesized, and they were investigating for their activity in cell-based assays [152].

5 Anti-inflammatory and Analgesic Activity of Pyrazole Derivatives

Among the azole family, pyrazoles hold an arena of interest in synthetic chemistry and natural products as well as the pharmaceutical industry. This area has witnessed continuous exploration, and the current study is aimed to enhance the knowledge about pyrazole derivatives as anti-inflammatory and analgesic agents. In this portion, modern advancements in the development of pyrazole-based pharmacologically active anti-inflammatory and analgesic compounds are highlighted.

Cyclooxygenase-2 (COX-2) inhibitory properties of pyrazole derivatives were evaluated to witness compound **142** as a selective COX-2 inhibitor. The compound **143** showed comparable anti-inflammatory activity and COX-2 inhibition to that of indomethacin among the screened 1*H*-pyrazolyl derivatives. The anti-inflammatory activity of compound **144** was observed to be higher than indomethacin among the tested pyrazolyl benzenesulfonamide derivatives with thiazolyl ring. In another, 3,5-diaryl pyrazole derivative **145** has potent anti-inflammatory agents against TNF-α and IL-6. The compound **146** displayed better analgesic and anti-inflammatory properties than pentazocine and diclofenac sodium among an innovative sequence of 1-acetyl/propyl-3-aryl-5-(5-chloro-3-methyl-1-phenyl-1*H*-pyrazol-4-yl)-2-pyrazolines examined. Additionally, using a carrageenan-induced rat paw oedema assay, pyrazolyl-pyrazoline derivative **147** was found to have greater anti-inflammatory activity (32%) than nimesulide (36%). Upon testifying the anti-inflammatory properties of 2,3-dihydro-imidazo[1,2-b]pyrazole analogues, compound **148** expressed a dual activity to inhibit both fMLP-OMe and IL-8-induced chemotaxis. The pyrazoles containing benzenesulfonamides were assessed for performing anti-inflammatory activities and the derivative **149** displayed better

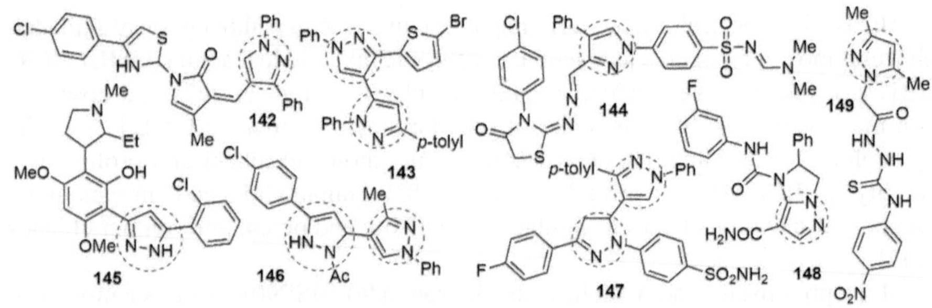

Fig. 23 Pyrazole derivatives with anti-inflammatory activity [156–161]

potency (anti-inflammation) (inhibition of oedema 62.67%) than indomethacin (inhibition of oedema 60.8%) (Fig. 23).

Pyrazole analogues were evaluated towards inhibiting ovine COX-1/COX-2 isozymes to realize compound **150** as the optimal COX-2 inhibitor with reference drug celecoxib among the testified compounds. The compound **151** was shown to express appreciable cyclooxygenase-2 (COX-2) inhibition among the dihydropyrazolyl-thiazolinone derivatives. In addition, 1,3,4-trisubstituted pyrazole derivatives anti-inflammatory activity have been tested as an extraordinary anti-inflammatory agent in comparison with diclofenac using carrageenans inducing paw oedema method to show compound **152**. Additionally, upon investigation of celecoxib analogues for COX-1/COX-2 inhibition and ulcerogenic liability, the derivative of 3-(pyridine 3-yl)pyrazole **153** was highest in power and showed an ulcerogenic potential of around 40 per cent reduction. In addition, the compound **154** has been seen in comparison with Indomethacin in the evaluated pyrazole-hydrazone derivatives as a potent medicinal product for inflammation (92.59% inhibition). The compound **155** contains 1-(4-substituted-phenyl)-3-phenyl-1*H*-pyrazole-4-carbaldehydes with maximum anti-inflammatory and analgesic activity. Compound **156** showed comparable anti-inflammatory activities similar to nimesulide among the tested compounds of pyrazole derivatives (Fig. 24).

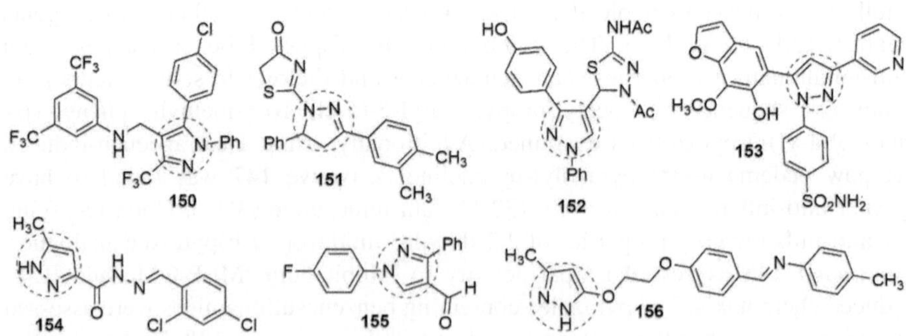

Fig. 24 Analgesic and anti-inflammatory activities of pyrazole derivatives [162–168]

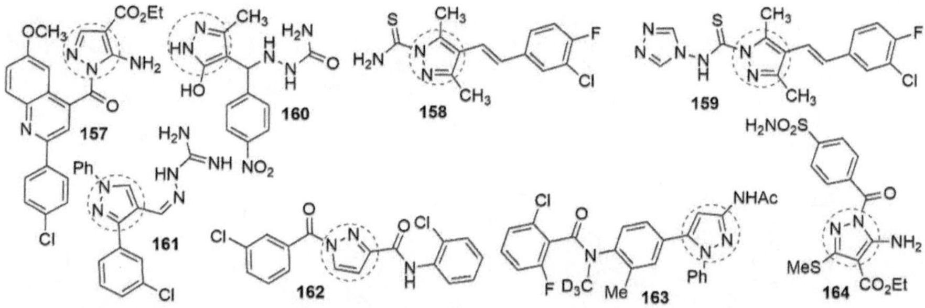

Fig. 25 Pyrazole derivatives with analgesic and anti-inflammatory activities [169–175]

Anti-inflammatory and ulcerogenic studies of pyrazole derivatives with a quinolone moiety revealed compound **157** to be a superior anti-inflammatory agent to celecoxib and to have the best COX-2 binding characteristics. Alongside, the compound **158** had the similar anti-inflammatory activities as that of reference drug ibuprofen. Similar analgesic properties were observed for compound **159** with good agreement to standard drugs. The anti-inflammatory activity of **160** was in comparison with diclofenac sodium which is a standard drug. The compound **161** had the best anti-inflammatory action (93.59% inhibition) when compared to the reference medications Ibuprofen (80.38% inhibition) and indomethacin in a study of 1,3-diaryl pyrazoles with aminoguanidine groups. The compound **162** was identified to inhibit human 15-lipoxygenase. In addition, the compound **163** was observed to be a more effective agent to act as RORγ inverse agonists among the pyrazole-containing benzamides investigated. In addition, analgesic and anti-inflammatory activities were checked for ethyl-5-amino-3-methyl-1*H*-pyrazole-4 carboxylates and the **164** compounds were performed as a superior analgesic and anti-inflammatory compound (Fig. 25).

In addition, anti-inflammatory activity of substituted pyrazoles was assessed using carrageenan-induced paw oedema standard technique to perceive compound **165** as the most effective anti-inflammatory agent. Among the 1*H*-pyrazole-4-acetates containing quinazolinone rings, compound **166** was shown to have significant analgesic and anti-inflammatory properties. Furthermore, compound **167** was discovered to be a brain penetrant molecule with efficacy in the CFA inflammatory pain model. Another study looked examined compound **168** as a non-steroidal anti-inflammatory medicine using pyrazole-containing tetrazole. The compound **169** was appeared to indicate significant analgesic and anti-inflammatory activities among the 1,3,4-trisubstituted pyrazoles screened. The benzoxazole and 1,2,4-triazole derivatives having pyrazole moieties were screened for analgesic activity to identify compound **170** as a significant analgesic agent. Furthermore, among the tested substituted pyrazoline derivatives, compound **171** was recognized as an anti-inflammatory and analgesic agent. When the anti-inflammatory and analgesic effects of 5-methyl-2-phenylthiazole-4-substituted pyrazole derivatives were examined, compound **172**

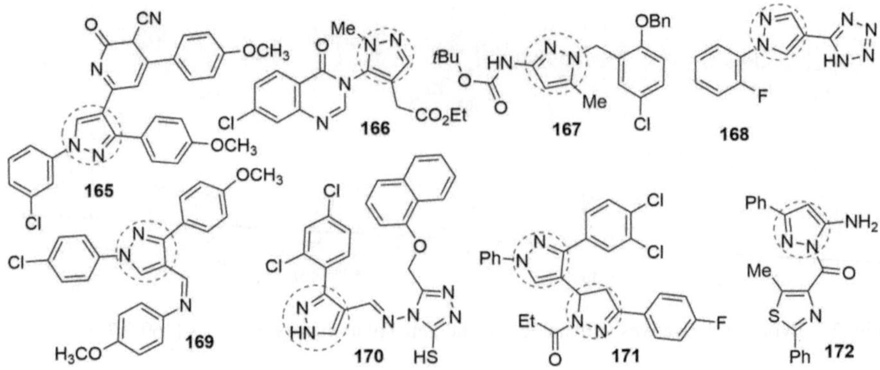

Fig. 26 Pyrazole derivatives with analgesic and anti-inflammatory activities [176–183]

was determined to be the most efficient anti-inflammatory and analgesic agent (Fig. 26).

The anti-inflammatory potential of pyrazole derivatives was evaluated using carrageenan-induced rat paw oedema assay to identify compound **173** as the most potent drug. The compound **174** exhibited admirable oral anti-inflammatory effectiveness in the pursuit of anti-inflammatory kinetics with modified and novel prodrugs of nitric oxide-releasing coxib. In another, COX inhibition was confirmed in flavone amalgamated pyrazoles, and acetic acid increased vascular permeability in mice and carrageenan-induced hind paw oedema in rats. Compound **175** was shown to have noteworthy inhibition contrary to COX-2, demonstrating that they are discriminating COX-2 inhibitors. Currently, anti-inflammatory against COX-1, COX-2 and 5-LOX enzymes inhibition was screened with pyrazole-hydrazone derivatives to witness compound **176** as the better COX-2 inhibitor than celecoxib. Upon exploration of pyrazole compounds as anti-inflammatory drugs for COX-2 inhibition, the hydroxymethyl group in compound **177**, which is ortho to the sulfonamide group, demonstrated good selectivity and inhibition. The canine whole blood (CWB) COX inhibition assay was also used to explore *N*-methanesulfonylpyridinyl-substituted trifluoromethylpyrazole analogues for canine selective COX-2 inhibition. The compound **178** had the better potency towards COX-2 inhibition with high selectivity index. The canine-specific COX-2 inhibitory activity of Pyrazole derivatives modified with *N*-methanesulfonyl pyridinyl and integrated with a heteroaryl motif at the 5-position (**179**) was discovered. The selectivity ratio of COX-1 to COX-2 was found to be 4000-fold, indicating a favourable efficacy profile for treating inflammation and pain. In addition, pyrazole derivatives were examined for canine selective COX-2 inhibition to realize compound **180** as the most potent COX-2 selective inhibitor (Fig. 27).

Additionally, the CWB assay was used to screen inhibitory activity against canine selective COX-2 for some of the *N*-methanesulfonylpyridinyl-substituted pyrazole compounds embedded with ether/thioether functionality (Fig. 28). The compound **181** exhibits reasonable selectivity and inhibitory potency against COX-2 enzyme. In a study, anti-inflammatory potency of 3/5-trifluoromethyl pyrazole derivatives

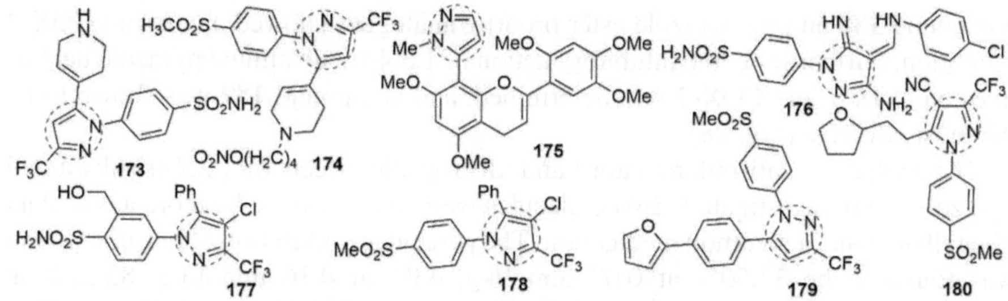

Fig. 27 Anti-inflammatory activity of pyrazole derivatives [184–191]

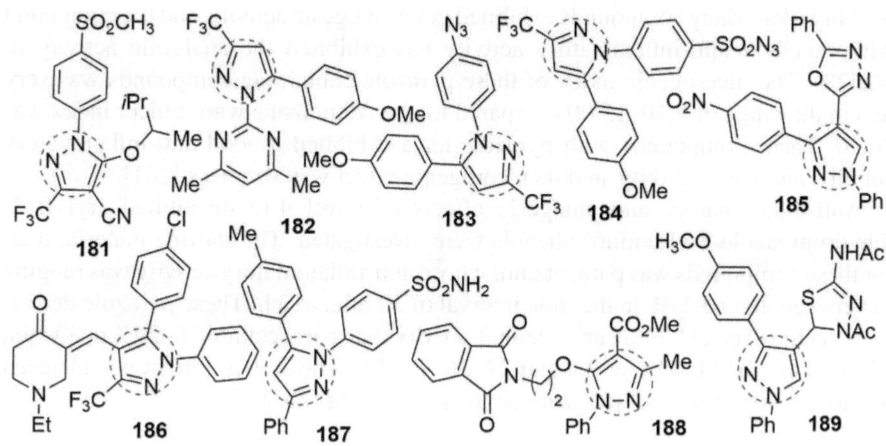

Fig. 28 Pyrazole derivatives with anti-inflammatory activities [192–200]

was verified and compound **182** possessed maximum potential of 78% which is comparable with indomethacin (78%) after 3 h of induction. In addition, Celecoxib analogues with an azido cluster in place of the SO_2NH_2 moiety were tested for their COX-1/COX-2 inhibitory ability, and the compound **183** demonstrated improved anti-inflammatory effects. In continuation, the compound **184** having SO_2N_3 group at para-position displayed good COX-1 enzyme inhibition among the series of celecoxib analogues which were assessed for their COX-1/COX-2 inhibition. In addition, the anti-inflammatory activity of 2-phenyl-5-(1,3-diphenyl-1*H*-pyrazol-4-yl)-1,3,4-oxadiazole analogues was investigated for discerning COX-2 inhibitory action and compound **185** has an optimal COX-2 inhibitory effect. In another, COX-2 inhibitory potential and anti-inflammatory activity of 4-substituted novel trifluoromethyl pyrazole derivatives were investigated and the compound **186** showcased an encouraging activity which is in comparison with diclofenac. In continuation, pyrazolyl benzenesulfonamides were also tested for COX-1 and COX-2 inhibition and anti-inflammatory activities, and derivative **187** has comparable activity with celecoxib and indomethacin among evaluated compounds. Additionally, the compound **188** among pyrazole ester prodrug analogues showed maximum COX-2 inhibition. Furthermore, the inhibitory action of 1,3,4-trisubstituted pyrazole derivatives on COX-1 and COX-2 was determined, and compound **189** was shown to be the most effective (Fig. 28).

The analgesic, anti-inflammatory and ulcerogenic effects of 1,3,4-trisubstituted pyrazoles were investigated. Two molecules were discovered to be more active than Phenylbutazone, a common medication. The percentage inhibition of one compound was found to be 37.50% at 0.08 mmol/kg, 43% at 0.16 mmol/kg, 82.22% at 0.32 mmol/kg, whereas the phenylbutazone exhibited 20.00% at 0.08 mmol/kg, 42.10% at 0.16 mmol/kg and 70.55% at 0.32 mmol/kg. The IC_{50} value of this compound was found to be 0.20 mmol/kg, whereas phenylbutazone was found to be o.22 mml/kg. Many compounds exhibited good analgesic activity, and the compound with excellent anti-inflammatory activity has exhibited the analgesic activity of 66.67%. The ulcerogenic index of these pyrazole-containing compounds was very less in the range of 5.40–15.00 compared to phenylbutazone whose ulcer index was 30.20. These compounds with pyrazole have exhibited a good anti-inflammatory activity, analgesic activity, and its ulcerogenic effect was very less [201].

Anti-inflammatory and analgesic effects of novel 4-(5-substituted aryl-4, 5-dihydropyrazole-3-yl-amino) phenols were investigated. The starting material used for these compounds was paracetamol whose anti-inflammatory activity was ranging in between 26 and 28% in the time interval of 30 min. to 3 h. These pyrazole derivatives synthesized exhibited an excellent activity than paracetamol 26–29% at 30 min, 32–37% at 1 h, 34–43% at 2 h and 27–36% at 3 h. The activity might be enhanced due to the substitution of pyrazole on the phenol ring [201].

6 Anti-tubercular Activity of Pyrazole Derivatives

Tuberculosis has become a serious threat worldwide due to infections arising from *mycobacterium tuberculosis,* and according to information, it occupies a one-third of the world's population. The medications to cure mycobacterial infections in particular tuberculosis have evolved into a huge concern because of the emergence *M. tuberculosis* strains which are monodrug and multidrug-resistant. Hence, the requirement is for establishing new drugs with structure benefits and modern mechanisms of action which are far distinct from pyrazinamide (PZA), rifampicin (RIF) and isoniazid (INH). In this respect, the discoveries towards designing novel anti-tubercular materials have been highly prioritized in medicinal chemistry research over the decades. Hence, we decided to provide an overview of modern advancements in the development of pyrazole-based pharmacologically active anti-tubercular compounds.

Compound **190**, the most active drug, was discovered to be the novel inhibitor of *Mycobacterium tuberculosis.* Antimicrobial activity of 3-substituted 5-hydroxy-5-trifluoro[chloro]methyl-1*H*-1-isonicotinoyl-4,5-dihydropyrazoles against non-tuberculous mycobacteria, INH-resistant clinical *M. tuberculosis* isolates and

Mycobacterium tuberculosis H37Rv was tested. Compound **191** was chosen because of its superior efficacy against *M. tuberculosis* and a variety of INH-resistant strains. The inhibitory action of pyrazole derivatives against M. tuberculosis H37Rv was tested, and The N1 position of compound **192** with a *p*-bromophenyl group was shown to be highly effective. In continuation, *M. tuberculosis* pantothenate synthetase inhibition of 5-*tert*-butyl-*N*-pyrazol-4-yl-4,5,6,7-tetrahydrobenzo[*d*]isoxazole-3-carboxamide derivatives was tested and the compound **193** showed the maximum activity. In addition, two sets of rigid pyrazolone derivatives were assessed for *M. tuberculosis* inhibition and the results were observed to be in favour of compound **194** with morpholine and *N*-Me-piperazine moieties. Similarly, among the series of 3a,4-dihydro-3*H*-indeno[1,2-c]pyrazole-2-carboxamide analogues, compound **195** was found to be particularly effective against isoniazid-resistant *M. tuberculosis* and *M. tuberculosis* H37Rv. *N*-aryl-1,4-dihydropyridine analogues embedded with a 1*H*-pyrazole ring were also tested towards anti-tubercular efficacy. The compound **196** had the smallest MIC of 0.02 μg/mL and was shown to be more powerful than isoniazid. The anti-mycobacterial activity of a series of *N*-phenyl-3-(4-fluorophenyl)-4-substituted pyrazoles was tested against *Mycobacterium tuberculosis* H37Rv, with compound **197** demonstrating the recommended potency. Anti-tubercular action was also examined in a series of 3-(4-chlorophenyl)-4-substituted pyrazoles against *M. tuberculosis* H37Rv strain to create novel anti-tubercular agents, with compound **198** exhibiting exceptional anti-tubercular activity among the validated analogues (Fig. 29).

Anti-tubercular activities of fluorinated pyrazoles were investigated against *M. tuberculosis* H37Rv and reports indicated that compound **199** exhibited notable anti-tubercular activities. In continuation, the compound **200** among 1-[(4-benzyloxy phenyl)-but-3-enyl]-1*H*-azole series has been recognized as a potent drug to act against *M. tuberculosis* with reasonable comparison with standard drugs. In addition, 3,5-diaryl-1*H*-pyrazoles were screened for inhibitory potencies against prokaryotic arylamine *N*-acetyltransferase enzymes and the compound **201** showed decent anti-mycobacterial movement on the expansion of MTB (*M. tuberculosis*). The anti-tubercular activity of hybrid furoxanyl *N*-acylhydrazones bearing pyrazole moiety

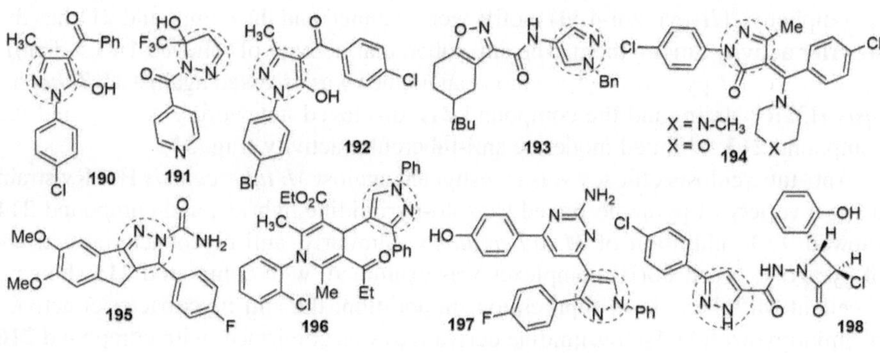

Fig. 29 Anti-tubercular activity of pyrazole derivatives [202–209]

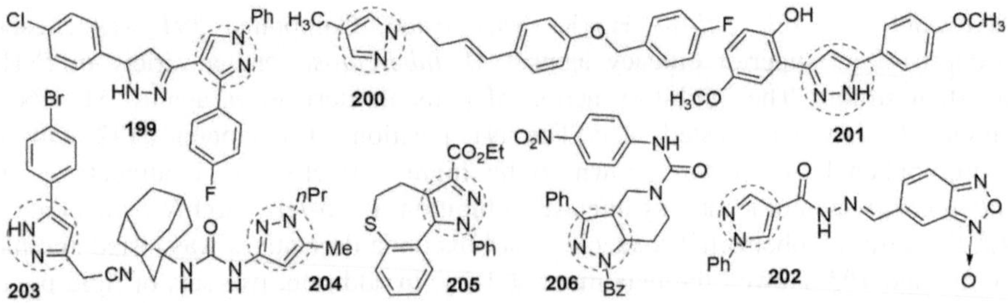

Fig. 30 Anti-tubercular activity of pyrazole analogues [210–217]

on *M. tuberculosis* H37Rv was screened and compound **202** displayed better activity against *M. tuberculosis* among them. Furthermore, anti-tubercular activity of substituted pyrazole derivatives has been evaluated against *M. tuberculosis* H37Rv strain and compound **203** was excellent among them. The compound **204** showed superior anti-mycobacterial activity among the series of pyrazoles containing 1-adamantyl-3-heteroaryl ureas against MTB H37Rv. Similarly, compound **205** exhibited maximum anti-tuberculosis activity among the series of pyrazoles embedded with thiochromeno and benzothiepino motifs against MTB. In a study, 3-phenyl-4,5,6,7-tetrahydro-1*H*-pyrazolo[4,3-*c*]pyridine analogues' bioactivity in contrast to *M. tuberculosis* (MTB) pantothenate synthetase trypanosoma, compound **206** was shown to be the most effective against MTB PS (Fig. 30).

The pyrazoles with a methylthiazole were investigated for anti-mycobacterial activities and compounds **207** exhibited comparable *M. tuberculosis* inhibition. In addition, in vitro anti-tuberculosis effectiveness of 5-imidazopyrazoles embedded with 2-amino-3-cyano pyridine analogues was investigated, and compound **208** outperformed reference medications. In addition, the anti-tubercular movement of polyhydroquinoline analogues was tested against the MTB H37Rv strain, and derivative **209** was showed activity similar to first-line medicines. The anti-tuberculosis activity of a series of pyrazolyl pyrazolines with fluoro-substitution was examined against *M. tuberculosis* H37Rv, and the compound **210** showed good anti-tubercular activity. Further, anti-tubercular activities of quinazolin-4(3*H*)-one analogues with (1,3-diphenyl-1*H*-pyrazol-4-yl) motifs were scanned and the compound **211** has the superior activity among them. The anti-tubercular activity of relieved 1-(3,5-diaryl-4,5-dihydro-1*H*-pyrazol-1-yl)ethanone analogues was assessed against *M. Tuberculosis* H37Rv strain, and the compound **212** displayed noteworthy activity and the compound **213** exhibited moderate anti-tubercular activity (Fig. 31).

Anti-tuberculosis efficacy was investigated against *M. tuberculosis* H37Rv strain using a variety of pyrazole-linked triazolo-pyrimidine hybrids, and compound **214** showed 99% inhibition of *M. tuberculosis.* Similarly, anti-mycobacterial activity of pyrazolyl-based Pd(II) complexes was examined, with compound **215** showing excellent M. tuberculosis suppression. In addition, the anti-mycobacterial activity of aminopyrazolo[1,5-a]pyrimidine derivatives was confirmed, with compound **216** showing the most activity among the compounds tested. Additionally, formononetin

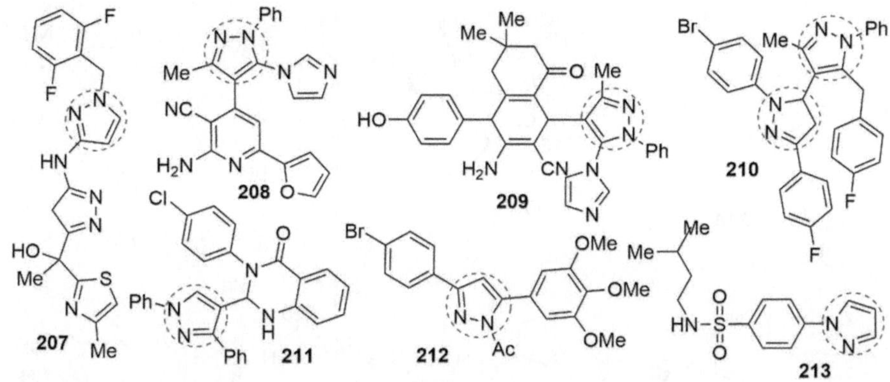

Fig. 31 Anti-tubercular activity of pyrazole analogues [218–222]

analogues of pyrazole ring were elucidated for anti-mycobacterial activity and the compound **217** exhibited H37Rv strain of *M. tuberculosis* inhibition in 40%. In comparison with the reference medicine ethambutol, the INH-pyrazole analogues were tested for anti-tubercular activity, and compound **218** demonstrated greater action with a MIC of 0.8 μg/mL. Furthermore, 1,2,4-oxadiazole/pyrazole analogues were evaluated for the same action; the studied compounds exhibited mild to sensible movement, with compound **219** having the highest efficacy against *M. tuberculosis* H37Rv strain. In addition, high-throughput screening revealed InhA inhibition of *N*-benzyl-4-((heteroaryl)methyl)-benzamides. The screened compounds showed greater activity versus MTB, preserving activity against KatG mutant clinical strains and developing as a prospective equipment against XDR-TB and MDR-TB. Furthermore, the pyrazole derivative **220** is a direct InhA inhibitor with moderate whole-cell activity and an appealing pharmacological profile, but it was ineffective in a TB infection model in mice. The MTB enzyme CYP121 inhibitory activity of pyrazole derivatives was tested, and the inhibitory potency of compound **221** against human P450s was excellent among the series of CYP121 inhibitors (Fig. 32).

For anti-tubercular activity, bi heterocyclic derivatives such as benzofuran, 1*H*-indole, pyrazolo[1,5-a]pyrimidine, pyrazolo[1,5-*a*]pyrimidin-5-(4*H*)-one, pyrazolo[5,1-*b*]thiazole, and imidazo[2,1-*b*]thiazole were tested. With various degrees of activity, the pyrazolo[5,1-*b*]thiazoles and imidazo[2,1-*b*]thiazoles outperformed the imidazo[2,1-*b*]thiazoles. With 2,6-dimethylpyrazolo[5,1-*b*]thiazole **222**, significant suppression of the H37Ra strain was observed. The quinoline-pyrazole derivatives with fluorine group are investigated for their anti-tubercular activity. The high selectivity index and low toxicity profile were observed for compound **223** and could serve as an anti-TB lead. In addition, it was screened and identified compound **224** as the most powerful antifungal agent with a superior selectivity index as the anti-tuberculous property of Sila analogues in Rimonabant. The following anti-tuberculosis activity was reported in the treatment M. Tuberculosis H37Rv line and the efficacy of MTB PS compound **225** were greater for inhibiting

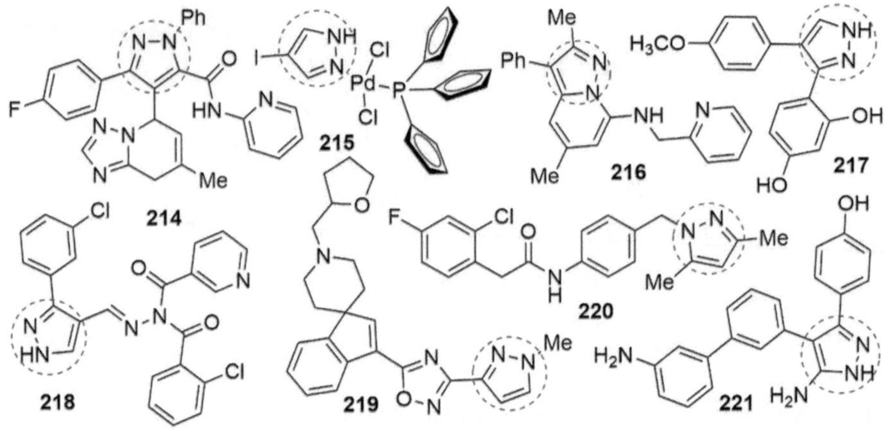

Fig. 32 Anti-tubercular activity of pyrazole analogues [223–230]

1-((1-(substituted)-1*H*-1,2,3-triazol-4-yl)methyl-*N*,3-diphenyl-6,7-dihydro-1*H*-pyrazolo[4,3-c]pyridine-5(4*H*)-carboxamides. Furthermore, hydrazide–hydrazone embedded 2-aroyl-[1]benzopyrano[4,3-c] showed anti-mycobacterial action. When pyrazol-4(1*H*)-one derivatives were tested contrary to the MTB H37Rv reference strain, compound **226** showed activity comparable to isoniazid. Furthermore, pyrazoles with bedaquiline derivatives were screened for inhibition of ATP synthesis in mycobacteria and the compound **227** was observed to inhibit ATP synthesis. A resazurin MIC assay was used to examine the anti-mycobacterial activity of pyrazole derivatives that are interlinked with isoniazid pharmacophores related with coumarin scaffold in contrast to *M. tuberculosis* H37Rv strain. Regarding tested analogues, compound **228** showed greater potency in inhibiting *M. tuberculosis* H37Rv development by 80%. However, pyrazole derivatives were tested for activity against MTB H37Rv, and compound **229** showed a MIC of 3.13 μg/mL, which is equivalent to ethambutol (3.25 μg/mL) and pyrazinamide (50 μg/mL) (Fig. 33).

The anti-tubercular activity of 8-trifluoromethyl quinoline substituted pyrazole-3-carboxamides was examined, and compound **230** exhibited important inhibitory activity in contrast to MTB H37Rv strain in comparison with ethambutol. Furthermore, the anti-tubercular activity of phenothiazine clubbed pyrazolo[3,4-*d*]pyrimidines was tested against *M. tuberculosis*, with the outcomes revealing that analogue **231** had the best anti-tubercular movement along with 96% inhibition. In addition, benzopyran-annulated pyrano[2,3-*c*]pyrazoles analogues were assessed for anti-tubercular action, and the results revealed that compound **232** has the most anti-tubercular activity, with 93% inhibition. Further research into the activity of benzofuran-pyrazole derivatives against MTB H37Rv revealed that compound **233** is more potent than MIC90. In addition, the anti-mycobacterial activity of quinolinyl heterocycles was tested contrary to *M. smegmatis*, and quinolinyl pyrazole hybrids like **234** exhibited superior movement that was comparable to that of isoniazid. The quinazolinone pyrazole derivatives have been tested for their ability to inhibit the

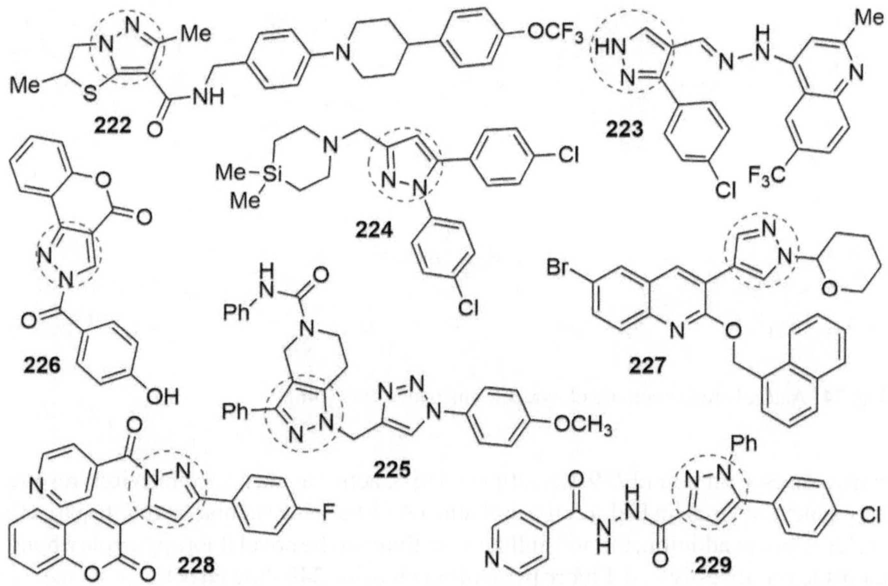

Fig. 33 Anti-tubercular activity of pyrazole analogues [231–238]

growth of MTB H37Rv, and the results are shown to be considerable for all the presented hybrids and in particular for compound **235**. The anti-tubercular of dihydropyrimidine pyrazole derivatives are screened to inhibit MTB H37Rv, and the compound **236** was observed to be the most potent compound with the maximum SI > 500, as it was more potent than INH. Additionally, anti-tubercular activity of 4-aminoquinolone piperidine amides was screened against drug-resistant strains of MTB and non-replicating phase (NRP). The maximum activity against DprE1 overexpressed (OE), InhA OE, MTB H37Rv, BTZ043 (C387S), PimA OE, TopA OE, moxifloxacin-resistant mutant clone 4.1 and TMC207-resistant mutant clone 8.1 strains was observed in case of compound **237** (Fig. 34).

7 Antiviral Activity of Pyrazole Derivatives

Antiviral activities of pyrazole and pyrazolo[4,3-*d*]-1,2,3-triazine-4-one ribonucleosides were substantiated against polio, vesicular stomatitis virus (VSV), African swine fever (ASFV), (HSV-1), HIV-1 and coxsackie. Compound **238** showed a selective inhibition of HIV-1 proliferation in acutely infected C8166 cells under pyrazole nucleosides. In identifying pyrazofurin derivatives that may have antiviral properties, the antiviral activity of 5′-deoxy pyrazofurin derivatives against a wide range of viruses has been studied, including arena-, myxo-, pox-, rhabdo-, herpes-, picorna-, reo-, toga- and

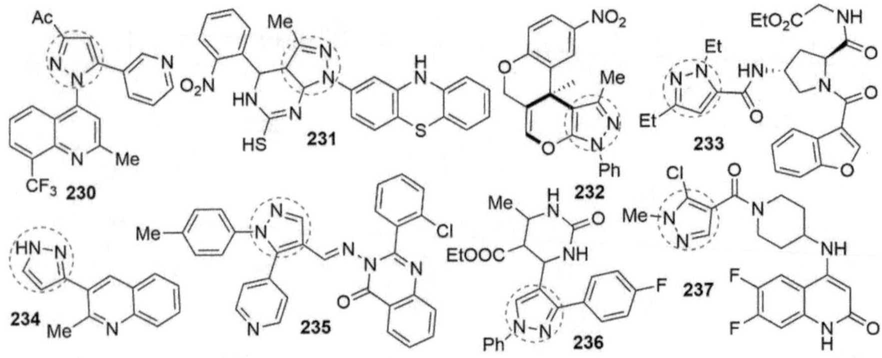

Fig. 34 Anti-tubercular activity of pyrazole analogues [239–246]

retroviruses. Compound **239** was estimated to be active against vaccinia virus, respiratory syncytial virus (in HeLa cells), influenza A virus and vesicular stomatitis virus (in HeLa cells). In addition, the anti-influenza activity of the novel fluoropyrazole ribonucleoside was investigated. Fluoro pyrazole nucleoside **240** showed better movement in contrast to influenza A and B. Furthermore, the 1,5-diphenylpyrazolea class of HIV-1 non-nucleoside reverse transcriptase inhibitors (NNRTIs) and derivative **241** exhibited high action against both wild-type and delavirdine-resistant P236L reverse transcriptase (RT). Furthermore, employing the RT-PCR technique, 1-(4-chlorophenyl)-4-hydroxy-1*H*-pyrazole-3-carboxylic acid hydrazide analogues were tested for their ability to inhibit hepatitis C virus (HCV) replication in a virus-infected HepG2 hepatocellular carcinoma cell line. The results showed that compound **242** was extremely effective at suppressing replication of both the (+) and (-) strands of HCV RNA. The antiviral activity of pyrazolo[4′,3′:5,6]pyrano[2,3-*d*]pyrimidine analogues in contrast to herpes simplex virus type-1 (HSV-1) was investigated; and as a result, pyrazolopyranopyrimidine **243** had the best antiviral activity among the compounds analysed. Alongside, 1-methyl-3-(trifluoromethyl)-*N*-[4-(pyrrolidinylsulfonyl)phenyl]-1*H*-pyrazole-5-carboxamide has been evaluated as an innovative agent that can inhibit several major strains of different measles virus (MeV) genotypes. The piperidine derivative **244** expressed workable activities against live MV without cytotoxic effects during treatment with a secondary virus titter reduction test. In addition, antiviral activities of pyrazolaldoxime ester derivatives were assessed against TMV and the results indicated scrawny to decent anti-TMV bioactivity for the above-mentioned compounds. The compound **245** has a greater affinity for TMV CP and has exhibited superior biological activity. Antiviral activity of pyrazole and pyrazolo[3,4-*d*]pyrimidine analogues against herpes simplex virus type-1 (HSV-1) and hepatitis A virus (HAV) was also investigated, with compound **246** showing the best anti-HAV activity among the compounds examined (Fig. 35).

The antiviral activity of phenyl-substituted 1*H*-pyrazole-3-carboxylic acids was evaluated in terms of the effect on HIV replication and IN inhibition, and the main antiviral activity was of 5 -(4-nitrophenyl)-1*H*-pyrazole-3-carboxylic acid **247**

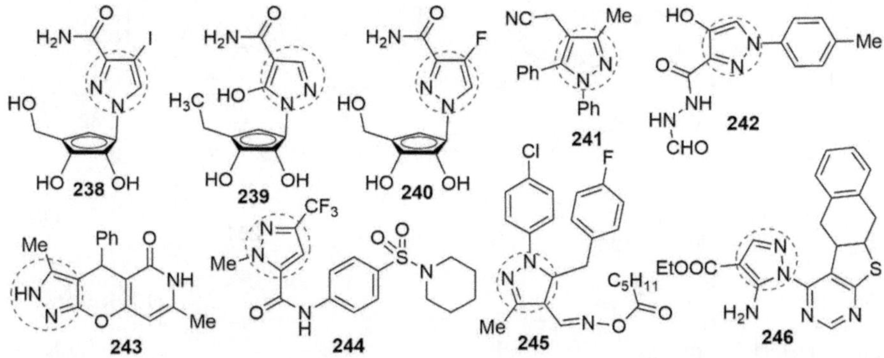

Fig. 35 Antiviral activity of pyrazole analogues [247–255]

and 3-(3 (benzyloxy) phenyl)isoxazole-5-carboxylic acid **248**. Alongside, anti-HIV activity of novel class of *N*-hydroxyethyl pyrazoles analogues was assessed and the compound **249** discloses greater activities against wide panels of drug-resistant and wild-type HIV and demonstrated compatible pharmacological profile versus isolated RT enzymes. According to another study, non-nucleoside HIV reverse transcriptase inhibitors (NNRTIs) based on a pyrazole template have been discovered to have clinically substantial action against maintaining mutations when tested against wild-type reverse transcriptase (RT). In continuation, 3,5-diethylpyrazole **250** is one of the most potent compounds among the 3,5-disubstituted pyrazole series. In addition, 3-substituted pyrazole ester analogues that could serve as an allosteric inhibitor of West Nile virus NS2B-NS3 protease were tested and compound **251** was found to be very effective. In addition, The antiviral activity of the 4,4' (arylmethylene) bis (1H-pyrazol-5-ol) derivative was selected to fight the peste des petits ruminant virus (PPRV). Compound **252** showed higher activity than the standard drug ribavirin. Additionally, antiviral activity of pyrazole derivatives was verified against HCV and the compound **253** was proved to be having remarkable anti-HCV activity. The pyrazole compounds are then cleared for selective anti-influenza virus action using a comparable cell neutralization assay (virus-induced cytopathic effects with inhibition). Compound **254** has shown potent inhibitory activity from among 20,800 randomly selected library compounds (Fig. 36).

In continuation, novel pyrazole derivatives for HIV-1 RT inhibition with nanomolar intrinsic motion on the key mutant enzymes, WT and potent antiviral activity in infected cells. Among this family of compounds, compound **255** exhibited good intrinsic antiviral activity versus Y181C, K103N, WT mutants and showed superior action in the cell-based assay in contrast to the above-mentioned mutants with little shift in its action. The antiviral activity of pyrazole- and isoxazole-based heterocycles against herpes simplex type-1 (HSV-1) was also examined and compound **256** showed the highest activity in reducing the HSV-1 plaque count among the compounds tested. The antiviral activity of a series of pyrazole amides embedded in an α-aminophosphonate moiety was assessed, and compound **257** showed partial

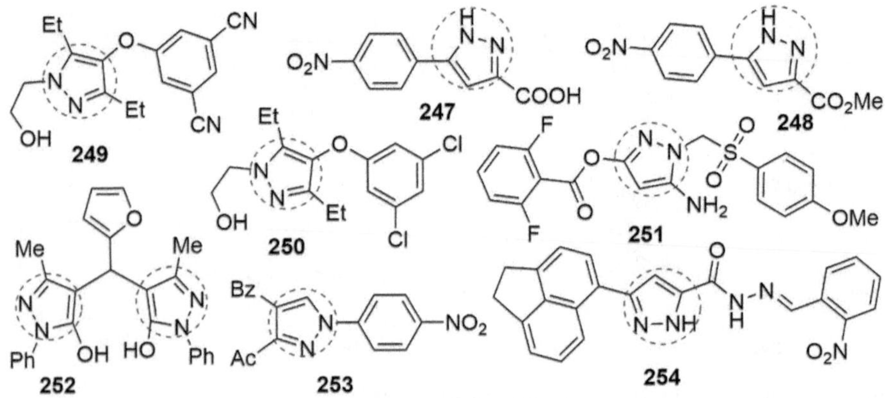

Fig. 36 Pyrazole analogues with antiviral activity [256–262]

healing properties (50%) against tobacco mosaic virus. Additionally, pyrazole rings were investigated for inhibiting nucleosides of Hepatitis C virus (HCV) NS5B polymerase and the pyrazole motif **258** displayed notable intrinsic potency with the inhibition of NS5B polymerase with NTP and exhibited engrossing antiviral properties in the replication assay. The properties of non-nucleoside reverse transcriptase inhibitors (NNRTIs) against human immunodeficiency virus (HIV) were then examined for aryl-substituted pyrazole compounds, and it was found that compound **259** has excellent activity against HIV-1 wild species as well as viruses with resistance mutations have K103N and Y181C on the back transcriptase gene. However, the further discovery on pyrazole derivatives by means of optimization of potency and aqueous solubility portray pyrazole **260** as a potent measles virus (MeV) inhibitor (Fig. 37).

Similarly, the antiviral activity of 4-substituted 3-methyl-1,5-diphenyl-1-*H*-pyrazoles was assessed against herpes simplex virus type-1. Viruses were identified in

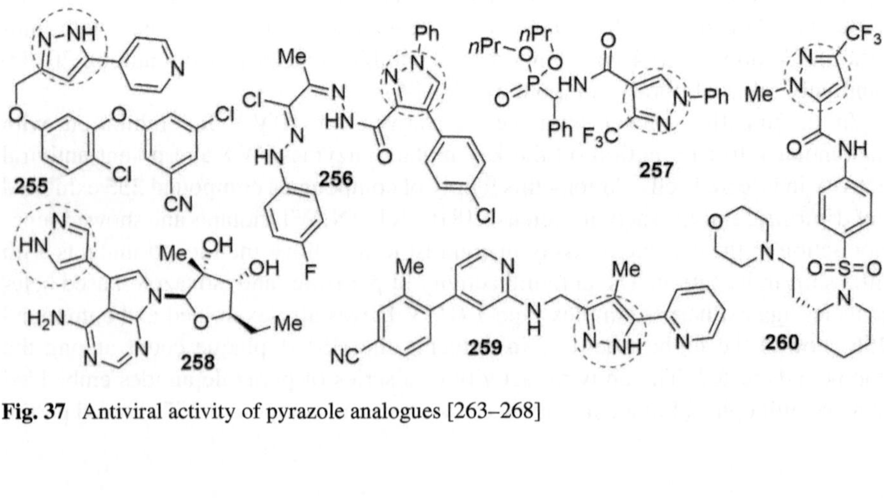

Fig. 37 Antiviral activity of pyrazole analogues [263–268]

kidney cells of the African Vero green monkey by plaque decrease assay phenomenon employ acyclovir as cultivated optimistic control. The obtained outcomes suggested that the compound **261** displayed stronger antiviral activity in comparison to standard drugs. In addition, the antiviral activity of bis-pyrazole compounds counter to tobacco mosaic virus (TMV) was evaluated, and compound **262** presented superior action over ningnanmycin. In another case, hepatitis C virus (HCV) inhibitory properties of 1,3,4-trisubstituted pyrazoles were evaluated by phenotypic high-throughput assessment with the help of infectious HCVcc. The compound **263** was shown to have higher potency among the evaluated compounds. Additionally, on studying anti-HIV properties of phenylpyrazole derivatives, 3,4-dichloro derivative **264** displays greater anti-HIV activity. Because of the increased resistance to hepatitis C virus (HCV) NS3 protease inhibitors at the current clinical stage, the need for an effective inhibitor with a superior resistance barrier is highly desirable. The invention of macrocyclic acylsulfonamides embedded with pyrazole moiety are screened for HCV protease inhibition by addressing the drawbacks with clinically trialled variants issued the compound **265** with notable antiviral activity against mutant D168V 1b and R155K 1a. In continuation, assuming pyrazole scaffold bearing diarylaniline analogues as non-nucleoside reverse transcriptase inhibitors (NNRTIs), the compound 2,4-dihydropyrano[2,3-*c*]pyrazole **266** works as anti-HIV-1 chemotherapeutic agent as NNRTIs. In addition, cytotoxicity and antiviral activity of *N*-((1,3-diphenyl-1*H*-pyrazol-4-yl)methyl)anilines were screened against a large panel of viruses and compound **267** replied with RSV replication in micromolar concentrations. Further, heterocyclic fused pyrazole analogues were tested for anti-BVDV activities and catalytic DNA cleavage abilities, and compound **268** showed tenfold higher activity than positive control ribavirin (Fig. 38).

Next, the HCV inhibitory properties of pyrazolecarboxamide derivatives were examined (Fig. 39) and compound **269** showed good activity in contrast to HCV 1b and condensed the RNA replicas of the infectious Jc1 chimeric 2a clone. In addition, the anti-HBV activities of pyridine-pyrazole-sulfonates were evaluated and the structure–activity relationship (SAR) was found in HepG2 2.2.15 cells, and compound **270** shows the highest potency. Additionally, pyrazole derivatives with

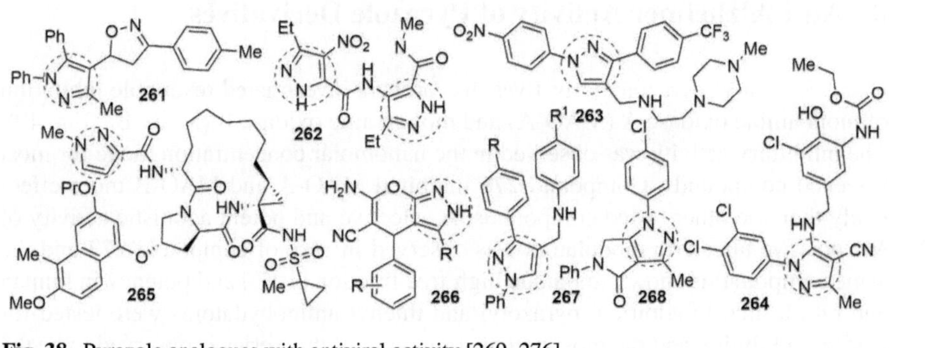

Fig. 38 Pyrazole analogues with antiviral activity [269–276]

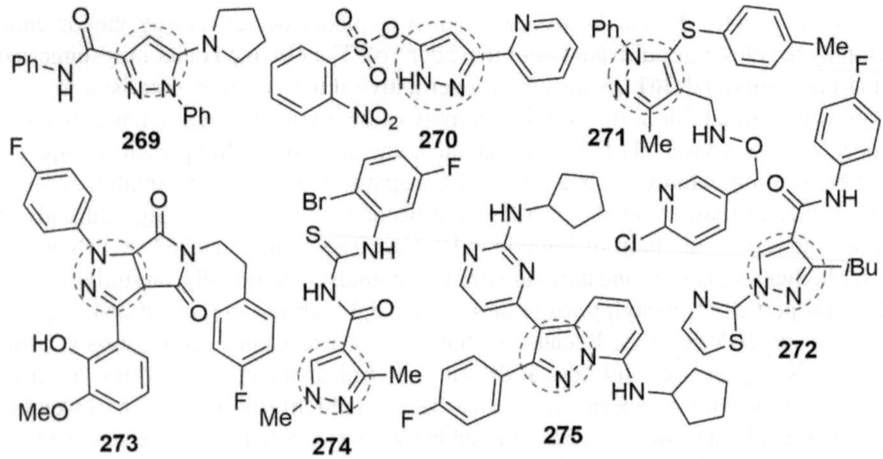

Fig. 39 Antiviral activity of pyrazole analogues [277–282]

oxime moiety were investigated for antiviral activities and the bioassay was identified to possess antiviral activities. The tobacco mosaic virus (TMV) was highly inactivated by compound **271** so as the commercial product ningnanmycin. Furthermore, non-nucleoside HBV inhibition by pyrazole derivatives via pharmacophore hybrid strategy and bioisosterism was investigated. The compound **272** was revealed to have higher potency against the secretion of HBeAg and HBsAg. In another, antiviral activity against HIV-1 was evaluated for pyrrolopyrazole derivatives and compound **273** had higher potency among the screened compounds. Similarly, anti-TMV activity of pyrazole acyl thiourea derivatives was verified and compound **274** showed maximum therapeutic activity among the products. However, antiviral action counter to herpes virus was assessed using pyrazolo[1,5-*a*]pyridine derivatives and the compound **275** was revealed to have the highest antiviral activity (Fig. 39).

8 Anti-Alzheimer Activity of Pyrazole Derivatives

Using 3,5-diaryl pyrazole derivatives, researchers investigated reversible inhibition of monoamine oxidase A (MAO-A) and monoamine oxidase B (MAO-B) (Fig. 40). The inhibitory activity was observed in the nanomolar concentration range for most screened compounds. Compound **276** inhibited MAO-A and MAO-B more effectively than the other tested compounds. A selective and potent agonistic activity of M1 positive allosteric modulators was observed in case of compound **277** and the same compound is shown to pursue high free fraction (10%) and potency in human and rat plasma. In addition, pyrazolyl and thienyl aminohydatoins were tested for BACE1 inhibition and the most potent compound of this series is observed to be the *n*-butyl analogue **278**. In another, inhibitors of metabolically stable γ-secretase were

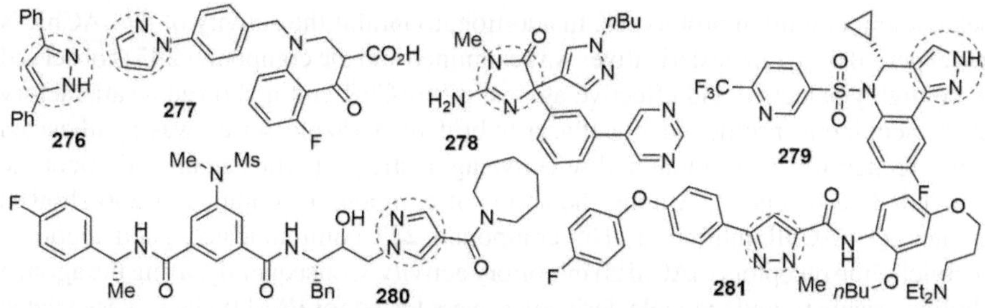

Fig. 40 Pyrazole analogues with anti-Alzheimer's activity [283–288]

very much selective towards inhibiting the production of amyloid-β over notch. In this regard, among the selected compounds, **279** were used in human clinical trials, and Aβ in the cerebrospinal fluid (CSF) of active volunteers was reduced. In addition, C-terminus Beta-secretase 1 (BACE1) inhibition can also be accomplished with pyrazole-based compounds. Moreover, upon modification over pyrazole moieties to identify a potent lead towards BACE1 inhibition, the compound **280** showed excellent potency. However, by identifying receptor inhibitors against advanced glycation end products (RAGE) in the control of Alzheimer's illness; the anti-Alzheimer's activity of pyrazole-5-carboxamide derivatives was assessed and the results indicated that analogue **281** had maximum inhibitory properties and Aβ-lowering effects in the brain (Fig. 40).

A novel series of pyrazolotacrine were investigated for their acetylcholinesterase (AChE) inhibition and the outcome showcases that the compound **282** found to have maximum AChE inhibition (Fig. 41). In addition, pyrano[2,3-*c*]pyrazole embedded with tetracyclic tacrine analogues were assessed for AChE inhibition and the compound **283** composed of 3,4-dimethoxyphenyl group showed maximum efficacy against AChE than the reference tacrine drug. In continuation, a α7-nicotinic acetylcholine receptors (α7 nAChR) exhibited optimistic properties for treating reasoning deficiency connected with schizophrenia and Alzheimer's disease (AD). The compound **284** was observed to be a strong and selective agonist against α7 nAChR which exhibited better brain levels, plasma stability and effectiveness in

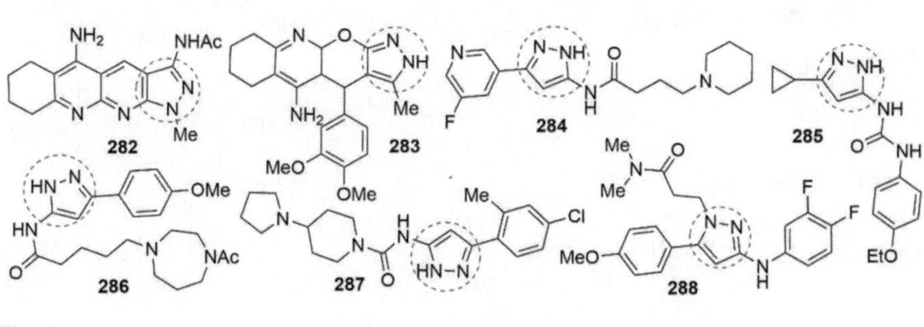

Fig. 41 Pyrazole derivatives as anti-Alzheimer's compounds [289–295]

behavioural cognition prototypes. In addition, to inhibit the activity of α7 nAChR, a novel period of pyrazole derivatives was examined and the compound **285** is observed to be highly selective and effective against α7 nAChR and it showed a satisfactory pharmacokinetic profile. Identically, a hybrid of pyrazole series was synthesized and screened for their potent and selective agonistic properties against α7 nicotinic acetylcholine receptors. Among the series of compounds, compound **286** showed greater α7 nAChR inhibition. The compound **287** communicated good nicotinic acetylcholine receptors (nAChRs) inhibitory activity. Consequently, using the agonist choline, trisubstituted pyrazole derivatives were tested for PAM types 1–4 associated with dynamic qualities in whole-cell voltage-clamp recordings, and the trisubstituted pyrazole **288** displayed incredible activity (Fig. 41).

9 Antidiabetic Activity of Pyrazole Derivatives

It was assessed the antidiabetic activity of substituted pyrazole 4-carboxylic acid (Fig. 42), and the results showed that compound **289** is the best hypoglycemic agent. Similarly, a novel sequence of 5-[(5-aryl-1*H*-pyrazol-3-yl)methyl]-1*H*-tetrazoles was tested in order to obtain anti-hyperglycemic action, and compound **290** showed 24.6 per cent blood glucose reducing activity out of the screened compounds. As a result, a variety of new 4-pyrazolyl-2-aminopyrimidines were developed as JNK inhibitors, including compound **291**, It shows high selectivity for many different proteins and lipid kinases. In addition, pyrazolopyrimidine has been tested as an inhibitor of dipeptidyl peptidase 4 (DPP4), with **292** demonstrating the highest efficacy and outstanding selectivity over the other dipeptidyl peptidases. Furthermore, utilizing an enhanced virtual screening strategy that combines ligand-centric and receptor-centric approaches, 1,3-diphenyl-1*H*-pyrazole derivatives were exhibited novel family of powerful PPAR limited agonists. Among the practical selections, the

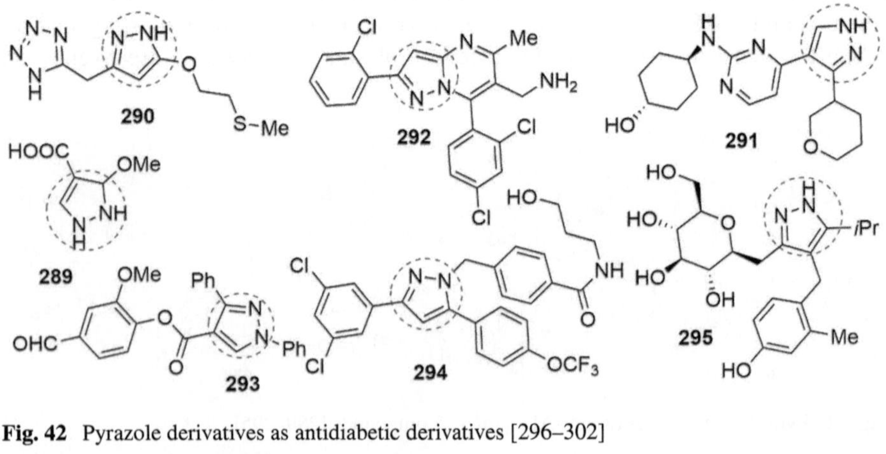

Fig. 42 Pyrazole derivatives as antidiabetic derivatives [296–302]

pyrazole-based drug **293** had comparatively significant binding abilities to PPAR. In addition, 1,3,5-pyrazoles were investigated for potent antagonists of the human glucagon receptor, with compound **294** being found as a potent antagonist with favourable pharmacokinetic contours in four preclinical class and great uttered pharmacodynamic efficiency It has been detected in monkeys and transgenic rats by delaying glucagon-induced hyperglycemia. Simultaneously, 4-benzyl-1*H*-pyrazol-3-yl-β-D-glucopyranoside analogues were screened for SGLT1 repressive action, and compound **295** was identified as a strong and selective SGLT1 inhibitor with increased intestinal stability over phlorizin (Fig. 42).

The structure–activity relationships of the new benzylpyrazole acylsulfonamides were investigated as agonists of the non-thiazolidinedione peroxisome proliferator-activated receptor γ (PPAR), which is not based on non-carboxylic acids (Fig. 43), with compound **296** exhibiting good metabolic stability and strong PPAR agonist shows activity. Also studied as an efficient, selective glucagon receptor antagonist was a new pyrazole molecule **297**, a competitive and reversible antagonist with strong binding affinity and functional activities of cAMP. In addition, Compound **298** (teneligliptin) was accepted for the handling of type 2 diabetes after it considerably reduced the rise in plasma glucose stages next to an oral glucose freight in Zucker fatty rats. GPR119 agonists based on pyrazoles were also studied, and compound **299** was shown to be around tenfold less powerful than exemplars from other series. In addition, acetyl-CoA carboxylase (ACC) inhibitors founded on a

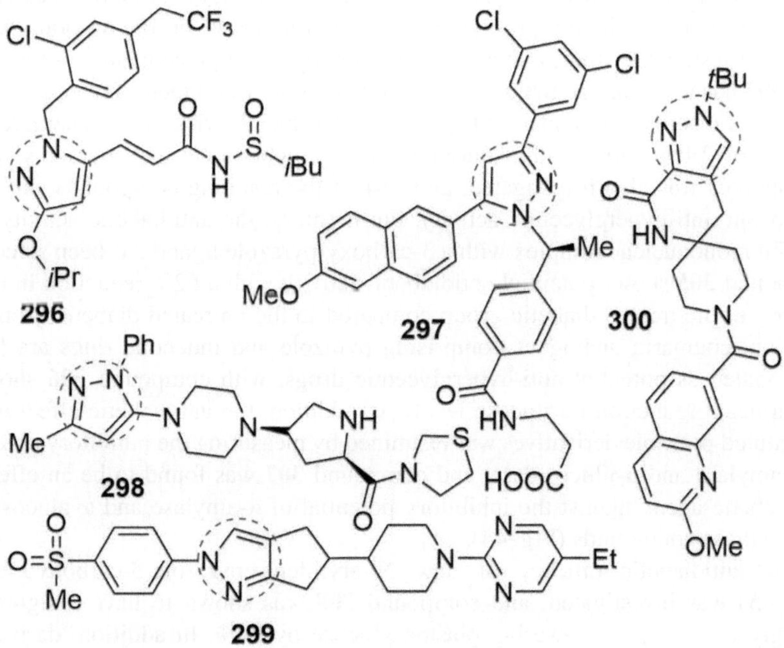

Fig. 43 Pyrazole derivatives as antidiabetic derivatives [303–307]

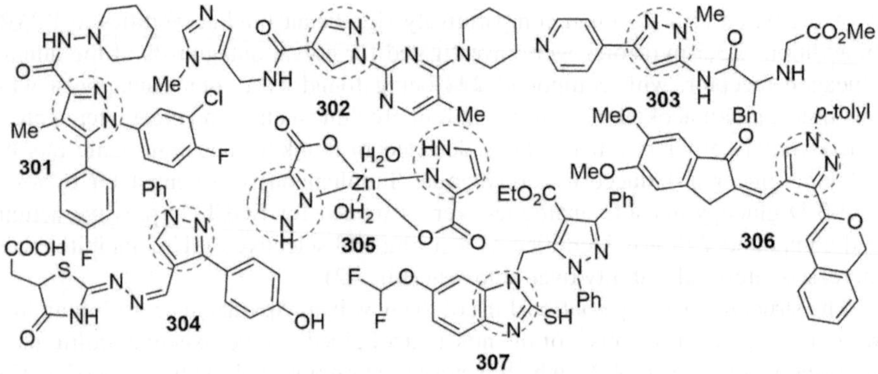

Fig. 44 Pyrazole derivatives as antidiabetic derivatives [308–314]

spirocyclic pyrazololactam core were investigated in pharmacokinetic studies in rats and Compound 300 was shown to have high oral bioavailability, moderate systemic clearance and reasonable exposure. Oral administration of compound **300** to rats resulted in a dose-proportional reduction of ACC activity (Fig. 43).

The hypoglycemic efficacy of original 1,5-diaryl pyrazole analogues was evaluated (Fig. 44), with compound **301** showing the greatest decrease in plasma glucose. New pyrazole compounds were also investigated as possible insulin secretagogues in order to obtain the medication of type 2 diabetes, wherein compound **302** shows significant glucose sinking properties in rats and monkeys during an oral glucose tolerance test. SAR studies showed that aminopyrazole-phenylalanine carboxylic acid **303** has a strong agonistic effect, excellent goal discernment, favourable pharmacokinetic belongings and no cytochrome P450 or hERG risk. Anti-hyperglycemic activity of 2-hydrazolyl-4-thiazolidinone-5-carboxylic acids with pyrazolyl pharmacophores was also investigated, and **304** of the resulting compounds exhibited significant anti-hyperglycemic activity. Furthermore, the antidiabetic activity of a new Zn mononuclear complex with a 3-carboxy-pyrazole ligand has been screened; compound **305** shows potential antidiabetic activity with a 62% reduction in blood glucose in the treated diabetic group compared to the untreated diabetic group. In addition, coumarin analogues comprising pyrazole and indenone rings are being investigated as potential anti-hyperglycemic drugs, with compound **306** showing a significant reduction in glucose levels. In addition, the antidiabetic effect of the substituted pyrazole derivatives was examined by measuring the inhibitory potential of α-amylase and α-glucosidase, and compound **307** was found to be an effective antidiabetic agent against the inhibitory potential of α-amylase and α-glucosidase among these compounds (Fig. 44).

The antidiabetic efficacy of new N′-arylidene pyrazole-3-carbohydrazides (Fig. 45) was investigated, and compound **308** was shown to have a significant hypoglycemic impact, lowering plasma glucose by 90%. In addition, derivatives of dihydropyrano[2,3-*c*]pyrazoles were examined for the inhibitory effect of α-glucosidase, and compound **309** was found to be the most effective of the group

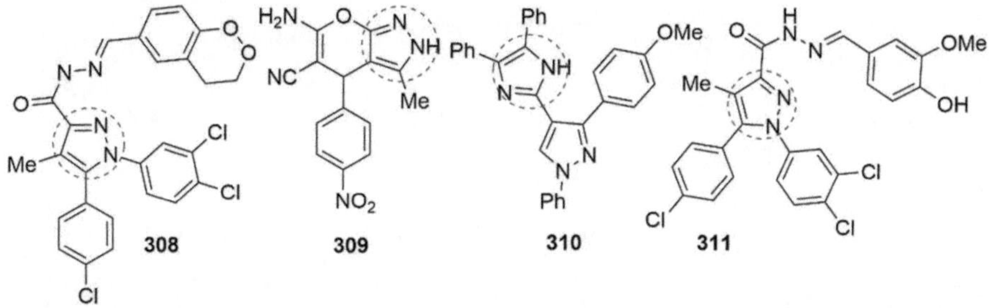

Fig. 45 Antidiabetic derivatives of pyrazole [315–318]

compared to the conventional drug acarbose. In addition, imidazolylpyrazoles were tested for their ability to inhibit α-glucosidase, and the enzyme inhibition revealed that compound **310** had significant inhibitory potentials and binding affinities when compared to reference acarbose. As a result, new 1,5-diarylpyrazole derivatives were investigated as antidiabetic agents, with the hybrid compound **311** showing both antidiabetic and antioxidant properties. Compound **311**, in particular, demonstrated a considerable anti-hyperglycemic impact in normoglycemic rats in a glucose tolerance test.

10 Anti-leishmanial Activity of Pyrazole Derivatives

The diversity of mammalian reservoirs (wild and domestic animals), vector species, and leishmania species contributes to the difficulty in controlling this parasite disease. Chemotherapy for leishmaniasis is usually ineffective, owing to the emergence of drug-resistant strains and the treatment drugs' toxicity. Pentavalent antimony drugs, such as sodium stibogluconate (pentostane) and meglumine antimonate (glucantim), are often used as first-line therapy; however, they can have serious side effects and lead to drug resistance. It is underlined that progress has been made in the creation of pyrazole-based pharmacologically efficacious anti-Leishmanial drugs.

As part of a programme to investigate prospective anti-Leishmania medicines, a sequence of 4-anilino-1*H*-pyrazolo[3,4-*b*]pyridine-5-carboxylic esters (Fig. 46) was confirmed against promastigote forms of *Leishmania amazonensis*. The compound **312** was shown to be the most effective against Leishmania amazonensis. The compound **313** was shown to be the utmost potent in contrast to *Leishmania amazonensis*, *Leishmania chagasi*, and *Leishmania braziliensis* species when the leishmanicidal activities of 1*H*-pyrazole-4-carbohydrazides derivatives were investigated. Furthermore, the anti-leishmanial activity of a novel pyrazole derivative **314** was studied, and this molecule inhibited *Leishmania tropica*, *Leishmania major*, and *Leishmania infantum* proliferation. Furthermore, anti-leishmanial properties of 1-aryl-1*H*-pyrazole-4-carboximidamide derivatives were evaluated, and compound **315** showed an activity profile that can be improved using medicinal chemistry

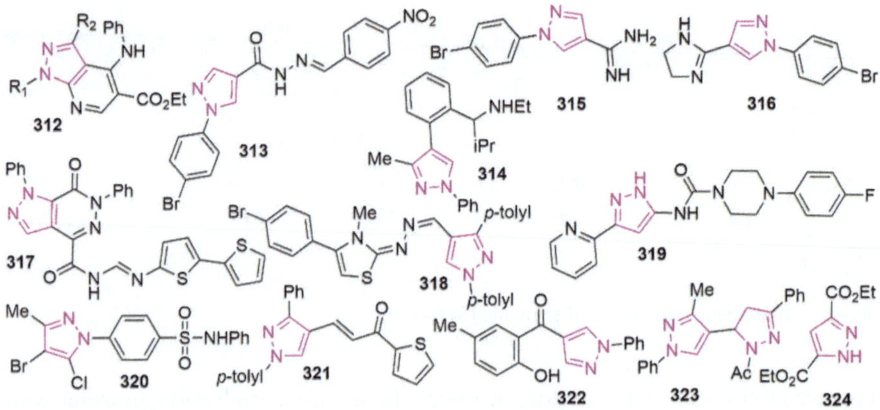

Fig. 46 Anti-leishmanial activity of pyrazole derivatives [319–331]

techniques. 1-aryl-4-(4,5-dihydro-1*H*-imidazol-2-yl)-1*H*-pyrazoles were assessed counter to 3 Leishmania types: *Leishmania amazonensis*, *Leishmania braziliensis* and *Leishmania infantum* (*Leishmania chagasi syn.*) and the compound **316** proved to be the most active against promastigotes forms of *L. amazonensis* among the derivatives studied. The anti-Leishmanial activity of pyrazolo[3,4-*d*]pyridazine-7-one analogues has also been studied against the promastigote and sterile amastigote forms of *Leishmania amazonensis* with data indicating that compound **317** showed better anti-Leishmanial action against both promastigote and axenic amastigote forms. Another study examined the anti-Leishmania activity of pyrazole derivatives against the *promastigotes* and *amastigotes* of *Leishmania aethiopica* and found that compound **318** has better anti-Leishmania activity. Similarly, efficient anti-leishmanial aminopyrazole ureas were investigated, and compound **319** showed significant levels of effectiveness (>90%) against *Leishmania infantum*. Furthermore, The anti-ileishman spectrum of 4-(1*H*-pyrazol-1-yl)-benzenesulfonamides was assessed in contrast to *Leishmania infantum* and *Leishmania amazonensis* Among them, **320** are the most active forms of infection against infectious *L. amazonensis promastigotes* and *L. infantum*. Apart from that, anti-leishmanial activities of 1*H*-pyrazole derivatives were tested against *L. aethiopica promastigotes*, with compound **321** demonstrating the greatest anti-leishmanial activity. Using a modified MTT assay. The antiparasitic activities of pyrazole derivatives on the promastigotes of *Leishmania mexicoides* (Bel21) and epimastigotes of *Trypanosoma cruzi* (DM28) were studied. Compound **322** showed selectivity for *L. Mexicana*. In addition, anti-leishmanial activity of pyrazole derivatives was tested, and compound **323** was shown to be more active than the standard miltefosine and amphotericin B deoxycholate for *Leishmania donovani*. Certain simple dialkyl pyrazole-3,5-dicarboxylates were tested for antiprotozoal action against *Leishmania infantum*, *Trypanosoma cruzi* and *Leishmania braziliensis*, and the diethyl ester **324** exhibited exceptional efficacy in contradiction of all three protozoa (Fig. 46).

11 Antimalarial Activity of Pyrazole Derivatives

Due to their physiological and pharmacological advantages, the pyrazole compounds have interested chemists and pharmacologists in consideration. Malaria remains a critical problem worldwide, despite tremendous improvement in the management of malaria during the previous 20 years. Malaria kills between 1.1 and 2.7 million people globally each year, according to a study, and more than 2400 million people are still at risk. In recent decades, parasites resistant to typical pharmacological treatments have emerged and spread. Plasmodium falciparum-resistant to chloroquine has now spread to most malarial regions, and resistance to other antimalarial medications such as mefloquine and sulfadoxine–pyrimethamine has become a serious concern in almost every country on the planet. A review of the pharmacologically active antimalarial drugs is required to solve this challenge. Antimalarial pyrazole-containing drugs have been provided in this section.

As part of attempts to investigate several powerful antimalarial drugs, a series of pyrazole moiety-containing compounds were screened (Fig. 47). Due to the presence of pyrazole derivative, compound **325** has excellent antimalarial efficacy. Curcumin analogues with pyrazole rings were also tested for antimalarial effectiveness in contrast to CQ-S and CQ-R parasites. In *Plasmodium falciparum* culture, compound **326** was revealed to be the utmost effective counterpart compared CQ-S and CQ-R. In the *P. berghei* mouse model, an aminomethylthiazole pyrazolecarboxamide lead **327** with high antiplasmodial action and microsomal metabolic constancy was discovered using whole-cell broadcast of a Soft Focus kinase library. The antimalarial characteristics of platinum(II) and palladium(II) complexes with pyrazole-derived ligands were also examined, with compound **328** demonstrating very limited antimalarial activity against two *P. falciparum* strains. Furthermore, antimalarial activity of polyhydroquinoline derivatives with pyrazole moieties was tested against *Plasmodium falciparum*, and compound **329** was shown to have high antimalarial activity among the verified compounds. In addition, inhibitors of *Plasmodium falciparum* calcium-dependent protein kinase 1 (PfCDPK1) called imidazopyridazine have been studied.

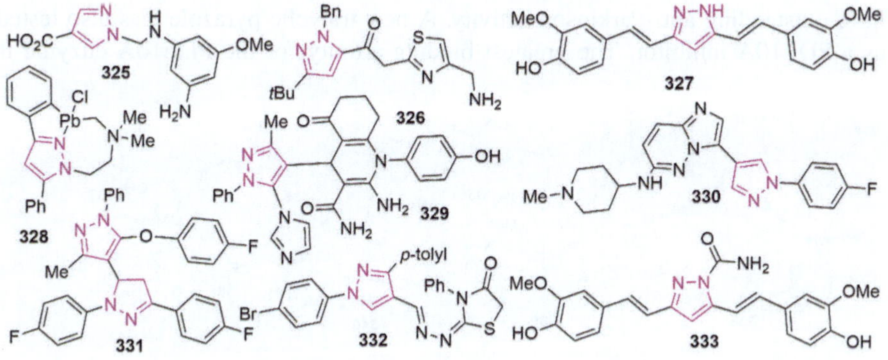

Fig. 47 Antimalarial pyrazole derivatives [332–337]

Against *Plasmodium falciparum*, Diaminocyclohexane **330** has a good potency and metabolic stability profile. Furthermore, antimalarial activity of fluoro-substituted pyrazolylpyrazolines against *Plasmodium falciparum* was tested, and compound **331** showed outstanding action Compared *P. falciparum* stain when against to quinine. In another study, pyrazole derivatives were tested for antimalarial efficacy counter to *Plasmodium berghei*-infected mice, with the utmost dynamic derivatives then tested against the chloroquine-resistant (RKL9) strain of *Plasmodium falciparum*. Furthermore, when compared to chloroquine phosphate, the antimalarial reference standard drug, compound **332** inhibited parasites by more than 90% and had lower IC_{50} values. The antimalarial capabilities of curcumin analogues with pyrazole rings (compound **333**) were investigated since it has the highest schizonticidal and parasiticidal activity (Fig. 47).

12 Anti-Parkinson Activity of Pyrazole Derivatives

A novel derivative of pyrazolo[3,4-*d*] pyrimidine (Fig. 48) has been studied as a positive allosteric modulator of metabotropic glutamate receptor subtype 4 (mGluR4). The compound **334** was shown to have significant anti-Parkinson action. It has also been confirmed that the activity of *N*1-thiocarbamoyl-3,5-di(hetero)aryl-4,5-dihydro- (1*H*)-pyrazole derivatives inhibits the activities of isoforms A and B of human monoamine oxidase (hMAO), with compound **335** being the utmost potent of the series. Similarly, the capacity of a pyrazole structure **336** to promote the activity of Ca^{2+}/calmodulin-dependent protein kinase II (CaMKII) was examined, and the findings revealed that the compound **336** increased the induction of both LTP and memory in rat hippocampal slices. In addition, a novel aminopyrazole was interpreted as a leucine-rich repeat kinase 2 (LRRK2) inhibitor, and compound **337** generated significant reductions in brain pLRRK2 levels after intravenous dosing. In addition, the Lundbeck chemical collection was screened for 4-(1-phenyl-1*H*-pyrazol-4-yl)quinoline **338**, which was shown to have mGlu4 receptor positive allosteric modulator characteristics. In addition, compound **338** has been found to have outstanding anti-Parkinson activity. A new tricyclic pyrazole was also tested as a PDE10A inhibitor. The greatest binding affinity for the PDE10A enzyme is

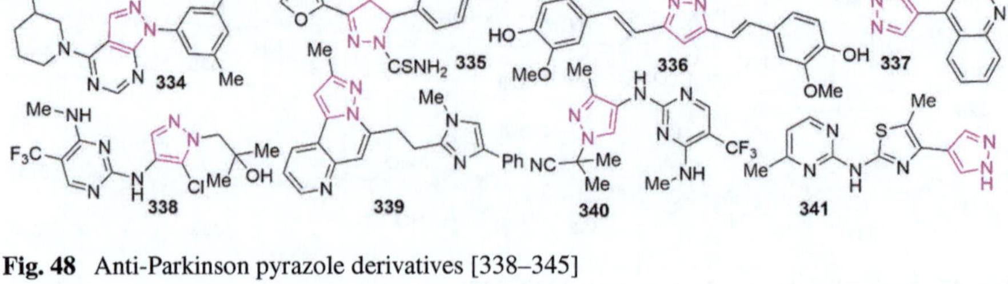

Fig. 48 Anti-Parkinson pyrazole derivatives [338–345]

Pyrazolo[5,1-*f*] [1,6]naphthyridine **339**. Aminopyrazoles were used to investigate the inhibition of leucine-rich repeat kinase 2 (LRRK2), and compound **340** was discovered to be a very effective and specific LRRK2 inhibitor. In addition, 5-methyl-*N*-(4-[11C]methylpyrimidine-2-yl)-4-(1*H*-pyrazol-4-yl)thiazole-2-amine (**341**), a type of metabotropic glutamate receptor subtype 4 (mGluR4) was synthesized and evaluated the new type of radioligand.

13 Agrochemical Activity of Pyrazole Derivatives

Due to its proven efficiency as an intermediary in the synthesis of novel biological resources, attention has improved in pyrazole derivatives in recent decades. Precisely in the agrochemical business, pyrazole derivatives have a long history of applications such as herbicides, insecticides, fungicides and acaricides. Pyrazole is a ring found in many agrochemical meaningful compounds such as tolfenpyrad [345], tolfenpyrad [346], and fenpyroximate [346]. The ring is present in a wide number of agrochemical relevant compounds (Fig. 49).

Pyrazole-based pharmacologically active agrochemical chemicals are included in this section (Fig. 50). Compound **342** has a high level of insecticidal action against *Nilaparvata lugens, Diabrotica undecimpunctata Howardi* and *Nephotettix cincticeps*, to name a few. Furthermore, a variety of new *N*-pyridylpyrazolecarboxamides showed outstanding insecticidal activity against the *Oriental armyworm* (*Mythimna separata*) and the *diamond-back moth* (*Plutella xylostella*). Compound **343** had potent larvicidal activity against P. xylostella, while compound **344** had excellent

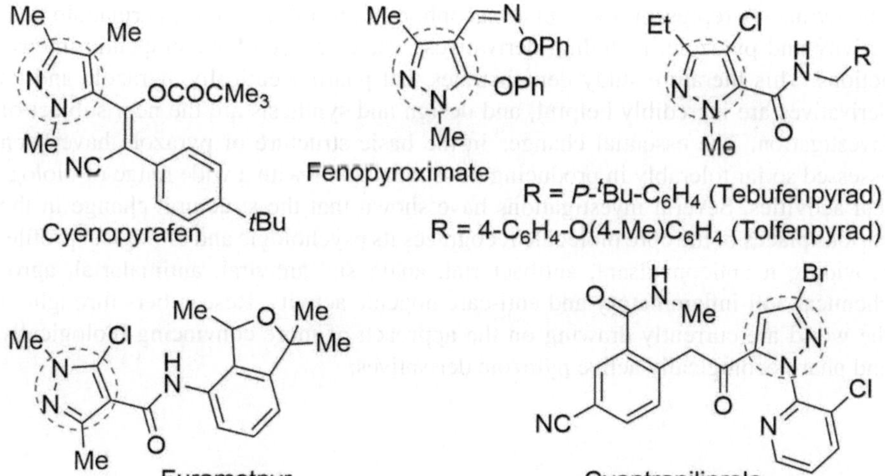

Fig. 49 Agrochemicals containing pyrazole scaffold [345, 346]

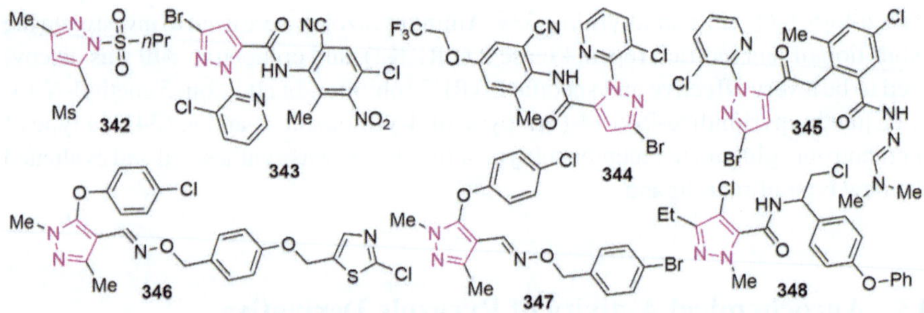

Fig. 50 Agrochemicals using pyrazole scaffolds [347–352]

activity against *M. separata.* Insecticidal activity of new pyrazole amide derivatives with hydrazine substructures (compound **345**) was good against a variety of insect species, including *P. xylostella, H. armigera, N. lugens, C. pipiens pallens* and *R. maidis*. In another study, the insecticidal and acaricidal properties of pyrazole oximes with a substituted thiazole ring were investigated. Compound **346** proved more effective against *Plutella xylostella* and *Tetranychus cinnabarinus* than the other compounds. Furthermore, compound **348** has a strong insecticidal efficacy against the cotton bollworm (*Helicoverpa armigera*). Insecticidal properties of new pyrazole oxime ether derivatives (compound **347**) were also shown to be effective.

14 Summary/Conclusion

The pyrazoles represent a key pharmacophore with different bio-pharmacological activity and pyrazole, including derivatives, was employed for therapeutic intense actions. This literature study demonstrates that pharmaceutically, pyrazole and its derivatives are incredibly helpful, and design and synthesis are the next subject of investigation. The essential changes in the basic structure of pyrazole have been assessed so far tolerably in producing novel analogues with a wide range of biological activities. Several investigations have shown that the structural change in the various places of the core molecule recognizes its psychologic and biological profile, providing it anticonvulsant, antibacterial, analgesic, antiviral, antimalarial, agrochemical anti-inflammatory and anti-carcinogenic activity. Researchers throughout the world are currently drawing on the approach of more convincing biologically and pharmacologically active pyrazole derivatives.

References

1. Naim MJ, Alam O, Nawaz F, Alam MJ, Alam P (2016) Current status of pyrazole and its biological activities. J Pharm Bioallied Sci 8:2–17
2. Faisal M, Saeed A, Hussain S, Dar P, Larik FA (2019) Recent developments in synthetic chemistry and biological activities of pyrazole derivatives. J Chem Sci 131:1–30
3. Karrouchi K, Radi S, Ramli Y, Taoufik J, Mabkhot YN, Al-aizari FA, Ansar M (2018) Synthesis and pharmacological activities of pyrazole derivatives: a review. Molecules 23:134–219
4. Fustero S, Sánchez-Roselló M, Barrio P, Simón-Fuentes A (2011) From 2000 to Mid-2010: a fruitful decade for the synthesis of pyrazoles. Chem Rev 111:6984–7034
5. Ansari A, Ali A, Asif M, Samsuzzaman (2017) Review: biologically active pyrazole derivatives. New J Chem 41:16–41
6. Steinbach G, Lynch PM, Robin KSP, Wallace MH, Hawk E, Gordon GB, Wakabayashi N, Saunders B, Shen Y, Fujimura T, Su L-K, Levin AB (2000) The effect of celecoxib, a cyclooxygenase-2 inhibitor, in familial adenomatous polyposis. N Engl J Med 342:1946–1952
7. Uslaner JM, Parmentier-Batteur S, Flick RB, Surles NO, Lam JS, McNaughton CH (2009) Dose-dependent effect of CDPPB, the mGluR5 positive allosteric modulator, on recognition memory is associated with GluR1 and CREB phosphorylation in the prefrontal cortex and hippocampus. Neuropharmacology 57:531–538
8. Friedrich G, Rose T, Rissler K (2002) Determination of lonazolac and its hydroxy and *O*-sulfated metabolites by on-line sample preparation liquid chromatography with fluorescence detection. J Chromatogr B 766:295–305
9. Hampp C, Hartzema AG, Kauf TL (2008) Cost-utility analysis of rimonabant in the treatment of obesity. Value Health 11:389–399
10. Spitz I, Novis B, Ebert R, Trestian S, LeRoith D, Creutzfeld W (1982) Betazole induced GIP secretion is not mediated by gastric HCl. Metabolism 31:380–382
11. Luttinger D, Hlasta DJ (1987) Antidepressant agents. Annu Rep Med Chem 22:21–30
12. Tsutomu K, Toshitaka N (1978) Effects of 1,3-diphenyl-5-(2 dimethylaminopropionamide)-pyrazole [difenamizole] on a conditioned avoidance response. Neuropharmacology 17:249–256
13. García-Lozano J, Server-Carrió J, Escrivà E, Folgado J-V, Molla C, Lezama L (1997) X-ray crystal structure and electronic properties of chlorobis (mepirizole) copper (II) tetrafluoroborate (mepirizole = 4-methoxy-2-(5-methoxy-3-methyl-1H-pyrazol-1-yl)-6 methylpyrimidine). Polyhedron 16:939–944
14. Alam MJ, Alam O, Khan SA, Naim MJ, Islamuddin M, Deora GS (2016) Synthesis, anti-inflammatory, analgesic, COX1/2-inhibitory activity, and molecular docking studies of hybrid pyrazole analogues. Drug Des Dev Ther 10:3529–3543
15. Wishart DS, Feunang YD, Guo AC, Lo EJ, Marcu A, Grant JR, Sajed T, Johnson D, Li C, Sayeeda Z, Assempour N, Iynkkaran I, Liu Y, Maciejewski A, Gale N, Wilson A, Chin L, Cummings R, Le D, Pon A, Knox C, Wilson M (2018) DrugBank 5.0: a major update to the DrugBank database for 2018. Nucleic Acids Res 46:D1074-D1082
16. Prasath R, Bhavana P, Sarveswari S, Ng SW, Tiekink ERT (2015) Efficient ultrasound-assisted synthesis, spectroscopic, crystallographic and biological investigations of pyrazole-appended quinolinyl chalcones. J Mol Struct 1081:201–210
17. Mert S, Kasimogullari R, Ica T, Colak F, Altun A, Ok S (2014) Synthesis, structure-activity relationships, and in vitro antibacterial and antifungal activity evaluations of novel pyrazole carboxylic and dicarboxylic acid derivatives. Eur J Med Chem 78:86–96
18. Rahimizadeh M, Pordel M, Bakavoli M, Rezaeian S, Sadeghian A (2010) Synthesis and antibacterial activity of some new derivatives of pyrazole. World J Microbiol Biotechnol 26:317–321
19. Chandrakantha B, Isloor A, Shetty P, Isloor S, Malladi S, Fun HK (2012) Synthesis, characterization and antimicrobial activity of novel ethyl 1-(N-substituted)-5-phenyl-1H-pyrazole-4-carboxylate derivatives. Med Chem Res 21:2702–2708

20. Sharma PK, Kumar S, Kumar P, Kaushik P, Kaushik D, Dhingra Y (2010) Synthesis and biological evaluation of some pyrazolylpyrazolines as anti-inflammatory-antimicrobial agents. Eur J Med Chem 45:2650–2655
21. Kendre BV, Landge MG, Bhusare SR (2019) Synthesis and biological evaluation of some novel pyrazole, isoxazole, benzoxazepine, benzothiazepine and benzodiazepine derivatives bearing an aryl sulfonate moiety as antimicrobial and anti-inflammatory agents. Arab J Chem 12:2091–2097
22. B'Bhatt H, Sharma S (2017) Synthesis and antimicrobial activity of pyrazole nucleus containing 2-thioxothiazolidin-4-one derivatives. Arab J Chem 10:1590–1596
23. Abunada NM, Hassaneen HM, Kandile NG, Miqdad OA (2008) Synthesis and antimicrobial activity of some new pyrazole, fused pyrazolo[3,4-d]-pyrimidine and pyrazolo[4,3-e][1,2,4]-triazolo[1,5-c]pyrimidine derivatives. Molecules 13:1501–1517
24. Padmaja A, Payani T, Reddy GD, Padmavathi V (2009) Synthesis, antimicrobial and antioxidant activities of substituted pyrazoles, isoxazoles, pyrimidine and thioxopyrimidine derivatives. Eur J Med Chem 44:4557–4566
25. Ragavan RV, Vijayakumar V, Kumari NS (2010) Synthesis and antimicrobial activities of novel 1,5-diaryl pyrazoles. Eur J Med Chem 45:1173–1180
26. Sharma PK, Chandak N, Kumar P, Sharma C, Aneja KR (2011) Synthesis and biological evaluation of some 4-functionalized-pyrazoles as antimicrobial agents. Eur J Med Chem 46:1425–1432
27. Pathak RB, Chovatia PT, Parekh HH (2012) Synthesis, antitubercular and antimicrobial evaluation of 3-(4-chlorophenyl)-4-substituted pyrazole derivatives. Bioorg Med Chem Lett 22:5129–5133
28. Vijesh AM, Isloor AM, Telkar S, Peethambar SK, Rai S, Isloor N (2011) Synthesis, characterization and antimicrobial studies of some new pyrazole incorporated imidazole derivatives. Eur J Med Chem 46:3531–3536
29. Wu J, Shi Q, Chen Z, He M, Jin L, Hu D (2012) Synthesis and bioactivity of pyrazole acyl thiourea derivatives. Molecules 17:5139–5150
30. Wu Z, Hu D, Kuangn J, Cain H, Wun S, Xuen W (2012) Synthesis and antifungal activity of N-(substituted pyridinyl)-1-methyl(phenyl)-3-(trifluoromethyl)-1H-pyrazole-4-carboxamide derivatives. Molecules 17:14205–14218
31. Xu L-L, Zheng C-J, Sun L-P, Miao J, Piao H-R (2012) Synthesis of novel 1,3-diaryl pyrazole derivatives bearing rhodanine-3-fatty acid moieties as potential antibacterial agents. Eur J Med Chem 48:174–178
32. Desai NC, Rajpara KM, Joshi VV (2013) Synthesis of pyrazole encompassing 2-pyridone derivatives as antibacterial agents. Bioorg Med Chem Lett 23:2714–2717
33. Jardosh HH, Sangani CB, Patel MP, Patel RG (2013) One step synthesis of pyrido [1,2-a] benzimidazoles derivatives of aryloxypyrazole and their antimicrobial evaluation. Chin Chem Lett 24:123–126
34. Malladi S, Isloor AM, Isloor S, Akhila D, Fun H-K (2013) Synthesis, characterization and antibacterial activity of some new pyrazole based Schiff bases. Arab J Chem 6:335–340
35. Nagamallu R, Kariyappa AK (2013) Synthesis and biological evaluation of novel formyl-pyrazoles bearing coumarin moiety as potent antimicrobial and antioxidant agents. Bioorg Med Chem Lett 23:6406–6409
36. Song M-X, Zheng C-J, Deng X-Q, Sun L-P, Wu Y, Hong L, Li Y-J, Liu Y, Wei Z-Y, Jin M-J, Piao H-R (2013) Synthesis and antibacterial evaluation of rhodanine-based 5-aryloxy pyrazoles against selected methicillin resistant and quinolone-resistant Staphylococcus aureus (MRSA and QRSA). Eur J Med Chem 60:376–385
37. Sayed GH, Azab ME, Anwer KE, Raouf MA, Negm NA (2018) Pyrazole, pyrazolone and enaminonitrile pyrazole derivatives: synthesis, characterization and potential in corrosion inhibition and antimicrobial applications. J Mol Liq 252:329–338
38. Kalaria PN, Satasia SP, Avalani JR, Raval DK (2014) Ultrasound-assisted one-pot four-component synthesis of novel 2-amino-3-cyanopyridine derivatives bearing 5-imidazopyrazole scaffold and their biological broadcast. Eur J Med Chem 83:655–664

39. Kumar R, Arora J, Ruhil S, Phougat N, Chhillar AK, Prasad AK (2014) Synthesis and antimicrobial studies of pyrimidine pyrazole heterocycles. Adv Chem 2014:1–12
40. Liu X-R, Wu H, He Z-Y, Ma Z-Q, Feng J-T, Zhang X (2014) Design, synthesis and fungicidal activities of some novel pyrazole derivatives. Molecules 19:14036–14051
41. Malladi S, Isloor AM, Peethambar S, Fun HK (2014) Synthesis and biological evaluation of newer analogues of 2,5-disubstituted 1,3,4-oxadiazole containing pyrazole moiety as antimicrobial agents. Arab J Chem 7:1185–1191
42. Mehta HB, Dixit BC, Dixit RB (2014) L-Proline catalyzed one-pot multi-component synthesis of 2-(1,3-diphenyl- 1H-pyrazol-4-yl)quinazolin-4(3H)-one derivatives and their biological studies. Chin Chem Lett 25:741–744
43. Ningaiah S, Bhadraiah UK, Doddaramappa SD, Keshavamurthy S, Javarasetty C (2014) Novel pyrazole integrated 1,3,4-oxadiazoles: synthesis, characterization and antimicrobial evaluation. Bioorg Med Chem Lett 24:245–248
44. Prakash TB, Reddy GD, Padmaja A, Padmavathi V (2014) Synthesis and antimicrobial activity of amine linked bis-and tris-heterocycles. Eur J Med Chem 82:347–3454
45. Du S, Tian Z, Yang D, Li X, Li H, Jia C, Che C, Wang M, Qin Z (2015) Synthesis, antifungal activity and structure-activity relationships of novel 3-(Difluoromethyl)-1-methyl-1H-pyrazole-4-carboxylic acid amides. Molecules 20:8395–8408
46. Miniyar PB, Barmade MA, Mahajan AA (2015) Synthesis and biological evaluation of 1-(5-(2-chloroquinolin- 3-yl)-3-phenyl-1H-pyrazol-1-yl) ethanone derivatives as potential antimicrobial agents. J Saudi Chem Soc 19:655–660
47. Radi S, Tighadouini S, Feron O, Riant O, Bouakka M, Benabbes R, Mabkhot YN (2015) Synthesis of novel-keto-enol derivatives tethered pyrazole, pyridine and furan as new potential antifungal and anti-breast cancer agents. Molecules 20:19684–19694
48. Rizk HF, Ibrahim SA, El-Borai MA (2015) Synthesis, fastness properties, color assessment and antimicrobial activity of some azo reactive dyes having pyrazole moiety. Dyes Pigm 112:86–92
49. Sun J, Zhou Y (2015) Synthesis and antifungal activity of the derivatives of novel pyrazole carboxamide and isoxazolol pyrazole carboxylate. Molecules 20:4383–4394
50. Yu L-G, Ni T-F, Gao W, He Y, Wang Y-Y, Cui H-W, Yang C-G, Qiu W-W (2015) The synthesis and antibacterial activity of pyrazole-fused tricyclic diterpene derivatives. Eur J Med Chem 90:10–20
51. Elshaier YAMM, Barakat A, Al-Qahtany BM, Al-Majid AM, Al-Agamy MH (2016) Synthesis of pyrazole-thiobarbituric acid derivatives: antimicrobial activity and docking studies. Molecules 21:1337–1353
52. Hafez HN, El-Gazzar A-RBA (2016) Synthesis and biological evaluation of N-pyrazolyl derivatives and pyrazolopyrimidine bearing a biologically active sulfonamide moiety as potential antimicrobial agent. Molecules 21:1156–1172
53. Mabkhot YN, Alatibi F, El-Sayed NNE, Kheder NA, Al-Showiman SS (2016) Synthesis and structure-activity relationship of some new thiophene-based heterocycles as potential antimicrobial agents. Molecules 21:1036–1044
54. Mu J-X, Shi Y-X, Yang M-Y, Sun Z-H, Liu X-H, Li B-J, Sun N-B (2016) Design, synthesis, DFT study and antifungal activity of pyrazolecarboxamide derivatives. Molecules 21:68–78
55. Nagamallu R, Srinivasan B, Ningappa MB, Kariyappa AK (2016) Synthesis of novel coumarin appended bis(formylpyrazole) derivatives: studies on their antimicrobial and antioxidant activities. Bioorg Med Chem Lett 26:690–694
56. Abrigach F, Bouchal B, Riant O, Macé Y, Takfaoui A, Radi S, Oussaid A, Bellaoui M, Touzani R (2016) New N, N, N0, N0-tetradentate pyrazoly agents: synthesis and evaluation of their antifungal and antibacterial activities. Med Chem 12:83–89
57. El Shehry MF, Ghorab MM, Abbas SY, Fayed EA, Shedid SA, Ammar YA (2018) Quinoline derivatives bearing pyrazole moiety: synthesis and biological evaluation as possible antibacterial and antifungal agents. Eur J Med Chem 143:1463–1473
58. Ahn M, Gunasekaran P, Rajasekaran G, Kim EY, Lee S-J, Bang G, Cho K, Hyun J-K, Lee H-J, Jeon YH, Kim N-H, Ryu EK, Shin SY, Bang JK (2017) Pyrazole derived ultra-short

antimicrobial peptidomimetics with potent anti-biofilm activity. Eur J Med Chem 125:551–564
59. Nada A, Al-Moghazy M, Soliman AAF, Rashwan GMT, Eldawy THA, Hassan AAE, Sayed GH (2018) Pyrazole-based compounds in chitosan liposomal emulsion for antimicrobial cotton fabrics. Int J Biol Macromol 107:585–594
60. Sharma A, Ghabbour H, Khan ST, de la Torre BG, Albericio F, El-Faham A (2017) Novel pyrazolyl-s-triazine derivatives, molecular structure and antimicrobial activity. J Mol Struct 1145:244–253
61. Zhou L, Wang P-Y, Zhou J, Shao W-B, Fang H-S, Wu Z-B, Yang S (2017) Antimicrobial activities of pyridinium-tailored pyrazoles bearing 1,3,4-oxadiazole scaffolds. J Saudi Chem Soc 21:852–860
62. Mandal S, Mondal M, Biswas JK, Cordes DB, Slawin AMZ, Butcher RJ, Saha M, Saha NC (2018) Synthesis, characterization and antimicrobial activity of some nickel, cadmium and mercury complexes of 5-methyl pyrazole-3yl-N-(20-methylthiophenyl) methyleneimine, (MPzOATA) ligand. J Mol Struct 1152:189–198
63. Mor S, Mohil R, Kumar D, Ahuja M (2012) Synthesis and antimicrobial activities of some isoxazolyl thiazolyl pyrazoles. Med Chem Res 21:3541–3548
64. Nitulescu GM, Draghici C, Chifiriuc MC, Marutescu L, Bleotu C, Missir AV (2012) Synthesis and antimicrobial screening of *N*-(1-methyl-1H-pyrazole-4-carbonyl)-thiourea derivatives. Med Chem Res 21:308–314
65. Tanitame A, Oyamada Y, Ofuji K, Fujimoto M, Iwai N, Hiyama Y, Suzuki K, Ito H, Terauchi H, Kawasaki M, Nagai K, Wachi M, Yamagishi J-I (2004) Synthesis and antibacterial activity of a novel series of potent DNA gyrase inhibitors pyrazole derivatives. J Med Chem 47:3693–3696
66. Çetin A, Bildirici İ (2016) A study on synthesis and antimicrobial activity of 4-acyl-pyrazoles. J Saudi Chem Sci 22:279–296
67. Dayakar C, Kumar BS, Sneha G, Sagarika G, Meghana K, Ramakrishna S, Prakasham RS, China RB (2017) Synthesis, pharmacological activities and molecular docking studies of pyrazolyltriazoles as anti-bacterial and anti-inflammatory agents. Bioorg Med Chem 25:5678–5691
68. Dai H, Chen J, Li G, Ge S, Shi Y, Fang Y, Ling Y (2017) Design, synthesis, and bioactivities of novel oxadiazole-substituted pyrazole oximes. Bioorg Med Chem Lett 27:950–953
69. Barakat A, Al-Majid AM, Al-Qahtany BM, Ali M, Teleb M, Al-Agamy MH, Naz S, Ul-Haq Z (2018) Synthesis, antimicrobial activity, pharmacophore modeling and molecular docking studies of new pyrazole-dimedone hybrid architectures. Chem Cent J 12:29
70. Ardiansah B (2018) A recent update: antimicrobial agents containing pyrazole nucleus. Asian J Pharm Clin Res 11:88–94
71. Kamal A, Shaik AB, Jain N, Kishor C, Nagabhushana A, Supriya B, Kumar GB, Chourasiya SS, Suresh Y, Mishra RK, Addlagatta A (2015) Design and synthesis of pyrazole–oxindole conjugates targeting tubulin polymerization as new anticancer agents. Eur J Med Chem 92:501–513
72. Xu Y, Liu X-H, Saunders M, Pearce S, Foulks JM, Parnell KM, Clifford A, Nix RN, Bullough J, Hendrickson TF, Wright K, McCullar MV, Kanner SB, Ho K-K (2014) Discovery of 3-(trifluoromethyl)-1H-pyrazole-5-carboxamide activators of the M2 isoform of pyruvate kinase (PKM2). Bioorg Med Chem Lett 24:515–519
73. Viale M, Anzaldi M, Aiello C, Fenoglio C, Albicini F, Emionite L, Gangemi R, Balbi A (2013) Evaluation of the anti-proliferative activity of three new pyrazole compounds in sensitive and resistant tumor cell lines. Pharmacol Rep 65:717–723
74. Tzanetou E, Liekens S, Kasiotis KM, Fokialakis N, Haroutounian SA (2012) Novel pyrazole and indazole derivatives: synthesis and evaluation of their anti-proliferative and anti-angiogenic activities. Arch Pharm 345:804–811
75. Zheng L-W, Shao J-H, Zhao B-X, Miao J-Y (2011) Synthesis of novel pyrazolo[1,5-a]pyrazin-4(5H)-one derivatives and their inhibition against growth of A549 and H322 lung cancer cells. Bioorg Med Chem Lett 21:3909–3913

76. Balbi A, Anzaldi M, Macciò C, Aiello C, Mazzei M, Gangemi R, Castagnola P, Miele M, Rosano C, Viale M (2011) Synthesis and biological evaluation of novel pyrazole derivatives with anticancer activity. Eur J Med Chem 46:5293–5309
77. Nitulescu GM, Draghici C, Missir AV, Synthesis of new pyrazole derivatives and their anticancer evaluation. Eur J Med Chem 45:4914–4919
78. Bandgar BP, Totre JV, Gawande SS, Khobragade CN, Warangkar SC, Kadam PD (2010) Synthesis of novel 3,5-diaryl pyrazole derivatives using combinatorial chemistry as inhibitors of tyrosinase as well as potent anticancer, anti-inflammatory agents. Bioorg Med Chem 18:6149–6155
79. Zheng L-W, Wu L-L, Zhao B-X, Dong W-L, Miao J-Y (2009) Synthesis of novel substituted pyrazole-5-carbohydrazide hydrazone derivatives and discovery of a potent apoptosis inducer in A549 lung cancer cells. Bioorg Med Chem 17:1957–1962
80. Xia Y, Dong Z-W, Zhao B-X, Ge X, Meng N, Shin D-S, Miao J-Y (2007) Synthesis and structure-activity relationships of novel 1-arylmethyl-3-aryl-1H-pyrazole-5-carbohydrazide derivatives as potential agents against A549 lung cancer cells. Bioorg Med Chem 15:6893–6899
81. Lian S, Su H, Zhao B-X, Liu W-Y, Zheng L-W, Miao J-Y (2009) Synthesis and discovery of pyrazole-5- carbohydrazide N-glycosides as inducer of autophagy in A549 lung cancer cells. Bioorg Med Chem 17:7085–7092
82. Wei F, Zhao B-X, Huang B, Zhang L, Sun C-H, Dong W-L, Shin D-S, Miao J-Y (2006) Design, synthesis, and preliminary biological evaluation of novel ethyl 1-(20-hydroxy-30-aroxypropyl)-3-aryl- 1H-pyrazole-5-carboxylate. Bioorg Med Chem Lett 16:6342–6347
83. Ding X-L, Zhang H-Y, Qi L, Zhao B-X, Lian S, Lv H-S, Miao J-Y (2009) Synthesis of novel pyrazolo carboxamide derivatives and discovery of modulators for apoptosis or autophagy in A549 lung cancer cells. Bioorg Med Chem Lett 19:5325–5328
84. Zheng L-W, Li Y, Ge D, Zhao B-X, Liu Y-R, Lv H-S, Ding J, Miao J-Y (2010) Synthesis of novel oxime-containing pyrazole derivatives and discovery of regulators for apoptosis and autophagy in A549 lung cancer cells. Bioorg Med Chem Lett 20:4766–4770
85. Liu Y-R, Luo J-Z, Duan P-P, Shao J, Zhao B-X, Miao J-Y (2012) Synthesis of pyrazole peptidomimetics and their inhibition against A549 lung cancer cells. Bioorg Med Chem Lett 22:6882–6887
86. Fan C-D, Zhao B-X, Wei F, Zhang G-H, Dong W-L, Miao J-Y (2008) Synthesis and discovery of autophagy inducers for A549 and H460 lung cancer cells, novel 1-(20-hydroxy-30-aroxypropyl)-3-aryl-1H-pyrazole-5-carbohydrazide derivatives. Bioorg Med Chem Lett 18:3860–3864
87. Shen S-L, Zhu J, Li M, Zhao B-X, Miao J-Y (2012) Synthesis of ferrocenyl pyrazole-containing chiral aminoethanol derivatives and their inhibition against A549 and H322 lung cancer cells. Eur J Med Chem 54:287–294
88. Farag AM, Mayhoub AS, Barakat SE, Bayomi AH (2008) Regioselective synthesis and antitumor screening of some novel N-phenylpyrazole derivatives. Bioorg Med Chem 16:881–889
89. Jin CH, Krishnaiah M, Sreenu D, Subrahmanyam VB, Rao KS, Mohan AVN, Park C-Y, Son J-Y, Sheen YY (2011) Synthesis and biological evaluation of 1-substituted-3-(6-methylpyridin-2-yl)-4-([1,2,4] triazolo[1,5-a]pyridin-6-yl) pyrazoles as transforming growth factor-β type 1 receptor kinase inhibitors. Bioorg Med Chem Lett 21:6049–6053
90. Newhouse BJ, Hansen JD, Grina J, Welch M, Topalov G, Littman N, Callejo M, Martinsona M, Galbraitha S, Laird ER, Brandhuber BJ, Vigers G, Morales T, Woessner R, Randolph N, Lyssikatos J, Olivero A (2011) Non-oxime pyrazole based inhibitors of B-Raf kinase. Bioorg Med Chem Lett 21:3488–3492
91. Zhang D, Wang G, Zhao G, Xu W, Huo L (2011) Synthesis and cytotoxic activity of novel 3-(1H-indol-3-yl)-1H-pyrazole-5-carbohydrazide derivatives. Eur J Med Chem 46:5868–5877
92. Baciu-Atudosie L, Ghinet A, Farce A, Dubois J, Belei D, Bîcu E (2012) Synthesis and biological evaluation of new phenothiazine derivatives bearing a pyrazole unit as protein farnesyltransferase inhibitors. Bioorg Med Chem Lett 22:6896–6902

93. Bai X-G, Yu D-K, Wang J-X, Zhang H, He H-W, Shao R-G, Li X-M, Wang Y-C (2012) Design, synthesis and anticancer activity of 1-acyl-3-amino-1,4,5,6-tetrahydropyrrolo[3,4-c]pyrazole derivatives. Bioorg Med Chem Lett 22:6947–6951
94. Bavetsias V, Crumpler S, Sun C, Avery S, Atrash B, Faisal A, Moore AS, Kosmopoulou M, Brown N, Sheldrake PW, Bush K, Henley A, Box G, Valenti M, Brandon ADH, Raynaud FI, Workman P, Eccles SA, Bayliss R, Linardopoulos S, Blagg J (2012) Optimization of imidazo[4,5-b]pyridine-based kinase inhibitors: identification of a dual FLT3/Aurora kinase inhibitor as an orally bioavailable preclinical development candidate for the treatment of acute myeloid leukemia. J Med Chem 55:8721–8734
95. Bondock S, Adel S, Etman HA, Badria FA (2012) Synthesis and antitumor evaluation of some new 1,3,4-oxadiazole-based heterocycles. Eur J Med Chem 48:192–199
96. Cui JJ, McTigue M, Nambu M, Tran-Dubé M, Pairish M, Shen H, Jia L, Cheng H, Hoffman J, Le P, Jalaie M, Goetz GH, Ryan K, Grodsky N, Deng Y-l, Parker M, Timofeevski S, Murray BW, Yamazaki S, Aguirre S, Li Q, Zou H, Christensen J (2012) Discovery of a novel class of exquisitely selective mesenchymal-epithelial transition factor (c-MET) protein kinase inhibitors and identification of the clinical candidate 2-(4-(1-(Quinolin-6-ylmethyl)-1H-[1,2,3]triazolo[4,5-b]pyrazin-6-yl)-1H-pyrazol-1-yl)ethanol (PF-04217903) for the treatment of cancer. J Med Chem 55:8091–8109
97. El Hamid MKA, Mihovilovic MD, El-Nassan HB (2012) Synthesis of novel pyrazolo[3,4-d]pyrimidine derivatives as potential anti-breast cancer agents. Eur J Med Chem 57:323–328
98. El-Borai MA, Rizk HF, Abd-Aal MF, El-Deeb IY (2012) Synthesis of pyrazolo[3,4-b]pyridines under microwave irradiation in multi-component reactions and their antitumor and antimicrobial activities-Part 1. Eur J Med Chem 48:92–96
99. Hanan EJ, Van Abbema A, Barrett K, Blair WS, Blaney J, Chang C, Eigenbrot C, Flynn S, Gibbons P, Hurley CA, Kenny JR, Kulagowski J, Lee L, Magnuson SR, Morris C, Murray J, Pastor RM, Rawson T, Siu M, Ultsch M, Zhou A, Sampath D, Lyssikatos JP (2012) Discovery of potent and selective pyrazolopyrimidine janus kinase 2 inhibitors. J Med Chem 55:10090–100107
100. Huang Y-Y, Wang L-Y, Chang C-H, Kuo Y-H, Kaneko K, Takayama H, Kimura M, Juang S-H, Wong FF (2012) One-pot synthesis and antiproliferative evaluation of pyrazolo[3,4-d]pyrimidine derivatives. Tetrahedron 68:9658–9664
101. Li X, Lu X, Xing M, Yang X-H, Zhao T-T, Gong H-B, Zhu H-L, Synthesis, biological evaluation, and molecular docking studies of N,1,3-triphenyl-1H-pyrazole-4-carboxamide derivatives as anticancer agents. Bioorg Med Chem Lett 22:3589–3593
102. Metwally M, Gouda M, Harmal AN, Khalil A (2012) Synthesis, antitumor, cytotoxic and antioxidant evaluation of some new pyrazolotriazines attached to antipyrine moiety. Eur J Med Chem 56:254–262
103. Mohareb RM, Al-Omran F (2012) Reaction of pregnenolone with cyanoacetylhydrazine: novel synthesis of hydrazide–hydrazone, pyrazole, pyridine, thiazole, thiophene derivatives and their cytotoxicity evaluations. Steroids 77:1551–1559
104. Mohareb RM, Wardakhan WW, Elmegeed GA, Ashour RM (2012) Heterocyclizations of pregnenolone: novel synthesis of thiosemicarbazone, thiophene, thiazole, thieno[2,3-b]pyridine derivatives and their cytotoxicity evaluations. Steroids 77:1560–1569
105. Puthiyapurayil P, Poojary B, Chikkanna C, Buridipad SK (2012) Design, synthesis and biological evaluation of a novel series of 1,3,4-oxadiazole bearing N-methyl-4-(trifluoromethyl) phenyl pyrazole moiety as cytotoxic agents. Eur J Med Chem 53:203–210
106. Strocchi E, Fornari F, Minguzzi M, Gramantieri L, Milazzo M, Rebuttini V, Breviglieri S, Camaggi CM, Locatelli E, Bolondi L, Comes-Franchini M (2012) Design, synthesis and biological evaluation of pyrazole derivatives as potential multi-kinase inhibitors in hepatocellular carcinoma. Eur J Med Chem 48:391–401
107. Vujasinović I, Paravić-Radičević A, Mlinarić-Majerski K, Brajša K, Bertoša B (2012) Synthesis and biological validation of novel pyrazole derivatives with anticancer activity guided by 3D-QSAR analysis. Bioorg Med Chem 20:2101–2110

108. Yamamoto S, Tomita N, Suzuki Y, Suzaki T, Kaku T, Hara T, Yamaoka M, Kanzaki N, Hasuoka A, Baba A, Ito M (2012) Design, synthesis, and biological evaluation of 4-arylmethyl-1-phenylpyrazole and 4-aryloxy-1-phenylpyrazole derivatives as novel androgen receptor antagonists. Bioorg Med Chem 20:2338–2352
109. Al-Adiwish WM, Tahir M, Siti-Noor-Adnalizawati A, Hashim SF, Ibrahim N, Yaacob WA (2013) Synthesis, antibacterial activity and cytotoxicity of new fused pyrazolo[1,5-a]pyrimidine and pyrazolo[5,1-c] [1,2,4]triazine derivatives from new 5-aminopyrazoles. Eur J Med Chem 64:464–476
110. Bavetsias V, Faisal A, Crumpler S, Brown N, Kosmopoulou M, Joshi A, Atrash B, Pérez-Fuertes Y, Schmitt JA, Boxall KJ, Burke R, Sun C, Avery S, Bush K, Henley A, Raynaud FI, Workman P, Bayliss R, Linardopoulos S, Blagg J (2013) Aurora isoform selectivity: design and synthesis of imidazo[4,5-b]pyridine derivatives as highly selective inhibitors of Aurora-A kinase in cells. J Med Chem 56:9122–9135
111. Desai B, Dixon K, Farrant E, Feng Q, Gibson KR, Van Hoorn WP, Mills J, Morgan T, Parry DM, Ramjee MK, Selway CN, Tarver GJ, Whitlock G, Wright AG, Rapid discovery of a novel series of Abl kinase inhibitors by application of an integrated microfluidic synthesis and screening platform. J Med Chem 56:3033–3047
112. Kamal A, Tamboli JR, Vishnuvardhan M, Adil S, Nayak VL, Ramakrishna S, Synthesis and anticancer activity of heteroaromatic linked 4-amido podophyllotoxins as apoptotic inducing agents. Bioorg Med Chem Lett 23:273–280
113. Koca İ, Özgür A, Coşkun KA, Tutar Y (2013) Synthesis and anticancer activity of acyl thioureas bearing pyrazole moiety. Bioorg Med Chem 21:3859–3865
114. Miyamoto N, Sakai N, Hirayama T, Miwa K, Oguro Y, Oki H, Okada K, Takagi T, Iwata H, Awazu Y, Yamasaki S, Takeuchi T, Miki H, Hori A, Imamura S (2013) Discovery of N-[5-({2-[(cyclopropylcarbonyl) amino]imidazo[1,2-b]pyridazin-6-yl}oxy)-2- methylphenyl]-1,3-dimethyl-1H-pyrazole-5-carboxamide (TAK-593), a highly potent VEGFR2 kinase inhibitor. Bioorg Med Chem 21:2333–2345
115. Shen S-L, Shao J-H, Luo J-Z, Liu J-T, Miao J-Y, Zhao B-X (2013) Novel chiral ferrocenylpyrazolo[1,5-a][1,4]diazepin-4-one derivatives–synthesis, characterization and inhibition against lung cancer cells. Eur J Med Chem 63:256–268
116. Zhu S-L, Wu Y, Liu C-J, Wei C-Y, Tao J-C, Liu H-M (2013) Synthesis and in vitro cytotoxic activity evaluation of novel heterocycle bridged carbothioamide type isosteviol derivatives as antitumor agents. Bioorg Med Chem Lett 23:1343–1346
117. Zhu S-L, Wu Y, Liu C-J, Wei C-Y, Tao J-C, Liu H-M (2013) Design and stereoselective synthesis of novel isosteviol-fused pyrazolines and pyrazoles as potential anticancer agents. Eur J Med Chem 65:70–82
118. Ali AR, El-Bendary ER, Ghaly MA, Shehata IA (2014) Synthesis, in vitro anticancer evaluation and in silico studies of novel imidazo[2,1-b]thiazole derivatives bearing pyrazole moieties. Eur J Med Chem 75:492–500
119. Sun J, Lv X-H, Qiu H-Y, Wang Y-T, Du Q-R, Li D-D, Yang Y-H, Zhu H-L (2013) Synthesis, biological evaluation and molecular docking studies of pyrazole derivatives coupling with a thiourea moiety as novel CDKs inhibitors. Eur J Med Chem 68:1–9
120. Zheng Y, Zheng M, Ling X, Liu Y, Xue Y, An L, Gu N, Jin M (2013) Design, synthesis, quantum chemical studies and biological activity evaluation of pyrazole–benzimidazole derivatives as potent Aurora A/B kinase inhibitors. Bioorg Med Chem Lett 23:3523–3530
121. Pirol ŞC, Çalışkan B, Durmaz İ, Atalay R, Banoglu E (2014) Synthesis and preliminary mechanistic evaluation of 5-(p-tolyl)-1-(quinolin-2-yl)pyrazole-3-carboxylic acid amides with potent antiproliferative activity on human cancer cell lines. Eur J Med Chem 87:140–149
122. Hopa C, Yildirim H, Kara H, Kurtaran R, Alkan M (2014) Synthesis, characterization and anti-proliferative activity of Cd (II) complexes with NNN type pyrazole-based ligand and pseudohalide ligands as coligand. Spectrochim Acta Part A 121:282–287
123. Jin CH, Krishnaiah M, Sreenu D, Subrahmanyam VB, Park H-J, Park S-J, Sheena YY, Kim D-K (2014) 4-([1,2,4]Triazolo[1,5-a]pyridin-6-yl)-5(3)-(6-methylpyridin-2-yl) imidazole and-pyrazole derivatives as potent and selective inhibitors of transforming growth factor-β type I receptor kinase. Bioorg Med Chem 22:2724–2732

124. Li S, Xu S, Tang Y, Ding S, Zhang J, Wang S, Zhou G, Zhou C, Li X (2014) Synthesis, anticancer activity and DNA-binding properties of novel 4-pyrazolyl-1,8-naphthalimide derivatives. Bioorg Med Chem Lett 24:586–590
125. Park BS, Al-Sanea MM, Abdelazem AZ, Park HM, Roh EJ, Park H-M, Yoo KH, Sim T, Tae JS, Lee SH (2014) Structure-based optimization and biological evaluation of trisubstituted pyrazole as a core structure of potent ROS1 kinase inhibitors. Bioorg Med Chem 22:3871–3878
126. Reddy GL, Guru SK, Srinivas M, Pathania AS, Mahajan P, Nargotra A, Bhushan S, Vishwakarma RA, Sawantad SD (2014) Synthesis of 5-substituted-1H-pyrazolo[4,3-d]pyrimidin-7(6H)-one analogs and their biological evaluation as anticancer agents: MTOR inhibitors. Eur J Med Chem 80:201–208
127. Wang S-F, Zhu Y-L, Zhu P-T, Makawana JA, Zhang Y-L, Zhao M-Y, Lv P-C, Zhu H-L (2014) Design, synthesis and biological evaluation of novel 5-phenyl-1H-pyrazole derivatives as potential BRAF V600E inhibitors. Bioorg Med Chem 22:6201–6208
128. Yao Y, Liao C, Li Z, Wang Z, Sun Q, Liu C, Yang Y, Tu Z, Jiang S (2014) Design, synthesis, and biological evaluation of 1,3-disubstituted-pyrazole derivatives as new class I and IIb histone deacetylase inhibitors. Eur J Med Chem 86:639–652
129. Zhang J-F, Li M, Miao J-Y, Zhao B-X (2014) Biological activities of novel pyrazolyl hydroxamic acid derivatives against human lung cancer cell line A549. Eur J Med Chem 83:516–525
130. Abd E-KS, Anwar MM, Mohamed NA, Nasr T, Elseginy SA (2015) Design, synthesis, biological evaluation and molecular docking studies of novel benzofuran–pyrazole derivatives as anticancer agents. Bioorg Chem 63:1–12
131. Aware V, Gaikwad N, Chavan S, Manohar S, Bose J, Khanna S, B-Rao C, Dixit N, Singh KS, Damre A, Sharma R, Patil S, Roychowdhury A (2015) Cyclopentyl-pyrimidine based analogues as novel and potent IGF-1R inhibitor. Eur J Med Chem 92:246–256
132. Ibrahim HS, Abou-Seri SM, Tanc M, Elaasser MM, Abdel-Aziz HA, Supuran CT (2015) Isatin-pyrazole benzenesulfonamide hybrids potently inhibit tumor-associated carbonic anhydrase isoforms IX and XII. Eur J Med Chem 103:583–593
133. Kamal A, Shaik AB, Polepalli S, Kumar GB, Reddy VS, Mahesh R, Garimella S, Jain N (2015) Synthesis of arylpyrazole linked benzimidazole conjugates as potential microtubule disruptors. Bioorg Med Chem 23:1082–1095
134. Khloya P, Ceruso M, Ram S, Supuran CT, Sharma PK (2015) Sulfonamide bearing pyrazolylpyrazolines as potent inhibitors of carbonic anhydrase isoforms I, II, IX and XII. Bioorg Med Chem Lett 25:3208–3212
135. Maggio B, Raimondi MV, Raffa D, Plescia F, Cascioferro S, Cancemi G, Tolomeo M, Grimaudo S, Daidone G (2015) Synthesis and antiproliferative activity of 3-(2-chloroethyl)-5-methyl-6-phenyl-8- (trifluoromethyl)-5,6-dihydropyrazolo[3,4-f][1,2,3,5]tetrazepin-4-(3H)-one. Eur J Med Chem 96:98–104
136. Nitulescu GM, Draghici C, Olaru OT, Matei L, Ioana A, Dragu LD, Bleotu C (2015) Synthesis and apoptotic activity of new pyrazole derivatives in cancer cell lines. Bioorg Med Chem 23:5799–5808
137. Nitulescu GM, Matei L, Aldea IM, Draghici C, Olaru OT, Bleotu C (2019) Ultrasound-assisted synthesis and anticancer evaluation of new pyrazole derivatives as cell cycle inhibitors. Arab J Chem 12:816–824
138. Rai US, Isloor AM, Shetty P, Pai K, Fun H-K (2015) Synthesis and in vitro biological evaluation of new pyrazole chalcones and heterocyclic diamides as potential anticancer agents. Arab J Chem 8:317–321
139. Reddy TS, Kulhari H, Reddy VG, Bansal V, Kamal A, Shukla R (2015) Design, synthesis and biological evaluation of 1,3-diphenyl-1H-pyrazole derivatives containing benzimidazole skeleton as potential anticancer and apoptosis inducing agents. Eur J Med Chem 101:790–805
140. Siddiqui T, Alam MG, Dar AM (2015) Synthesis, characterization and anticancer studies of new steroidal oxadiazole, pyrrole and pyrazole derivatives. J Saudi Chem Soc 19:387–391

141. Shi JB, Tang WJ, Li R, Liu XH (2015) Novel pyrazole-5-carboxamide and pyrazole–pyrimidine derivatives: synthesis and anticancer activity. Eur J Med Chem 90:889–896
142. Alam R, Wahi D, Singh R, Sinha D, Tandon V, Grover A (2016) Design, synthesis, cytotoxicity, HuTopoII inhibitory activity and molecular docking studies of pyrazole derivatives as potential anticancer agents. Bioorg Chem 69:77–90
143. Lv X-H, Ren Z-L, Zhou B-G, Li Q-S, Chu M-J, Liu D-H, Zhang L-S, Yao X-K, Cao H-Q (2016) Discovery of N-(benzyloxy)-1,3-diphenyl-1H-pyrazole-4-carboxamide derivatives as potential antiproliferative agents by inhibiting MEK. Bioorg Med Chem 24:4652–4659
144. Wang F-Q, Yang H, He B, Jia Y-K, Meng S-Y, Zhang C, Liu H-M, Liu F-W (2016) A novel domino approach for synthesis of indolyl tetrahydropyrano[4,3-c]pyrazole derivatives as anticancer agents. Tetrahedron 72:5769–5775
145. Wen J, Bao Y, Niu Q, Yang J, Fan Y, Li J, Jing Y, Zhao L, Liu D (2016) Identification of *N*-(6-mercaptohexyl)-3-(4-pyridyl)-1H-pyrazole-5-carboxamide and its disulfide prodrug as potent histone deacetylase inhibitors with in vitro and in vivo anti-tumor efficacy. Eur J Med Chem 109:350–359
146. Galal SA, Abdelsamie AS, Shouman SA, Attia YM, Ali HI, Tabll A, El-Shenawye R, El Abd YS, Ali MM, Mahmoud AE, Abdel-Halim AH, Fyiad AA, Girgis AS, El-Diwani HI (2017) Part I: design, synthesis and biological evaluation of novel pyrazole-benzimidazole conjugates as checkpoint kinase 2 (Chk2) inhibitors with studying their activities alone and in combination with genotoxic drugs. Eur J Med Chem 134:392–405
147. Huang QP, Zhang SN, Zhang SH, Wang K, Xiao Y (2017) Solvent and copper ion-induced synthesis of pyridyl–pyrazole-3-one derivatives: crystal structure cytotoxicity. Molecules 22:1813–1823
148. Li J, Huo H, Guo R, Liu B, Li L, Dan W, Xiao X, Zhang J, Shi B (2017) Facile and efficient access to Androsten-17-(10,30,40)-pyrazoles and Androst-17β-(10,30,40)-pyrazoles via Vilsmeier reagents, and their antiproliferative activity evaluation in vitro. Eur J Med Chem 130:1–14
149. Shi W, Hu J, Bao N, Li D, Chen L, Sun J (2017) Design, synthesis and cytotoxic activities of scopoletin-isoxazole and scopoletin-pyrazole hybrids. Bioorg Med Chem Lett 27:147–151
150. Wang M, Xu S, Lei H, Wang C, Xiao Z, Jia S, Zhi J, Zheng P, Zhu W (2017) Design, synthesis and antitumor activity of Novel Sorafenib derivatives bearing pyrazole scaffold. Bioorg Med Chem 25:5754–5763
151. Daidone G, Maggio B, Raffa D, Plescia S, Schillaci D, Raimondi MV (2004) Synthesis and in vitro antileukemic activity of new 4-triazenopyrazole derivatives. Farmaco 59:413–417
152. Chauhan S, Paliwal S, Chauhan R (2014) Anticancer activity of pyrazole via different biological mechanisms. Synth Commun 44:1333–1374
153. Vijayaraghavan S, Moulder S, Keyomarsi K, Layman RM (2018) Inhibiting CDK in cancer therapy: current evidence and future directions. Target Oncol 13:21–38
154. Chieffi P (2018) Aurora B: a new promising therapeutic target in cancer. Intractable Rare Dis Res 7:141–144
155. Marcucci G, Perrotti D, Caligiuri MA (2003) Understanding the molecular basis of imatinib mesylate therapy in chronic myelogenous leukemia and the related mechanisms of resistance. Clin Cancer Res 9:1333–1337
156. Kamble RD, Meshram RJ, Hese SV, More RA, Kamble SS, Gacche RN, Dawane BS (2016) Synthesis and in silico investigation of thiazoles bearing pyrazoles derivatives as anti-inflammatory agents. Comput Biol Chem 61:86–96
157. Bekhit AA, Abdel-Aziem T (2004) Design, synthesis and biological evaluation of some pyrazole derivatives as anti-inflammatory-antimicrobial agents. Bioorg Med Chem 12:1935–1945
158. Bekhit AA, Ashour HMA, Abdel Ghany YS, Bekhit A-A, Baraka A, Synthesis and biological evaluation of some thiazolyl and thiadiazolyl derivatives of 1H-pyrazole as anti-inflammatory antimicrobial agents. Eur J Med Chem 43:456–463
159. Girisha KS, Kalluraya B, Narayana V, Padmashree V, Synthesis and pharmacological study of 1-acetyl/propyl-3-aryl-5-(5-chloro-3-methyl-1-phenyl-1H-pyrazol-4-yl)-2-pyrazoline. Eur J Med Chem 45:4640–4644

160. Brullo C, Spisani S, Selvatici R, Bruno O (2012) N-Aryl-2-phenyl-2,3-dihydro-imidazo[1,2-b]pyrazole-1-carboxamides 7-substituted strongly inhibiting both fMLP-OMe-and IL-8-induced human neutrophil chemotaxis. Eur J Med Chem 47:573–579
161. El-Moghazy S, Barsoum F, Abdel-Rahman H, Marzouk A (2012) Synthesis and anti-inflammatory activity of some pyrazole derivatives. Med Chem Res 21:1722–1733
162. El-Sayed MA-A, Abdel-Aziz NI, Abdel-Aziz AA-M, El-Azab AS, ElTahir KEH (2012) Synthesis, biological evaluation and molecular modeling study of pyrazole and pyrazoline derivatives as selective COX-2 inhibitors and anti-inflammatory agents. Part 2. Bioorg Med Chem 20:3306–3316
163. Qiu K-M, Yan R, Xing M, Wang H-H, Cui H-E, Gong H-B, Zhu H-L (2012) Synthesis, biological evaluation and molecular modeling of dihydro-pyrazolyl-thiazolinone derivatives as potential COX-2 inhibitors. Bioorg Med Chem 20:6648–6654
164. Alegaon SG, Alagawadi KR, Garg MK, Dushyant K, Vinod D (2014) 1,3,4-Trisubstituted pyrazole analogues as promising anti-inflammatory agents. Bioorg Chem 54:51–59
165. Hassan GS, Abou-Seri SM, Kamel G, Ali MM (2014) Celecoxib analogs bearing benzofuran moiety as cyclooxygenase-2 inhibitors: design, synthesis and evaluation as potential anti-inflammatory agents. Eur J Med Chem 76:482–493
166. Karrouchi K, Chemlal L, Doudach L, Taoufik J, Cherrah Y, Radi S, Fouzi MA, Ansar M (2014) Synthesis, anti-inflammatory and antioxidant activities of some new pyrazole derivatives. J Pharm Res 8:1171–1177
167. Selvam TP, Kumar PV, Saravanan G, Prakash CR (2014) Microwave-assisted synthesis, characterization and biological activity of novel pyrazole derivatives. J Saudi Chem Soc 18:1015–1021
168. Tewari AK, Singh VP, Yadav P, Gupta G, Singh A, Goel RK, Shinde P, Mohan CG (2014) Synthesis, biological evaluation and molecular modeling study of pyrazole derivatives as selective COX-2 inhibitors and anti-inflammatory agents. Bioorg Chem 56:8–15
169. El-Feky SAH, El-Samii ZKA, Osman NA, Lashine J, Kamel MA, Thabet HK (2015) Synthesis, molecular docking and anti-inflammatory screening of novel quinoline incorporated pyrazole derivatives using the Pfitzinger reaction II. Bioorg Chem 58:104–116
170. Hussain S, Kaushik D (2015) Noval 1-substituted-3,5-dimethyl-4-[(substituted phenyl)diazenyl] pyrazole derivatives: synthesis and pharmacological activity. J Saudi Chem Soc 19:274–281
171. Surendra KR, Arif IA, Ahamed A, Idhayadhulla A (2016) Anti-inflammatory and antimicrobial activities of novel pyrazole analogues. Saudi J Biol Sci 23:614–620
172. Li Y-R, Li C, Liu J-C, Guo M, Zhang T-Y, Sun L-P, Zheng C-J, Piao H-R (2015) Synthesis and biological evaluation of 1,3-diaryl pyrazole derivatives as potential antibacterial and anti-inflammatory agents. Bioorg Med Chem Lett 25:5052–5057
173. Pelcman B, Sanin A, Nilsson P, Schaal W, Olofsson K, Krog-Jensen C, Forsell P, Hallberg A, Larhed M, Boesen T, Kromann H, Claesson H-E (2015) N-Substituted pyrazole-3-carboxamides as inhibitors of human 15-lipoxygenase. Bioorg Med Chem Lett 25:3017–3023
174. Wang T, Banerjee D, Bohnert T, Chao J, Enyedy I, Fontenot J, Guertina K, Jones H, Lina EY, Marcotte D, Talreja T, Vloten KV (2015) Discovery of novel pyrazole-containing benzamides as potent RORγ inverse agonists. Bioorg Med Chem Lett 25:2985–2990
175. Thore S, Gupta SV, Baheti KG (2016) Novel ethyl-5-amino-3-methylthio-1H-pyrazole-4-carboxylates: synthesis and pharmacological activity. J Saudi Chem Sci 20:259–264
176. Nossier ES, Fahmy HH, Khalifa NM, El-Eraky WI, Baset MA (2017) Design and synthesis of novel pyrazole-substituted different nitrogenous heterocyclic ring systems as potential anti-inflammatory agents. Molecules 22:512–527
177. Maggio B, Daidone G, Raffa D, Plescia S, Mantione L, Cutuli VMC, Mangano NG, Caruso A (2001) Synthesis and pharmacological study of ethyl 1-methyl-5-(substituted 3,4-dihydro-4-oxoquinazolin-3-yl)- 1H-pyrazole-4-acetates. Eur J Med Chem 36:737–742
178. Hall A, Billinton A, Brown SH, Clayton NM, Chowdhury A, Giblin GMP, Goldsmith P, Hayhow TG, Hurst DN, Kilford IR, Naylor A, Passingham B, Winyard L (2008) Non-acidic pyrazole EP1 receptor antagonists with in vivo analgesic efficacy. Bioorg Med Chem Lett 18:3392–3399

179. De Moura SS, de Ávila RI, Brito LB, de Oliveira R, de Oliveira GAR, Pazini F, Aline RM, Cesar CB, Marize KG, Valadares C (2017) In vitro genotoxicity and in vivo subchronic evaluation of the anti-inflammatory pyrazole compound LQFM021. Chem Biol Interact 277:185–194
180. Ragab FA, Abdel Gawad NM, Georgey HH, Said MF (2013) Synthesis of novel 1,3,4-trisubstituted pyrazoles as anti-inflammatory and analgesic agents. Eur J Med Chem 63:645–654
181. Vijesh AM, Isloor AM, Shetty P, Sundershan S, Fun HK (2013) New pyrazole derivatives containing1,2,4-triazoles and benzoxazoles as potent antimicrobial and analgesic agents. Eur J Med Chem 62:410–415
182. Viveka S, Dinesha SP, Nagaraja GK, Ballav S, Kerkar S (2015) Design and synthesis of some new pyrazolyl-pyrazolines as potential anti-inflammatory, analgesic and antibacterial agents. Eur J Med Chem 101:442–451
183. Thore SN, Gupta SV, Baheti KG (2016) Synthesis and pharmacological evaluation of 5-methyl-2- phenylthiazole-4-substituted heteroazoles as a potential anti-inflammatory and analgesic agents. J Saudi Chem Soc 20:6–52
184. Chowdhury MA, Abdellatif KR, Dong Y, Knaus EE (2008) Synthesis of new 4-[2-(4-methyl(amino)sulfonylphenyl)-5-trifluoromethyl-2H-pyrazol-3-yl]-1,2,3,6-tetrahydropyridines: a search for novel nitric oxide donor anti-inflammatory agents. Bioorg Med Chem 16:8882–8888
185. Chowdhury MA, Abdellatif KR, Dong Y, Yu G, Huang Z, Rahman M, Das D, Velázquez CA, Suresh MR, Knaus EE (2010) Celecoxib analogs possessing a N-(4-nitrooxybutyl) piperidin-4-yl or N-(4-nitrooxybutyl)-1,2,3,6-tetrahydropyridin-4-yl nitric oxide donor moiety: synthesis, biological evaluation and nitric oxide release studies. Bioorg Med Chem Lett 20:1324–1329
186. Chavan HV, Bandgar BP, Adsul LK, Dhakane VD, Bhale PS, Thakare VN, Masand V (2013) Design, synthesis, characterization and anti-inflammatory evaluation of novel pyrazole amalgamated flavones. Bioorg Med Chem Lett 23:1315–1321
187. Abdelgawad MA, Labib MB, Abdel-Latif M (2017) Pyrazole-hydrazone derivatives as anti-inflammatory agents: design, synthesis, biological evaluation, COX-1,2/5-LOX inhibition and docking study. Bioorg Chem 74:212–220
188. Singh SK, Reddy PG, Rao KS, Lohray BB, Misra P, Rajjak SA, Rao YK, Venkateswarlu A (2004) Polar substitutions in the benzenesulfonamide ring of celecoxib afford a potent 1,5-diarylpyrazole class of COX-2 inhibitors. Bioorg Med Chem Lett 14:499–504
189. Li J, DeMello KML, Cheng H, Sakya SM, Bronk BS, Rafka RJ, Jaynes BH, Ziegler CB, Kilroy C, Mann DW, Nimz EL, Lynch MP, Haven ML, Kolosko NL, Minich ML, Li C, Dutra JK, Rast B, Crosson RM, Morton BJ, Kirk GW, Callaghan KM, Koss DA, Shavnya A, Lund LA, Seibel SB, Petras CF, Silvia A (2004) Discovery of a potent, selective and orally active canine COX-2 inhibitor, 2-(3-difluoromethyl-5-phenyl-pyrazol-1-yl)-5-methanesulfonyl-pyridine. Bioorg Med Chem Lett 14:95–98
190. Cheng H, DeMello KML, Li J, Sakya SM, Ando K, Kawamura K, Kato T, Rafka RJ, Jaynes BH, Ziegler CB, Stevens R, Lund LA, Mann DW, Kilroy C, Haven ML, Nimz EL, Dutra JK, Li C, Minich ML, Kolosko NL, Petras C, Silvia AM, Seibel SB (2006) Synthesis and SAR of heteroaryl-phenyl-substituted pyrazole derivatives as highly selective and potent canine COX-2 inhibitors. Bioorg Med Chem Lett 16:2076–2080
191. Sakya SM, DeMello KML, Minich ML, Rast B, Shavnya A, Rafka RJ (2006) 5-Heteroatom substituted pyrazoles as canine COX-2 inhibitors. Part 1: structure-activity relationship studies of 5-alkylamino pyrazoles and discovery of a potent, selective, and orally active analog. Bioorg Med Chem Lett 16:288–292
192. Sakya SM, Cheng H, DeMello KML, Shavnya A, Minich ML, Rast B (2006) 5-Heteroatom-substituted pyrazoles as canine COX-2 inhibitors: Part 2. Structure-activity relationship studies of 5-alkylethers and 5-thioethers. Bioorg Med Chem Lett 16:1202–1206
193. Aggarwal R, Bansal A, Rozas I, Kelly B, Kaushik P, Kaushik D (2013) Synthesis, biological evaluation and molecular modeling study of 5-trifluoromethyl-D2-pyrazoline and isomeric 5/3-trifluoromethylpyrazole derivatives as anti-inflammatory agents. Eur J Med Chem 70:350–357

194. Habeeb AG, Rao PNP, Knaus EE (2001) Design and synthesis of celecoxib and rofecoxib analogues as selective cyclooxygenase-2 (COX-2) inhibitors: replacement of sulfonamide and methylsulfonyl pharmacophores by an azido bioisostere. J Med Chem 44:3039–3042
195. Uddin MJ, Rao PNP, Knaus EE (2003) Design and synthesis of novel celecoxib analogues as selective cyclooxygenase-2 (COX-2) inhibitors: replacement of the sulfonamide pharmacophore by a sulfonylazide bioisostere. Bioorg Med Chem 11:5273–5280
196. Bansal S, Bala M, Suthar SK, Choudhary S, Bhattacharya S, Bhardwaj V, Singla S, Joseph A (2014) Design and synthesis of novel 2-phenyl-5-(1,3-diphenyl-1H-pyrazol-4-yl)-1,3,4-oxadiazoles as selective COX-2 inhibitors with potent anti-inflammatory activity. Eur J Med Chem 80:167–174
197. El-Sayed MAA, Abdel-Aziz NI, Abdel-Aziz AAM, El-Azab AS, Asiri YA, ElTahir KEH (2011) Design, synthesis, and biological evaluation of substituted hydrazone and pyrazole derivatives as selective COX-2 inhibitors: Molecular docking study. Bioorg Med Chem 19:3416–3424
198. Bekhit AA, Ashour HM, Bekhit AE-DA, Abdel-Rahman HM, Bekhit SA (2009) Synthesis of some pyrazolyl benzenesulfonamide derivatives as dual anti-inflammatory antimicrobial agents. J Enzym Inhib Med Chem 24:296–309
199. Tewari AK, Srivastava P, Singh VP, Singh A, Goel RK, Mohan CG (2010) Novel anti-inflammatory agents based on pyrazole based dimeric compounds; design, synthesis, docking and in vivo activity. Chem Pharm Bull 58:634–638
200. Alegaon SG, Hirpara MB, Alagawadi KR, Hullatti KK, Kashniyal K (2014) Synthesis of novel pyrazole–thiadiazole hybrid as potential potent and selective cyclooxygenase-2 (COX-2) inhibitors. Bioorg Med Chem Lett 24:5324–5329
201. Miyata Y (2003) Molecular chaperone HSP90 as a novel target for cancer chemotherapy. Folia Pharmacol Jpn 121:33–42
202. Manetti F, Magnani M, Castagnolo D, Passalacqua L, Botta M, Corelli F, Saddi M, Deidda D, Logu AD (2006) Ligand-based virtual screening, parallel solution-phase and microwave-assisted synthesis as tools to identify and synthesize new inhibitors of mycobacterium tuberculosis. Chem Med Chem 1:973–989
203. Almeida PE, Ramos DF, Bonacorso HG, de la Iglesia AI, Oliveira MR, Coelho T, Navarini J, Morbidoni HR, Zanatta N, Martins MAP (2008) Synthesis and in vitro antimycobacterial activity of 3-substituted 5-hydroxy-5-trifluoro[chloro]methyl-4,5-dihydro-1H-1-(isonicotinoyl)pyrazoles. Int J Antimicrob Agents 32:139–144
204. Castagnolo D, De Logu A, Radi M, Bechi B, Manetti F, Magnani M, Supino S, Meleddu R, Chisu L, Botta M, Synthesis, biological evaluation and SAR study of novel pyrazole analogues as inhibitors of Mycobacterium tuberculosis. Bioorg Med Chem 16:8587–85891
205. Velaparthi S, Brunsteiner M, Uddin R, Wan B, Franzblau SG, Petukhov PA (2008) 5-tert-Butyl-Npyrazol-4-yl-4,5,6,7-tetrahydrobenzo[d]isoxazole-3-carboxamide derivatives as novel potent inhibitors of Mycobacterium tuberculosis pantothenate synthetase: initiating a quest for new antitubercular drugs. J Med Chem 51:1999–2002
206. Castagnolo D, Manetti F, Radi M, Bechi B, Pagano M, De Logu A, Meleddu R, Saddi M, Botta M (2009) Synthesis, biological evaluation, and SAR study of novel pyrazole analogues as inhibitors of Mycobacterium tuberculosis: Part 2. Synthesis of rigid pyrazolones. Bioorg Med Chem 17:5716–5721
207. Ahsan MJ, Samy JG, Soni S, Jain N, Kumar L, Sharma LK, Yadav H, Saini L, Kalyansing RG, Devenda NS, Prasad R, Jain CB (2011) Discovery of novel antitubercular 3a,4-dihydro-3H-indeno[1,2-c] pyrazole-2-carboxamide/carbothioamide analogues. Bioorg Med Chem Lett 21:5259–5261
208. Trivedi AR, Dodiya DK, Dholariya BH, Kataria VB, Bhuva VR, Shah VH (2011) Synthesis and biological evaluation of some novel N-aryl-1,4-dihydropyridines as potential antitubercular agents. Bioorg Med Chem Lett 21:5181–5183
209. Khunt RC, Khedkar VM, Chawda RS, Chauhan NA, Parikh AR, Coutinho EC (2012) Synthesis, antitubercular evaluation and 3D-QSAR study of N-phenyl-3-(4-fluorophenyl)-4-substituted pyrazole derivatives. Bioorg Med Chem Lett 22:666–678

210. Shelke SN, Mhaske GR, Bonifácio VD, Gawande MB (2012) Green synthesis and anti-infective activities of fluorinated pyrazoline derivatives. Bioorg Med Chem Lett 22:5727–5730
211. Anand N, Ramakrishna K, Gupt MP, Chaturvedi V, Singh S, Srivastava KK, Sharma P, Rai N, Ramachandran R, Dwivedi A, Gupta V, Kumar B, Pandey S, Shukla PK, Pandey SK, Lal J, Tripathi RP (2013) Identification of 1-[4-benzyloxyphenyl)-but-3-enyl]-1H-azoles as new class of antitubercular and antimicrobial agents. ACS Med Chem Lett 4:958–963
212. Fullam E, Talbot J, Abuhammed A, Westwood I, Davies SG, Russell AJ, Sim E (2013) Design, synthesis and structure-activity relationships of 3,5-diaryl-1H-pyrazoles as inhibitors of arylamine N-acetyltransferase. Bioorg Med Chem Lett 23:2759–2764
213. Hernández P, Rojas R, Gilman RH, Sauvain M, Lima LM, Barreiro EJ, González M, Cerecetto H (2013) Hybrid furoxanyl N-acylhydrazone derivatives as hits for the development of neglected diseases drug candidates. Eur J Med Chem 59:64–74
214. Maurya HK, Verma R, Alam S, Pandey S, Pathak V, Sharma S, Srivastava KK, Negi AS, Gupta A (2013) Studies on substituted benzo[h]quinazolines, benzo[g]indazoles, pyrazoles, 2,6-diarylpyridines as anti-tubercular agents. Bioorg Med Chem Lett 23:5844–5849
215. North EJ, Scherman MS, Bruhn DF, Scarborough JS, Maddox MM, Jones V, Grzegorzewicz A, Yang L, Hess T, Morisseau C, Jackson M, McNeil MR, Lee RE (2013) Design, synthesis and anti-tuberculosis activity of 1-adamantyl-3-heteroaryl ureas with improved in vitro pharmacokinetic properties. Bioorg Med Chem 21:2587–2599
216. Palanisamy P, Jenniefer SJ, Muthiah PT, Kumaresan S (2013) Synthesis, characterization, antimicrobial, anticancer, and anti-tuberculosis activity of some new pyrazole, isoxazole, pyrimidine and benzodiazepine derivatives containing thiochromeno and benzothiepino moieties. RSC Adv 3:19300–19310
217. Samala G, Devi PB, Nallangi R, Yogeeswari P, Sriram D (2013) Development of 3-phenyl-4,5,6,7-tetrahydro-1H-pyrazolo[4,3-c]pyridine derivatives as novel Mycobacterium tuberculosis pantothenate synthetase inhibitors. Eur J Med Chem 69:356–364
218. Shirude PS, Madhavapeddi P, Naik M, Murugan K, Shinde V, Nandishaiah R, Bhat J, Kumar A, Hameed S, Holdgate G, Davies G, McMiken H, Hegde N, Ambady A, Venkatraman J, Panda M, Bandodkar B, Sambandamurthy VK, Read JA (2013) Methyl-thiazoles: a novel mode of inhibition with the potential to develop novel inhibitors targeting InhA in Mycobacterium tuberculosis. J Med Chem 56:8533–8542
219. Kalaria PN, Satasia SP, Raval DK (2014) Synthesis, characterization and pharmacological screening of some novel 5-imidazopyrazole incorporated polyhydroquinoline derivatives. Eur J Med Chem 78:207–216
220. Karad SC, Purohit VB, Raval DK, Design, synthesis and characterization of fluoro substituted novel pyrazolylpyrazolines scaffold and their pharmacological screening. Eur J Med Chem 84:51–58
221. Pathak V, Maurya HK, Sharma S, Srivastava KK, Gupta A (2014) Synthesis and biological evaluation of substituted 4,6-diarylpyrimidines and 3,5-diphenyl-4,5-dihydro-1H-pyrazoles as anti-tubercular agents. Bioorg Med Chem Lett 24:2892–2896
222. Villemagne B, Flipo M, Blondiaux N, Crauste C, Malaquin S, Leroux F, Piveteau C, Villeret V, Brodin P, Villoutreix BO, Sperandio O, Soror SH, Wohlkönig A, Wintjens R, Deprez B, Baulard AR, Willand N (2014) Ligand efficiency driven design of new inhibitors of mycobacterium tuberculosis transcriptional repressor EthR using fragment growing, merging, and linking approaches. J Med Chem 57:4876–4888
223. Bhatt JD, Chudasama CJ, Patel KD (2015) Pyrazole clubbed triazolo[1,5-a]pyrimidine hybrids as an anti-tubercular agents: synthesis, in vitro screening and molecular docking study. Bioorg Med Chem 23:7711–7716
224. Da Silva C, Ribeiro LB, Furuno CC, da Cunha GA, de Souza RF, Netto AV, Mauro AE, Frem RC, Fernandes JA, Paz FAA, Marino LB, Pavan FR, Leite CQF (2015) Pyrazolyl Pd (II) complexes containing triphenylphosphine: synthesis and antimycobacterial activity. Polyhedron 100:10–16

225. De Melo Candice S, Feng T-S, van der Westhuyzen R, Gessner RK, Street LJ, Morgans GL, Warner DF, Moosa A, Naran K, Lawrence N, Boshoff HIM, Barry CE III, Harris CJ, Gordon R, Chibale K (2015) Aminopyrazolo[1,5 a]pyrimidines as potential inhibitors of Mycobacterium tuberculosis: structure activity relationships and ADME characterization. Bioorg Med Chem 23:7240–7250
226. Mutai P, Pavadai E, Wiid I, Ngwane A, Baker B, Chibale K (2015) Synthesis, antimycobacterial evaluation and pharmacophore modeling of analogues of the natural product formononetin. Bioorg Med Chem Lett 25:2510–2513
227. Nayak N, Ramprasad J, Dalimba U (2015) New INH-pyrazole analogs: design, synthesis and evaluation of antitubercular and antibacterial activity. Bioorg Med Chem Lett 25:5540–5545
228. Tantry SJ, Degiacomi G, Sharma S, kumar Jena L, Narayan A, Guptha S, Shanbhag G, Menasinakai S, Mallya M, Awasthy D, Balakrishnan G, Kaur P, Bhattacharjee D, Narayan C, Reddy J, Naveen Kumar CN, Shandil R, Boldrin F, Ventura M, Manganelli R, Hartkoorn RC, Cole ST, Panda M, Markad SD, Ramachandran V, Ghorpade SR, Dinesh N (2015) Whole cell screen based identification of spiropiperidines with potent antitubercular properties. Bioorg Med Chem Lett 25:3234–3245
229. Guardia A, Gulten G, Fernandez R, Gómez J, Wang F, Convery M, Blanco D, Martínez M, Pérez-Herrán E, Alonso M, Ortega F, Rullás J, Calvo D, Mata L, Young R, Sacchettini JC, Mendoza-Losana A, Remuiñán M, Ballell PL, Castro-Pichel J (2016) N-Benzyl-4-((heteroaryl)methyl) benzamides: a new class of direct NADH-dependent 2-trans Enoyl-Acyl Carrier Protein Reductase (InhA) inhibitors with antitubercular activity. Chem Med Chem 11:687–701
230. Kavanagh ME, Coyne AG, McLean KJ, James GG, Levy CW, Marino LB, De Carvalho LPS, Chan DS, Hudson SA, Surade S, Leys D, Munro AW, Abell C (2016) Fragment-based approaches to the development of Mycobacterium tuberculosis CYP121 inhibitors. J Med Chem 59:3272–3302
231. Lu X, Tang J, Liu Z, Li M, Zhang T, Zhang X, Ding K (2016) Discovery of new chemical entities as potential leads against Mycobacterium tuberculosis. Bioorg Med Chem Lett 26:5916–5919
232. Nayak N, Ramprasad J, Dalimba U (2016) Synthesis and antitubercular and antibacterial activity of some active fluorine containing quinoline–pyrazole hybrid derivatives. J Fluor Chem 183:59–68
233. Ramesh R, Shingare RD, Kumar V, Anand A, Swetha B, Veeraraghavan S, Viswanadha S, Ummanni R, Gokhale R, Reddy DS (2016) Repurposing of a drug scaffold: identification of novel sila analogues of rimonabant as potent antitubercular agents. Eur J Med Chem 122:723–730
234. Amaroju S, Kalaga MN, Srinivasarao S, Napiorkowska A, Augustynowicz-Kopec E, Murugesan S, Chander S, Krishnan R, Chandra Sekhar KVG (2017) Identification and development of pyrazolo[4,3-c]pyridine carboxamides as Mycobacterium tuberculosis pantothenate synthetase inhibitors. New J Chem 41:347–357
235. Angelova VT, Valcheva V, Pencheva T, Voynikov Y, Vassilev N, Mihaylova R, Momekov G, Shivachev B (2017) Synthesis, antimycobacterial activity and docking study of 2-aroyl-[1]benzopyrano[4,3-c]pyrazol-4(1H)-one derivatives and related hydrazide-hydrazones. Bioorg Med Chem Lett 27:2996–3002
236. He C, Preiss L, Wang B, Fu L, Wen H, Zhang X, Cui H, Meier T, Yin D (2017) Structural simplification of bedaquiline: the discovery of 3-(4-(N, N-Dimethylaminomethyl)phenyl) quinoline-derived antitubercular lead compounds. Chem Med Chem 12:106–119
237. Aragade P, Palkar M, Ronad P, Satyanarayana D, Coumarinyl pyrazole derivatives of INH: promising antimycobacterial agents. Med Chem Res 22:2279–2283
238. Gajbhiye JM, More NA, Patil MD, Ummanni R, Kotapalli SS, Yogeeswari P, Sriram D, Masand VH (2015) Discovery of rimonabant and its potential analogues as anti-TB drug candidates. Med Chem Res 24:2960–2971
239. Nayak N, Ramprasad J, Dalimba U (2015) Design, synthesis, and biological evaluation of new 8-trifluoromethylquinoline containing pyrazole-3-carboxamide derivatives. J Heterocycl Chem 54:171–182

240. Trivedi AR, Dholariya BH, Vakhariya CP, Dodiya DK, Ram HK, Kataria VB, Siddiqui AB, Shah VH (2012) Synthesis and anti-tubercular evaluation of some novel pyrazolo[3,4-d]pyrimidine derivatives. Med Chem Res 21:1887–1891
241. Labana BM, Brahmbhatt GC, Sutariya TR, Parmar NJ, Padrón JM, Kant R, Gupta VK (2017) Efficient synthesis and biological evaluation of new benzopyran-annulated pyrano[2,3-c]pyrazole derivatives. Mol Divers 21:339–354
242. Encinas L, O'Keefe H, Neu M, Remuiñán MJ, Patel AM, Guardia A, Davie CP, Pérez-Macías N, Yang H, Convery MA, Messer JA, Pérez-Herrán E, Centrella PA, Álvarez-Gómez D, Clark MA, Huss S, O'Donovan GK, Ortega-Muro F, McDowell W, Castañeda P, Arico-Muendel CC, Pajk S, Rullás J, Angulo-Barturen I, Álvarez-Ruíz E, Mendoza-Losana A, Pages LB, Castro-Pichel J, Evindar G (2014) Encoded library technology as a source of hits for the discovery and lead optimization of a potent and selective class of bactericidal direct inhibitors of Mycobacterium tuberculosis InhA. J Med Chem 57:1276–1288
243. Rachakonda V, Alla M, Kotipalli SS, Ummani R (2013) Design, diversity-oriented synthesis and structure activity relationship studies of quinolinyl heterocycles as antimycobacterial agents. Eur J Med Chem 70:536–547
244. Pandit U, Dodiya A (2013) Synthesis and antitubercular activity of novel pyrazole–quinazolinone hybrid analogs. Med Chem Res 22:3364–3371
245. Mistry PT, Kamdar NR, Haveliwala DD, Patel SK (2012) Synthesis, characterization, and in vitro biological studies of some novel pyran fused pyrimidone derivatives. J Heterocycl Chem 49:349–357
246. Naik M, Humnabadkar V, Tantry SJ, Panda M, Narayan A, Guptha S, Panduga V, Manjrekar P, Jena LK, Koushik K, Shanbhag G, Jatheendranath S, Manjunatha MR, Gorai G, Bathula C, Rudrapatna S, Achar V, Sharma S, Ambady A, Hegde N, Mahadevaswamy J, Kaur P, Sambandamurthy VK, Awasthy D, Narayan C, Ravishankar S, Madhavapeddi P, Reddy J, Prabhakar KR, Saralaya R, Chatterji M, Whiteaker J, McLaughlin B, Chiarelli LR, Riccardi G, Pasca MR, Binda C, Neres J, Dhar N, Signorino-Gelo F, McKinney JD, Ramachandran V, Shandil R, Tommasi R, Iyer PS, Narayanan S, Hosagrahara V, Kavanagh S, Dinesh N, Ghorpade SR (2014) 4-aminoquinolone piperidine amides: noncovalent inhibitors of DprE1 with long residence time and potent antimycobacterial activity. J Med Chem 57:5419–5434
247. Manfredini S, Bazzanini R, Baraldi PG, Guarneri M, Simoni D, Marongiu ME, Pani A, Tramontano E, La Colla P (1992) Pyrazole-related nucleosides. Synthesis and antiviral/antitumor activity of some substituted pyrazole and pyrazolo[4,3-d]-1,2,3-triazin-4-one nucleosides. J Med Chem 35:917–924
248. Chen X, Schneller SW, Ikeda S, Snoeck R, Andrei G, Balzarini J, De Clercq E (1993) Synthesis and antiviral activity of 5'-deoxypyrazofurin. J Med Chem 36:3727–3730
249. Storer R, Ashton CJ, Baxter AD, Hann MM, Marr CLP, Mason AM, Mo C-L, Myers PL, Noble SA, Penn CR, Weir NG, Woods JM, Coe PL (1999) The synthesis and antiviral activity of 4-fluoro-1-β-D-ribofuranosyl-1H-pyrazole-3-carboxamide. Nucleosides Nucleotides 18:203–216
250. Genin MJ, Biles C, Keiser BJ, Poppe SM, Swaney SM, Tarpley WG, Yagi Y, Romero DL (2000) Novel 1,5-diphenylpyrazole nonnucleoside HIV-1 reverse transcriptase inhibitors with enhanced activity versus the delavirdine-resistant p236l mutant: lead identification and SAR of 3- and 4-substituted derivatives. J Med Chem 43:1034–1040
251. Rostom SAF, Shalaby MA, El-Demellawy MA (2003) Polysubstituted pyrazoles, part 5. Synthesis of new 1-(4-chlorophenyl)-4-hydroxy-1H-pyrazole-3-carboxylic acid hydrazide analogs and some derived ring systems. A novel class of potential antitumor and anti-HCV agents. Eur J Med Chem 38:959–974
252. Shamroukh AH, Zaki ME, Morsy EM, Abdel-Motti FM, Abdel-Megeid FM (2007) Synthesis of pyrazolo [4′,3′:5,6] pyrano[2,3-d]pyrimidine derivatives for antiviral evaluation. Arch Pharm 340:236–243
253. Sun A, Chandrakumar N, Yoon J-J, Plemper RK, Snyder JP (2007) Non-nucleoside inhibitors of the measles virus RNA-dependent RNA polymerase complex activity: synthesis and in vitro evaluation. Bioorg Med Chem Lett 17:5199–5203

254. Ouyang G, Chen Z, Cai X-J, Song B-A, Bhadury PS, Yang S, Jin L-H, Xue W, Hu D-Y, Zeng S (2008) Synthesis and antiviral activity of novel pyrazole derivatives containing oxime esters group. Bioorg Med Chem 16:9699–9707
255. Rashad AE, Hegab MI, Abdel-Megeid RE, Micky JA, Abdel-Megeid FM (2008) Synthesis and antiviral evaluation of some new pyrazole and fused pyrazolopyrimidine derivatives. Bioorg Med Chem 16:7102–7106
256. Zeng L-F, Zhang H-S, Wang Y-H, Sanchez T, Zheng Y-T, Neamati N, Long Y-Q (2008) Efficient synthesis and utilization of phenyl-substituted heteroaromatic carboxylic acids as aryl diketo acid isosteres in the design of novel HIV-1 integrase inhibitors. Bioorg Med Chem Lett 18:4521–4524
257. Mowbray CE, Burt C, Corbau R, Gayton S, Hawes M, Perros M, Tran I, Price DA, Quinton FJ, Selby MD, Stupple PA, Webster R, Wood A (2009) Pyrazole NNRTIs 4: selection of UK-453,061 (lersivirine) as a development candidate. Bioorg Med Chem Lett 19:5857–5860
258. Mowbray CE, Burt C, Corbau R, Perros M, Tran I, Stupple PA, Webster R, Wood A (2009) Pyrazole NNRTIs 1: design and initial optimisation of a novel template. Bioorg Med Chem Lett 19:5599–5602
259. Sidique S, Shiryaev SA, Ratnikov BI, Herath A, Su Y, Strongin AY, Cosford ND (2009) Structure-activity relationship and improved hydrolytic stability of pyrazole derivatives that are allosteric inhibitors of West Nile Virus NS2B-NS3 proteinase. Bioorg Med Chem Lett 19:5773–5777
260. Sujatha K, Shanthi G, Selvam NP, Manoharan S, Perumal PT, Rajendran M (2009) Synthesis and antiviral activity of 4,40-(arylmethylene)bis(1H-pyrazol-5-ols) against peste des petits ruminant virus (PPRV) Bioorg. Med Chem Lett 19:4501–4503
261. Riyadh SM, Farghaly TA, Abdallah MA, Abdalla MM, El-Aziz MRA (2010) New pyrazoles incorporating pyrazolylpyrazole moiety: synthesis, anti-HCV and antitumor activity. Eur J Med Chem 45:1042–1050
262. Shih S-R, Chu T-Y, Reddy GR, Tseng S-N, Chen H-L, Tang W-F, Wu M-S, Yeh J-Y, Chao Y-S, Hsu JT, Hsieh H-P, Horng J-T (2010) Pyrazole compound BPR1P0034 with potent and selective anti-influenza virus activity. J Biomed Sci 17:13–21
263. Su D-S, Lim JJ, Tinney E, Tucker TJ, Saggar S, Sisko JT, Wan B-L, Young MB, Anderson KD, Rudd D, Munshi V, Bahnck C, Felock PJ, Lu M, Lai M-T, Touch S, Moyer G, DiStefano DJ, Flynn JA, Liang Y, Sanchez R, Perlow-Poehnelt R, Miller M, Vacca JP, Williams TM, Anthony NJ (2010) Biaryl ethers as potent allosteric inhibitors of reverse transcriptase and its key mutant viruses: Aryl substituted pyrazole as a surrogate for the pyrazolopyridine motif. Bioorg Med Chem Lett 20:4328–4332
264. Dawood KM, Abdel-Gawad H, Mohamed HA, Badria FA (2011) Synthesis, anti-HSV-1, and cytotoxic activities of some new pyrazole-and isoxazole-based heterocycles. Med Chem Res 20:912–919
265. Wu L, Song B, Bhadury PS, Yang S, Hu D, Jin L (2011) Synthesis and antiviral activity of novel pyrazolo amides containing-aminophosphonate moiety. J Heterocycl Chem 48:389–396
266. Di Francesco ME, Avolio S, Pompei M, Pesci S, Monteagudo E, Pucci V, Giuliano C, Fiore F, Rowley M, Summa V, Synthesis and antiviral properties of novel 7-heterocyclic substituted 7-deaza-adenine nucleoside inhibitors of Hepatitis C NS5B polymerase. Bioorg Med Chem 20:4801–4811
267. Kim J, Lee D, Park C, So W, Jo M, Ok T, Kwon J, Kong S, Jo S, Kim Y, Choi J, Kim HC, Ko Y, Choi I, Park Y, Yoon J, Ju MK, Kim J, Han S-J, Kim T-H, Cechetto J, Nam J, Sommer P, Liuzzi M, Lee J, No Z (2012) Discovery of phenylaminopyridine derivatives as novel HIV-1 non-nucleoside reverse transcriptase inhibitors. ACS Med Chem Lett 3:678–682
268. Ndungu JM, Krumm SA, Yan D, Arrendale RF, Reddy GP, Evers T, Howard R, Natchus MG, Saindane MT, Liotta DC, Plemper RK, Snyder JP, Sun A (2012) Non-nucleoside inhibitors of the measles virus RNA-dependent RNA polymerase: synthesis, structure-activity relationships, and pharmacokinetics. J Med Chem 55:4220–4230
269. Tantawy AS, Nasr MN, El-Sayed MA, Tawfik SS, Synthesis and antiviral activity of new 3-methyl-1,5-diphenyl-1H-pyrazole derivatives. Med Chem Res 21:4139–4149

270. Zhang DQ, Xu GF, Fan ZJ, Wang DQ, Yang XL, Yuan DK (2012) Synthesis and anti-TMV activity of novel N-(3-alkyl-1H-pyrazol-4-yl)-3-alkyl-4-substituted-1H-pyrazole-5-carboxamides. Chin Chem Lett 23:669–672
271. Hwang JY, Kim H-Y, Park D-S, Choi J, Baek SM, Kim K, Kim S, Seong S, Choi I, Lee H-G, Windisch MP, Lee J (2013) Identification of a series of 1,3,4-trisubstituted pyrazoles as novel hepatitis C virus entry inhibitors. Bioorg Med Chem Lett 23:6467–6473
272. Mizuhara T, Kato T, Hirai A, Kurihara H, Shimada Y, Taniguchi M, Maeta H, Togami H, Shimura K, Matsuoka M, Okazaki S, Takeuchi T, Ohno H, Oishi S, Fujii N (2013) Structure-activity relationship study of phenylpyrazole derivatives as a novel class of anti-HIV agents. Bioorg Med Chem Lett 23:4557–4561
273. Moreau B, O'Meara JA, Bordeleau JE, Garneau M, Godbout C, Gorys V, Leblanc ML, Villemure E, White PW, Llinàs-Brunet M (2013) Discovery of hepatitis C virus NS3-4A protease inhibitors with improved barrier to resistance and favorable liver distribution. J Med Chem 57:1770–1776
274. Bhadoriya KS, Sharma MC, Jain SV (2015) 2,4-Dihydropyrano[2,3-c]pyrazole: discovery of new lead as through pharmacophore modelling, atom-based 3D-QSAR, virtual screening and docking strategies for improved anti-HIV-1 chemotherapy. J Taibah Univ Sci 9:521–530
275. Fioravanti R, Desideri N, Biava M, Droghini P, Atzori EM, Ibba C, Collu G, Sanna G, Delogu I, Loddo R (2015) *N*-((1,3-Diphenyl-1H-pyrazol-4-yl)methyl) anilines: a novel class of anti-RSV agents. Bioorg Med Chem Lett 25:2401–2404
276. Han C, Guo Y-C, Wang D-D, Dai X-J, Wu F-J, Liu H-F, Dai G-F, Tao J-C (2015) Novel pyrazole fused heterocyclic ligands: synthesis, characterization, DNA binding/cleavage activity and anti-BVDV activity. Chin Chem Lett 26:534–538
277. Manvar D, Pelliccia S, La Regina G, Famiglini V, Coluccia A, Ruggieri A, Anticoli S, Lee J-C, Basu A, Cevik O, Nencioni L, Palamara AT, Zamperini C, Botta M, Neyts J, Leyssen P, Kaushik-Basu N, Silvestri R (2015) New 1-phenyl-5-(1H-pyrrol-1-yl)-1H-pyrazole-3-carboxamides inhibit hepatitis C virus replication via suppression of cyclooxygenase-2. Eur J Med Chem 90:497–506
278. Chuang H, Huang L-CS, Kapoor M, Liao Y-J, Yang C-L, Chang C-C, Wu C-Y, Hwu JR, Huang T-J, Hsu M-H (2016) Design and synthesis of pyridine-pyrazole-sulfonate derivatives as potential anti-HBV agents. Med Chem Comm 7:832–836
279. Ouyang G, Cai X-J, Chen Z, Song B-A, Bhadury PS, Yang S, Jin L-H, Xue W, Hu D-Y, Zeng S (2008) Synthesis and antiviral activities of pyrazole derivatives containing an oxime moiety. J Agric Food Chem 56:10160–10167
280. Jia H, Bai F, Liu N, Liang X, Zhan P, Ma C, Jiang X, Liu X (2016) Design, synthesis and evaluation of pyrazole derivatives as non-nucleoside hepatitis B virus inhibitors. Eur J Med Chem 123:202–210
281. Liu G-N, Luo R-H, Zhou Y, Zhang X-J, Li J, Yang L-M, Zheng Y-T, Liu H (2016) Synthesis and Anti-HIV-1 Activity Evaluation for Novel 3a,6a-Dihydro-1H-pyrrolo[3,4-c]pyrazole-4,6-dione Derivatives. Molecules 21:1198
282. Johns BA, Gudmundsson KS, Allen SH (2007) Pyrazolo[1,5-a]pyridine antiherpetics: effects of the C3 substituent on antiviral activity. Bioorg Med Chem Lett 17:2858–2862
283. Chimenti F, Fioravanti R, Bolasco A, Manna F, Chimenti P, Secci D, Befani O, Turini P, Ortuso F, Alcaro S (2007) Monoamine oxidase isoform-dependent tautomeric influence in the recognition of 3,5-diaryl pyrazole inhibitors. J Med Chem 50:425–428
284. Kuduk SD, Di Marco CN, Cofre V, Pitts DR, Ray WJ, Ma L, Wittmann M, Veng L, Seager MA, Koeplinger K, Thompson CD, Hartman GD, Bilodeau MT (2010) *N*-Heterocyclic derived M 1 positive allosteric modulators. Bioorg Med Chem Lett 20:1334–1337
285. Malamas MS, Erdei J, Gunawan I, Barnes K, Hui Y, Johnson M, Robichaud A, Zhou P, Yan Y, Solvibile W, Turner J, Fan KY, Chopra R, Bard J, Pangalos MN (2011) New pyrazolyl and thienyl aminohydantoins as potent BACE1 inhibitors: exploring the S20 region. Bioorg Med Chem Lett 21:5164–5170
286. Probst G, Aubele DL, Bowers S, Dressen D, Garofalo AW, Hom RK, Konradi AW, Marugg JL, Mattson MN, Neitzel ML, Semko CM, Sham HL, Smith J, Sun M, Truong AP, Ye V,

Xu Y-Z, Dappen MS, Jagodzinski JJ, Keim PS, Peterson B, Latimer LH, Quincy D, Wu J, Goldbach E, Ness DK, Quinn KP, Sauer J-M, Wong K, Zhang H, Zmolek W, Brigham EF, Kholodenko D, Hu K, Kwong GT, Lee M, Liao A, Motter RN, Sacayon P, Santiago P, Willits C, Bard F, Bova MP, Hemphill SS, Nguyen L, Ruslim L, Tanaka K, Tanaka P, Wallace W, Yednock TA, Basi GS (2013) Discovery of (R)-4-Cyclopropyl-7,8-difluoro-5-(4-(trifluoromethyl)phenylsulfonyl) -4,5-dihydro-1H-pyrazolo[4,3-c]quinoline (ELND006) and (R)-4-Cyclopropyl- 8-fluoro-5-(6-(trifluoromethyl) pyridin-3-ylsulfonyl)-4,5-dihydro-2H-pyrazolo[4,3-c]quinoline (ELND007): metabolically stable-secretase inhibitors that selectively inhibit the production of Amyloid-β over Notch. J Med Chem 56:5261–5274

287. Zou Y, Xu L, Chen W, Zhu Y, Chen T, Fu Y, Li L, Ma L, Xiong B, Wang X, Li J, He J, Zhang H, Xu Y, Li J, Shen J (2013) Discovery of pyrazole as C-terminus of selective BACE1 inhibitors. Eur J Med Chem 68:270–283
288. Han YT, Kim K, Choi G-I, An H, Son D, Kim H, Ha H-J, Son J-H, Chung S-J, Park H-J, Lee J, Suh Y-G (2014) Pyrazole-5-carboxamides, novel inhibitors of receptor for advanced glycation end products (RAGE). Eur J Med Chem 79:128–142
289. Silva D, Chioua M, Samadi A, Carreiras MC, Jimeno M-L, Mendes E, de los Ríos C, Romero A, Villarroya M, López MG, Marco-Contelles J (2011) Synthesis and pharmacological assessment of diversely substituted pyrazolo[3,4-b]quinoline, and benzo[b]pyrazolo[4,3-g][1,8]naphthyridine derivatives. Eur J Med Chem 46:4676–4681
290. Khoobi M, Ghanoni F, Nadri H, Moradi A, Hamedani MP, Moghadam FH, Emami S, Vosooghi M, Zadmard R, Foroumadi A, A Shafiee (2015) New tetracyclic tacrine analogs containing pyrano[2,3-c]pyrazole: efficient synthesis, biological assessment and docking simulation study. Eur J Med Chem 89:296–303
291. Zanaletti R, Bettinetti L, Castaldo C, Ceccarelli I, Cocconcelli G, Comery TA, Dunlop J, Genesio E, Ghiron C, Haydar SN, Jow F, Maccari L, Micco I, Nencini A, Pratelli C, Scali C, Turlizzi E, Valacchi M (2012) *N*-[5-(5-Fluoropyridin-3-yl)-1H-pyrazol-3-yl]-4-piperidin-1-ylbutyramide (SEN78702, WYE-308775): a medicinal chemistry effort toward an 7 nicotinic acetylcholine receptor agonist preclinical candidate. J Med Chem 55:10277–10281
292. Zanaletti R, Bettinetti L, Castaldo C, Cocconcelli G, Comery T, Dunlop J, Gaviraghi G, Ghiron C, Haydar SN, Jow F, Maccari L, Micco I, Nencini A, Scali C, Turlizzi E, Valacchi M (2012) Discovery of a novel alpha-7 nicotinic acetylcholine receptor agonist series and characterization of the potent, selective, and orally efficacious agonist 5-(4-acetyl[1,4]diazepan-1-yl)pentanoic acid[5-(4-methoxyphenyl)-1H-pyrazol-3-yl] amide (SEN15924, WAY-361789). J Med Chem 55:4806–4823
293. Nencini A, Castaldo C, Comery TA, Dunlop J, Genesio E, Ghiron C, Haydar S, Maccari L, Micco I, Turlizzi E, Zanaletti R, Zhang J (2014) Design and synthesis of a hybrid series of potent and selective agonists of 7 nicotinic acetylcholine receptor. Eur J Med Chem 78:401–418
294. Ernst G, Frietze W, Simpson T (2008) Novel pyrazole derivatives and their use as modulators of nicotinic acetylcholine receptors. Google Patent, WO2006068591A1
295. Thuring JWJF, MacDonald GJ, Zhuang W (2014) Trisubstituted pyrazoles as acetylcholine receptor modulators. Google Patents
296. Cottineau B, Toto P, Marot C, Pipaud A, Chenault J (2002) Synthesis and hypoglycemic evaluation of substituted pyrazole-4-carboxylic acids. Bioorg Med Chem Lett 12:2105–2108
297. Sharon A, Pratap R, Tiwari P, Srivastava A, Maulik PR, Ram VJ (2005) Synthesis and in vivo antihyperglycemic activity of 5-(1H-pyrazol-3-yl)methyl-1H-tetrazoles. Bioorg Med Chem Lett 15:2115–2117
298. Humphries PS, Lafontaine JA, Agree CS, Alexander D, Chen P, Do Q-QT, Li LY, Lunney EA, Rajapakse RJ, Siegel K, Timofeevski SL, Wang T, Wilhite DM (2009) Synthesis and SAR of 4-substituted-2-aminopyrimidines as novel c-Jun N-terminal kinase (JNK) inhibitors. Bioorg Med Chem Lett 19:2099–2102
299. Brigance RP, Meng W, Fura A, Harrity T, Wang A, Zahler R, Kirby MS, Hamann LG (2010) Synthesis and SAR of azolopyrimidines as potent and selective dipeptidyl peptidase-4 (DPP4) inhibitors for type 2 diabetes. Bioorg Med Chem Lett 20:4395–4398

300. Choi J, Park Y, Lee HS, Yang Y (2010) Yoon S (2010) 1,3-Diphenyl-1H-pyrazole derivatives as a new series of potent PPAR partial agonists. Bioorg Med Chem 18:8315–8323
301. Shen D-M, Brady EJ, Candelore MR, Dallas-Yang Q, Ding VDH, Feeney WP, Jiang G, McCann ME, Mock S, Qureshi SA, Saperstein R, Shen X, Tong X, Tota LM, Wright MJ, Yang X, Zheng S, Chapman KT, Zhang BB, Tata JR, Parmee ER (2011) Discovery of novel, potent, selective, and orally active human glucagon receptor antagonists containing a pyrazole core. Bioorg Med Chem Lett 21:76–81
302. Fushimi N, Fujikura H, Shiohara H, Teranishi H, Shimizu K, Yonekubo S, Ohno K, Miyagi T, Itoh F, Shibazaki T, Tomae M, Ishikawa-Takemura Y, Nakabayashi T, Kamada N, Ozawa T, Kobayashi S, Isaji M (2012) Structure-activity relationship studies of 4-benzyl-1H-pyrazol-3-yl β–d glucopyranoside derivatives as potent and selective sodium glucose co-transporter 1 (SGLT1) inhibitors with therapeutic activity on postprandial hyperglycemia. Bioorg Med Chem 20:6598–6612
303. Rikimaru K, Wakabayashi T, Abe H, Imoto H, Maekawa T, Ujikawa O, Murase K, Matsuo T, Matsumoto M, Nomura C, Tsuge H, Arimura N, Kawakami K, Sakamoto J, Funami M, Mol CD, Snell GP, Bragstad KA, Sang BC, Dougan DR, Tanaka T, Katayama N, Horiguchi Y, Momose Y (2012) A new class of non-thiazolidinedione, non-carboxylic-acid-based highly selective peroxisome proliferator-activated receptor (PPAR) agonists: design and synthesis of benzylpyrazole acylsulfonamides. Bioorg Med Chem 20:714–733
304. Xiong Y, Guo J, Candelore MR, Liang R, Miller C, Dallas-Yang Q, Jiang G, McCann PE, Qureshi SA, Tong X, Xu SS, Shang J, Vincent SH, Tota LM, Wright MJ, Yang X, Zhang BB, Tata JR, Parmee ER (2012) Discovery of a Novel Glucagon Receptor Antagonist N-[(4-{(1S)-1-[3-(3,5-Dichlorophenyl)-5-(6-methoxynaphthalen-2-yl)-1H-pyrazol-1-yl]ethyl}phenyl)carbonyl]-β-alanine (MK-0893) for the treatment of type II diabetes. J Med Chem 55:6137–6148
305. Yoshida T, Akahoshi F, Sakashita H, Kitajima H, Nakamura M, Sonda S, Takeuchi M, Tanaka Y, Ueda N, Sekiguchi S, Ishige T, Shima K, Nabeno M, Abe Y, Anabuki J, Soejima A, Yoshida K, Takashina Y, Ishii S, Kiuchi S, Fukuda S, Tsutsumiuchi R, Kosaka K, Murozono T, Nakamaru Y, Utsumi H, Masutomi N, Kishida H, Miyaguchi I, Hayashi Y (2012) Discovery and preclinical profile of teneligliptin (3-[(2S,4S)-4-[4-(3-methyl-1- phenyl-1H-pyrazol-5-yl)piperazin-1-yl]pyrrolidin-2-ylcarbonyl]thiazolidine): a highly potent, selective, long-lasting and orally active dipeptidyl peptidase IV inhibitor for the treatment of type 2 diabetes. Bioorg Med Chem 20:5705–5719
306. Futatsugi K, Mascitti V, Guimarães CR, Morishita N, Cai C, DeNinno MP, Gao H, Hamilton MD, Hank R, Harris AR, Kung DW, Lavergne SY, Lefker BA, Lopaze MG, McClure KF, Munchhof MJ, Preville C, Robinson RP, Wright SW, Bonin PD, Cornelius P, Chen Y, Kalgutkar AS (2013) From partial to full agonism: identification of a novel 2,4,5,6-tetrahydropyrrolo[3,4-c]pyrazole as a full agonist of the human GPR119 receptor. Bioorg Med Chem Lett 23:194–197
307. Griffith DA, Dow RL, Huard K, Edmonds DJ, Bagley SW, Polivkova J, Zeng D, Garcia-Irizarry CN, Southers JA, Esler W, Amor P, Loomis K, McPherson K, Bahnck KB, Préville C, Banks T, Moore DE, Mathiowetz AM, Menhaji-Klotz E, Smith AC, Doran SD, Beebe DA, Dunn MF (2013) Spirolactam-based acetyl-CoA carboxylase inhibitors: toward improved metabolic stability of a chromanone lead structure. J Med Chem 56:7110–7119
308. Hernández-Vázquez E, Aguayo-Ortiz R, Ramírez-Espinosa JJ, Estrada-Soto S, Hernández-Luis F (2013) Synthesis, hypoglycemic activity and molecular modeling studies of pyrazole-3-carbohydrazides designed by a CoMFA model. Eur J Med Chem 69:10–21
309. Toda N, Hao X, Ogawa Y, Oda K, Yu M, Fu Z, Chen Y, Kim Y, Lizarzaburu M, Lively S, Lawlis S, Murakoshi M, Nara F, Watanabe N, Reagan JD, Tian H, Fu A, Motani A, Liu Q, Lin Y-J, Zhuang R, Xiong Y, Fan P, Medina J, Li L, Izumi M, Okuyama R, Shibuya S (2013) Potent and orally bioavailable GPR142 agonists as novel insulin secretagogues for the treatment of type 2 diabetes. ACS Med Chem Lett 4:790–794
310. Yu M, Lizarzaburu M, Motani A, Fu Z, Du X, Liu J, Jiao X, Lai S, Fan P, Fu A, Liu Q, Murakoshi M, Nara F, Oda K, Okuyama R, Reagan JD, Watanabe N, Yamazaki M, Xiong

Y, Zhang Y, Zhuang R, Lin DC-H, Houze JB, Medina JC, Li L (2013) Aminopyrazole-phenylalanine based GPR142 agonists: discovery of tool compound and in vivo efficacy studies. ACS Med Chem Lett 4:829–834
311. Bhosle MR, Mali JR, Pal S, Srivastava AK, Mane RA (2014) Synthesis and antihyperglycemic evaluation of new 2-hydrazolyl-4-thiazolidinone-5-carboxylic acids having pyrazolyl pharmacophores. Bioorg Med Chem Lett 24:2651–2654
312. López-Viseras ME, Fernández B, Hilfiker S, González CS, González JL, Calahorro AJ, Colacio E, Rodríguez-Diéguez A (2014) In vivo potential antidiabetic activity of a novel zinc coordination compound based on 3-carboxy-pyrazole. J Inorg Biochem 131:64–67
313. Kenchappa R, Bodke YD, Chandrashekar A, Sindhe MA, Peethambar S (2017) Synthesis of coumarin derivatives containing pyrazole and indenone rings as potent antioxidant and antihyperglycemic agents. Arab J Chem 10:3895–3906
314. Doddaramappa SD, Lokanatha Rai KM, Srikantamurthy N, Chandra, Chethan J (2015) Novel 5-functionalizedpyrazoles: synthesis, characterization and pharmacological screening. Bioorg Med Chem Lett 25:3671–3675
315. Hernández-Vázquez E, Salgado-Barrera S, Ramírez-Espinosa JJ, Estrada-Soto S, Hernández-Luis F (2016) Synthesis and molecular docking of N0-arylidene-5-(4-chlorophenyl)-1-(3,4-dichlorophenyl)-4-methyl-1Hpyrazole- 3-carbohydrazides as novel hypoglycemic and antioxidant dual agents. Bioorg Med Chem 24:2298–2306
316. Kashtoh H, Muhammad MT, Khan JJA, Rasheed S, Khan A, Perveen S, Javaid K, Atia Tul W, Khan KM, Choudhary MI (2016) Dihydropyrano[2,3-c]pyrazole: novel in vitro inhibitors of yeast α-glucosidase. Bioorg Chem 65:61–72
317. Chaudhry F, Naureen S, Huma R, Shaukat A, Al-Rashida M, Asif N, Ashraf M, Munawar MA, Khan MA (2017) In search of new α-glucosidase inhibitors: imidazolylpyrazole derivatives. Bioorg Chem 71:102–109
318. Hernández-Vázquez E, Ocampo-Montalban H, Cerón-Romero L, Cruz M, Gómez-Zamudio J, Hiriart-Valencia G, Villalobos-Molina R, Flores-Flores A, Estrada-Soto S (2017) Antidiabetic, antidyslipidemic and toxicity profile of ENV-2: a potent pyrazole derivative against diabetes and related diseases. Eur J Pharmacol 803:159–166
319. De Mello H, Echevarria A, Bernardino AM, Canto-Cavalheiro M, Leon LL (2004) Antileishmanial pyrazolopyridine derivatives: synthesis and structure-activity relationship analysis. J Med Chem 47:5427–5432
320. Bernardino AMR, Gomes AO, Charret KS, Freitas ACC, Machado GMC, Canto-Cavalheiro MM, Leon LL, Amaral VF (2006) Synthesis and leishmanicidal activities of 1-(4-X-phenyl)-N'-[(4-Y-phenyl)methylene]-1H-pyrazole-4-carbohydrazides. Eur J Med Chem 41:80–87
321. Dardari Z, Lemrani M, Sebban A, Bahloul A, Hassar M, Kitane S, Berrada M, Boudouma M (2006) Antileishmanial and antibacterial activity of a new pyrazole derivative designated 4-[2-(1-(Ethylamino)-2-methyl-propyl)phenyl]-3-(4-methyphenyl)-1-phenylpyrazole. Arch Pharm 339:291–298
322. Dos Santos MS, Gomes AO, Bernardino AM, Souza MCD, Khan MA, Brito MAD, Castro HC, Abreu PA, Rodrigues CR, de Léo RMM, Leon LL, Canto-Cavalheiro MM (2011) Synthesis and antileishmanial activity of new 1-aryl-1H-pyrazole-4-carboximidamides derivatives. J Braz Chem Soc 22:352–358
323. Dos Santos MS, Oliveira ML, Bernardino AM, de Léo RM, Amaral VF, de Carvalho FT, Leon LL, Canto-Cavalheiro MM (2011) Synthesis and antileishmanial evaluation of 1-aryl-4-(4,5-dihydro-1H-imidazol-2-yl)-1H-pyrazole derivatives. Bioorg Med Chem Lett 21:7451–7454
324. Jacomini AP, Silva MJV, Silva RGM, Gonçalves DS, Volpato H, Basso EA, Paula FR, Nakamura CV, Sarragiotto MH, Rosa FA (2016) Synthesis and evaluation against Leishmania amazonensis of novel pyrazolo[3,4-d]pyridazinone-N-acylhydrazone-(bi)thiophene hybrids. Eur J Med Chem 124:340–349
325. Bekhit AA, Hassan AMM, Abd El Razik HA, El-Miligy MMM, El-Agroudy EJ, Bekhit AE-DA (2015) New heterocyclic hybrids of pyrazole and its bioisosteres: design, synthesis and biological evaluation as dual acting antimalarial-antileishmanial agents. Eur J Med Chem 94:30–44

326. Mowbray CE, Braillard S, Speed W, Glossop PA, Whitlock GA, Gibson KR, Mills JEJ, Brown AD, Gardner JMF, Cao Y, Hua V, Morgans GL, Feijens P-B, Matheeussen A, Maes LJ (2015) Novel Amino-pyrazole ureas with potent in vitro and in vivo antileishmanial activity. J Med Chem 58:9615–9624
327. Marra R, Bernardino A, Proux T, Charret K, Lira M-L, Castro H, Souza A, Oliveira C, Borges J, Rodrigues C, Canto-Cavalheiro MM, Leon LL, Amaral VF (2012) 4-(1H-Pyrazol-1-yl) benzenesulfonamide derivatives: identifying new active antileishmanial structures for use against a neglected disease. Molecules 17:12961–12973
328. Bekhit AA, Haimanot T, Hymete A (2014) Evaluation of some 1H-pyrazole derivatives as a dual acting antimalarial and anti-leishmanial agent. Pak J Pharm Sci 27:1767–1773
329. Figarella K, Marsiccobetre S, Galindo-Castro I, Urdaneta N, Herrera JC, Canudas N, Galarraga E (2017) Antileishmanial and antitrypanosomal activity of synthesized hydrazones, pyrazoles, pyrazolo [1,5-a]-pyrimidines and pyrazolo[3,4-b]-pyridine. Curr Bioact Compd 13:000–000
330. Tuha A, Bekhit AA, Seid Y (2017) Screening of some pyrazole derivatives as promising antileishmanial agent. Afr J Pharm Pharmacol 11:32–37
331. Reviriego F, Olmo F, Navarro P, Marín C, Ramírez-Macías I, García-España E, Albelda MT, Gutiérrez-Sánchez R, Sánchez-Moreno M, Arán VJ (2017) Simple dialkyl pyrazole-3,5-dicarboxylates show in vitro and in vivo activity against disease-causing trypanosomatids. Parasitology 144:1133–1143
332. Domínguez JN, Charris JE, Caparelli M, Riggione F (2002) Synthesis and antimalarial activity of substituted pyrazole derivatives. Arzneimittelforschung 52:482–488
333. Mishra S, Karmodiya K, Surolia N, Surolia A (2008) Synthesis and exploration of novel curcumin analogues as anti-malarial agents. Bioorg Med Chem 16:2894–2902
334. González CD, Douelle F, Feng T-S, Nchinda AT, Younis Y, White KL, Wu Q, Ryan E, Burrows JN, Waterson D, Witty MJ, Wittlin S, Charman SA, Chibale K (2011) Novel orally active antimalarial thiazoles. J Med Chem 54:7713–7719
335. Quirante J, Ruiz D, Gonzalez A, López C, Cascante M, Cortés R, Messeguer R, Calvis C, Baldomà L, Pascual A, Guérardel Y, Pradines B, Font-Bardía M, Calvet T, Biot C (2011) Platinum (II) and palladium (II) complexes with (N, N0) and (C, N, N0)–ligands derived from pyrazole as anticancer and antimalarial agents: synthesis, characterization and in vitro activities. J Inorg Biochem 105:1720–1728
336. Large JM, Osborne SA, Smiljanic-Hurley E, Ansell KH, Jones HM, Taylor DL, Clough B, Green JL, Holder AA (2013) Imidazopyridazines as potent inhibitors of Plasmodium falciparum calcium-dependent protein kinase 1 (PfCDPK1): preparation and evaluation of pyrazole linked analogues. Bioorg Med Chem Lett 23:6019–6024
337. Balaji SN, Ahsan MJ, Jadav SS, Trivedi V (2015) Molecular modelling, synthesis, and antimalarial potentials of curcumin analogues containing heterocyclic ring. Arab J Chem 12:2492–2500
338. Niswender CM, Lebois EP, Luo Q, Kim K, Muchalski H, Yin H, Conn PJ, Lindsley CW (2008) Positive allosteric modulators of the metabotropic glutamate receptor subtype 4 (mGluR4): Part I. Discovery of pyrazolo[3,4-d]pyrimidines as novel mGluR4 positive allosteric modulators. Bioorg Med Chem Lett 18:5626–5630
339. Chimenti F, Carradori S, Secci D, Bolasco A, Bizzarri B, Chimenti P, Granese A, Yanez M, Orallo F (2010) Synthesis and inhibitory activity against human monoamine oxidase of N1-thiocarbamoyl-3,5-di(hetero)aryl-4,5-dihydro-(1H)-pyrazole derivatives. Eur J Med Chem 45:800–804
340. Maher P, Akaishi T, Schubert D, Abe K (2010) A pyrazole derivative of curcumin enhances memory. Neurobiol Aging 31:706–709
341. Chan BK, Estrada AA, Chen H, Atherall J, Baker-Glenn C, Beresford A, Burdick DJ, Chambers M, Dominguez SL, Drummond J, Gill A, Kleinheinz T, Le Pichon CE, Medhurst AD, Liu X, Moffat JG, Nash K, Scearce-Levie K, Sheng Z, Shore DG, Van de Poël H, Zhang S, Zhu H, Sweeney ZK (2012) Discovery of a highly selective, brain-penetrant aminopyrazole LRRK2 inhibitor. ACS Med Chem Lett 4:85–90

342. Jimenez HN, Liu KG, Hong S-P, Reitman MS, Uberti MA, Bacolod MD, Cajina M, Nattini M, Sabio M, Doller D (2012) 4-(1-Phenyl-1H-pyrazol-4-yl) quinolines as novel, selective and brain penetrant metabotropic glutamate receptor 4 positive allosteric modulators. Bioorg Med Chem Lett 22:3235–3239
343. Dore A, Asproni B, Scampuddu A, Pinna GA, Christoffersen CT, Langgård M, Kehler J (2014) Synthesis and SAR study of novel tricyclic pyrazoles as potent phosphodiesterase 10A inhibitors. Eur J Med Chem 84:181–193
344. Estrada AA, Chan BK, Baker-Glenn C, Beresford A, Burdick DJ, Chambers M, Chen H, Dominguez SL, Dotson J, Drummond J, Flagella M, Fuji R, Gill A, Halladay J, Harris SF, Heffron TP, Kleinheinz T, Lee DW, Le Pichon CE, Liu X, Lyssikatos JP, Medhurst AD, Moffat JG, Nash K, Scearce-Levie K, Sheng Z, Shore DG, Wong S, Zhang S, Zhang X, Zhu H, Sweeney ZK (2014) Discovery of highly potent, selective, and brain-penetrant aminopyrazole leucine-rich repeat kinase 2 (LRRK2) small molecule inhibitors. J Med Chem 57:921–936
345. Fujinaga M, Yamasaki T, Nengaki N, Ogawa M, Kumata K, Shimoda Y, Yui J, Xie L, Zhang Y, Kawamura K, Zhang MR (2016) Radiosynthesis and evaluation of 5-methyl-N-(4-[11C]methylpyrimidin-2-yl)-4-(1H-pyrazol-4-yl)thiazol-2-amine ([11C]ADX88178) as a novel radioligand for imaging of metabotropic glutamate receptor subtype 4 (mGluR4). Bioorg Med Chem Lett 26:370–374
346. Kim M, Si C, Shin D, Suh E, Cho K (2006) Residual and sublethal effects of fenpyroximate and pyridaben on the instantaneous rate of increase of Tetranychus urticae. Crop Prot 25:542–548
347. Finkelstein BL, Strock CJ (1997) Synthesis and insecticidal activity of novel pyrazole methanesulfonates. Pest Sci 50:324–328
348. Mao M, Li Y, Liu Q, Zhou Y, Zhang X, Xiong L, Li Y, Li Z (2013) Synthesis and insecticidal evaluation of novel N-pyridylpyrazolecarboxamides containing cyano substituent in the ortho-position. Bioorg Med Chem Lett 23:42–46
349. Wu J, Song BA, Hu DY, Yue M, Yang S (2012) Design, synthesis and insecticidal activities of novel pyrazole amides containing hydrazone substructures. Pest Manag Sci 68:801–810
350. Dai H, Xiao Y-S, Li Z, Xu X-Y, Qian X-H (2014) The thiazoylmethoxy modification on pyrazole oximes: synthesis and insecticidal biological evaluation beyond acaricidal activity. Chin Chem Lett 25:1014–1016
351. Song H, Liu Y, Xiong L, Li Y, Yang N, Wang Q (2012) Design, synthesis, and insecticidal activity of novel pyrazole derivatives containing α-hydroxymethyl-N-benzyl carboxamide, α-chloromethyl-N-benzyl carboxamide, and 4,5-dihydrooxazole moieties. J Agric Food Chem 60:1470–1479
352. Fu C, Pei J, Ning Y, Liu M, Shan P, Liu J (2014) Synthesis and insecticidal activities of novel pyrazole oxime ether derivatives with different substituted pyridyl rings. Pest Manag Sci 70:1207–1214

Chapter 8
An Overview on Biological Evaluation of Tetrazole Derivatives

Arup K. Kabi, Sattu Sravani, Raghuram Gujjarappa, Aakriti Garg, Nagaraju Vodnala, Ujjawal Tyagi, Dhananjaya Kaldhi, Ravichandiran Velayutham, Virender Singh, Sreya Gupta, and Chandi C. Malakar

Abbreviations

U.S. FDA	United States Food and Drug Administration
HIV	Human Immunodeficiency Virus
ARB	Angiotensin II receptor blocker
DNA	Deoxyribonucleic Acid
HMG-CoA	3-hydroxy-3-methyl-glutaryl-coenzyme A
COX	Cyclooxygenase
GABA	Gamma-Aminobutyric Acid
ABSSSI	Acute Bacterial Skin and skin structure infection
ADP	Adenosine di-phosphate
AMP	Adenosine monophosphate
PARP	poly(ADP-ribose)polymerase
RNA	Ribonucleic Acid
SAR	Structure–Activity Relationship
NS3	Non-structural protein 3
AIDS	Acquired Immune Deficiency Syndrome
MRSA	Methicillin-resistant Staphylococcus aureus
WHO	World Health Organization

A. K. Kabi · R. Gujjarappa · N. Vodnala · U. Tyagi · D. Kaldhi · C. C. Malakar (✉)
Department of Chemistry, National Institute of Technology Manipur, Langol, Imphal 795004, Manipur, India
e-mail: cmalakar@nitmanipur.ac.in

S. Sravani · A. Garg · R. Velayutham · S. Gupta (✉)
Department of Medicinal Chemistry, National Institute of Pharmaceutical Education and Research (NIPER), Chunilal Bhawan, 168, Maniktala Main Road, Kolkata 700054, India

V. Singh
Department of Chemistry, Department of Chemistry, Central University of Punjab, Bathinda 151001, Punjab, India

B. P. Swain (ed.), *Nanostructured Biomaterials*, Materials Horizons: From Nature to Nanomaterials, https://doi.org/10.1007/978-981-16-8399-2_8

HSV	Herpes Simplex Viruses
ER	Endoplasmic Reticulum
DNMT	DNA methyltransferases
HepG-2	Human liver cancer cell line
OGTT	Oral Glucose Tolerance Test

1 Introduction

Tetrazoles are one of the most influential and stable five-membered heterocyclic compounds because they possess interesting applications in pharmaceuticals [1], agriculture [2–4], photography [5] and as components of explosives [6]. Additionally, tetrazoles have a suitable position in the modern pharmaceutical fields (Fig. 1). The tetrazole compounds display various biological activities and are broadly employed in clinical practice. Tetrazoles occupy a centre stage of investigations and drug design carried out by leading pharmaceutical companies and scientific institutes. Tetrazole is stated to be a very significant structural unit by the World Health Organization (WHO) and U.S. FDA and it is effectively used for the strategy of drugs of the twenty-first century. Owing to their multidimensional features, the medicinal chemistry and drug discovery field has witnessed a steep increase in the number of patents and publications. Diverse substituted Tetrazoles are abridged in this chapter to understand the chemistry as well as biological and pharmacological activities.

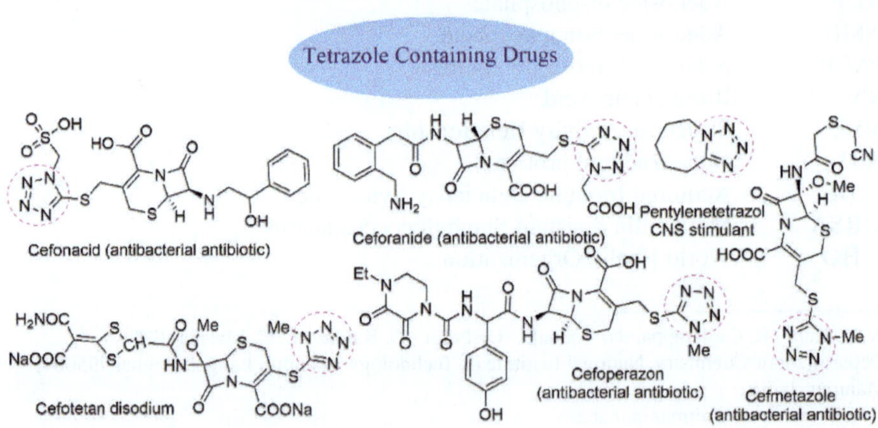

Fig. 1 Tetrazole containing marketed drugs [1–4]

1.1 Mechanism of Action of Cefotetan

It acts by preventing the synthesis of bacterial cell walls, resulting in cell lysis. The β-lactam antibiotics are structurally comparable to D-alanyl-D-alanine's (component of peptidoglycan), they are capable of reacting with penicillin-binding proteins. Cephalosporins interrupt the synthesis of peptidoglycan layers of the bacterial cell wall.

The role of tetrazole ring in medicinal chemistry is confined to work as a metabolically stable carboxylic acid isostere. It is considered as bioisosteres of amides and carboxylic acids having high lipophilicity. They are utilized as plant herbicides, fungicides, growth regulators and have antiallergic, antibiotic, CCK-B antagonist, anti-hypertensive, antiviral activities. In recent times, aryl thiotetrazolyl acetanilide functionality of tetrazole motif (**1–5**) is largely employed for binding with HIV-1 reverse transcriptase [6]. The compound BMS-183920 (**2**) is a potent angiotensin II receptor antagonist that is poorly absorbed from the gut presumably because of the di-acidic nature of the molecule. Olmesartanmedoxonil **3**, the tetrazole analogues, is accessible for treatment of hypertension. This drug plays a vibrant role in inhibiting angiotensin-converting enzymes and to block AT1 and AT2 receptors present in brain, heart, kidney, adrenal glands and vascular smooth muscle cells. Candesartan **4** is an angiotensin II receptor antagonist employed widely for the treatment of congestive heart failure, hypertension, diabetic nephropathy and myocardial infarction. It is an orally active lipophilic drug with rapid oral absorption. Cilostazol **5** is bound to inhibit phosphodiesterase III with additional pharmacological properties such as inhibition of platelet activation, inhibition of thrombosis and improved blood flow to the limbs. Cilostazol **5** can also perform vasodilation, aggregation, perfection in serum lipids with lowering of triglycerides, the reserve of vascular smooth muscle cell growth and boost of high-density lipoprotein cholesterol (Fig. 2) [7–10].

1.2 The Arrangements of the Existence of the Tetrazole Ring

Tetrazole is a heterocycle with two nitrogen atoms that are capable of exhibiting simultaneous intermolecular hydrogen bonding and it is quite comparable with purine and pyrimidine bases. The nitrogen atoms in tetrazole actively participate in multi-centre intermolecular interactions with the pyridine ring and hydrogen atoms from the surrounding molecules, which includes the H-atoms at the active site pockets of enzymes [2]. Tetrazoles can exist in diverse tautomeric forms and also as cations and anions [11]. Here we discussed the various forms of the tetrazole ring. The structure **A** is the mono-substituted NH-tetrazole and **B** is its anion. Disubstituted derivatives are characterized by two regioisomers **C** and **D**.

The tetrazole ring annulated to other rings (**E** and **F**) can vary in character and aromaticity. The synthesis of tetrazolium salts can be performed with the sharing of 1,4,5 (**G**), 1,3,5 (**H**), or 2,3,5-trisubstituted tetrazolium (**I**) cations. The examples of

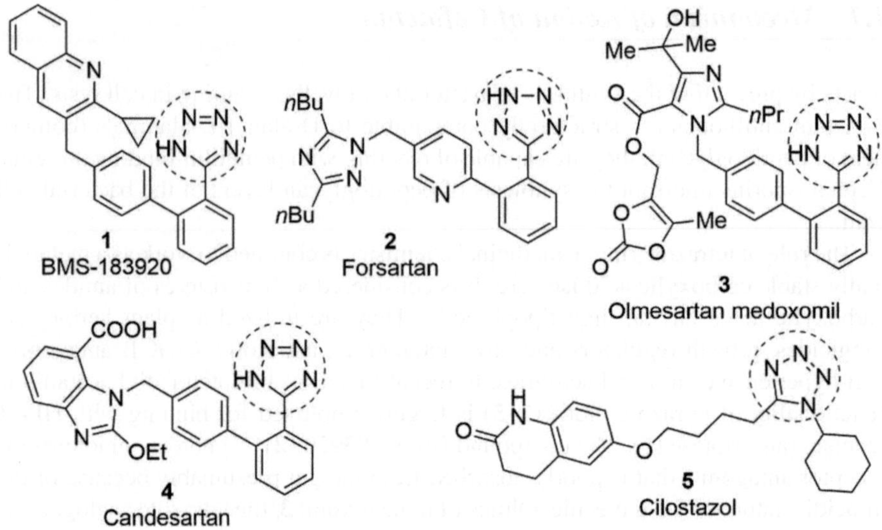

Fig. 2 Tetrazole containing marketed drugs [7–10]

partially hydrogenated tetrazoles are considered to be 1,4-dihydro tetrazoles **J** and **K** are considered (Fig. 3).

Tetrazole containing compounds exhibit various pharmacological activities like hypotensive, antiviral, antimicrobial, antiallergic, nootropic, cryostatic and also some more biological activities (Table 1). These also act as a component of medical purposes and diagnostic complexes.

The list of FDA-approved *N*-heterocycle containing drugs has been portrayed in Table 1 [9].

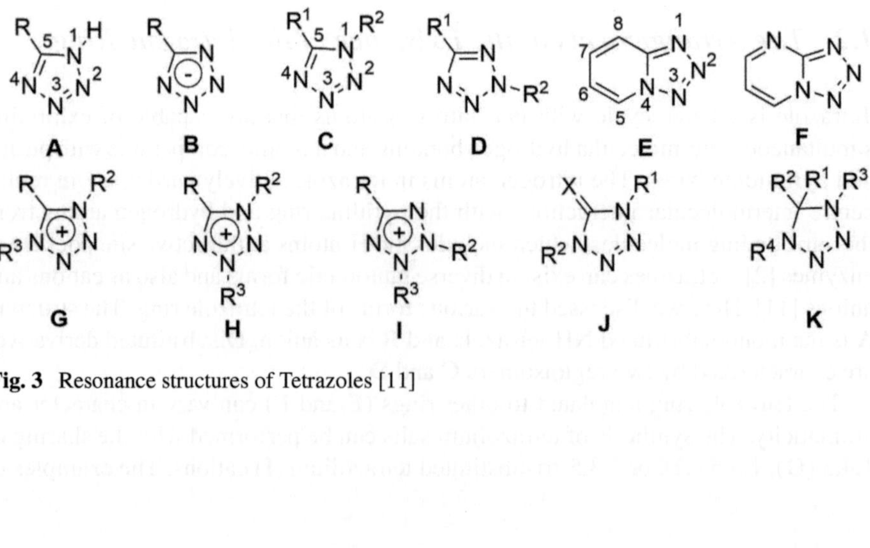

Fig. 3 Resonance structures of Tetrazoles [11]

Table 1 Tetrazoles containing drugs and their mode of action [9]

Structure of the drug Name of the drug (Drug Bank ID)	Category/indication	Mechanism of action
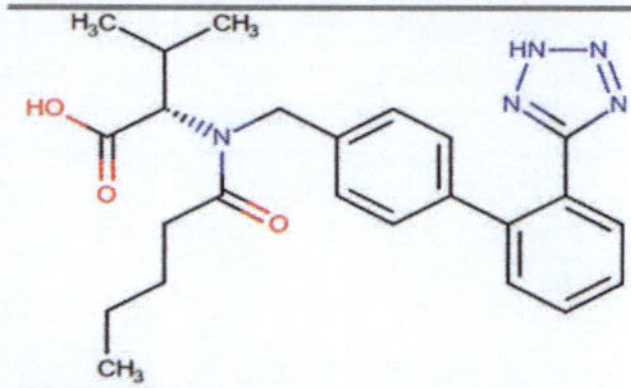 Valsartan (DB00177)	Anti-hypertensive	Angiotensin-converting enzyme inhibitor (ACEI)
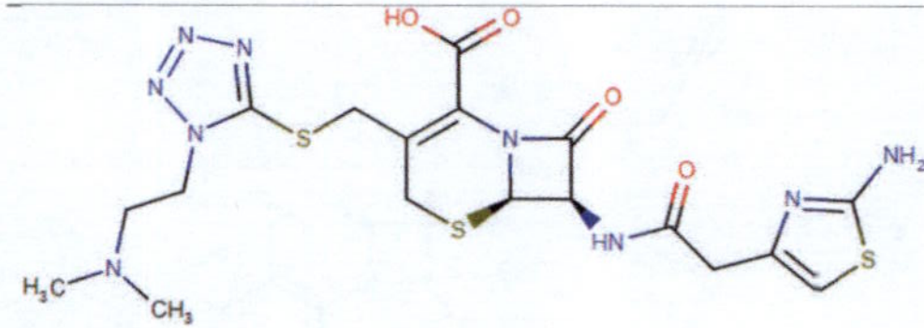 Cefotiam (DB00299)	Bactericidal	The cell wall synthesis was prohibited by its affinity towards penicillin-binding proteins (PBPs) inhibits

(continued)

Table 1 (continued)

Structure of the drug Name of the drug (Drug Bank ID)	Category/indication	Mechanism of action
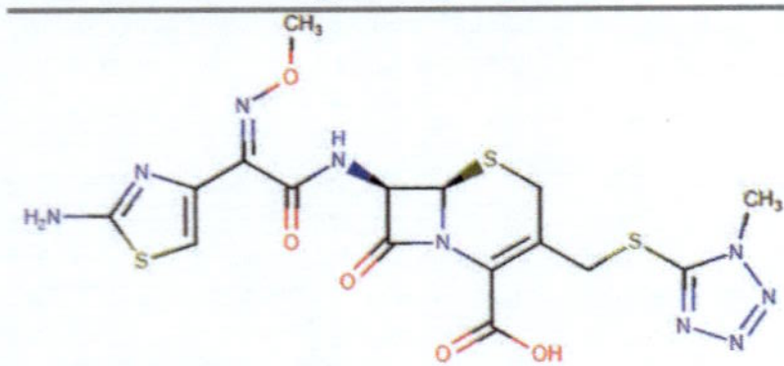 Cefmenoxime (DB00267)	Cephalosporin antibiotic	The cell wall synthesis was prohibited by its affinity towards penicillin-binding proteins (PBPs) inhibits
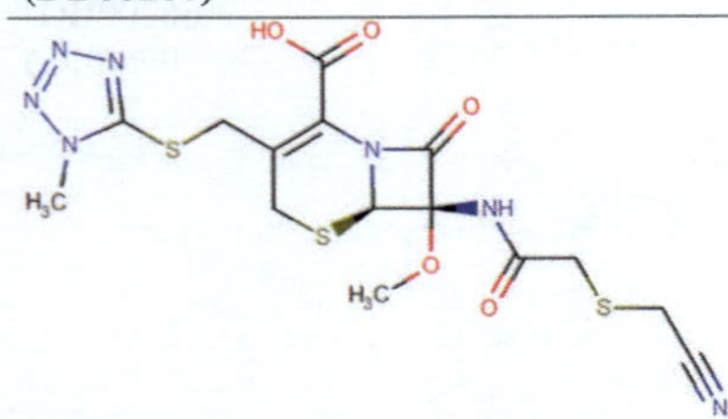 Cefmetazole (DB00274)	Cephamycin antibiotic	Acts against both gram-positive and gram-negative microorganisms

(continued)

Table 1 (continued)

Structure of the drug Name of the drug (Drug Bank ID)	Category/indication	Mechanism of action
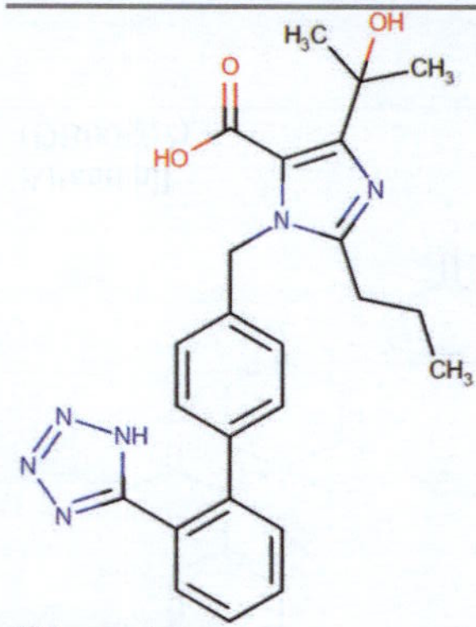 Olmesartan (DB00275)	Anti-hypertensive	Angiotensin II receptor blocker (ARB)
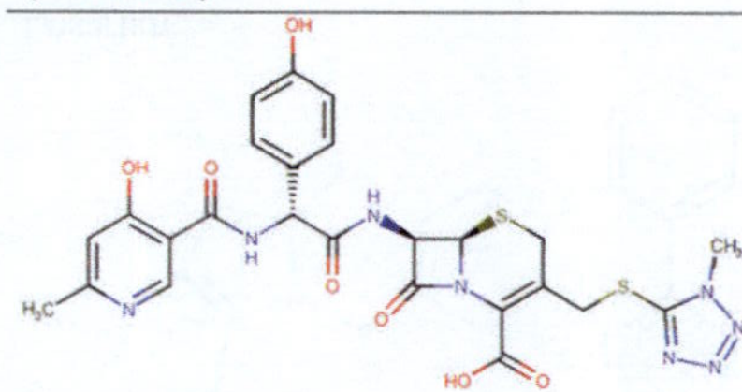 Cefpiramide (DB00430)	Cephalosporin antibiotic	Inhibiting bacterial cell wall biosynthesis against *Pseudomonas aeruginosa*

(continued)

Table 1 (continued)

Structure of the drug Name of the drug (Drug Bank ID)	Category/indication	Mechanism of action
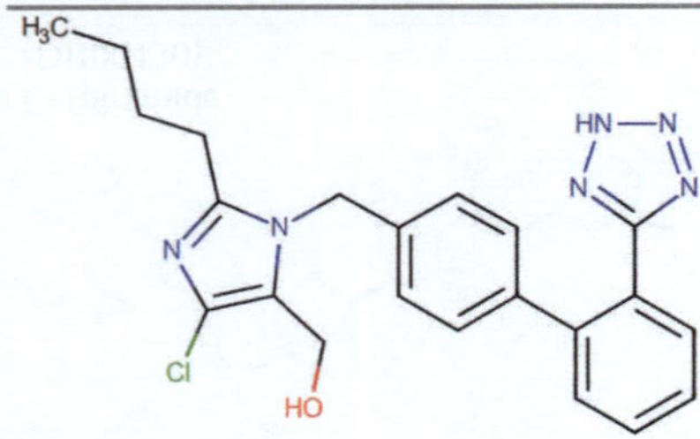 Losartan (DB00678)	Anti-hypertensive and to reduce the risk of stroke	HMG-CoA inhibitor
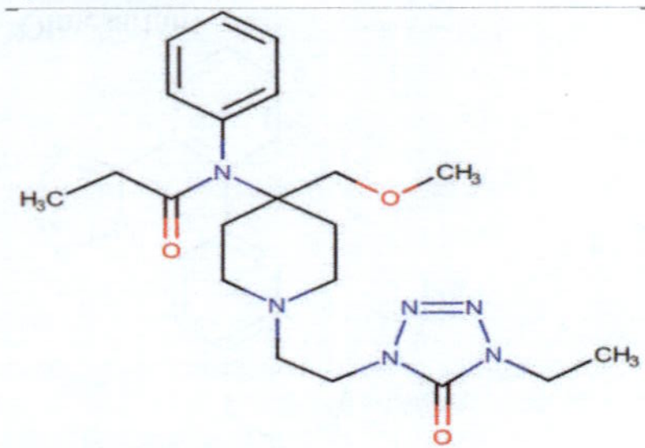 Alfentanil (DB00802)	Postoperative pain and the maintenance of general anaesthesia	Mu-type opioid receptor in humans

(continued)

Table 1 (continued)

Structure of the drug Name of the drug (Drug Bank ID)	Category/indication	Mechanism of action
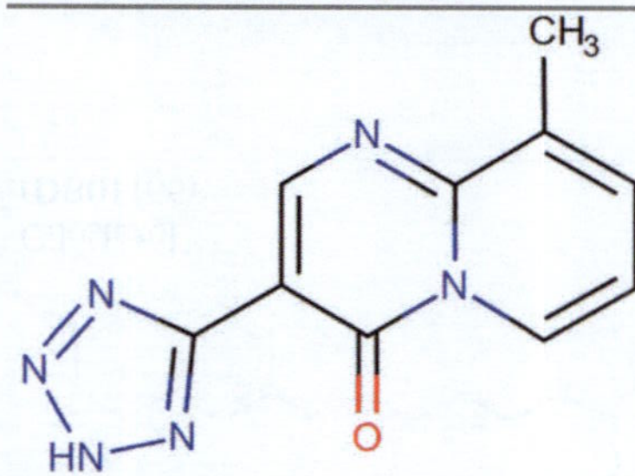 Pemirolast (DB00885)	It acts to prevent itching of the eyes arisen due to allergies such as allergic conjunctivitis and hay fever	Histamine H1 antagonist
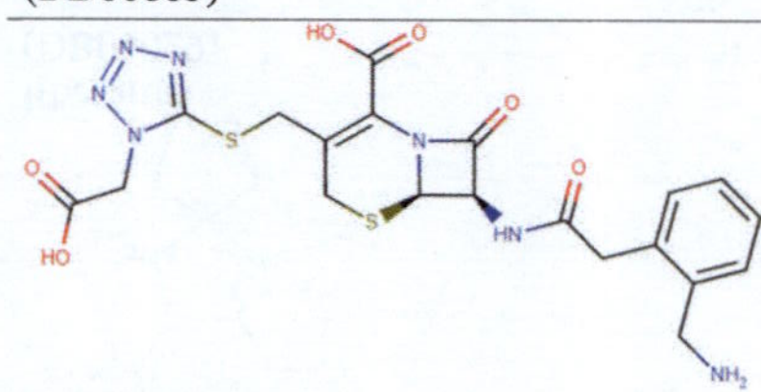 Ceforanide (DB00923)	Anti-infective	The cell wall synthesis was prohibited by its affinity towards penicillin-binding proteins (PBPs) inhibits

(continued)

Table 1 (continued)

Structure of the drug Name of the drug (Drug Bank ID)	Category/indication	Mechanism of action
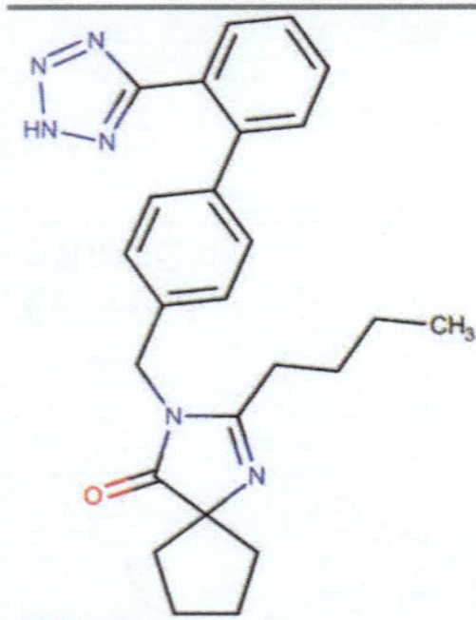 Irbesartan (DB01029)	Used for treating diabetic nephropathy and hypertension in hypertensive patients associated with type 2 diabetes	An angiotensin-receptor blocker (ARB)
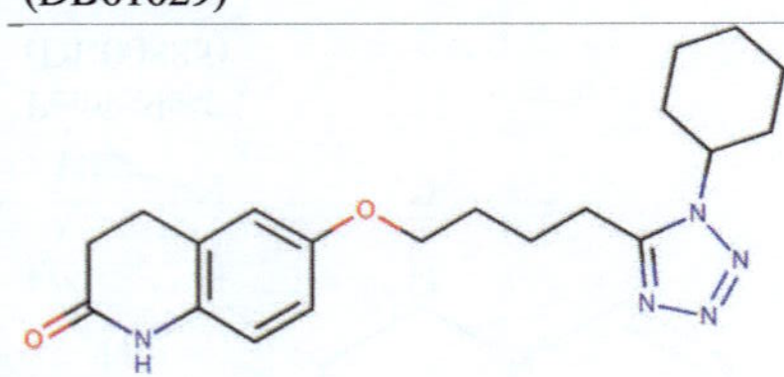 Cilostazol (DB01166)	Recommended for the treating symptoms aroused out of intermittent claudication	cyclic AMP (cAMP), phosphodiesterase III inhibitors (PDE III inhibitors)

(continued)

Table 1 (continued)

Structure of the drug Name of the drug (Drug Bank ID)	Category/indication	Mechanism of action
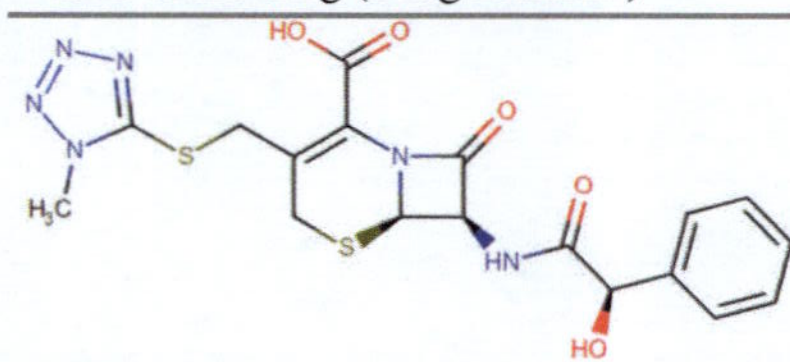 Cefamandole (DB01326)	Used to treat infections originated by susceptible strains of microorganisms	Binds to specific penicillin-binding proteins (PBPs) present inside the cell wall of a bacteria
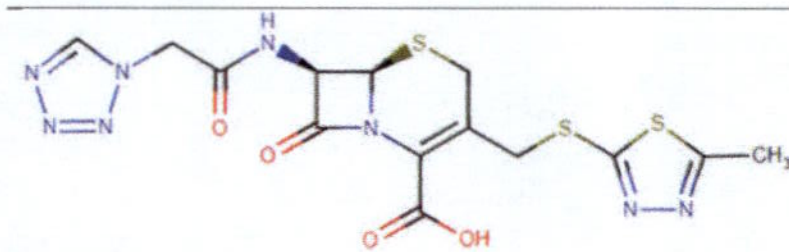 Cefazolin (DB01327)	To treat bacterial infections of the skin	The cell wall synthesis was prohibited by its affinity towards penicillin-binding proteins (PBPs) inhibits
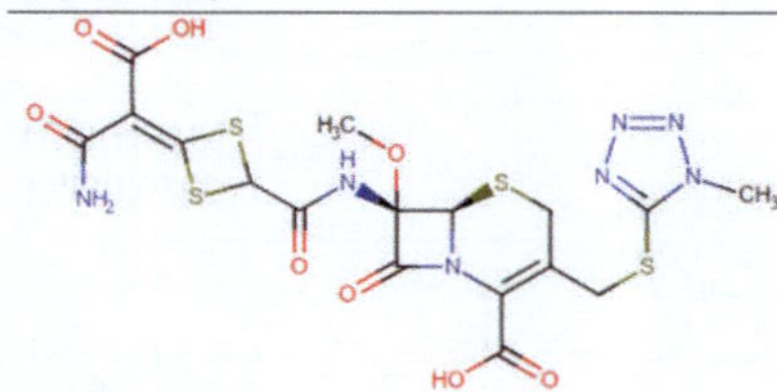 Cefotetan (DB01330)	For prophylaxis and treatment of bacterial infections	The cell wall synthesis was inhibited by binding and also inhibits the bacterial penicillin-binding proteins which are useful in cell wall biosynthesis

(continued)

Table 1 (continued)

Structure of the drug Name of the drug (Drug Bank ID)	Category/indication	Mechanism of action
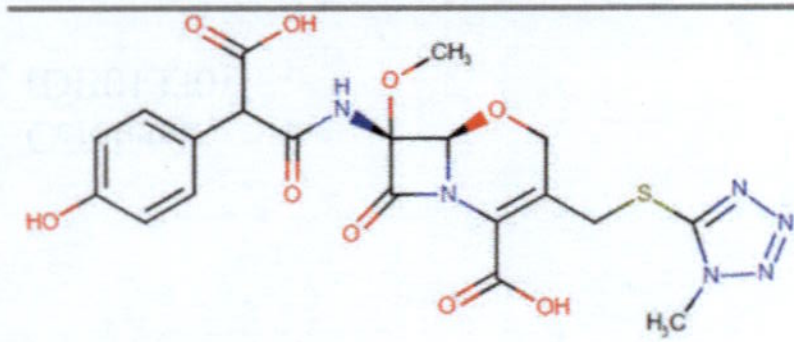 Latamoxef (DB04570)	Oxacephem antibiotic	It acts by acylating the transpeptidase C-terminal domain which is sensitive to penicillin—by lactam ring opening

(continued)

Table 1 (continued)

Structure of the drug Name of the drug (Drug Bank ID)	Category/indication	Mechanism of action
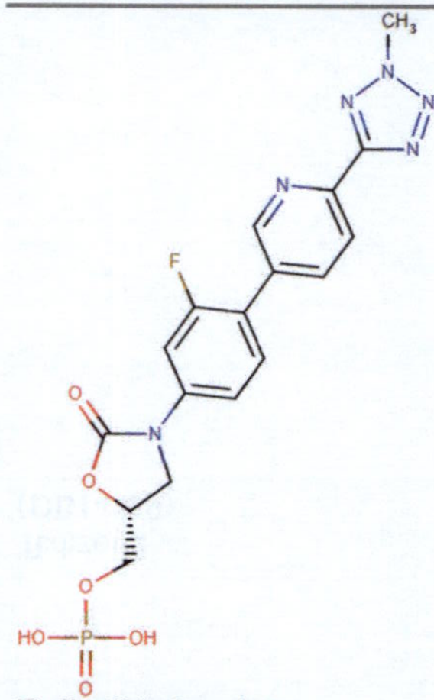 Tedizolid phosphate (DB09042)	Oxazolidinone-class antibiotic prodrug	Recommended for the treatment of acute bacterial skin and skin structure infections (ABSSSI)

(continued)

Table 1 (continued)

Structure of the drug Name of the drug (Drug Bank ID)	Category/indication	Mechanism of action
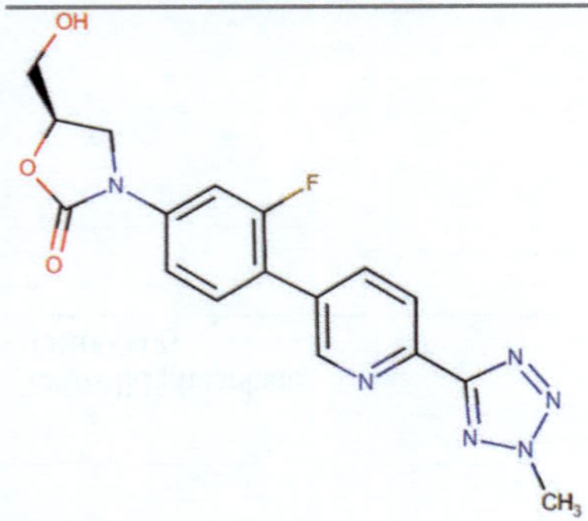 Tedizolid (DB14569)	No information	No information

(continued)

Table 1 (continued)

Structure of the drug Name of the drug (Drug Bank ID)	Category/indication	Mechanism of action
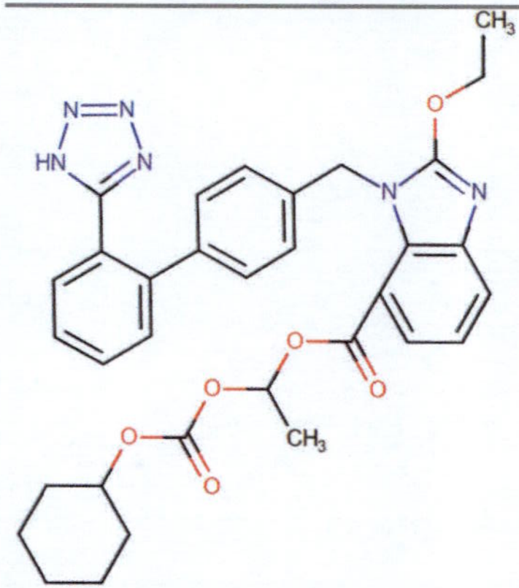 Candesartan cilexetil (DB00796	Anti-hypertensive	An angiotensin-receptor blocker (ARB)
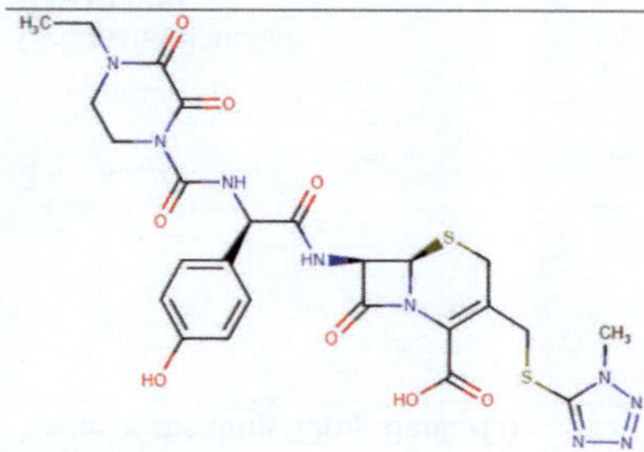 Cefoperazone (DB01329)	Broad-spectrum cephalosporin	Used to treat various bacterial infections

(continued)

Table 1 (continued)

Structure of the drug Name of the drug (Drug Bank ID)	Category/indication	Mechanism of action
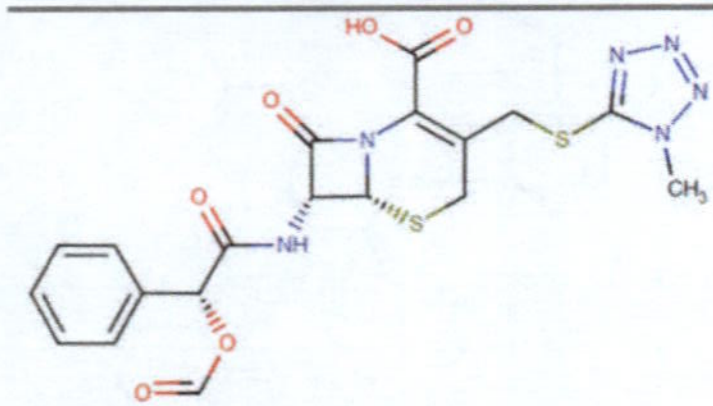 Cefamandolenafate (DB14725)	No information	No information

2 Tetrazoles Drugs and Their Hypotensive Action

Renin Angiotensin System (RAS) plays a crucial role in regulating blood pressure and homeostasis. Angiotensin II (AII) is an octapeptide, which is constituted of angiotensin I within the RAS due to the angiotensin-converting enzyme (ACE) catalyzed reactions and it is also a powerful vasoconstrictor. The control of RAS can be most capably done by inhibiting the activity of AII by blocking its active sites. Losartan **6** is used as a hypertensive drug that consists of a 5-aryl-1*H*-tetrazole moiety and candesartan cilexetil **9** which is used as an angiotensin II receptor blocker (ARB) also consists of 5-aryl-1*H*-tetrazole. The first representative of non-peptide AII antagonists considered was Losartan (Dup-753, Cozaar) [12]. The AII receptor antagonists (**6**–**12**) contain tetrazole as a common structural fragment (Fig. 4) [13].

A novel AT1 receptor antagonist which is non-peptide angiotensin has been delineated. It was evident from pharmacological results that compound **13** inhibited the binding of AT1 receptors to angiotensin II in rat liver membranes without causing any interplay with AT2 receptors in bovine cerebellar membranes. The AT1 receptor subtype-selective inhibition was described by an angiotensin II receptor antagonist. Compound **14** displayed exceptional inhibitory properties in comparison with ARBs such as losartan and candesartan in the diminution of isolated rabbit thoracic aorta. Additionally, the structure–activity relationships of cyano ester dihydropyridines were also reported. The compound **15** exhibited improved solubility and metabolic stability without CYP inhibition liability (Fig. 5) [14–18].

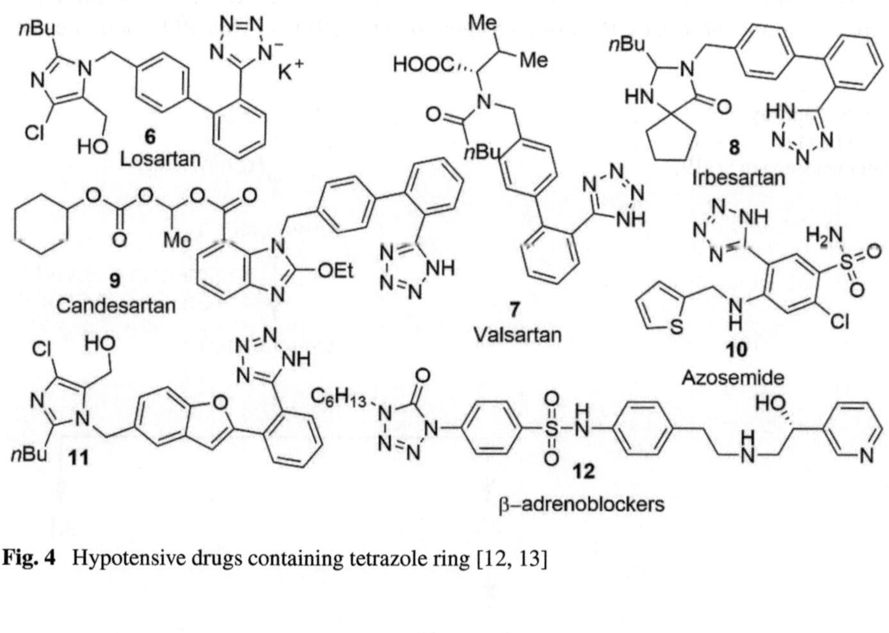

Fig. 4 Hypotensive drugs containing tetrazole ring [12, 13]

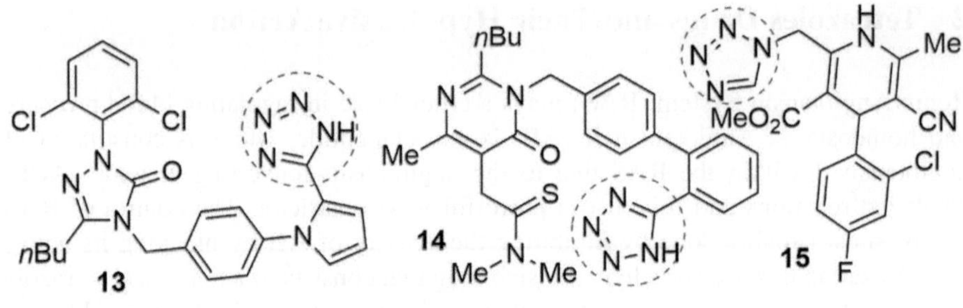

Fig. 5 Anti-hypertensive activity of tetrazole ring [14–18]

2.1 Mechanism of Action

Hypertension is a type of cardiovascular ailment that is troubling a larger portion of elderly people [7]. Angiotensinogen is a serpin family protein that is converted into an inactive form i.e. Angiotensin I by Renin. It is further transformed by Angiotensin-Converting Enzymes into Angiotensin II (an active form). It has a vital role in the preservation of arterial blood pressure and fluid and electrolyte balance. The Angiotensin II binds to the AT-I receptor which is involved in cardiovascular effects like oxidative stress, vasoconstriction, sympathetic stimulation, cell proliferation, etc. (Fig. 6) [19].

The conditions like Hypercholesterolemia, Hypertension, Diabetes, Heart failure activities the Angiotensin II in higher levels which causes AT1 receptor increased expression and produces Reactive oxygen species (ROS). This ROS activates the

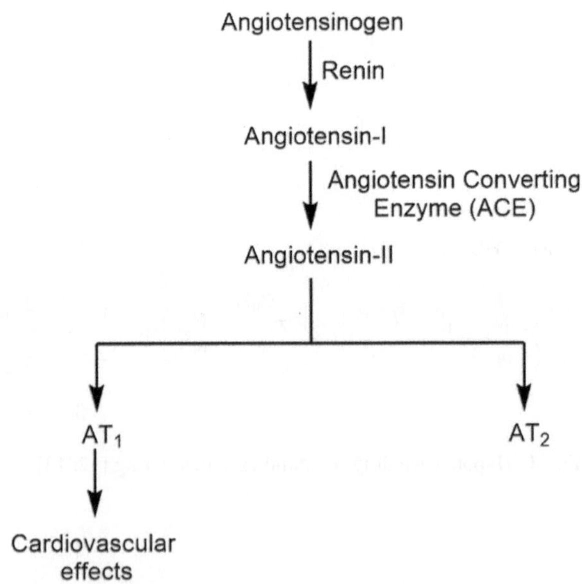

Fig. 6 Pathway of conversion of Angiotensinogen [7, 19]

redox-sensitive genes and reduces the bioavailability of NO. The activated redox genes may result in the release of Proinflammatory mediators and it also initiates and progresses Atherosclerosis condition. The reduced bioavailability of NO also worsens the condition of Atherosclerosis. So, inhibition of the Angiotensin II binding to the AT-1 receptor plays a vital role in the case of the above-mentioned conditions [19].

Angiotensin II is capable of stimulating the release of aldosterone and also acts as a vasoconstrictor. Upon the liberation of aldosterone, water and sodium will be reabsorbed and it finally results in the elevation of blood pressure. Angiotensin II receptor blockers (ARBs) are known to act as anti-hypertensive agents (Fig. 7) [7].

ARBs, selective non-peptide antagonists, are utilized for the curing of high blood pressure and other cardiovascular disorders. There are 8 sartans namely Azilsartan, Eprosartan, Candesartan, Irbesartan, Losartan, Telmisartan, Olmesartan and Valsartan [19, 20]. All the ARBs except Telmisartan contain tetrazole in their chemical structures and only it crosses BBB [21].

Candesartan, a tetrazole derivative binds to the AT1 angiotensin II receptor and results in the prevention of binding between angiotensin II and the receptor. This

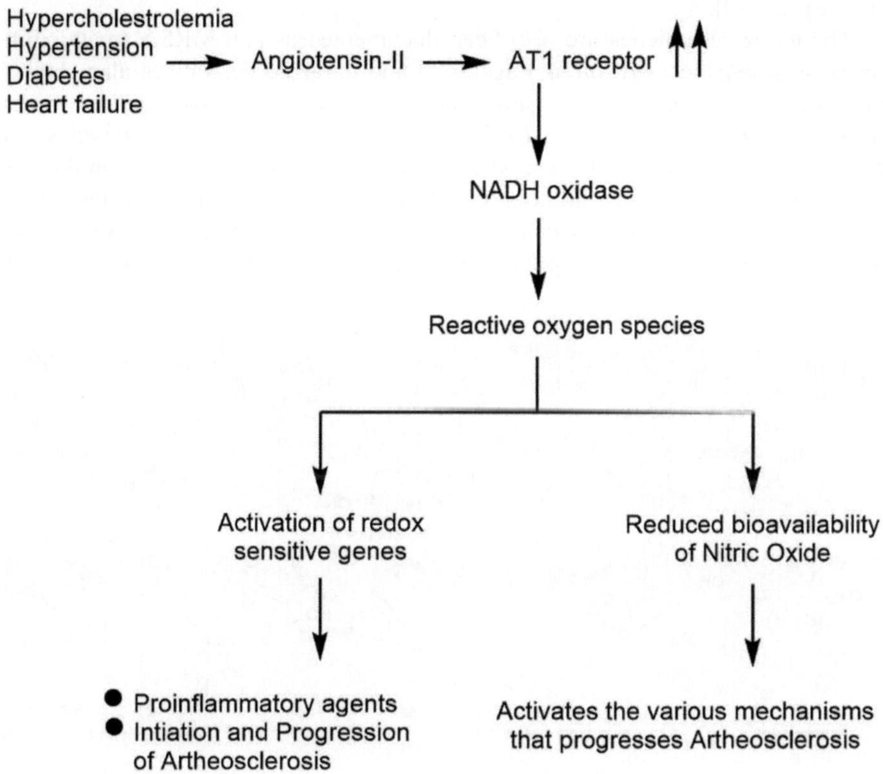

Fig. 7 Angiotensin II mechanism of action [19]

binding causes a blocking in the vasoconstriction and the ability of angiotensin II to secrete aldosterone [7]. The angiotensin II receptor belongs to a G-protein-coupled receptor superfamily. ARBs act by blocking the diverse effects of Ang II and are extremely selective for the AT1 receptor. The tetrazole moiety binds to the AT1 receptor by multiple interactions with Lys[199] and His[256] residues that comprise an identical substitute to the ligand-binding pocket [22].

3 Antimicrobial and Anti-inflammatory Activity of Tetrazole Derivatives

The first-generation cephalosporin antibiotics kefzol **16** and its demethylated analogue ceftezole **17** show a broad range of activities as both drugs are vigorously used as veterinary medicines. Cefamandole **18** consists of tetrazole moiety that goes to the second-generation cephalosporin antibiotics. The third-generation cephalosporin antibiotic is Latamoxef **19**. The equivalent of the macrolide antibiotic Rapamycin **20** and the analogue of Nocathiacin I **21** are remarkable for their activity (Fig. 8) [23–30].

The tetrazole moieties are also been documented as anti-MRSA (methicillin-resistant *Staphylococcus aureus*) agents. Some tetrazole derivatives also showed high potency against gram-negative and gram-positive microorganisms which include *Staphylococcus aureus*, *Escherichia coli*, *Streptococcus Pneumoniae* and *Enterococcus faecalis* which were otherwise resistant to penicillin series antibiotics. Tetrazole containing oxazolidinone moiety exhibited high activity against gram-positive bacteria. The tetrazole containing a terminal of carbazole derivatives showed antiseptic properties. The phenothiazine derivatives with tetrazole moiety in the side

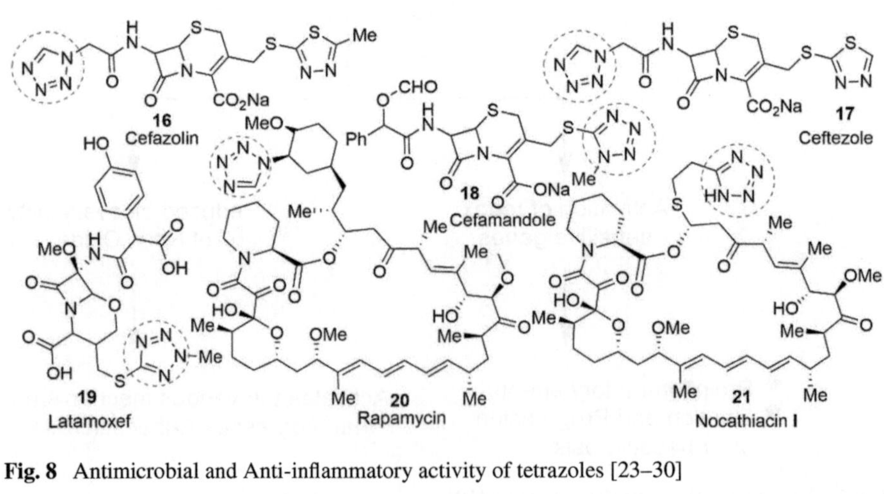

Fig. 8 Antimicrobial and Anti-inflammatory activity of tetrazoles [23–30]

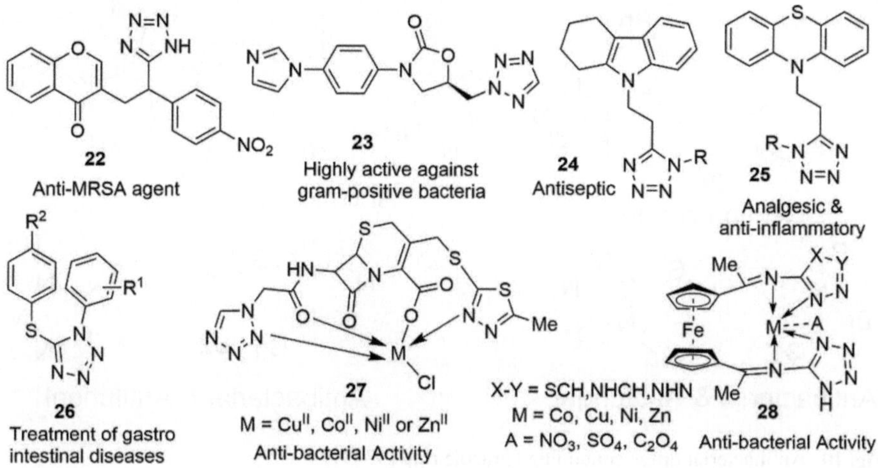

Fig. 9 Different biological activity of tetrazoles [31–36]

chain are effective analgesics and can act as anti-inflammatory drugs (**22–28**) (Fig. 9) [31–36].

The better activity was shown by compound **29** against all the tested cultures, except *Staphylococcus aureus* in comparison with ampicillin. An in vitro antimicrobial activity using disc diffusion method measuring zones of inhibition was exhibited by a series of tetrazole compounds. The compound 2-methyl-3-{4-[2-(1*H*-tetrazol-5-yl-ethylamino]phenyl}-3*H*-quinazolin-4-one (**30**) is shown to have considerable antimicrobial activity versus the tested microorganisms. The exhibited activity was less in comparison to the reference drugs such as fluconazole and ciprofloxacin. The moderate activity was reported by 5-thio-substituted tetrazole derivatives against the tested organisms. The compounds **31** and **32** were found to exhibit antifungal and antibacterial activities (Fig. 10) [37–39].

The compounds **33–38** were reported to have minimal inhibitory concentration (MIC) values ranging from 23.40 to 46.87 μg/L to act as an antimicrobial. The Society for Artistic Research reported that the pyran derivatives were proved to be more effective with an order of activity for the -R substituent: 4-OMe > 4-Me > 3-OH > H > 4-Cl > 4-NO_2 than pyridine derivatives (Fig. 11) [40].

A group of synthesized oxazolidinone derivatives were (Fig. 12) and were tested on clinically pertinent resistant gram-positive organisms such as *Moraxellacatarrhalis* and *Haemophilus influenza* (*Bacillus influenzae*) for their substituent effects on antibacterial activity (**39–43**) [41, 42].

The tetrazolo[1,5-*c*]quinazoline-5-thione **44** which has a *para*-methoxyphenyl (PMB) group at the 5 positions emerged to have the foremost antimicrobial

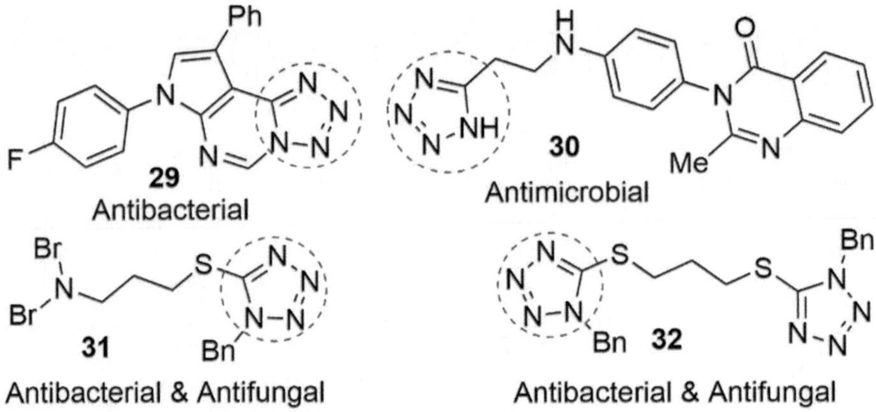

Fig. 10 Antibacterial drugs containing tetrazole ring [37–39]

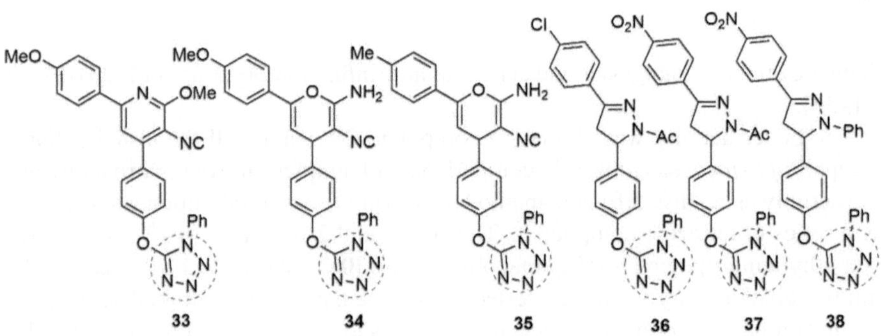

Fig. 11 Tetrazole ring containing antimicrobial drugs [40]

agent out of all the screened compounds. Additionally, the growth of *Pseudomonas aeruginosa* (gram-negative opportunistic pathogen) and *Klebsiellapneumoniae* (gram-negative rod-shaped bacterium) showed moderate inhibition. In addition, the compounds *N*-(4-(2-(2*H*-tetrazol-5-yl)ethyl)-5-thioxo-4,5-dihydro-1,3,4-thiadiazol-2-yl)benzamide **45** tend to exhibit antimicrobial activities with a MIC value of 100 μg/mL against *Bacillus subtilis* in comparison to penicillin (31 μg/mL). It also exhibits a MIC value of 125 μg/mL against *Pseudomonas aeruginosa* in comparison to a penicillin (46 μg/mL) and a MIC value of 125 μg/mL against *Streptomyces* species paralleled with penicillin (33 μg/mL). A Group of substituted 5,6,7,8-tetrahydro-3*H*-benzo[4,5]thieno[2,3-*d*]pyrimidine-4-one derivatives showed antimicrobial activity against *Aspergillus Niger, Staphylococcus aureus*, *Bacillus subtilis*, *Escherichia coli and Candida albicans*. The group of substituted thienopyrimidine derivative **46** is found to be the most highly active compound (Fig. 13) [43–45].

Tetrazole derivatives embedded with azetidinone have been derived and investigated for their antitubercular activity. The compound **47** is recommended to have

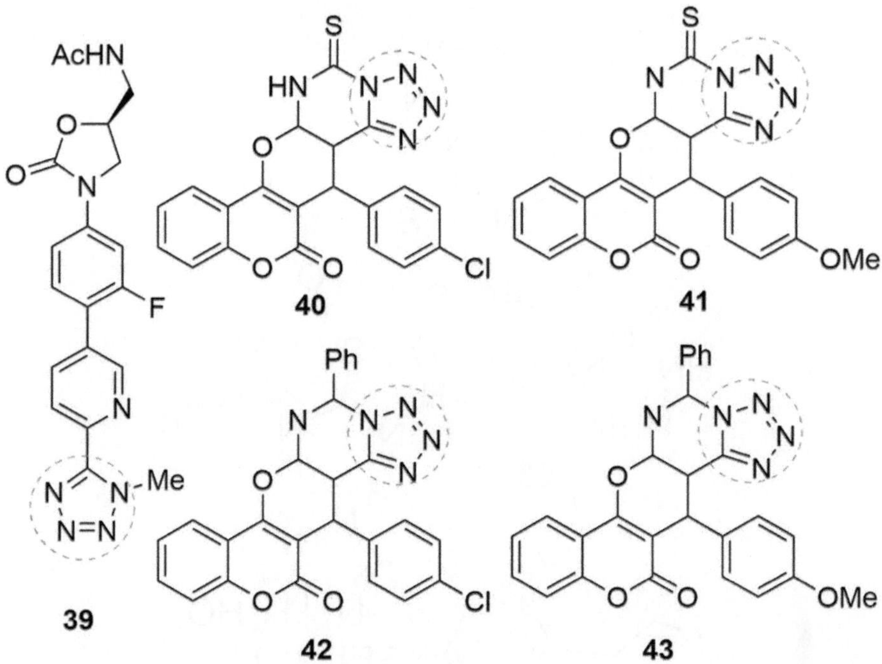

Fig. 12 Tetrazole ring containing antibacterial drugs [41, 42]

exhibited maximum activity and is regarded as a lead for further discoveries towards an effective class of antimicrobial drugs. Compound **48** showed strong activity against *Bacillus subtilis, Pseudomonas aeruginosa, Escherichia coli* and *Streptococcus lactis* in comparison with ciprofloxacin and showcases greater activities versus the fungi when compared to fusidic acid. Meanwhile, compounds **49** and **50** exhibited greater activity against the gram-negative bacteria *Pseudomonas aeruginosa* and *Escherichia coli.* New compounds such as α-acetoxyphosphonate and α-hydroxyphosphonate **51** derivatives **52** have shown antimicrobial activities as well (Fig. 14) [46–48].

1-substituted tetrazoles mostly exhibit antibiotic activity. The compounds 3-[(1*H*-tetrazol-5-yl) methoxy]-2-(4-fluorophenyl)-4*H*-chromen-4-ones were synthesized and their potency was tested for antibacterial and antifungal activity. The tetrazole substituted compounds exhibited excellent activity versus several bacterial strains like *Pseudomonas aeruginosa, Staphylococcus aureus, Bacillus subtilis, Escherichia coli* and fungal strains like *Candida glabrata; Candida tropicalis; Candida albicans.* All the tetrazole substituted compounds exhibited excellent activity compared to the standard drugs like Streptomycin for antibacterial activity against *Escherichia coli, Bacillus subtilis, Pseudomonas aeruginosa and* Miconazole for antifungal activity. The MIC value of the tetrazole compounds against *Bacillus subtilis* ranges between 50 and 200 μg/ml, *E. coli* in the range of 100–400 μg/ml, Pseudomonas aeruginosa in the range of 50–200 μg/ml. The antifungal activity *Candida albicans* ranges

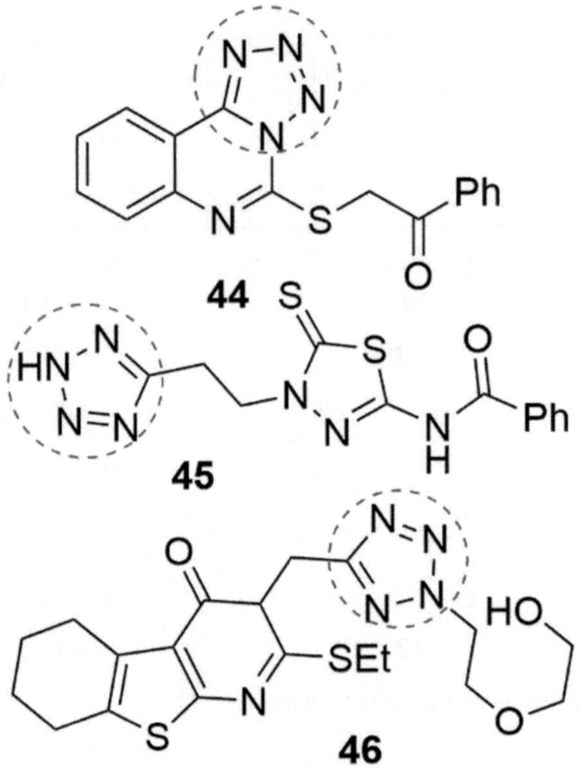

Fig. 13 Different biological activity of Tetrazoles [43–45]

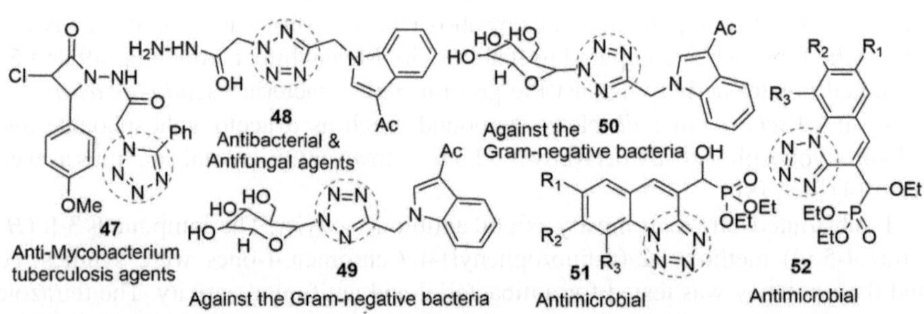

Fig. 14 Various antibacterial and antimicrobial activity of Tetrazoles [46–48]

between 12.5 and 50 μg/ml, *Candida glabrata* ranges from 12.5 to 50 μg/ml and *Candida tropicalis* in the range of 12.5–50 μg/ml [49].

Tetrazole and/or cyanamide-containing compounds were synthesized and screened for anti-inflammatory activity, ulcerogenic activity. Some of the compounds have exhibited good COX-2 inhibitory potency with the IC_{50} value of 0.11 and

0.14 μM whereas the IC_{50} value of the standard drug, Celecoxib was found to be 0.16 μM. The ulcerogenic index of these compounds was studied and they have exhibited a very low ulcer index in the range of 0.25–2.00 compared to Indomethacin whose ulcer index was 22.50 and Celecoxib is 0.50 [50].

4 Antifungal Activity of Tetrazole Derivatives

Over the past 50 years, the diseases arising due to fungal infections are largely increasing owing to the increase in the number of weakened immune systems. The patients diagnosed with organ transplants, AIDS, anticancer chemotherapy and long antimicrobial treatments are immunosuppressed and are endangered to get affected with systemic fungal infections such as cryptococcosis, aspergillosis and candidiasis which are life-threatening. The dugs itraconazole and fluconazole (triazole antifungal medication) have been widely recommended in antifungal chemotherapy and they are orally active and can strongly inhibit lanosterol 14α-demethylase (cytochrome P-45014DM). Oteseconazole **53** also known as VT-1161, is an antifungal agent consisting of tetrazole moiety and is potentially used for the treatment of candida vaginal infection [51].

In 2004, *Upadhayaya* et al. designed tetrazole based triazole derivatives (**54–57**) which are showing strong growth inhibitory activity against *Aspergillus spp. Candida spp.* and *Cryptococcus neoformance*. Furthermore, they designed and manufactured a group of new compounds with antifungal potency (Fig. 15) [52].

In 2010, *Movie popat. B* et al. have synthesized 3-aryl 1-(5-phenyl-1*H*-tetrazol-1-yl)prop-2-en-1-one (**58**) and classified it for its antifungal activity utilizing cup and plate method in which the presence of -Cl group shows high activity (Fig. 15) [53].

Additionally, the presence of the combination of the tetrazole and 1,2,4-triazole rings in antifungal agent TAK-456 (**59**) showed promising results and, remarkably, the above compound does not contain the 2,4-difluorophenyl moiety which is a characteristic of active substances of many antifungal agents [54].

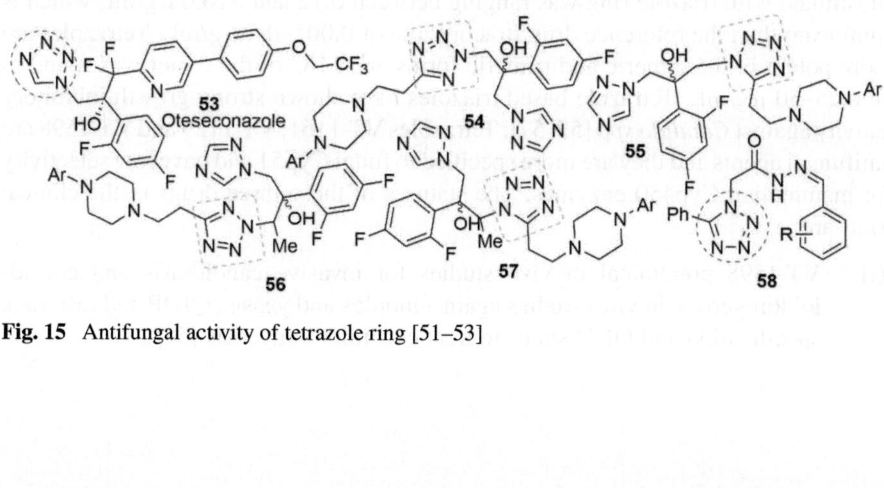

Fig. 15 Antifungal activity of tetrazole ring [51–53]

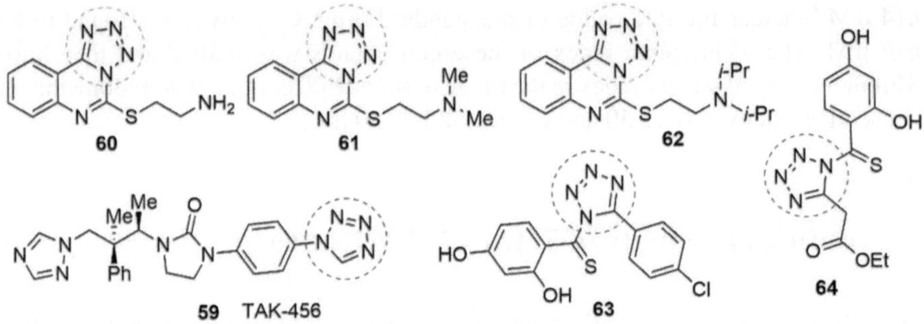

Fig. 16 Tetrazole moieties showing antifungal activity [54–56]

A series of synthesized tetrazolo[1,5-*c*]quinazoline-5-thione derivatives were screened at a concentration of 100 μg for antifungal and antibacterial activities. The SAR has shown that the shortening of the dialkylamino group in substances **60–62** moderately decreased the antimicrobial activity against *Enterococcus faecalis*. Otherwise, it has led to an increment in activity against *Escherichia coli* and *Staphylococcus aureus* and also antifungal activity versus *Candida albicans* [55]. In addition, tetrazole derivatives such as compounds **63** and **64** were the most active compounds against *Candida albicans* (pathogenic yeast). Moreover, maximum compounds exhibited higher activities than reference drugs fluconazole and itraconazole (Fig. 16) [56].

4.1 Mechanism of Tetrazoles as Antifungal Agents

Antifungal activities of tetrazole have been examined and some of its derivatives displayed the activity against both drug-resistant and drug-susceptible fungi. Tetrazoles along with the azoles provide more effective antifungal activity as the azole moiety can exercise diverse non-covalent interactions. The MIC of the tetrazole substituted with triazole ring was ranging between 0.12 and >16.0 μg/mL which is significant than the reference drug Itraconazole of 0.007–0.5 μg/mL. Tetrazoles are more potent in the dimeric and trimeric forms and MIC of the dimer was found to be 6.25–50 μg/mL. Tetrazole based triazoles have shown strong growth inhibitory activity against *Candida spp* [52, 57], Tetrazoles VT-1161, VT-1129 and VT-1598 are antifungal agents and they are more specified to fungi Cyp51 and have low selectivity for mammalian Cyp450 enzymes. The statuses of these three drugs in the clinical trials are:

(i) VT-1598 preclinical in vivo studies for invasive candidiasis and coccidioidomycosis; in vitro studies against moulds and yeasts; QIDP and fast track specification and ODD status to treat coccidioidomycosis [58].

(ii) Preclinical in vitro and in vivo studies of VT-1129 versus Cryptococcus and Candida.

(iii) VT-1161 has finished phase 2 clinical trials to treat acute vulvovaginal candidiasis and tinea pedis and began the studies towards curing recurrent vulvovaginal candidiasis and onychomycosis; It has obtained QIDP status and fast track specification by U.S. FDA for curing recurrent vulvovaginal candidiasis.

5 Antiviral Activity of Tetrazole Derivatives

In the last 15 years, an active search towards designing an antiviral drug embedded with tetrazole ring has been demonstrated. The activity of hepatitis C virus protease inhibitors was examined against the strains NS3 **65** and NS3/4A **66**. The vital activity of the enterovirus EV71 was suppressed by pyridylimidazolidinone derivative **67** (Fig. 17) [59–63].

Among the new anti-AIDS drugs, 1-(5-chloroindol-3-yl)-3-hydroxy-3-(2*H*-tetrazol-5-yl)propenone (5-CITEP) (**68**) is the 2nd generation HIV integrase

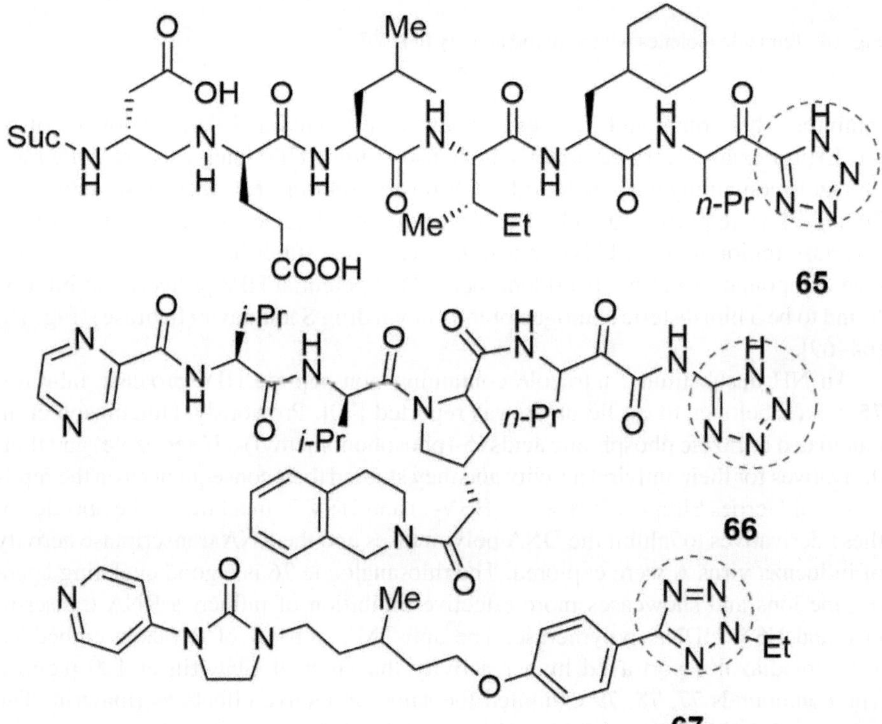

65

66

67

Fig. 17 Antiviral activity of Tetrazoles [59–63]

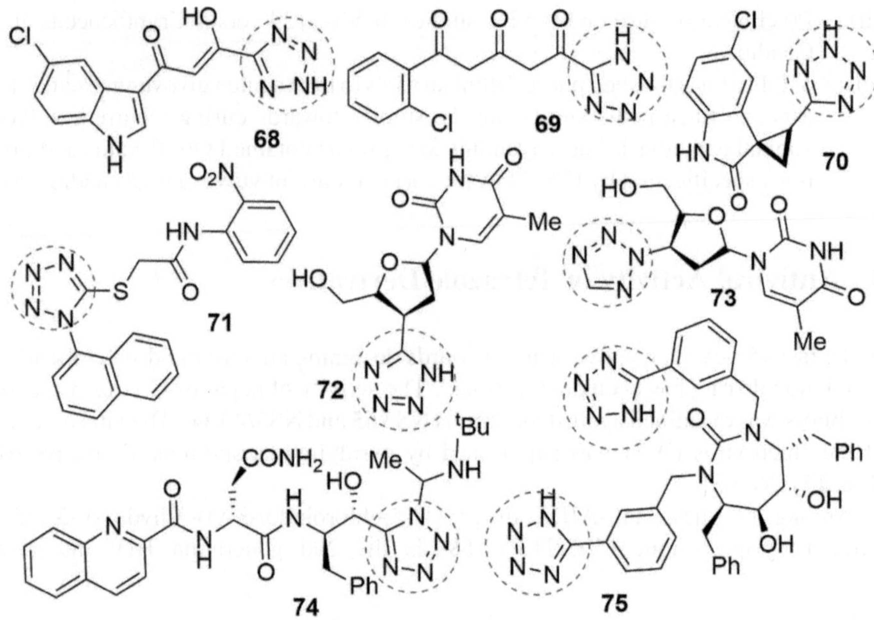

Fig. 18 Tetrazole moieties with antiviral activity [64–69]

inhibitor. This compound has been widely investigated and data analogues of 5-CITEP, for example, triketone **69**, which also inhibits HIV integrase, are quite rare. Tetrazole containing non-nucleoside HIV reverse transcriptase (revertase) inhibitors **70** and **71** were portrayed well. The above-disclosed nucleoside analogues of HIV revertase inhibitors (NARTIs) **72** and **73** were also established in a series of tetrazole compounds. Original peptidomimetic **74**, a potential HIV protease inhibitor, is found to be a bioisosteric analogue of the known drug Saquinavir (Invirase) (Fig. 18) [64–69].

An NH-unsubstituted tetrazole containing non-peptide HIV protease inhibitor **75**, which belongs to cyclic ureas was reported [70]. Previously, Hutchinson et al. elucidated tetrazole phosphonic acids (5-(phosphonomethyl)-1*H*-tetrazole) and their derivatives for their antiviral activity and they studied their consequences on the replication of Herpes Simplex Viruses-1 (HSV-1) and HSV-2. In addition, the abilities of these derivatives to inhibit the DNA polymerases and the RNA transcriptase activity of influenza virus A were explored. The thio-analogue **76** is a good chelating agent of zinc ions and showcases more effective inhibition of influenza RNA transcriptase and HSV-1 DNA polymerase. The anti-TMV activity of tetrazole embedded 1,2,3-thiadiazoles portrayed higher activity than that of ribavirin at 100 μg/mL. The compounds **77**, **78**, **79** exhibited the same protective effects as ribavirin. The compound 7-(2*H*-Tetrazol-5-yl)-1*H*-indole **80** exhibited potent inhibitory activity against HIV-1 attachment with a disadvantage of oral bioavailability in rats (Fig. 19) [71, 72].

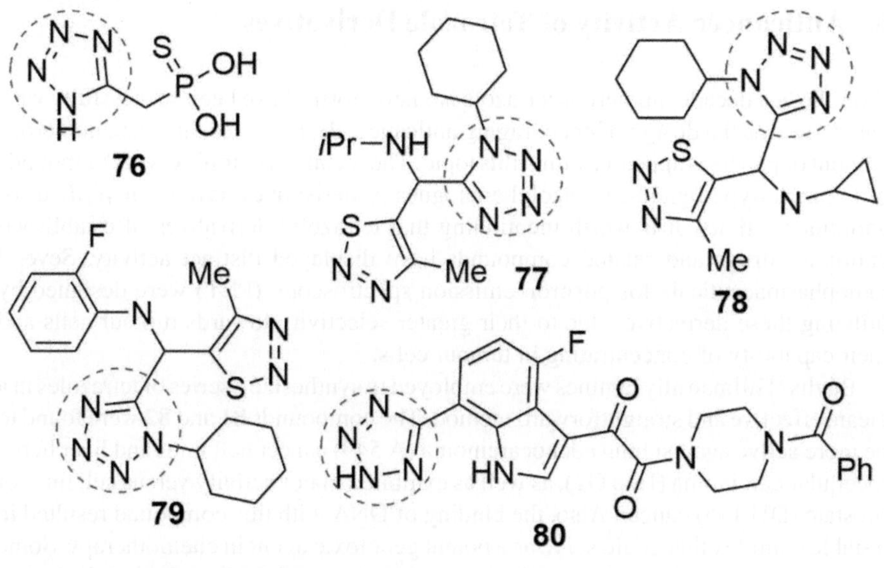

Fig. 19 Antiviral activity of tetrazole ring [70–72]

The 2-adamantyl-tetrazole derivatives were synthesized and screened for antiviral activity. These compounds have not exhibited very good antiviral activity but some of these compounds exhibited anti-influenza properties. On contrary, antiviral activity of 1-adamantyl tetrazoles exhibits higher activity than 2-adamantyl isomer [73]. This indicates that the presence of tetrazole increases the activity of the organic compounds.

In Immunocompromised patients, several diseases were caused by Herpes simplex viruses (HSV) like disseminated disease. It is also reported that genital herpes is enhancing the transmission of HIV-1 (human immunodeficiency virus) through sexual transmission. The serotypes of HSV-1 and HSV-2 destroy or activate epithelial cells, produce chemokines and proinflammatory cytokines and activate HIV-1 targets. So, HSV inhibition is essential in order to prevent the transmission of HIV-1.

There was a report that 5-(phosphomethyl)-1*H*-tetrazole was highly efficient in inhibiting the replication of HSV-2 when compared with Phosphonoacetic acid. The thio group substitution has made 5-(phosphomethyl)-1*H*-tetrazole a more effective inhibitor of Influenza RNA transcriptase and HSV-1 DNA polymerase. The phosphono group substitution has also slightly increased the activity of the compounds [74].

6 Anticancer Activity of Tetrazole Derivatives

Over the last decade, numerous tetrazole isomeric forms have been successfully validated towards the design of encouraging anticancer drugs. In recent years, numerous amount of papers is appeared citing this topic. The metal ion complexes of compounds of tetrazole-type ligands, tetrazolyl analogues or derivatives have been verified for antitumor activity. It is worth mentioning that tetrazolyl derivatives of established antitumor drugs and related compounds have displayed distinct activity. Several radiopharmaceuticals for positron emission spectroscopy (PET) were designed by utilizing these derivatives due to their greater selectivity towards tumour cells and their capability of concentrating in tumour cells.

Baylis–Hillman allyl amines were employed to synthesize a series of tetrazoles in a clean, effective and straightforward method. The compounds **81** and **82** were found to be more active against lung adenocarcinoma (A 549) cancer cell lines and liver hepatocellular carcinoma (Hep G2), as well as exhibited major activity versus cell lines of prostate (DU 145) cancer. Also, the binding of DNA with this compound resulted in a stable complex that could serve as a potent genotoxic agent in chemotherapy. Some compounds derived from 2-phenylindole such as (tetrazol-5-yl)methylindole derivatives were considered to act as anticancer agents on the human liver carcinoma cell line (Hep G2). The compound **83** exhibits higher activity among the series of screened compounds and it shows an IC_{50} value of 4.2 μM and cell viability was affected in a dose-dependent manner. The growth of cells overexpressed P-glycoprotein which is multidrug resistant was inhibited by some tetrazole derivatives and the compound **84** tend to show strong inhibition of tubulin polymerization with an $IC_{50} = 1.1$ μM and powerfully inhibits the attaching of colchicine and tubulin (78% inhibition). The compound **84** has proved to be a propitious tubulin-binding agent and stands a further appraisal on its chemotherapeutic properties (Fig. 20) [54, 75, 76].

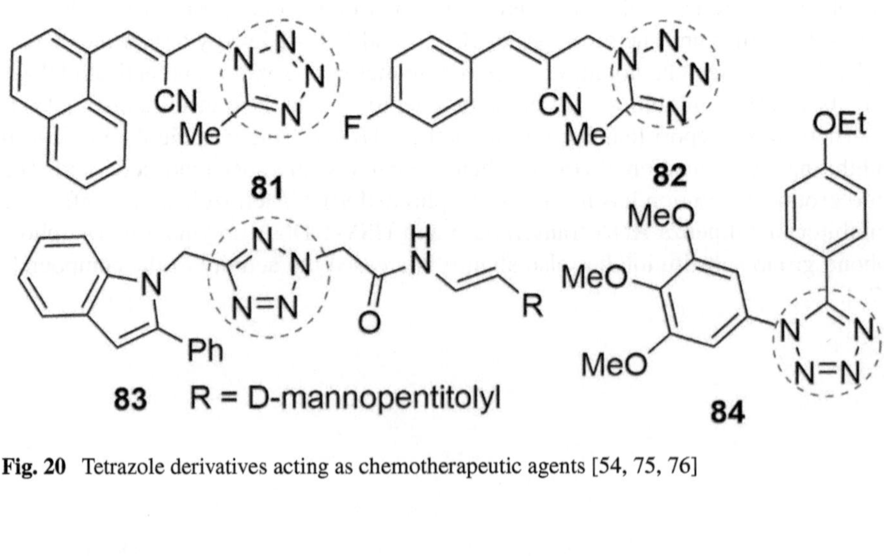

Fig. 20 Tetrazole derivatives acting as chemotherapeutic agents [54, 75, 76]

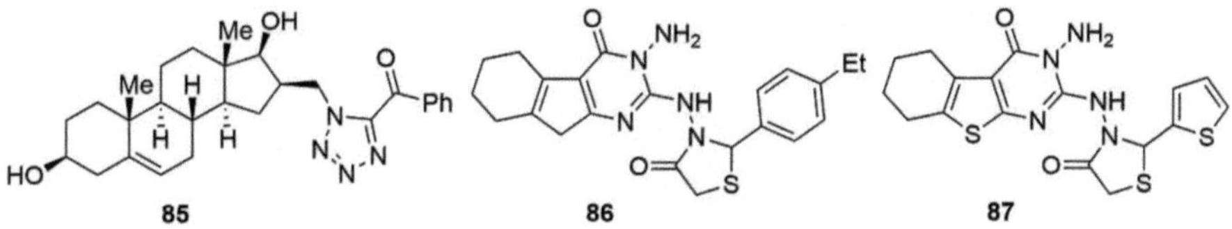

Fig. 21 Tetrazoles with chemotherapeutic activity [77, 78]

Several steroidal tetrazoles with D-ring-substitution were constructed by means of 1,3-dipolar cycloaddition and were chosen for their anticancer activity against three human gynaecological cancer cell lines (HeLa, MCF7 and A2780) and out of which compound **85** had a moderate effect [77]. Additionally, derivatives of spirothieno[2,3-*d*]pyrimidine, imidazolidine, substituted thiazolidine thieno[2,3-*d*]pyrimidines and substituted pyrimidinyl showed the potent anticancer activity. The combination of compounds **86** and **87** showed potent anticancer activity with lesser toxicity levels and the ease of synthesis marks them for future lead compounds towards cancer chemotherapy (Fig. 21) [78].

Jackman et al. reported a drug ZD9331 **88** which showed cytotoxic activity and potent growth inhibitor. The compound ZD9331[79] showed a distinct or at least an imperfectly overlapping spectrum of toxicity profile and antitumor activity, related with tomudex and possibly other TS inhibitors which are presently being clinically studied. In addition, 1,2-substituted tetrazoles were assessed for their activity against breast cancer cell lines such as MCF-7 (ER positive), MDA-MB-231 and ZR-75 (ER negative). The compounds **89**, **90** and **91** exhibited greater inhibitory potency over MCF-7 cells and compound **92** displayed higher inhibition of ZR-75 cells and MDA-MB-231 cells at a 10 − 5 M concentration level (Fig. 22) [80].

Moreover, combretastatin analogues tethered with 1,5-disubstituted tetrazole were classified for their anti-proliferative and antitubulin activity. The compounds **93** and **94**, with *ortho*- and *meta*-substituted 4-methoxyphenyl B rings with hydrogen-bonding donor groups, act as anti-proliferative agents and showed stronger inhibition of tubulin polymerization with IC50 values in micromolar concentrations. Also, it was reported that 7*a*-aza-B-homostigmast-5-eno[7*a*,7-*d*] tetrazole **95** was verified to act upon two human cancer cell lines: HepG2 and CT116 along with one non-cancerous HFL1 (human lung fibroblast) cell line. The IC_{50} values of all the compound were identified to be identical with doxorubicin. Several cancer cell lines were impeded in a dose-dependent manner and the growth of cancer cells was efficiently inhibited by compound **95** (Fig. 23) [81, 82].

Tetrazole based compounds were synthesized and their activity was evaluated for anticancer activity and their results were compared with the already reported DNA methyltransferases (DNMTs) inhibitors, i.e. RG108 and 1149. Almost all the compounds have exhibited a superior activity compared to the known DNMTs inhibitors. The GI_{50} value of these compounds was found to be in the range of 0.30 μM to 1.86 μM and one compound exhibited GI_{50} value of 6.15 μM whereas the GI_{50} value of the standards (RG108 and 1149) was found to be 127.90 μM and 7.31 μM

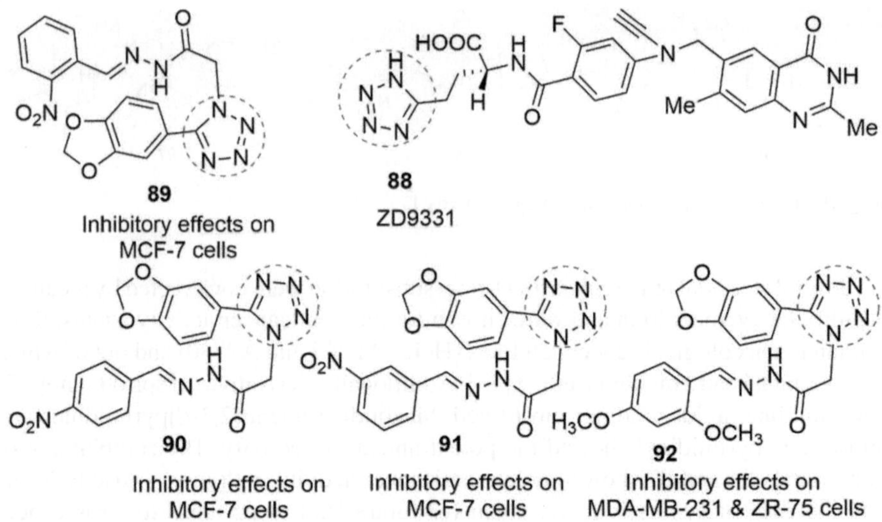

Fig. 22 Anticancer activity of substituted tetrazole ring [79, 80]

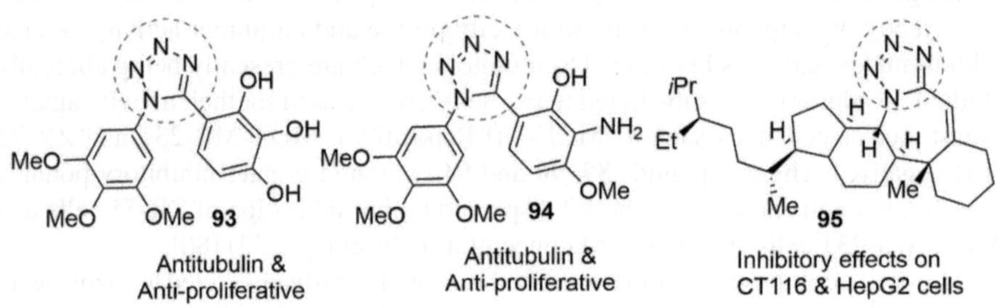

Fig. 23 Anticancer activity of 1,5-disubstituted Tetrazoles [81, 82]

respectively. These compounds act by inhibiting the DNA methyltransferases enzyme inhibitors [83].

DNA methyltransferases catalyze methylation of DNA at 5-position which is an epigenetic modification [83, 84]. It promotes normal DNA expression, stabilizes DNA–protein complexes and prevents gene expression. The process of DNA methylation by DNA methyltransferases takes place in a stepwise manner (Fig. 24).

(i) DNMTs binds to the 6th position of the Cytosine residue present in DNA
(ii) The methyl group transfer from S-adenosyl methionine to C-5 of Cytosine residue
(iii) Subsequent release of S-adenosyl-L-homocysteine (SAH) and the removal of proton from 5th position of Cytosine followed by release of the free enzyme via β-elimination and finally the generation of methylated Cytosine.

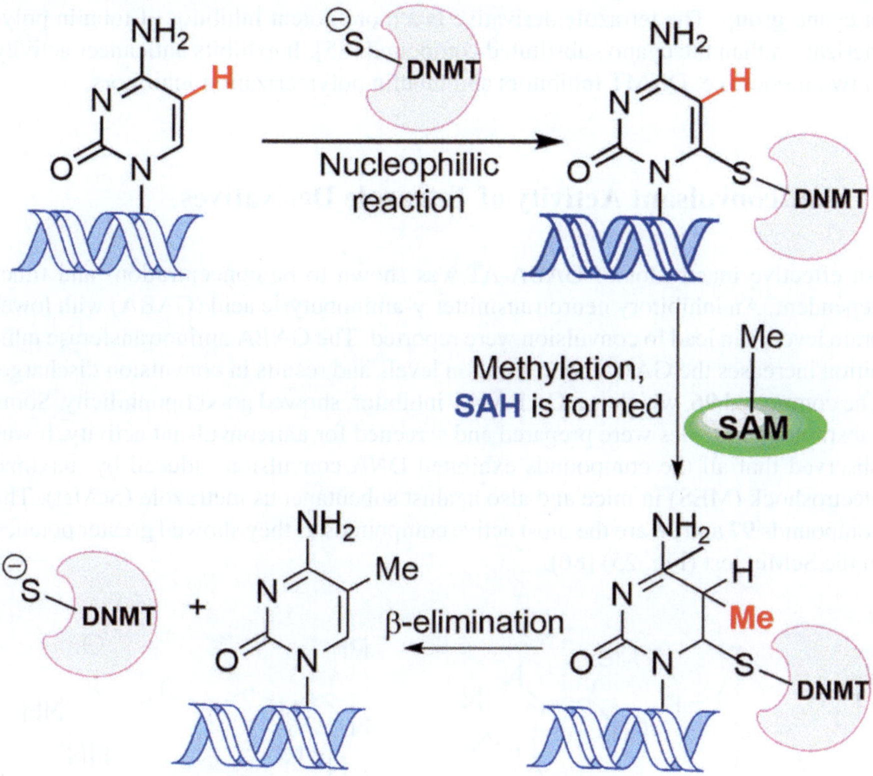

Fig. 24 Process of DNA methylation by DNA methyltransferases [83, 84]

Inhibition of these DNA methyltransferases leads to misleading of the above process which leads to apoptosis and cell death. These tetrazole moieties are good inhibitors of these DNA methyltransferases. The tetrazole ring in these compounds acts as a good copycat for the indole ring present in tryptophan moiety and it effectively binds to the active site of DNMT1 via various non-covalent interactions [83].

Tetrazole rings have drug-likeness properties and also possess high metabolic stability. This ring is a bioisosteric replacement of COOH group as it exhibits similar pKa value and similar biological properties. These are more lipophilic than the carboxylates which enhance their systemic bioavailability. So, tetrazole derivatives of Resveratrol were synthesized and assessed for their anticancer activity in various Cancer cell lines. Two tetrazolyl stilbene analogues have shown greater than 60% inhibition in almost all cell lines out of 60 cell lines. These tetrazole analogue and cyano analogue were docked at the Colchicine binding site of α/β-tubulin and their binding free energy was calculated and it was found to be −31.7 kcal/mol and −40.7 kcal/mol respectively. This proves that the tetrazole moiety can act as a mimic of cyano group. The tetrazole derivative is a more potent inhibitor of tubulin polymerization than the cyano substituted compound [85]. It exhibits anticancer activity in two modes, i.e. DNMT inhibitors and tubulin polymerization inhibitors.

7 Anticonvulsant Activity of Tetrazole Derivatives

An effective inactivator of GABA-AT was shown to be concentration- and time-dependent. An inhibitory neurotransmitter γ-aminobutyric acid (GABA) with lower brain levels can lead to convulsion were reported. The GABA aminotransferase inhibition increases the GABA concentration levels and results in convulsion discharge. The compound **96**, which is a GABA-AT inhibitor, showed good lipophilicity. Some substituted tetrazoles were prepared and screened for anticonvulsant activity. It was observed that all the compounds exhibited DNA convulsion induced by maximal electroshock (MES) in mice and also against subcutaneous metrazole (ScMet). The compounds **97** and **98** are the most active compounds as they showed greater potency in the ScMet test (Fig. 25) [86].

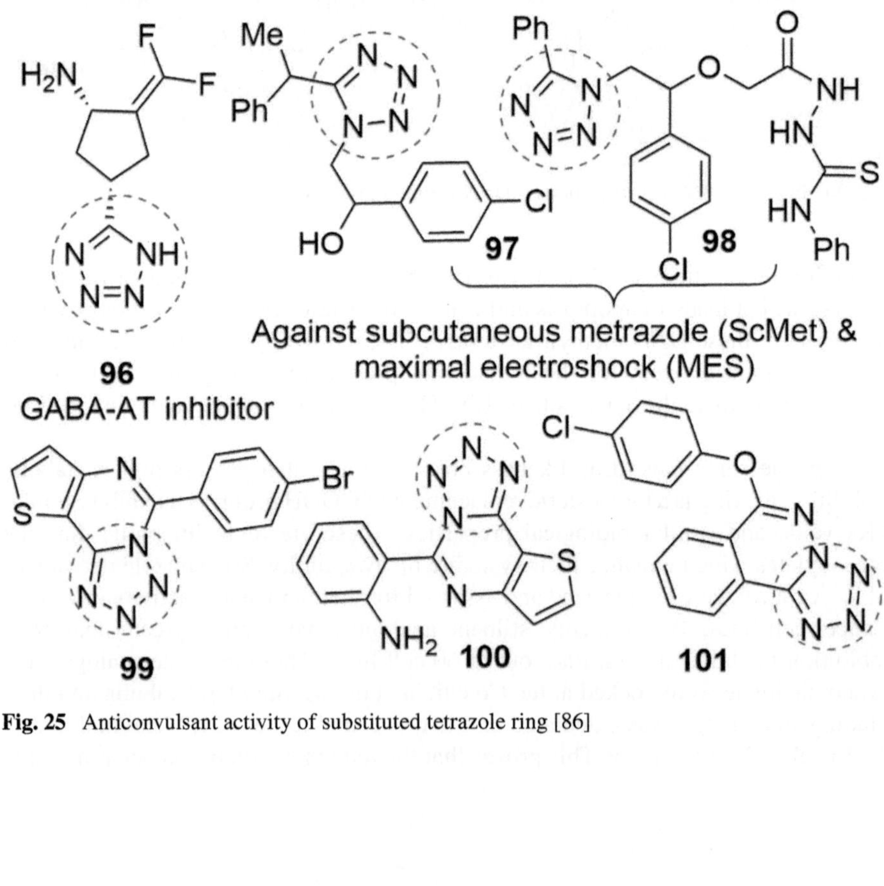

Fig. 25 Anticonvulsant activity of substituted tetrazole ring [86]

Moreover, anticonvulsant properties of 5-alkoxytetrazolo-[1,5-*c*]thieno[2,3-*e*]pyrimidine derivatives were examined [87]. In addition to that, anticonvulsant activities of 6-alkyoxytetrazolo[5,1-*a*]phthalazine derivatives are evaluated. The pharmacological outcomes exhibited that compound **101** has superior potency and moreover, the activity is quite higher than the reference drug carbamazepine [88].

A series of 5-substituted 1*H*-tetrazoles were also synthesized and they were examined for anticonvulsant activity by using two models namely scPTZ model and MES model. The ED50 value of one of the compounds was found to be 83.3 mg/kg which is very less compared to standard drug, Ethosuximide with 167.0 mg/kg in the scPTZ model. The protective index value of one of the compounds was found to be 19.7 which is higher than the standard drugs [89].

8 Hypoglycaemic Activity of Tetrazole Derivatives

Innovative *N*-glycosides tethered tetrazoles were proposed and constructed to act as SGLT2 inhibitors. Their hypoglycaemic activity was assessed by a mice oral glucose tolerance test (OGTT). Two compounds showed more powerful activity in comparison to positive control dapagliflozin (a drug of the gliflozin class). The compounds **102** and **103** showed 73.9% and 77.0% blood glucose levels inhibition in mice OGTT in comparison with dapagliflozin (68.3%) [90].

Additionally, 5-(4-alkoxyphenylalkyl)-1*H*-tetrazole derivatives were appraised in two genetically obese and diabetic animal models for their antidiabetic effects in wistar fatty rats and KKAy mice. Specifically, compound **104** had potent glucose lowering activity, which is 72 times more active in comparison to pioglitazone hydrochloride [91]. This compound also exhibited strong lowering effects for glucose and lipid in wistar fatty rats. The agonistic activity of this compound resulted in its antidiabetic effects. The compounds **105** and **106** are capable lead compounds in developing selective aldose reductase inhibitors, which can target the long-term complexities arising from diabetes mellitus (Fig. 26) [92].

Some triazolopyridine acetohydrazide conjugates were synthesized and they were explored for their antidiabetic activity. Few of the compounds have decreased the glucose concentration in the blood and their results were equated with the standard drug, Glibenclamide and they exhibited comparable activity. The Glucose concentration in the blood was found to be 60 mg/DL, 63 mg/DL of two tetrazole derivatives whereas the Glibenclamide treated was found to be 40 mg/DL [93].

A series of 5-(4-alkozyphenylalkyl)-1*H*-tetrazole derivatives were synthesized and screened for their antidiabetic activity. One tetrazole derivative has exhibited potent hypoglycaemic activity which is 72 times more active than Pioglitazone hydrochloride. The ED_{50} value of the tetrazole derivative was found to be 0.0839 mg/kg/d whereas the Pioglitazone hydrochloride was found to be 6.0 mg/kg/d [94].

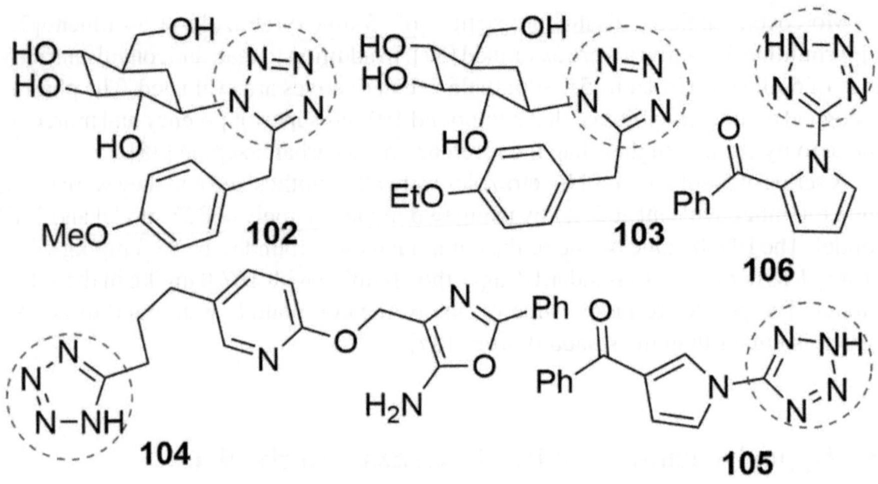

Fig. 26 Hypoglycaemic activity of substituted tetrazole ring [90–92]

9 Antihistaminic Activity of Tetrazole Derivatives

Since 1980, Tazanoplast **107** is effectively employed to treat acute reversible airway obstruction. The NH-unsubstituted tetrazole ring containing drugs such as Pemiroplast **108** and Pranlukast **109** corresponds to new generation of antihistaminic drugs to act upon both H1 and H2-receptors of mast cells (Fig. 27) [2].

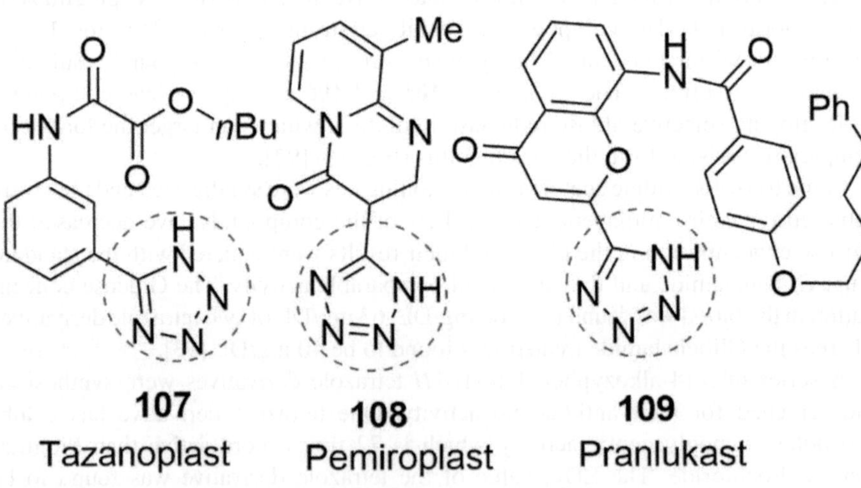

Fig. 27 Antihistaminic activity of tetrazole ring [2]

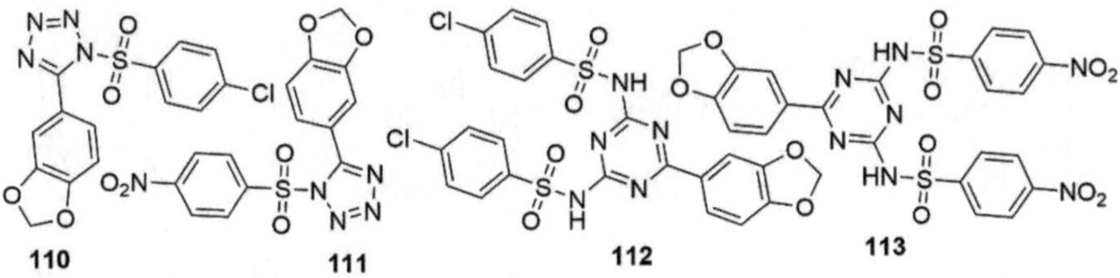

Fig. 28 Antiparasitic activity of tetrazole ring [95]

10 Antiparasitic Activity of Tetrazole Derivatives

Tetrazole derivatives have a satisfactory position in antiparasitic activity. The tetrazole derivatives with a SO_2NH function were verified for their antiamoebic activity and witnessed that the compounds **110** and **111** were least cytotoxic and exceptional entamoeba histolytica (an anaerobic parasitic amoebozoa) inhibitors. The replacement of tetrazole ring with a triazine ring in case of compounds **112** and **113** resulted in a fourfold improvement in the activity (Fig. 28) [95].

Also, pyrazoline derivatives were also screened to substantiate their effect towards the development of HM1:IMSS strain present in entamoeba histolytica. The compound **114** exhibited twice better potency as antiamoebic drugs and least cytotoxicity in comparison with standard metronidazole. The derivatives of 5-(1-aryl-3-methyl-1*H*-pyrazol-4-yl)-1*H*-tetrazole and their precursor 1-aryl-3-methyl-1*H*-pyrazole-4-carbonitriles were studied for their antileishmaniasis properties against Leishmania amazonensis Promastigotes and Leishmania Braziliensis. The cytotoxicity of compounds **115, 116** were assessed on the RAW 264.7 cell line and found to be most potent against L. braziliensis promastigote in reference to the drug pentamidine (Fig. 29) [96, 97].

11 Summary/Conclusion

Tetrazole and its derivatives have an ample position in the modern chemical community and drug discovery. Among the family of *N*-heterocycles, tetrazole derivatives exhibit diverse biological properties such as anticancer, hypotensive, antihistaminic, antimicrobial, antiallergic, antiviral, cytostatic, nootropic and are generally used in clinical practice. The pioneering of the tetrazole moiety into scaffolds of an organic substrate quite often accompany an improvement in the efficiency as well as an increased prolongation of drug action. However, the last few years have witnessed a considerable expansion in the number of patents and publications on new drugs as well as tetrazole containing biologically and pharmaceutically active scaffolds. The unique features of tetrazole such as different biological activities and applications are discussed in this chapter.

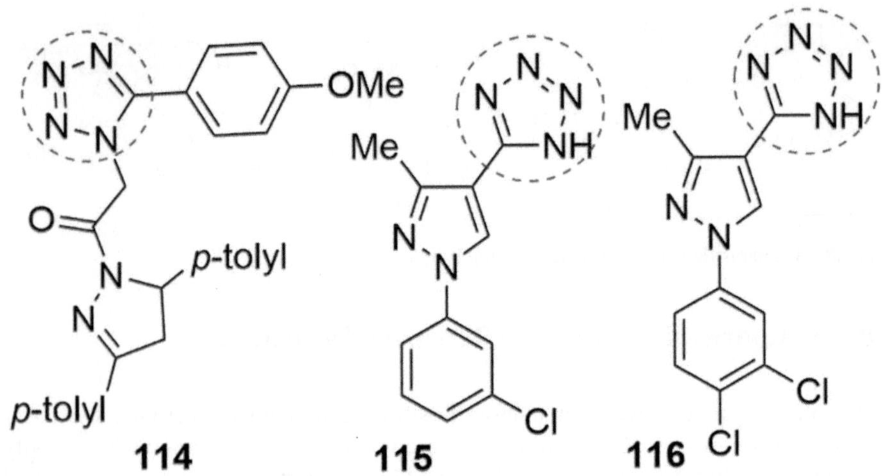

Fig. 29 Medicinally active tetrazole compounds [96, 97]

References

1. Mohite PB, Bhaskar VH (2001) Potential pharmacological activities of tetrazoles in the new millennium. Int J Pharm Tech Res 3:1557–1566
2. Ostrovskii VA, Trifonov RE, Popova EA (2012) Medicinal chemistry of tetrazoles. Russ Chem Bull 61:768–780
3. Herr RJ (2002) 5-Substituted-1*H*-tetrazoles as carboxylic acid isosteres: medicinal chemistry and synthetic methods. Bioorg Med Chem 10:3379–3393
4. Sureshbabu VV, Venkataramanarao R, Naik SA, Chennakrishnareddy G (2007) Synthesis of tetrazole analogues of amino acids using Fmoc chemistry: isolation of amino free tetrazoles and their incorporation into peptides. Tetrahedron Lett 48:7038–7041
5. Modarresi-Alam AR, Khamooshi F, Rostamizadeh M, Keykha H, Nasrollahzadeh M, Bijanzadeh HR, Kleinpeter E (2007) Dynamic 1H NMR spectroscopic study of the restricted SN rotation in aryl-N-(arylsulfonyl)-N-(triphenylphosphoranylidene) imidocarbamates. J Mol Struct 841:61–66
6. Manzoor S, Tariq Q, Yin X, Zhang J-G (2021) Nitro-tetrazole based high performing explosives: recent overview of synthesis and energetic properties. Def Technol. https://doi.org/10.1016/j.dt.2021.02.002
7. Husain A, Azim MS, Mitra M, Bhasin PS (2011) A review on candesartan: pharmacological and pharmaceutical profile. J Appl Pharm Sci 01:12–17
8. Weintraub WS (2006) The vascular effects of cilostazol. Can J Cardiol 22:56–60
9. Wishart DS, Feunang YD, Guo AC, Lo EJ, Marcu A, Grant JR, Sajed T, Johnson D, Li C, Sayeeda Z, Assempour N, Iynkkaran I, Liu Y, Maciejewski A, Gale N, Wilson A, Chin L, Cummings R, Le D, Pon A, Knox C, Wilson M (2018) DrugBank 5.0: a major update to the DrugBank database for 2018. Nucleic Acids Res 46:D1074–D1082
10. Basile J (2010) Critical appraisal of amlodipine and olmesartanmedoxomil fixed-dose combination in achieving blood pressure goals. Integr Blood Press Control 3:91–104
11. Ostrovskii VA, Koldobskii GI, Trifonov RE (2008) Reference module in chemistry, molecular sciences and chemical engineering. Comp HeterocyclChem III 6:257–423
12. Le Bourdonnec B, Meulon E, Yous S, Goossens JF, Houssin R, Hénichart JP (2000) Synthesis and pharmacological evaluation of new pyrazolidine-3, 5-diones as AT (1) angiotensin II receptor antagonists. J Med Chem 43:2685–2697

13. Kurup A, Gard R, Carini DJ, Hansch C (2001) Comparative QSAR: angiotensin II antagonists. Chem Rev 101:2727–2750
14. Wu J, Wang Q, Guo J, Hu Z, Yin Z, Xu J, Wu X (2008) Characterization of angiotensin II antagonism displayed by Ib, a novel nonpeptide angiotensin AT (1) receptor antagonist. Eur J Pharmacol 589:220–224
15. Yan B, Wang G, Sun J, Li X, Zheng Y, Ai H, Lv H, Wu X, Xu J (2007) Identification of the major metabolites of 5-n-butyl-4-{4-[2-(1H-tetrazole-5-yl)-1H-pyrrol-1-yl]phenylmethyl}-2,4-dihydro-2-(2,6-dichloridephenyl)-3H-1,2,4-triazol-3-one, a new angiotensin type 1 receptor antagonist, in rat bile by HPLC-diode array detection-MS and HPLC-MS/MS. Biomed Chromatogr 21:912–924
16. Kim JH, Lee JH, Paik SH, Kim JH, Chi YH (2012) Fimasartan, a novel angiotensin II receptor antagonist. Arch Pharmacal Res 35:1123–1126
17. Choi MJ, Kwon GH, Han NS, Yoo BW, Kim JH, Paik SH, Chi YH, Lee KT, Lee JY (2013) Development of 3D-QSAR CoMSIA models for 5-(biphenyl-2-yl)-1H-tetrazole derivatives as angiotensin II receptor type 1 (AT1) antagonists. Bioorg Med Chem Lett 23:4540–4546
18. Arhancet GB, Woodard SS, Iyanar K, Case BL, Woerndle R, Dietz JD, Garland DJ, Collins JT, Payne MA, Blinn JR, Pomposiello SI, Hu X, Heron MI, Huang H-C, Lee LF (2010) Discovery of novel cyanodihydropyridines as potent mineralocorticoid receptor antagonists. J Med Chem 53:5970–5978
19. Nickenig G (2004) Should angiotensin II receptor blockers and statins be combined? Circulation 110:1013–1020
20. Nakamura M, Anzai N, Jutabha P, Sato H, Sakurai H, Ichida K (2010) Concentration-dependent inhibitory effect of Irbesartan on renal UricAcid transporters. J Pharmacol Sci 114:115–118
21. Miura S, Okabe A, Matsuo Y, Karnik SS, Saku K (2013) Unique binding behavior of the recently approved angiotensin II receptor blocker azilsartan compared with that of candesartan. Hypertens Res 36:134–139
22. Noda K, Saad Y, Kinoshita A, Boyle TP, Graham RM, Husain A, Karnik SS (1995) Tetrazole and carboxylate groups of angiotensin receptor antagonists bind to the same subsite by different mechanisms. J Biol Chem 270:2284–2289
23. Martirosyan AO, Aleksanyan MV, Terzyan SS, Ter-Zakharyan YZ, Agababyan RV, Karapetyan AA, Mndzhoyan SL, Tamazyan RA (2001) Synthesis and molecular and crystal structure of 1-(5-Benzyl-2-tetrazolyl)-1-cyclopentanecarboxylic acid. Penicillin and cephalosporins based on this acid. Pharm Chem J 35:169–171
24. Kidwai M, Misra P, Bhushan KR, Saxena RK, Singh M (2000) Microwave-assisted solid-phase synthesis of cephalosporin derivatives with antibacterial activity. Monatsh Chem 131:937–943
25. Anacona JR, Alvarez P (2002) Synthesis and antibacterial activity of metal complexes of cefazolin. Transit Met Chem 27:856–860
26. Chohan ZH, Supuran CT, Scozzafava A (2004) Metalloantibiotics: synthesis and antibacterial activity of cobalt(II), copper(II), nickel(II) and zinc(II) complexes of kefzol. J Enzyme Inhib Med Chem 19:79–84
27. Tremblay LW, Xu H, Blanchard JS (2010) Structures of the michaelis complex (1.2 Å) and the covalent acyl intermediate (2.0 Å) of cefamandole bound in the active sites of the mycobacterium tuberculosis β-Lactamase K73A and E166A mutants. Biochemistry 49:9685–9687
28. Vilain S, Cosette P, Junter GA, Jouenne T (2002) Phosphate deprivation is associated with high resistance to latamoxef of gel-entrapped, sessile-like Escherichia coli cells. J Antimicrob Chemother 49:315–320
29. Wagner R, Mollison KW, Liu L, Henry CL, Rosenberg TA, Bamaung N, Tu N, Wiedeman PE, Or Y, Luly JR, Lane BC, Trevillyan J, Chen YW, Fey T, Hsieh G, Marsh K, Nuss M, Jacobson PB, Wilcox D, Carlson RP, Carter GW, Djuric SW (2005) Rapamycinanalogs with reduced systemic exposure. Bioorg Med Chem Lett 15:5340–5343
30. Naidu BN, Sorenson ME, Bronson JJ, Pucci MJ, Clark JM, Ueda Y (2005) Synthesis, in vitro, and in vivo antibacterial activity of nocathiacin I thiol-Michael adducts. Bioorg Med Chem Lett 15:2069–2072

31. Diwakar SD, Bhagwat SS, Shingare MS, Gill CH (2008) Substituted 3-((Z)-2-(4-nitrophenyl)-2-(1*H*-tetrazol-5-yl) vinyl)-4*H*-chromen-4-ones as novel anti-MRSA agents: synthesis, SAR, and in-vitro assessment. Bioorg Med Chem Lett 18:4678–4681
32. Renslo AR, Luehr GW, Gordeev MF (2006) Recent developments in the identification of novel oxazolidinone antibacterial agents. Bioorg Med Chem 14:4227–4240
33. Rajasekaran A, Thampi PP (2005) Synthesis and antinociceptive activity of some substituted-{5-[2-(1, 2, 3, 4-tetrahydrocarbazol-9-yl) ethyl] tetrazol-1-yl} alkanones. Eur J Med Chem 40:1359–1364
34. Rajasekaran A, Thampi PP (2004) Synthesis and analgesic evaluation of some 5-[b-(10-phenothiazinyl)ethyl]-1-(acyl)-1,2,3,4-tetrazoles. Eur J Med Chem 39:273–279
35. Adamec J, Waisser K, Kuneš J, Kaustová J (2005) A Note on the antitubercular activities of 1-Aryl-5-benzylsulfanyltetrazoles. Arch Pharm Chem Life Sci 338:385–389
36. Gaponik PN, Voitekhovich SV, Ivashkevich OA (2006) Metal derivatives of tetrazoles. Russ Chem Rev 75:507–539
37. Dave CG, Shah RD (2002) Annellation of triazole and tetrazole systems onto pyrrolo[2,3 *d*]pyrimidines: synthesis of tetrazolo[1,5-c]-pyrrolo[3,2-e]-pyrimidines and triazolo[1,5-c]pyrrolo[3,2-e]pyrimidines as potential antibacterial agents. Molecules 7:554–565
38. Arulmurugan S, Kavitha HP (2010) 2-Methyl-3-{4-[2-(1H-tetrazol-5-yl)ethylamino]phenyl}-3Hquinazolin-4-one. Molbank. https://doi.org/10.3390/M695
39. Dhayanithi V, Syed SS, Kumaran K, Reguraman K, Sankar J, Ragavan RV, Kumar GP, S., Kumari NS, Pati HN (2011) Synthesis of selected 5-thio-substituted tetrazole derivatives and evaluation of their antibacterial and antifungal activities. J Serb Chem Soc 76:165–175
40. Moustafa MA, El-Sherbeny MA, El-Sherbiny DT, El-Sayed SM (2012) Molecular modeling, synthesis and antimicrobial evaluation of new molecular hybrids of tetrazole derivatives. J Am Sci 8:973–986
41. Jo YW, Im WB, Rhee JK, Shim MJ, Kim WB, Choi EC (2004) Synthesis and antibacterial activity of oxazolidinones containing pyridine substituted with heteroaromatic ring. Bioorg Med Chem 12:5909–5915
42. Kanakaraju S, Kumar PSV, Prasanna B, Chandramouli GVP (2013) Design, synthesis, and in vitro antimicrobial evaluation of fused pyrano[3,2-e]tetrazolo[1,5-c]pyrimidines and diazepines. ISRN Org Chem 21:635384–635393
43. Antypenko LM, Kovalenko SI, Antypenko OM, Katsev AM, Achkasova OM (2013) Design and evaluation of novel antimicrobial and anticancer agents among tetrazolo[1,5-c]-quinazoline-5-thione S-derivatives. Sci Pharm 81:15–42
44. Ramiz MMM, Abdel-Rahman AA-H (2011) Antimicrobial activity of newly synthesized 2,5-disubstituted 1,3,4-thiadiaozle derivatives. Bull Korean Chem Soc 32:4227–4232
45. Wael AE-S, Omar MA, Ali OM, Zyada RA, Mohamed AA, Abdel-Rahman AA (2012) Synthesis and antimicrobial activity of new substituted thienopyrimidines, their tetrazolyl and sugar derivatives. Acta Pol Pharm 69:439–447
46. Mohite PB, Bhaskar VH (2012) In vitro evaluation of tetrazoles as a novel class of anti-mycobacterium tuberculosis agents. Adv Pharm Bull 2:31–36
47. El-Sayed WA, Abdel Megeid RE, Abbas H-AS (2011) Synthesis and antimicrobial activity of new 1-[(tetrazol-5-yl)methyl]indole derivatives, their 1,2,4-triazole thioglycosides and acyclic analogs. Arch Pharm Res 34:1085–1096
48. Kategaonkar AH, Pokalwar RU, Sonar SS, Gawali VU, Shingate BB, Shingare MS (2010) Synthesis, in vitro antibacterial and antifungal evaluations of new α-hydroxyphosphonate and new α-acetoxyphosphonate derivatives of tetrazolo[1,5-a]quinoline. Eur J Med Chem 45:1128–1132
49. Varadaraji D, Suban SS, Ramasam VR, Kubendiran K, Raguraman JSKG, Nalilu SK, Pati HN (2010) Synthesis and evaluation of a series of 1-substituted tetrazole derivatives as antimicrobial agents. Org Commun 3:45–56
50. Lamie PL, Philoppes JN, Azouz AA, Safwat NM (2017) Novel tetrazole and cyanamide derivatives as inhibitors of cyclooxygenase-2 enzyme: design, synthesis, anti-inflammatory evaluation, ulcerogenic liability and docking study. J Enzyme Inhib Med Chem 32:805–820

51. Warrilow AGS, Hull CM, Parker JE, Garvey EP, Hoekstra WJ, Moore WR, Schotzinger RJ, Kelly DE, Kelly SE (2014) Antimicrob Agents Chemother 58:7121–7127
52. Upadhayaya RS, Jain S, Sinha N, Kishore N, Chandra R, Arora SK (2004) Synthesis of novel substituted tetrazoles having antifungal activity. Eur J Med Chem 39:579–592
53. Mohite PB, Pandhare RB, Khanage SG, Bhaskar VH (2010) Synthesis and in vitro antimicrobial activity of some novel chalcones containing 5-phenyl tetrazole. Acta Pharm Sci 52:505–510
54. Kumar CNSSP, Parida DK, Santhoshi A, Kota AK, Sridhar B, Rao VJ (2011) Synthesis and biological evaluation of tetrazole containing compounds as possible anticancer agents. Med Chem Commun 2:486–492
55. Wei CX, Bian M, Gong GH (2015) Tetrazolium compounds: synthesis and applications in medicine. Molecules 20:5528–5553
56. Matysiak J, Niewiadomy A, Krajewska-Kułak E, Macik-Niewiadomy G (2003) Synthesis of some 1-(2,4-dihydroxythiobenzoyl)imidazoles, -imidazolines and -tetrazoles and their potent activity against Candida species. Farmaco 58:55–61
57. Nishimoto AT, Wiederhold NP, Flowers SA, Zhang Q, Kelly SL, Morschhäuser J, Yates CM, Hoekstra WJ, Schotzinger RJ, Garvey EP, Rogers PD (2019) Antimicrob Agents Chemother 63:e00341–e419
58. Wiederhold NP (2018) The antifungal arsenal: alternative drugs and future targets. Int J Antimicrob Agents 51:333–339
59. Colarusso S, Gerlach B, Koch U, Muraglia E, Conte I, Stansfield I, Matassa VG, Narjes F (2002) Evolution, synthesis and SAR of tripeptide α-ketoacid Inhibitors of the hepatitis C virus NS3/NS4A serine protease. Bioorg Med Chem Lett 12:705–708
60. Johansson A, Poliakov A, Akerblom EA, Wiklund K, Lindeberg G, Winiwarter S, Danielson UH, Samuelsson B, Hallberg A (2003) Acyl sulfonamides as potent protease inhibitors of the hepatitis C virus full-length NS3 (Protease-Helicase/NTPase): a comparative study of different C-terminals. Bioorg Med Chem 11:2551–2568
61. Poliakov A, Johansson A, Ekerblom E, Oscarsson K, Samuelsson B, Hallberg A, Danielson UH (2004) Structure-activity relationships for the selectivity of hepatitis C virus NS3 protease inhibitors. Biochim Biophys Acta 167:51–59
62. Perni RB, Pitlik J, Britt SD, Court JJ, Courtney LF, Deininger DD, Farmer LJ, Gates CA, Harbeson SL, Levin RB, Lin C, Lin K, Moon YC, Luong YP, O´Malley ET, Rao BG, Thomson JA, Tung RD, Van Drie JH, Wei Y (2004) Inhibitors of hepatitis C virus NS3·4A protease 2. Warhead SAR and optimization. Bioorg Med Chem Lett 14:1441–1446
63. Chang CS, Lin YT, Shih SR, Lee CC, Lee YC, Tai CL, Tseng SN, Chern JH (2005) Design, synthesis, and antipicornavirus activity of 1-[5-(4-arylphenoxy)alkyl]-3-pyridin-4-ylimidazolidin-2-one derivatives. J Med Chem 48:3522–3535
64. Sotriffer CA, Ni H, McCammon JA (2000) Active site binding modes of HIV-1 integrase inhibitors. J Med Chem 43:4109–4117
65. Walker MA, Johnson T, Ma Z, Banville J, Remillard R, Kim O, Zhang Y, Staab A, Wong H, Torri A, Samanta H, Lin Z, Deminie C, Terry B, Krystal M, Meanwell N (2006) Triketoacid inhibitors of HIV-integrase: a new chemotype useful for probing the integrasepharmacophore. Bioorg Med Chem Lett 16:2920–2924
66. Jiang T, Kuhen KL, Wolff K, Yin H, Bieza K, Caldwell J, Bursulaya B, Tuntland T, Zhang K, Karanewsky D, He Y (2006) Design, synthesis, and biological evaluations of novel oxindoles as HIV-1 non-nucleoside reverse transcriptase inhibitors. Part 2. Bioorg Med Chem Lett 16:2109–2112
67. Muraglia E, Kinzel OD, Laufer R, Miller MD, Moyer G, Munshi V, Orvieto F, Palumbi MC, Pescatore G, Rowley M, Williams PD, Summa V (2006) Tetrazole thioacetanilides: potent non-nucleoside inhibitors of WT HIV reverse transcriptase and its K103N mutant. Bioorg Med Chem Lett 16:2748–2752
68. Abell AD, Foulds GJ (1997) Synthesis of a cis-conformationally restricted peptide bondisostere and its application to the inhibition of the HIV-1 protease. J Chem Soc Perkin Trans 1:2475–2482

69. May BCH, Abell AD (2002) α-Methylene tetrazole-based peptidomimetics: synthesis and inhibition of HIV protease. J Chem Soc Perkin Trans 1:172–178
70. Han Q, Chang C-H, Li R, Ru Y, Jadhav PK, Lam PYS (1998) Cyclic HIV protease inhibitors: design and synthesis of orally bioavailable, pyrazole P2/P2' cyclic ureas with improved potency. J Med Chem 41:2019–2028
71. Wang SX, Fang Z, Fan ZJ, Wang D, Li YD, Ji XT, Hua XW, Huang Y, Kalinina TA, Bakulev VA, Morzherin YY (2013) Synthesis of tetrazole containing 1,2,3-hiadiazole derivatives via U-4CR and their anti-TMV activity. Chin Chem Lett 24:889–892
72. Yeung KS, Qiu Z, Yin Z, Trehan A, Fang H, Pearce B, Yang Z, Zadjura L, D'Arienzo CJ, Riccardi K, Shi PY, Spicer TP, Gong YF, Browning MR, Hansel S, Santone K, Barker J, Coulter T, Lin PF, Meanwell NA, Kadow JF (2013) Inhibitors of HIV-1 attachment. Part 8: the effect of C7-heteroaryl substitution on the potency, and in vitro and in vivo profiles of indole-based nhibitors. Bioorg Med Chem Lett 23:203–208
73. Zarubaev VV, Golod EL, Anfimov PM, Shtro AA, Saraev VV, Gavrilov AS, Logvinov AV, Kiselev OI (2010) Synthesis and anti-viral activity of azolo-adamantanes against influenza A virus. Bioorg Med Chem 18:839–848
74. Hutchinson DW, Naylor M (1985) The antiviral activity of tetrazole phosphomic acids and their analogues. Nucleic Acids Res 13:8519–8530
75. El-Sayed WA, El-Kosy SM, Ali OM, Emselm HM, Abdel-Rahman AA (2012) Anticancer activity of new (tetrazol-5-yl) methylindole derivatives and their acyclic c nucleoside analogs. Acta Pol Pharm 69:669–677
76. Romagnoli R, Baraldi PG, Salvador MK, Preti D, Tabrizi MA, Brancale A, Fu XH, Li J, Zhang SZ, Hamel E, Bortolozzi R, Basso G, Viola G (2011) Synthesis and evaluation of 1,5-disubstituted tetrazoles as rigid analogues of combretastatin A-4 with potent antiproliferative and antitumor activity. J Med Chem 55:475–488
77. Kádár Z, Kovács D, Frank É, Schneider G, Huber J, Zupkó I, Bartók T, Wölfling J (2011) Synthesis and in vitroantiproliferative activity of novel androst-5-ene triazolyl and tetrazolyl derivatives. Molecules 16:4786–4806
78. Al-Duaij OK, Hafez HN, el-Gazzar ARBA (2013) Synthesis of novel series of benzothieno [2,3-*d*] pyrimidine derivatives, promising anticancer agents. J Chem Eng 7:725–742
79. Jackman AL, Kimbell R, Aherne GW, Brunton L, Jansen G, Stephens TC, Smith MN, Wardleworth JM, Boyle FT (1997) Cellular pharmacology and in vivo activity of a new anticancer, agent, ZD9331: a water-soluble, nonpolyglutamatable, quinazoline-based inhibitor of thymidylate synthase. Clin Cancer Res 3:911–921
80. Arshad M, Bhat AR, Pokharel S, Kim JE, Lee EJ, Athar F, Choi I (2014) Synthesis, characterization and anticancer screening of some novel piperonyl-tetrazole derivatives. Eur J Med Chem 71:229–236
81. Jedhe GS, Paul D, Gonnade RG, Santra MK, Hamel E, Nguyen TL, Sanjayan GJ (2013) Correlation of hydrogen-bonding propensity and anticancer profile of tetrazole-tethered combretastatin analogues. Bioorg Med Chem Lett 23:4680–4684
82. Alam M, Nami SAA, Husain A, Lee D-U, Park S (2013) Synthesis, characterization, X-ray diffraction, antimicrobial and in vitro cytotoxicity studies of 7α-Aza-B-homostigmast-5-eno[7a,7-d]tetrazole. C R Chim 16:201–206
83. Zhu B, Ge J, Yao SQ (2015) Developing new chemical tools for DNA methyltransferase 1 (DNMT 1): a small-molecule activity-based probe and novel tetrazole-containing inhibitors. Bioorg Med Chem 23:2917–2927
84. Ghoshal K, Bai S (2007) DNA methyltransferases as targets for cancer therapy. Drugs Today 43:395–422
85. Bommagani S, Penthala NR, Balasubramaniam MS, Kuravi S, Caldas-Lopes E, Guzman ML, Balusu R, Crooks PA (2019) A novel tetrazole analogue of resveratrol is a potent anticancer agent. Bioorg Med Chem Lett 29:172–178
86. Rostom SA, Ashour HM, El Razik HA, Abd El Fattah H, El-Din NN (2009) Azole antimicrobial pharmacophore-based tetrazoles: synthesis and biological evaluation as potential antimicrobial and anticonvulsant agents. Bioorg Med Chem 17:2410–2422

87. Wang SB, Deng XQ, Zheng Y, Yuan YP, Quan ZS, Guan LP (2012) Synthesis and evaluation of anticonvulsant and antidepressant activities of 5-alkoxytetrazolo[1,5 c]thieno[2,3-e]pyrimidine derivatives. Eur J Med Chem 56:139–144
88. Sun XY, Wei CX, Deng XQ, Sun ZG, Quan ZS (2010) Synthesis and primary anticonvulsant activity evaluation of 6-alkyoxyl-tetrazolo [5,1-a]phthalazine derivatives. Arzneim Forsch 60:289–292
89. Liao AM, Wang T, Cai B, Jin Y, Cheon S, Chun C, Wang Z (2016) Design, synthesis and evaluation of 5-substituted 1-H-tetrazoles as potent anticonvulsant agents. Arch Pharm Res 40:435–443
90. Gao YL, Zhao GL, Liu W, Shao H, Wang YL, Xu WR, Tang LD, Wang JW (2010) Design, synthesis and *in vivo* hypoglycemic activity of tetrazole-bearing *N*-glycosides as SGLT2 inhibitors. Indian J Chem B 49:1499–1508
91. Momose Y, Maekawa T, Odaka H, Ikeda H, Sohda T (2002) Novel 5-substituted-1*H*-tetrazole derivatives as potent glucose and lipid lowering agents. Chem Pharm Bull (Tokyo) 50:100–111
92. Pegklidou K, Koukoulitsa C, Nicolaou I, Demopoulos VJ (2010) Design and synthesis of novel series of pyrrole based chemotypes and their evaluation as selective aldose reductase inhibitors. A case of bioisosterism between a carboxylic acid moiety and that of a tetrazole. Bioorg Med Chem 18:2107–2114
93. Arif M, Jabeen F, Saeed A, Qureshi IZ, Mushtaq N (2017) A new class of potential antidiabeticacetohydrazides: synthesis, *in vivo* antidiabetic activity and molecular docking studies. Bangladesh J Pharmacol 12:319–332
94. Momose Y, Maekawa T, Odaka H, Ikeda H, Sohda T (2002) Novel 5-substituted-1H-tetrazole derivatives as potent glucose and lipid lowering agents. Chem Pharm Bull 50:100–111
95. Wani MY, Bhat AR, Azam A, Choi I, Athar F (2012) Probing the antiamoebic and cytotoxicity potency of novel tetrazole and triazine derivatives. Eur J Med Chem 48:313–320
96. Wani MY, Bhat AR, Azam A, Lee DH, Choi I, Athar F (2012) Synthesis and in vitro evaluation of novel tetrazole embedded 1,3,5-trisubstituted pyrazoline derivatives as Entamoebahistolyticagrowth inhibitors. Eur J Med Chem 54:845–854
97. Faria JV, dos Santos MS, Bernardino AM, Becker KM, Machado GM, Rodrigues RF, Canto-Cavalheiro MM, Leon LL (2013) Synthesis and activity of novel tetrazole compounds andtheir pyrazole-4-carbonitrile precursors against Leishmaniaspp. Bioorg Med Chem Lett 23:6310–6312

Chapter 9
An Overview on Biological Activity of Benzimidazole Derivatives

Arup K. Kabi, Sattu Sravani, Raghuram Gujjarappa, Aakriti Garg, Nagaraju Vodnala, Ujjawal Tyagi, Dhananjaya Kaldhi, Virender Singh, Sreya Gupta, and Chandi C. Malakar

Abbreviations

NSAID	Non-steroidal anti-inflammatory drugs
HIV	Human immunodeficiency virus
ARB	Angiotensin II receptor blocker
DNA	Deoxyribonucleic acid
OLED	Organic light-emitting diodes
GERDL	Gastroesophageal reflux disease
EMA	European medicines agency
U.S. FDA	United State Food and Drug Administration
ADP	Adenosine di-phosphate
ACE	Angiotensin-converting enzyme
PARP	Poly(ADP-ribose)polymerase
RNA	Ribonucleic acid
SAR	Structure–activity relationship
GI	Gastrointestinal
AIDS	Acquired immuno deficiency syndrome
MRSA	Methicillin-resistant Staphylococcus aureus
WHO	World Health Organization
IBS-D	Irritable bowel syndrome with diarrhoea

A. K. Kabi · R. Gujjarappa · N. Vodnala · U. Tyagi · D. Kaldhi · C. C. Malakar (✉)
Department of Chemistry, National Institute of Technology Manipur, Imphal, Langol, Manipur 795004, India
e-mail: cmalakar@nitmanipur.ac.in

S. Sravani · A. Garg · S. Gupta (✉)
Department of Medicinal Chemistry, National Institute of Pharmaceutical Education and Research (NIPER), Chunilal Bhawan, 168, Maniktala Main Road, Kolkata 700054, India

V. Singh
Department of Chemistry, Central University of Punjab, Bathinda, Punjab 151001, India

B. P. Swain (ed.), *Nanostructured Biomaterials*, Materials Horizons: From Nature to Nanomaterials, https://doi.org/10.1007/978-981-16-8399-2_9

CLL	Chronic lymphocytic leukaemia
NHL	Non-Hodgkin lymphoma
HCV	Hepatitis C virus
UV	Ultraviolet

1 Introduction

Nitrogen-containing compounds occur extensively in synthetic drugs and natural products. Benzimidazoles are considered as a crucial class of compounds in the pharmaceutical industry. Moreover, synthetic drugs assembled with higher numbers of nitrogen atoms than natural products owing to their property of carrying a positive charge and can also perform as hydrogen bond acceptor or donor which can strongly influence the interactions between a medicinal agent and its target. The benzimidazole's structural motif is also a building block of pharmaceutics and functional materials. The opportunities of benzimidazole moieties in pharmaceutical chemistry includes the factor Xa (FXa) suppressors, poly(ADP-ribose)polymerase (PARP) preventors, human cytomegalovirus (HCMV) inhibitors, HIV reverse transcriptase inhibitor L-697, 661, oestrogen receptor-b agonist ERB-041, anticancer agent NSC-693638, and 5-HT3 receptor agonist [1, 2]. The functionalized benzimidazole scaffolds are abridged in this chapter to realize regarding the chemical as well as pharmacological and biological properties. Benzimidazole moieties are witnessed to be exercised for the organic light-emitting diodes (OLEDs). Benzimidazole motifs containing compounds exhibit a wide spectrum of biological properties such as antiviral, antibacterial, anticancer, antifungal, and anti-HIV activities (Fig. 1) [3].

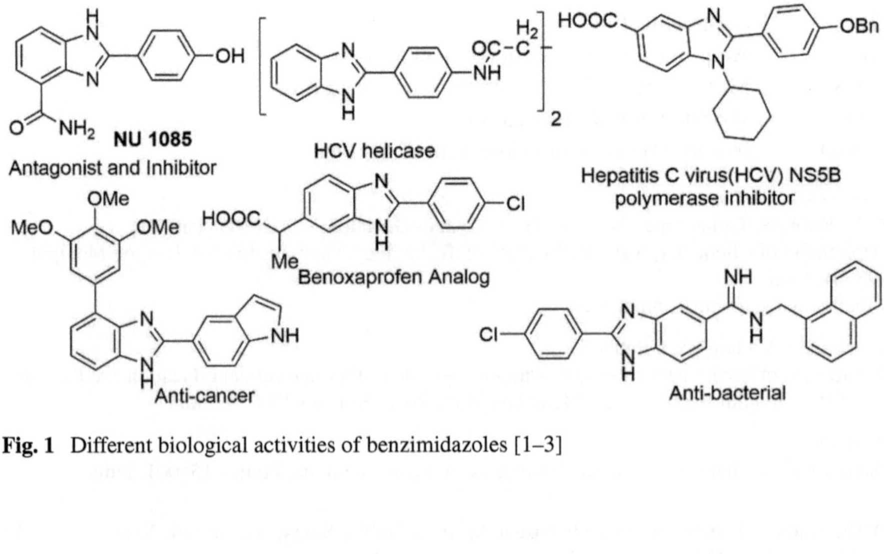

Fig. 1 Different biological activities of benzimidazoles [1–3]

1.1 Benzimidazoles and Their Pharmaceutical Drugs

Angiotensin II receptor blockers (ARBs) are extremely active antihypertensive agents. Commercially accessible ARBs, telmisartan consists of the prolonged half-life of approximately 24 h. That means, telmisartan should have a long period of action. Albendazole has a mechanism of activities related to alternative benzimidazoles. This is potent against a widespread domain of some protozoa and helminths. It acts actively on intestinal parasites. In addition, flubendazole is endorsed for treating gastrointestinal nematode infections in both veterinary and human medicine. In another, proton pump inhibitors (PPIs) are referred to the medications of preferences for the handling of gastroesophageal reflux disease (GERD). It is known that the esomeprazole is the modern proton pump inhibitor (PPI) and S-isomeric forms of omeprazole was industrialized. This effort refers to recover its pharmacokinetic properties. Omeprazole has been shown to be effective for the cure of HP infection, peptic ulcer disease, GERD, and several acid-relevant conditions in children. In another, bendamustine is known as the alkylating agent with several special mechanisms of action. It is permitted by the European Medicines Agency (EMA) and the U.S. Food and Drug Administration (FDA) for treating patients with rituximab-refractory indolent B-cell non-Hodgkin's lymphoma (NHL). Furthermore, astemizole is the first non-sedating antihistamine drug to be introduced for therapeutic use. Additionally, rabeprazole acts as a proton pump inhibitor (PPI) and can deactivate the gastric parietal cell proton pump (H^+/K^+-ATPase) by forming a covalent bond with it. This prevents the generation of gastric acid formation and increases gastric pH. Furthermore, maribavir is a novel antiviral agent. It exhibits an antiviral mechanism of action unlike those of currently available antiviral agents for cytomegalovirus (CMV). Maribavir directly prevents UL97 kinase, which is an early viral gene product intricated in viral DNA packaging, DNA elongation, and egress or flaking of capsids from viral nuclei. In addition, wide-spectrum anthelmintic-effective drugs such as thiabendazole are active against gastrointestinal nematodes in lungworms and ruminants in sheep. The drug can also exhibit useful fungicide properties in controlling pathogenic fungi, influencing field crops, stored vegetables, and fruits. Additionally, benomyl (carbendazim) is extensively used as a fungicide in agriculture and home gardening, and also used as an anthelmintic in veterinary medicine. In another, anti-helminthic drug mebendazole (known as Vermox) has been in clinical use and has anticancer properties that have been explained in a wide range of pre-clinical studies across several different cancer types. In addition, DMA acts as a non-cytotoxic radio protector in mammalian cells. It exhibits an exceptional radioprotection in mice at a single nontoxic oral dose (Fig. 2) [4–13].

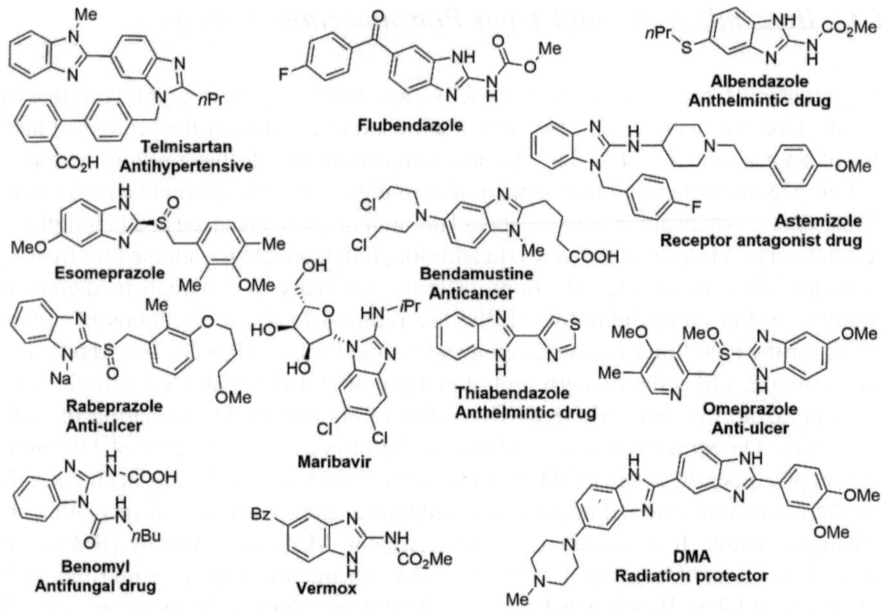

Fig. 2 Chemical structures of benzimidazole-related pharmaceutical drugs [4–13]

Few of the FDA-approved benzimidazole-containing drugs are listed in Table 1 [12].

2 Antimicrobial Activity of Benzimidazole Derivatives

Antibacterial activity in modern times has more importance because the rate of infection increases by way of antibiotic-resistant microorganisms. Compound **1** (**a**, **b**) having good antibacterial activity with benzimidazole rings having a halogen and imidazopyridine ring having bromine at phenyl moiety [14]. Additionally, Compound **2** showed good activity against *P. aeruginosa*, having 16 times more effectiveness than reference drug Chloramphenicol [15]. In addition, thiadiazolo-thiazolidinones **3** were selected for antimicrobial activity with the help of benzimidazole [16]. Indeed, compound **4** exhibited comparable activity against gram-positive bacteria *Bacillus subtilis* [17], and compound **5** shows antifungal activity against *B. ellipica* and *P. nicotianae* comparable with standard carbendazim [18]. In addition, compounds **6** and **7a** revealed antimicrobial activity against *B. subtilis.* Additionally, compounds **7** (**a**, **b**) controlled more activity than the fluconazole against F. oxysporum (Fig. 3) [19].

In continuation, compounds **8c**, **8i**, and **8j** having the heterocyclic groups at C_6 position so exhibited the highest inhibition against test organisms [20]. Additionally,

Table 1 FDA-approved benzimidazole-containing drugs [12]

Structure of the drug Name of the drug (Drug Bank ID)	Category/indication	Mechanism of action
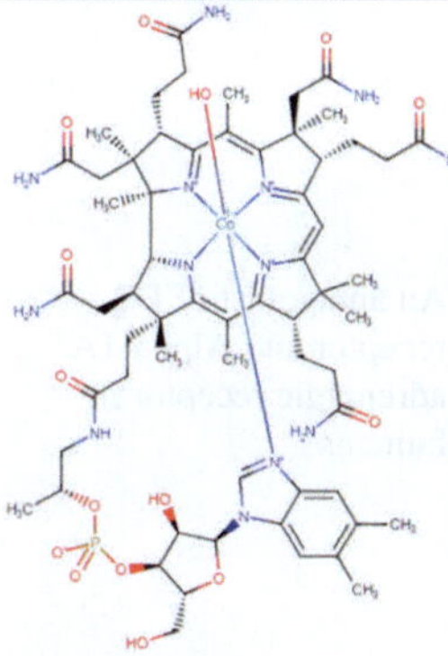 Hydroxocobalamin (DB00200)	For the prevention and treatment of vitamin B12 deficiency arising from various reasons and treatment of known or suspected cyanide poisoning, and treatment of pernicious anaemia	Supplement of Vitamin B12
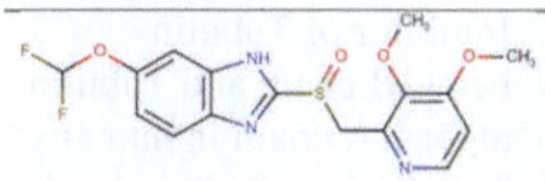 Pantoprazole (DB00213)	This drug is used for the management of GERD. It is also used to impede the reappearance of gastric problem or stomach ulcers from chronic utilization of NSAIDs and in medicating pathological hypersecretory disorders involving Zollinger–Ellison (ZE) Syndrome	Inhibitor of potassium-transporting ATPase alpha chain 1, N(G),N(G)-dimethyl arginine dimethyl amino-hydrolase 1
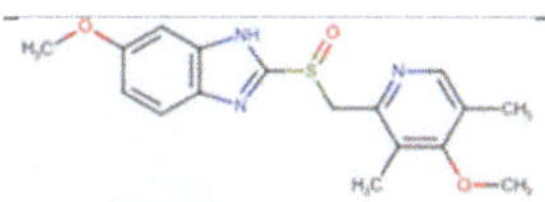 Omeprazole (DB00338)	Duodenal ulcers in adults, eradication of Helicobacter pylori were treated by employing the proton pump suppressor. Also used to cure benign gastric ulcer (active) in adults	Inhibitor of potassium-transporting ATPase alpha chain 1 and agonist of Aryl hydrocarbon receptor
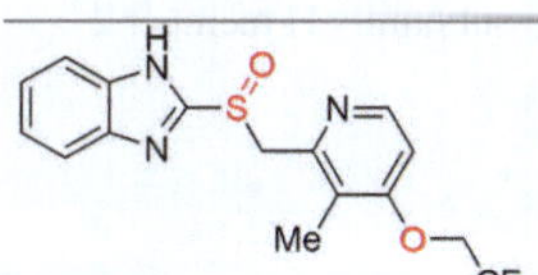 Lansoprazole (DB00448)	Used to treat short-span disorders such as erosive reflux oesophagitis, active gastric ulcers, symptomatic gastroesophageal reflux disease, active duodenal ulcers, duodenal ulcers execution of hypersecretory situations along with non-steroidal anti-inflammatory drug (NSAID)-induced gastric and Zollinger–Ellison syndrome. By using a combination of clarithromycin and amoxicillin, it eradicates H. pylori	Targets H^+, K^+-ATPase which promotes the ultimate process in the acid discharge mechanism in parietal cells

(continued)

Table 1 (continued)

Structure of the drug Name of the drug (Drug Bank ID)	Category/indication	Mechanism of action
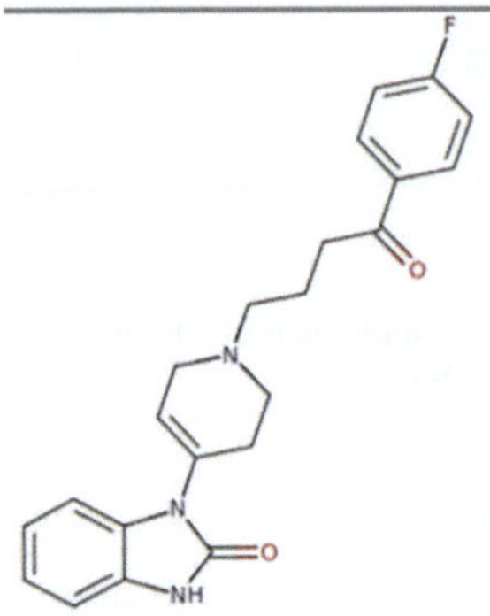 Droperidol (DB00450)	It acts to reduce the occurrence of nausea, vomiting in diagnostic and surgical procedures and also utilized as a tranquilizer	An antagonist of D2 receptor and Alpha-1A adrenergic receptor in humans
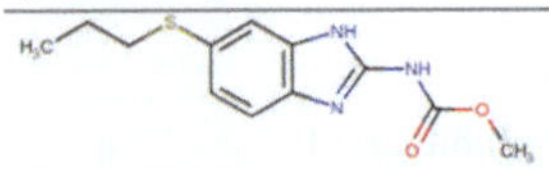 Albendazole (DB00518)	Recommended for the cystic hydatid disease of the liver, lung, and peritoneum, induced by the *Echinococcus granulosus (*larval form of the dog tapeworm) and parenchymal neurocysticercosis provoked by *Taenia (*larval forms of the pork tapeworm)	Inhibitor of Tubulin beta-4B chain and Tubulin alpha-1A chain in humans Tubulin beta-2 chain in pig roundworm, Fumarate reductase flavoprotein subunit in Shewanella oneidensis (strain MR-1)
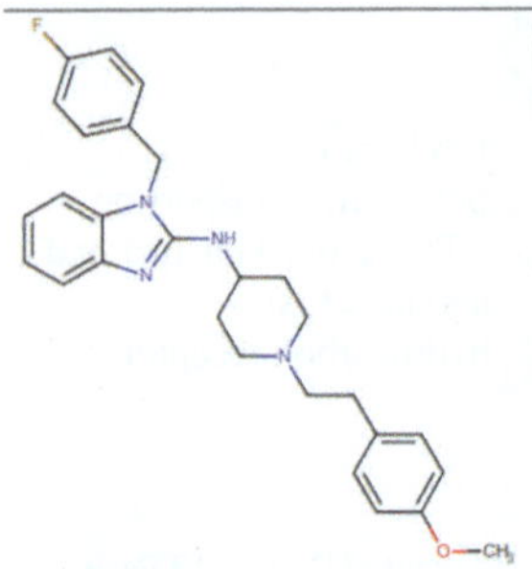 Astemizole (DB00637)	Used as an indicator in relieving allergy symptoms of rhinitis and conjunctivitis. Due to concerns about arrhythmias, it has been withdrawn from the market	H1 receptor antagonist and inhibitor of potassium voltage-gated channel subfamily H member 2

(continued)

compounds, **9 (a-d)**, exposed good antibacterial activity against most of the microorganisms [21]. Furthermore, sulfonyl-benzimidazoles **10 (a-j)/11 (a-j)** displayed antitubercular and antimicrobial activity. Additionally, compounds **10b**, **10d**, **10e,** and **10 h** have acceptable functions towards verified fungal and bacterial strains. In details, compounds (**10b**), (**10e**), and (**10 h**) exhibited significant activity against *Mycobacterium tuberculosis* (MTB) $H_{37}Rv$ strain [22]. Indeed, Compounds **13a and 13b**

Table 1 (continued)

Structure of the drug Name of the drug (Drug Bank ID)	Category/indication	Mechanism of action
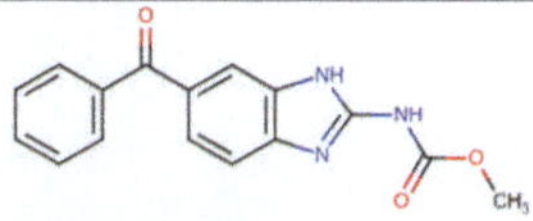 Mebendazole (DB00643)	Treatment of Enterobius vermicularis (pinworm), Trichuris trichiura (whipworm), Ancylostoma duodenale (common hookworm), Necator americanus (American hookworm), and Ascaris lumbricoides (common roundworm), in single or mixed infections	Inhibitor of tubulin alpha-1A chain, Tubulin beta-4B chain
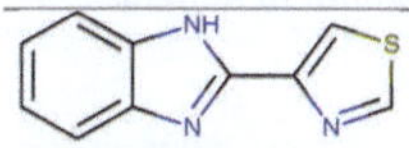 Thiabendazole (DB00730)	Used to treat trichinosis, visceral larva migrans, cutaneous larva migrans (creeping eruption), and strongyloidiasis (threadworm)	Perform the inhibition of helminth-specific enzyme fumarate reductase
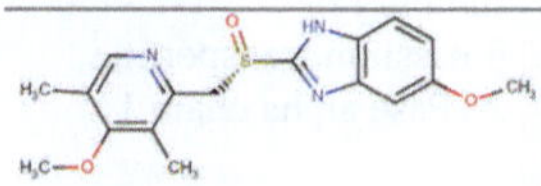 Esomeprazole (DB00736)	Used to treat disorders related to acid-reflux (control and curing of symptomatic gastroesophageal reflux disease (GERD), erosive esophagitis, H. pylori eradication, and peptic ulcer disease. For the treatment of pathological hypersecretory conditions along with Zollinger–Ellison Syndrome and to suppress gastrointestinal bleeds originating from utilization of NSAID	Inhibits potassium-transporting ATPase alpha chain 1 in humans
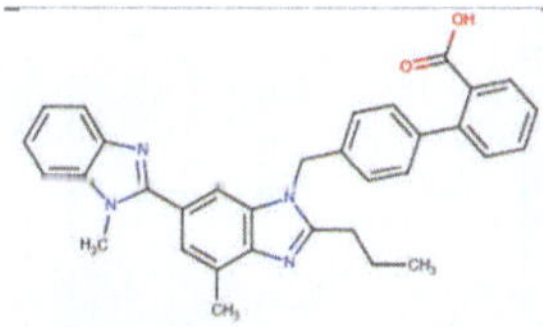 Telmisartan (DB00966)	Recommended for hypertensive patients towards diabetic nephropathy. It refers for the patients who suffers from type 2 diabetes mellitus, hypertension, congestive heart failure, but only for those who are unable tolerate ACE inhibitors	An antagonist of Type-1 angiotensin II receptor and partial agonist of PPAR gamma in humans
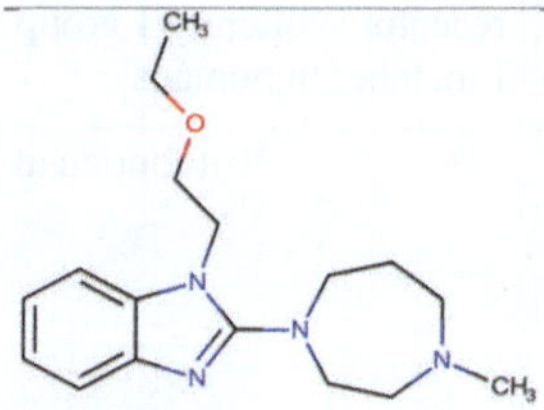 Emedastine (DB01084)	Helps in short-term solace for the signs and symptoms appeared due to allergic conjunctivitis	An antagonist of histamine H1 receptor in humans

(continued)

Table 1 (continued)

Structure of the drug Name of the drug (Drug Bank ID)	Category/indication	Mechanism of action
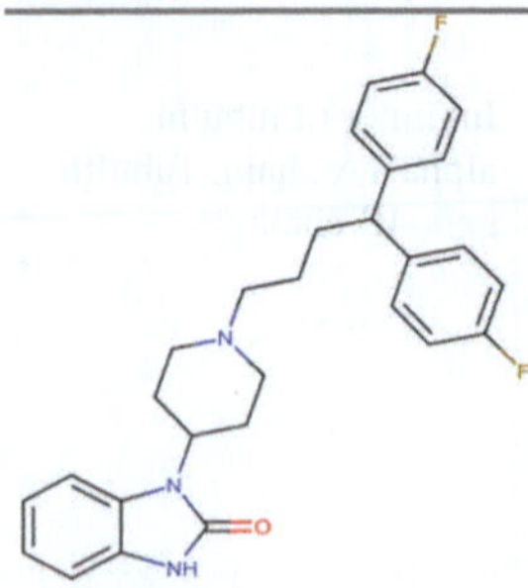 Pimozide (DB01100)	Used for patients who fail to respond to standard treatment of Tourette's disorder for suppression of motor and phonic tics	An antagonist of dopamine D2 and D3 receptors and inhibitor of potassium voltage-gated channel subfamily H member 2 and calmodulin in humans
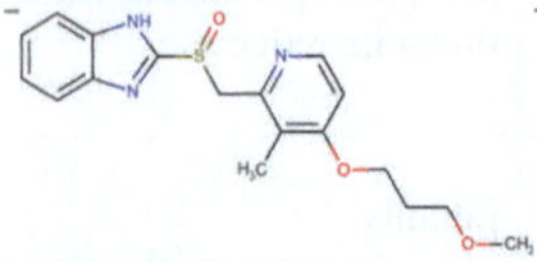 Rabeprazole (DB01129)	Used to prevent GI bleeds caused by excess NSAID use and acid-reflux disorders (GERD), H. pylori eradication, and peptic ulcer disease	Inhibitor of potassium-transporting ATPase alpha chain 1
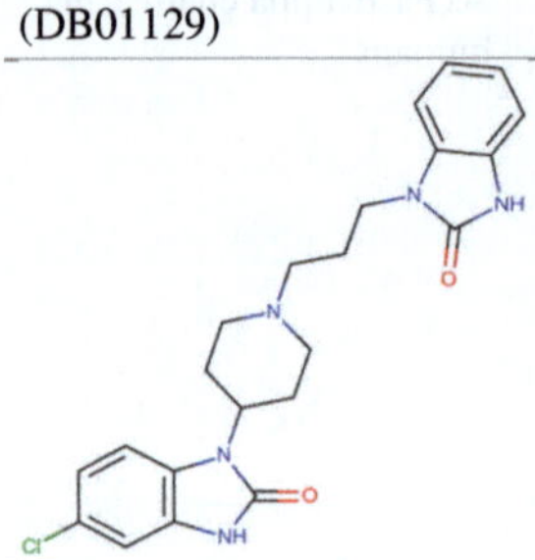 Domperidone (DB01184)	Used to treat vomiting, nausea, dyspepsia, heartburn, and epigastric pain	An antagonist of dopamine D2 and D3 receptors in humans
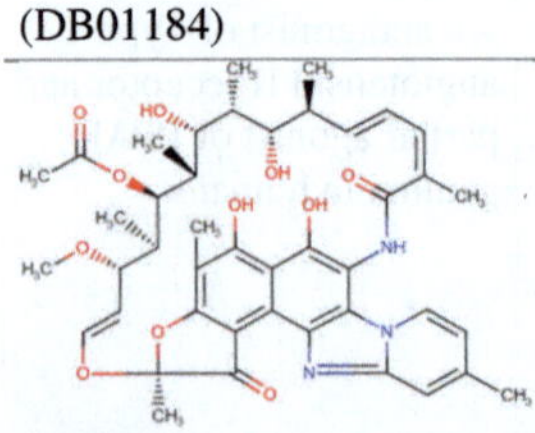 Rifaximin (DB01220)	Used to treat irritable bowel syndrome with diarrhoea (IBS-D), travellers' diarrhoea (caused by non-invasive strains of E. coli) and to reduce over-hepatic encephalopathy recurrence	Inhibitor of DNA-directed RNA polymerase subunit beta of E. coli (strain K12), agonist of nuclear receptor subfamily 1 group I member in humans

(continued)

Table 1 (continued)

Structure of the drug Name of the drug (Drug Bank ID)	Category/indication	Mechanism of action
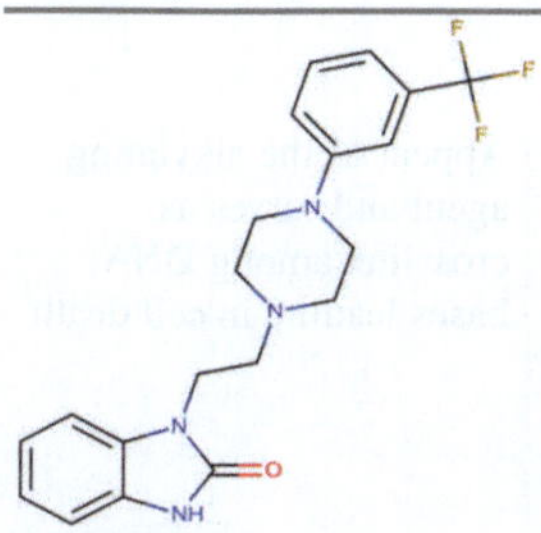 Flibanserin (DB04908)	Hypoactive sexual desire disorder (HSDD) in premenopausal women	Agonist of 5-hydroxytryptamine receptor 1A and antagonist of 5-hydroxytryptamine receptor 2A and dopamine D4 receptor
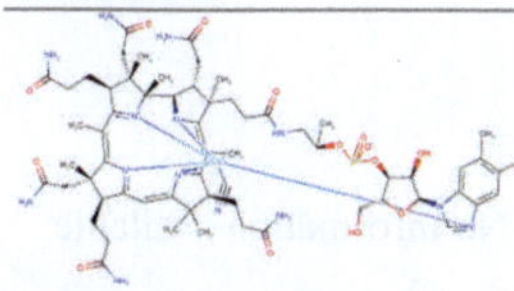 Cyanocobalamin (DB00115)	Vitamin B12 deficiency	Cofactor of methionine synthase, mitochondrial Methylmalonyl-CoA mutase, methionine synthase reductase methylmalonic aciduria and homocystinuria type C protein, methylenetetrahydrofolate reductase and binder of methylmalonic aciduria type A protein, mitochondrial
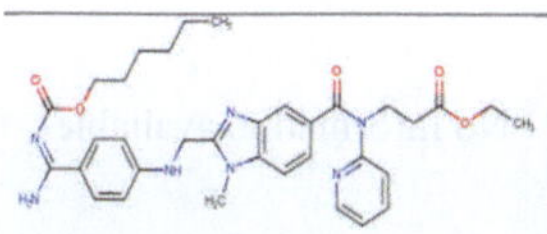 Dabigatran etexilate (DB06695)	It prevents venous thromboembolism events caused in patients going through hip or knee replacement surgery. It also prevents systemic embolism and stroke in patients with atrial fibrillation	Prothrombin inhibitor in humans
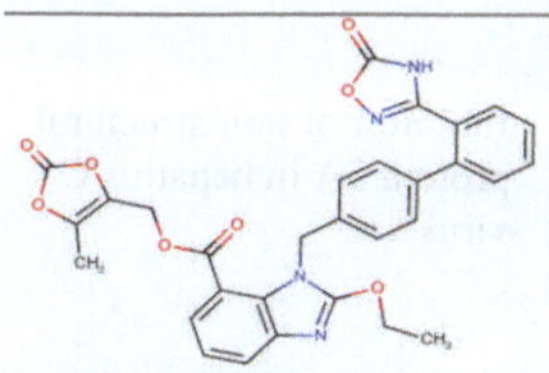 Azilsartan medoxomil (DB08822)	Hypertension treatment (alone or as an adjunct)	Type-1 angiotensin II receptor antagonist in humans

(continued)

Table 1 (continued)

Structure of the drug Name of the drug (Drug Bank ID)	Category/indication	Mechanism of action
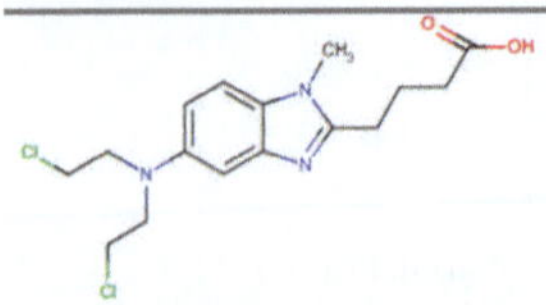 Bendamustine (DB06769)	CLL and NHL are two forms of chronic lymphocytic leukaemia that has progressed during or in a period of six months of therapy using rituximab or a rituximab-consisting regimen	Appear as the alkylating agent and serves as crosslink among DNA bases leading in cell death
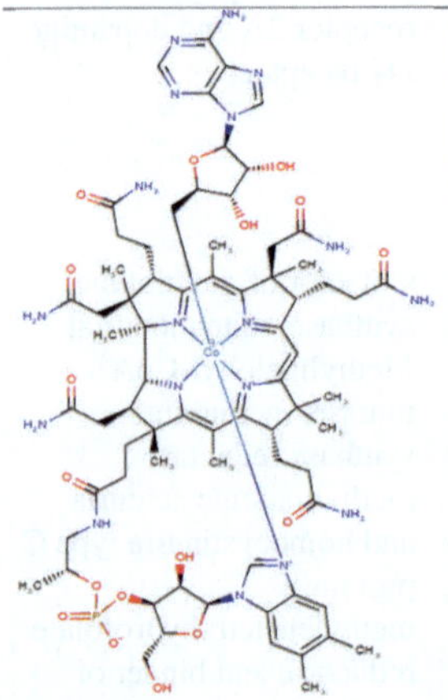 Cobamamide (DB11191)	No information available	No information available
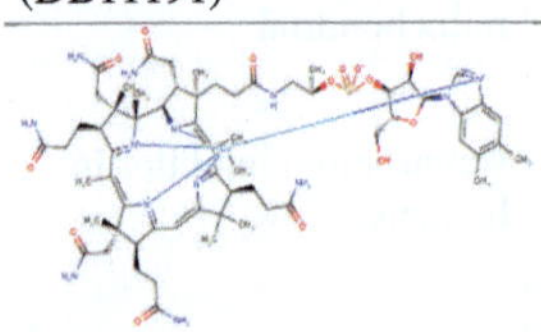 Methylcobalamin (DB03614)	No information available	No information available
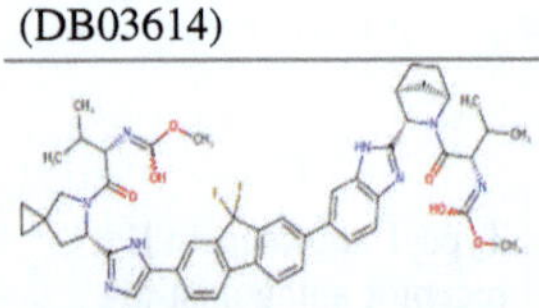 Ledipasvir (DB09027)	Used to treat HCV in patients co-infected with HIV and to treat HCV genotypes 1,4,5 and 6	Inhibitor of non-structural protein 5A in hepatitis C virus

(continued)

Table 1 (continued)

Structure of the drug Name of the drug (Drug Bank ID)	Category/indication	Mechanism of action
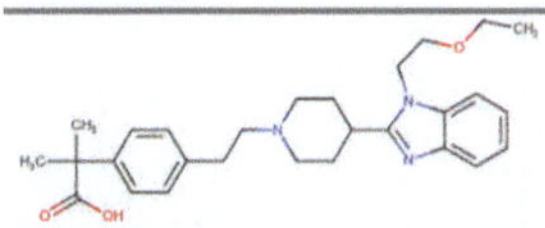 Bilastine (DB11591)	Symptomatic relief of seasonal rhinitis (≥12-year-old patients) and chronic spontaneous urticaria (≥18-year-old patients)	H1 receptor antagonist in humans
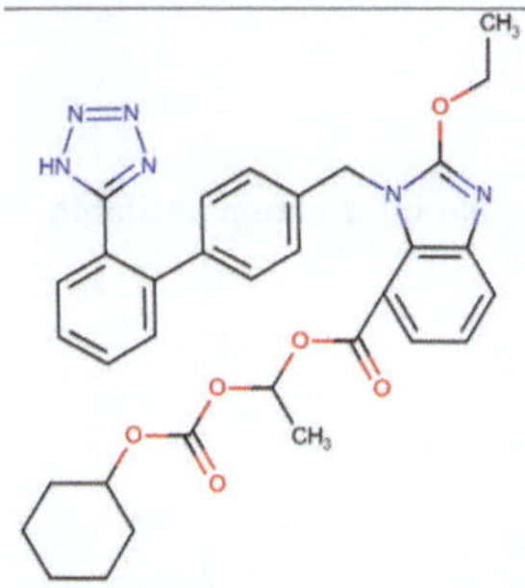 Candesartan cilexetil (DB00796)	Used to treat isolated systolic hypertension, left ventricular hypertrophy, hypertension, and delay the progression of diabetic nephropathy uncomplicate under the first-line agent to treat. Serves as a second-line factor to treat myocardial infarction, systolic dysfunction, coronary artery disease and congestive heart failure when patients are illiberal to ACE inhibitors	An antagonist of Type-1 angiotensin II receptor in humans
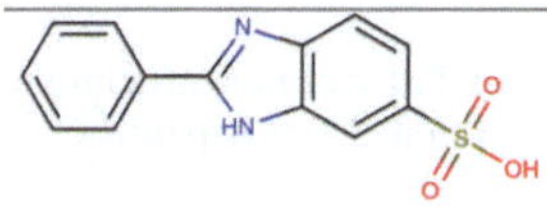 Ensulizole (DB11115)	The UV-B-absorbing molecule in sunscreen formulations	Instigates destruction of DNA, causes fragmentation of DNA strand and photosensitizes the formation of oxidized guanines via type I and II photosensitization mechanisms following UV-A or UV-B irradiation. Generates ROS, which includes singlet oxygen upon photoexcitation
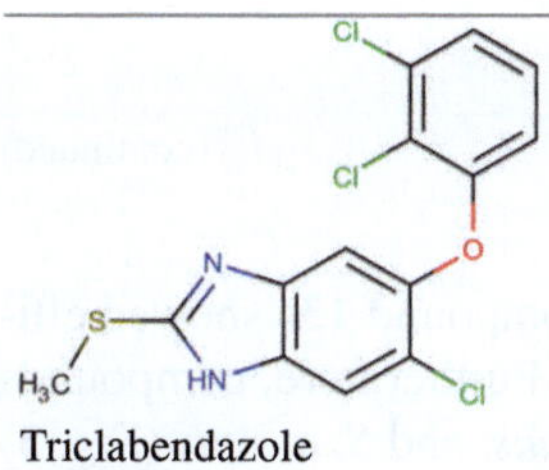 Triclabendazole (DB12245)	Treatment of fascioliasis (patients ≥6 years)	Inhibitor of excretory secretory (ES) proteins in Fasciola hepatica

(continued)

Table 1 (continued)

Structure of the drug Name of the drug (Drug Bank ID)	Category/indication	Mechanism of action
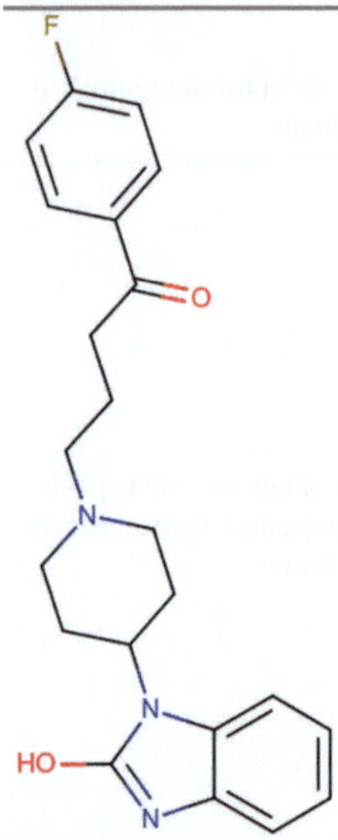 Benperidol (DB12867)	No information available	No information available
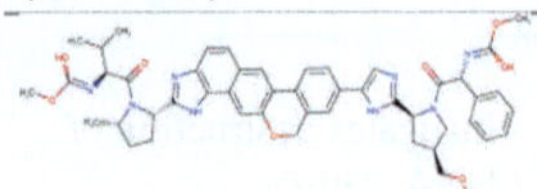 Velpatasvir (DB11613)	To treat chronic hepatitis C virus (HCV) genotype 1–6 infection without cirrhosis or with compensated cirrhosis in adult patients	Inhibitor of non-structural protein 5A of hepatitis C virus
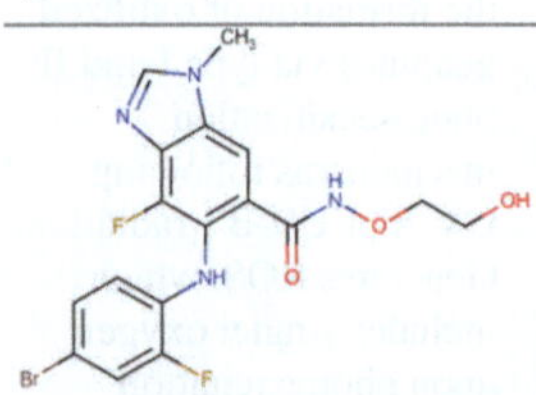 Binimetinib (DB11967)	The combination with encorafenib works well for metastatic and unresectable melanoma accompanying a BRAF V600E or V600K mutation	Inhibitor of MAPK kinase 1 and dual-specificity MAP kinase 2

(continued)

show good antimicrobial activity against *S. aureus*, and compound **13c** showed efficiently anti-oxidative activity in the cellular system [23]. Furthermore, compounds **12** (**a, b**) exhibited more activity against *E. coli, C. albicans*, and *S. aureus* (Fig. 3) [24].

Few notable benzimidazole scaffolds with antimicrobial activity are mentioned below (Fig. 4) [25–35].

Table 1 (continued)

Structure of the drug Name of the drug (Drug Bank ID)	Category/indication	Mechanism of action
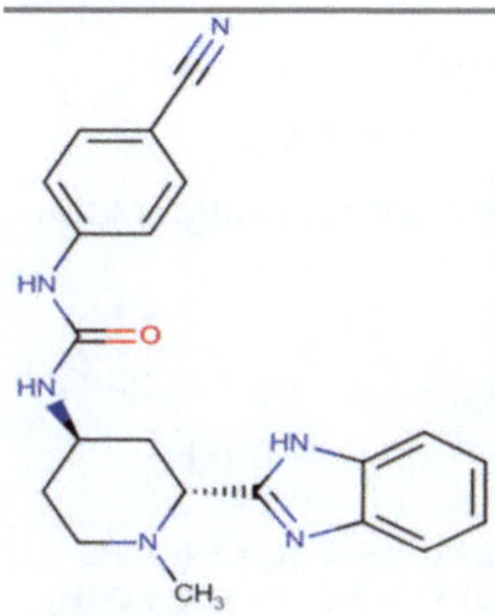 Glasdegib (DB11978)	Newly diagnosed acute myeloid leukaemia treatment	Inhibitor of smoothened homolog in humans
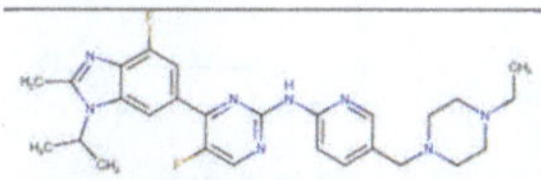 Abemaciclib (DB12001)	It is an antitumor agent recommended for the treatment of HER2-negative and HR-positive advanced or metastatic breast cancer that has progressed followed by an ineffective endocrine therapy	Inhibitor of CDK-4 and CDK-6 in Humans
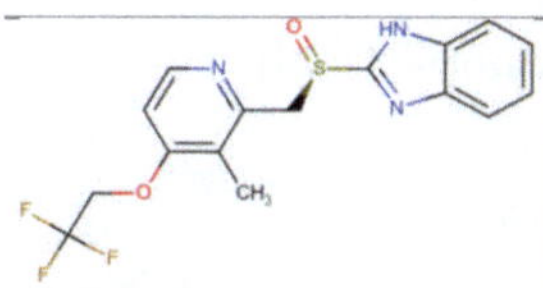 Dexlansoprazole (DB05351)	Used for managing symptoms allied with gastroesophageal reflux disease (GERD) and erosive esophagitis	New generation proton pump inhibitor (PPI)
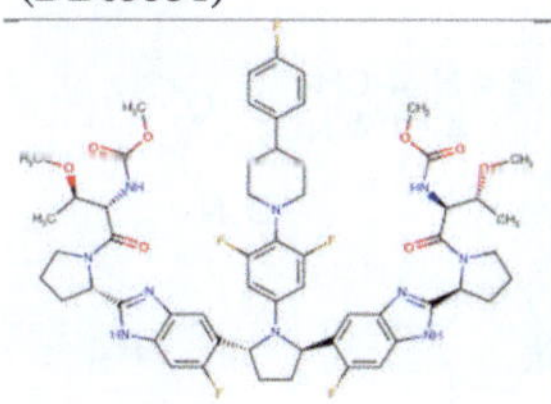 Pibrentasvir (DB13878)	To treat chronic hepatitis C virus (HCV) genotype 1–6 infection without cirrhosis or with compensated cirrhosis in adult patients	Known as hepatitis C virus (HCV) NS5A suppressors and antiviral agent which targets the viral RNA replication and virion assembly

The antibacterial effect exerted by the benzimidazole derivative is due to the azetidinone ring and functional group in their structure. It has shown the excellent antibacterial effect in gram-negative bacteria with the inhibition zone of around 15.3 mm and 19.5 mm, whereas the inhibizone of Gentamicin, a reference drug was found to be 9 mm and 9.3 mm in both *E. coli* and Staphylococcus aureus,

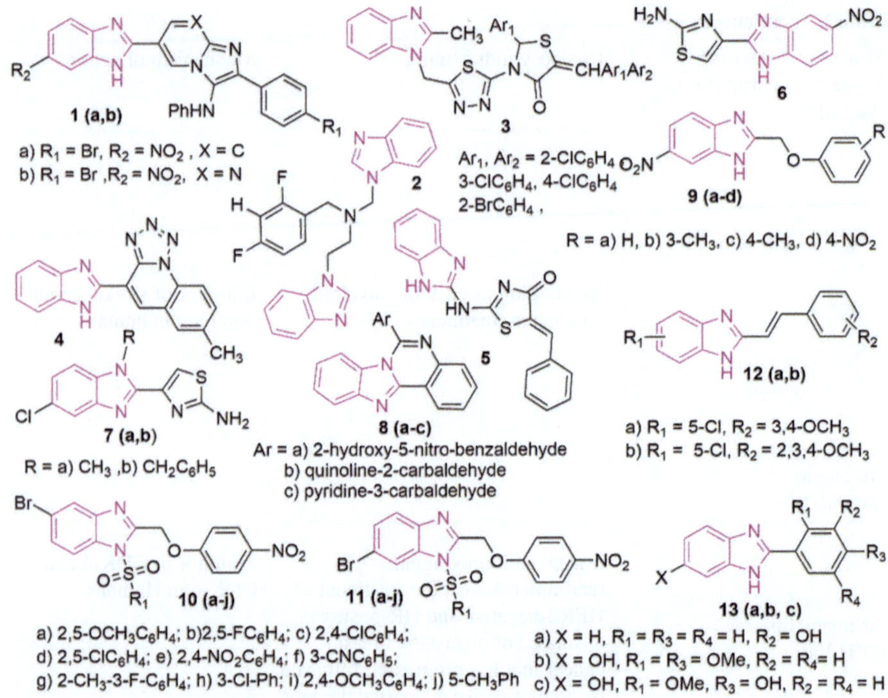

Fig. 3 Chemical structures of some biologically important benzimidazoles [14–24]

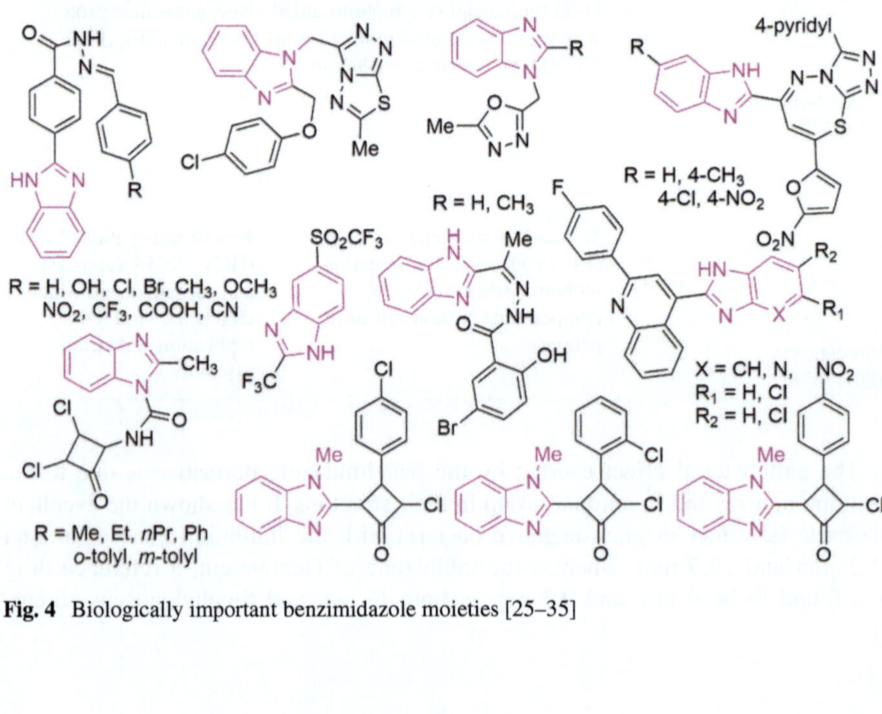

Fig. 4 Biologically important benzimidazole moieties [25–35]

respectively. It indicates that the benzimidazole derivatives exert a higher inhibitory effect compared to Gentamicin [36].

Benzimidazole derivatives bearing various (benz)azolylthio moieties have shown the antibacterial effect greater than the chloramphenicol which was indicated by their MIC values like 6.25 (μg/mL) and 12.5 (μg/mL), respectively, when tested in *E. coli* 35,218. Some of the derivatives exerted potent antibacterial effects with the MIC values ranging from 12.5 μg/mL to 25 μg/mL when tested in *P. vulgaris*.

Some of the benzimidazole compounds are tested for the antibacterial effect against gram (+) and gram (-) bacterial strains, and they have exerted the greater effect than the reference drugs with the MIC range from 2 μg/ mL to 8 μg/mL and in one case 16 μg/mL [37].

3 Anti-inflammatory and Analgesic Activity of Benzimidazole Derivatives

The compounds **18** (**a**,**b**) indicated for their analgesic and anti-inflammatory activities in comparison to standard drug nimesulide. Indeed, the SAR study indicated that the aniline ring embedded with chloro group at the *meta*-position enhances the anti-inflammatory and analgesic activities [38]. Additionally, phenyl pyrazolo benzimidazole quinoxaline derivatives **16** (**a**-**c**) exhibited antioxidant and anti-inflammatory activity [39]. Indeed, 2-styryl benzimidazole (**19**) more equipotent than paracetamol and compound **20** more potent than diclofenac [40]. The thiazolidinone derivatives **14** (**a**,**b**) showed decent anti-inflammatory and analgesic activity and were found to be more potent than diclofenac [41]. In addition, compound **15** is having more potent analgesic and anti-inflammatory activities than nimesulide. The SAR study revealed that compounds with phenyl ring having hydroxy (-OH) and methoxy (-OCH_3) substitution at the fourth position of azetidinone ring enhances the anti-inflammatory and analgesic activity [42]. Furthermore, compounds **17** inhibited the release of TNF-α and IL 6 in a dose-dependent manner and displays no cytotoxicity in hepatic cells (Fig. 5) [43].

Few notable benzimidazole scaffolds with anti-inflammatory and analgesic activity are mentioned below (Fig. 6) [44–50].

Due to the presence of versatile cores in several drugs, benzimidazole scaffolds reveal a wide spectrum of pharmacological activities. Novel coumarin and benzimidazole compounds were prepared and tested for analgesic and anti-inflammatory property. The roles of most of the molecules were comparable with indomethacin, and the activity was increased by the time being [51]. 2-methylamino benzimidazole derivatives have been prepared and screened for anti-inflammatory and analgesic activity. Some of these compounds have shown potent analgesic activity, i.e., 89% at 100 mg/kg body weight and anti-inflammatory activity of 100% at 50 mg/kg body weight compared to the reference drug nimesulide. These group of compounds exhibited superior anti-inflammatory and analgesic activities [38].

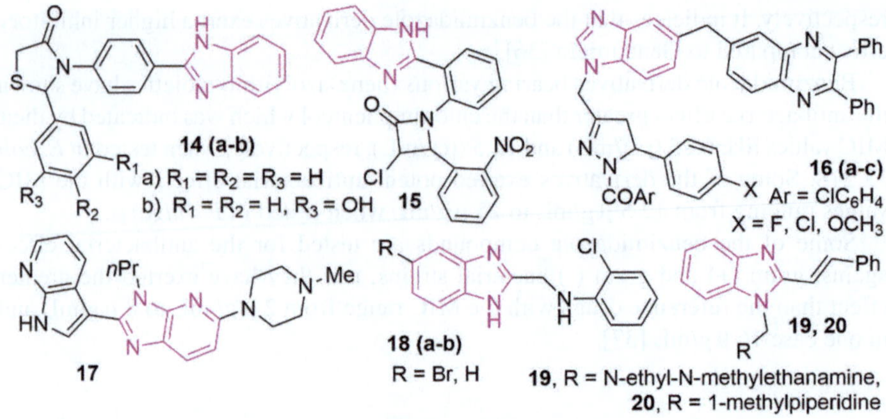

Fig. 5 Chemical structures of some biologically important benzimidazoles [38–43]

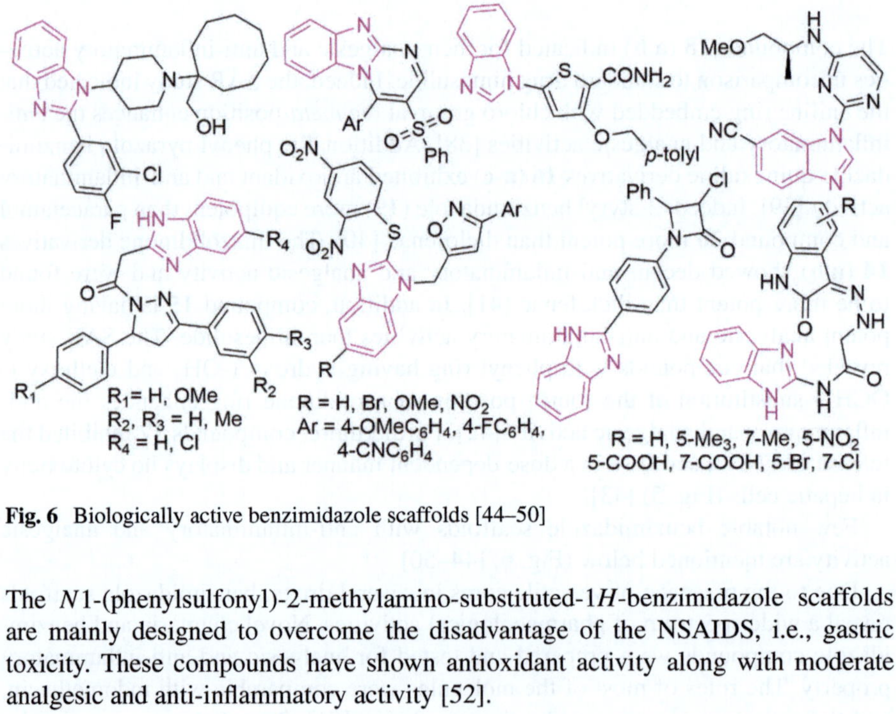

Fig. 6 Biologically active benzimidazole scaffolds [44–50]

The *N*1-(phenylsulfonyl)-2-methylamino-substituted-1*H*-benzimidazole scaffolds are mainly designed to overcome the disadvantage of the NSAIDS, i.e., gastric toxicity. These compounds have shown antioxidant activity along with moderate analgesic and anti-inflammatory activity [52].

4 Anticancer Activity of Benzimidazole Derivatives

At this time, a large range of the world population was affected by cancer due to which many anticancer agents were testified for treating them. However, these drugs suffer from drawbacks of selectivity while affecting abnormal cells; they cause damage to normal cells [53–57].

The derivatives of pyrazino-benzimidazoles and 2-arylbenzimidazoles were screened against diverse cancer cell lines. The compounds **21 and 22** exhibit significant anticancer activity containing methoxy and halogen which were more active than other substitutes (Fig. 7) [58].

Furthermore, benzimidazoles containing oxothiazolidine **33**, 2-[(4-fluorobenzylidene **24** (**a**, **b**), cyclo-alkylidene) cyanomethyl] benzimidazoles **25** (**a**, **b**), 2-[(4- or 5-oxothiazolidin-2-ylidene, 4-substituted thiazolyl-2-ylidene, and [1, 3] thiazin-2-ylidene)cyanomethyl] benzimidazoles (**26–28**, **30**, respectively) displayed anti-tumour properties towards human hepatoma (HEPG2), human breast adenocarcinoma (MCF-7) cell lines, and human colon carcinoma (HCT-116) [59]. Additionally, compound **31** (**a-e**) exhibited more activity against CNS cancer (SF-268), cervical carcinoma (KB), melanoma (G 361), and leukaemia (U937) cell lines [60]. In continuation, conjugated naphthyl-fused system compound **29** is

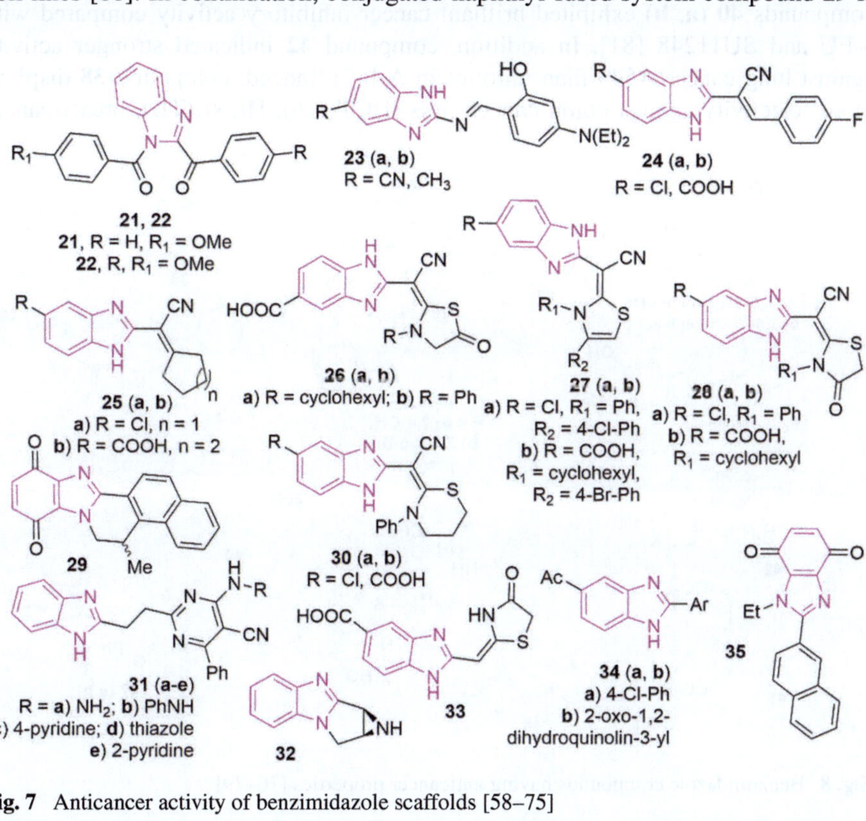

Fig. 7 Anticancer activity of benzimidazole scaffolds [58–75]

most specific towards human cancer cell lines overexpressing reductase enzymes. Compound 2-naphthyl-substituted benzimidazolequinone **35** was presented toxicity towards all human cell lines [61]. Furthermore, the first diazole molecule (**32**) of MMC, exhibited more cytotoxicity on the cell lines (HCC1937 and MCF-7) of human breast cancer in comparison to the normal cell line (GM00637) [62]. Additionally, compounds **23** (**a,b**) exhibited the most antiproliferative activities and also have a robust concentration-dependent effect on MCF-7 and HeLa cell lines [63]. Indeed, **34** (**a, b**) unveiled the maximum activity against a cancer cell lines such as melanoma, non-small cell lung cancer, and leukaemia with a noteworthy values (Fig. 7) [64–75].

In addition, compound **36** (**a, b**) and also compound **37** (showing anticancer activity against NCI 60 cell lines) combined as a lead compound on tumour cell lines with a wide spectrum of anticancer activities [76, 77]. In continuation, compound **39** worked as a potent protein kinase inhibitor which can inhibit cancer cell lines of the NCI panel in sub-micromolar concentration levels [78]. Compound **43** exhibited the maximum activity among the tested compound, according to SAR chloro substituent at *ortho*-position shows the highest anticancer activity (Fig. 8) [79].

According to the study, compound **44** showed less toxicity on human peripheral blood mononuclear cells (PBMC) than 5-Fluorouracil (FU) [80]. Furthermore, compounds **40** (**a, b**) exhibited brilliant cancer inhibitory activity compared with 5-FU and SU11248 [81]. In addition, compound **42** indicated stronger activity against lung cancer A549 than mitomycin A [82]. Indeed, compound **38** displays more selectivity against colon cancer cells (HCT-116), HL60 (TB), breast cancer

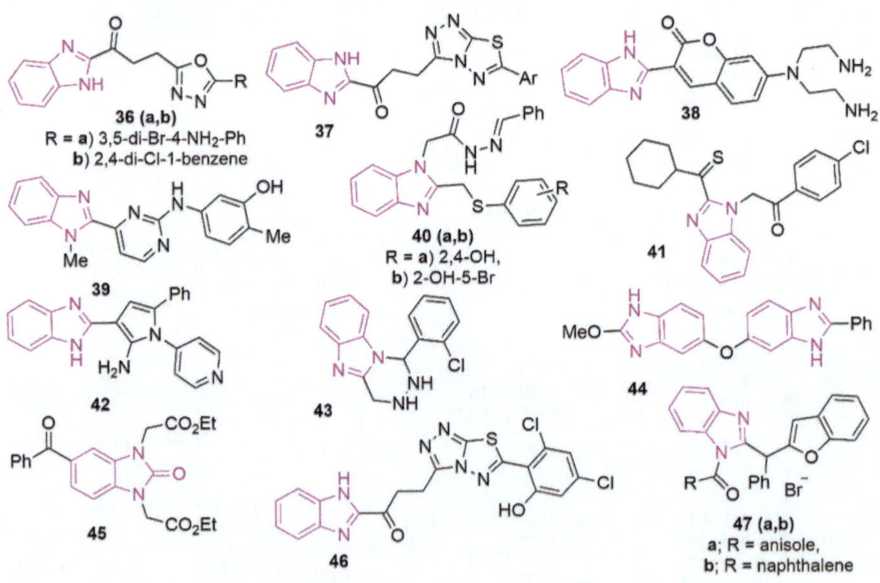

Fig. 8 Benzimidazole compounds having anticancer properties [76–79]

cells (MCF-7, T-47D), RPMI-8226), leukaemia cancer cells (CCRF-CEM, K-562, HCT-15), and melanoma cancer cells (LOX IMVI, UACC-257) [83]. In continuation, compound **46** displayed important growth inhibition and contains brilliant selectivity for leukaemia cell lines [84]. Additionally, compound **47a** and compound **47b** were effective derivatives against five strains of human tumour cell lines. The molecules displayed selective cytotoxic properties towards myeloid liver carcinoma (SMMC-7721) and breast carcinoma (MCF-7), respectively, in comparison to cisplatin (DDP) [85]. Indeed, compound **45** having relatively high antiproliferative effects against HT-29 [86]. In continuation, compound **41** containing 4-chlorophenyl substituent has more activity against the A549 cell line as well as exhibits more cytotoxic activity than cisplatin and doxorubicin [87].

Few notable benzimidazole scaffolds with anticancer activity are mentioned below (Fig. 9).

Benzimidazole acts as an anticancer agent by acting on various biotargets like topoisomerase I and II inhibitors, DNA intercalation and alkylating agents, androgen receptor antagonists, PARP inhibitors, protein tyrosine kinase inhibitors, tyrosine phosphate inhibitors, dihydrofolate reductase inhibitors, microtubule inhibitors, and beta-glucuronidase inhibitors. The benzimidazole ring is present in various

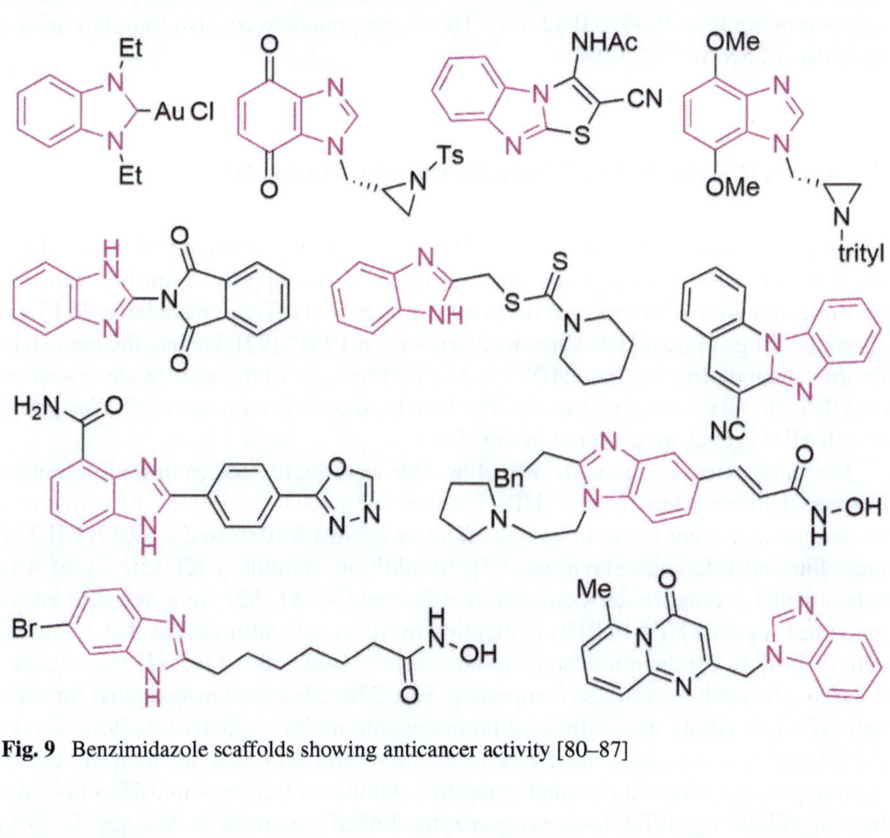

Fig. 9 Benzimidazole scaffolds showing anticancer activity [80–87]

anticancer drugs like Veliparib (PARP inhibitor), Carbendazim (inhibits microtubule functioning), Nocodazole (interferes with microtubule polymerization), and Bendamustine.

Topoisomerases have a role in the winding and unwinding of DNA structure. Inhibition of this topoisomerase leads to stabilization of the DNA complexes, and it may result in DNA strand breaks. Novel benzimidazole acridine compounds were synthesized, and one of the compounds of this class has shown the highest activity in comparison with Colchicine and Imatinib when tested in K-562 and HepG2 cells with IC_{50} of 2.68 mM and 8.11 mM, respectively.

CaPARP inhibitors sensitize the tumour cells towards the DNA damaging agents which leads to multiple strand breaks which cannot be repaired perfectly and results in cell death [88]. There are many PARP inhibitors but they were low-selective and low-potent molecules. The PARP inhibitors with benzimidazole-containing moieties have shown more intrinsic activity [89]. Microtubule inhibition leads to cell death by mitotic arrest during cell division. Some conjugates by Schiff base reaction were synthesized and tested for activity in human breast cancer cells (MCF-7 cells) with IC_{50} of 24.95, 26.36, and 22.59 nM. Some benzimidazole compounds with dehydroabietic acid were synthesized and verified for anticancer activity against numerous cancer cell lines. One of the compounds showed highly potent activity with the IC_{50} values ranging from 0.04 to 0.42 mM. These compounds were also found to induce apoptosis of SMMc-7721 [90].

5 Anti-HIV Activity of Benzimidazole Derivatives

"According to source of united nations World Health Organization (WHO) in 2019, there are 38 million individuals infected by HIV and approx. 2 million individuals infected with AIDS-related diseases per year" [91]. There have been 26 FDA-approved drugs to treat HIV since its discovery in 1981″ [92]. Hence, the persistent improvement of drugs to treat HIV-1 is essential because of the worldwide extension of AIDS and HIV-1 drug resistance. The benzimidazole derivatives useful for acting as anti-HIV agents are depicted in Fig. 10.

The most active compounds **48b**, **49a**, **49b** were highly active to inhibit human immuno deficiency virus type-1 (HIV-1) when compared to nevirapine. Nevertheless, the compounds were found to be less effective against K103N and L100I RT HIV-1 than **48a** and reference efavirenz [93]. In addition, inhibitory RT activity of *N*1-substituted 1,3-dihydro-2*H*-benzimidazol-2-ones (**50**, **51**, **52**) were assessed enzymatic tests against HIV-1 (IIIB) replication in MT-4 cell cultures and also cytotoxic activity [94]. In continuation, benzimidazole embedded with 4-Oxo-4*H*-quinolizine-3-carboxylic acid derivatives (compounds **53a**, **53b**, **53c**) were investigated for their anti-HIV IN activity and witnessed no noticeable inhibitory activities [95]. The *o*-substituted amine (compound **54**) showed more anti-HIV activity than the corresponding *m*- and *p*-substituted analogues [96]. Additionally, compound **55** inhibits the binding of HIV-1 gp120-CD4 capture enzyme-linked immunosorbent assay (ELISA)

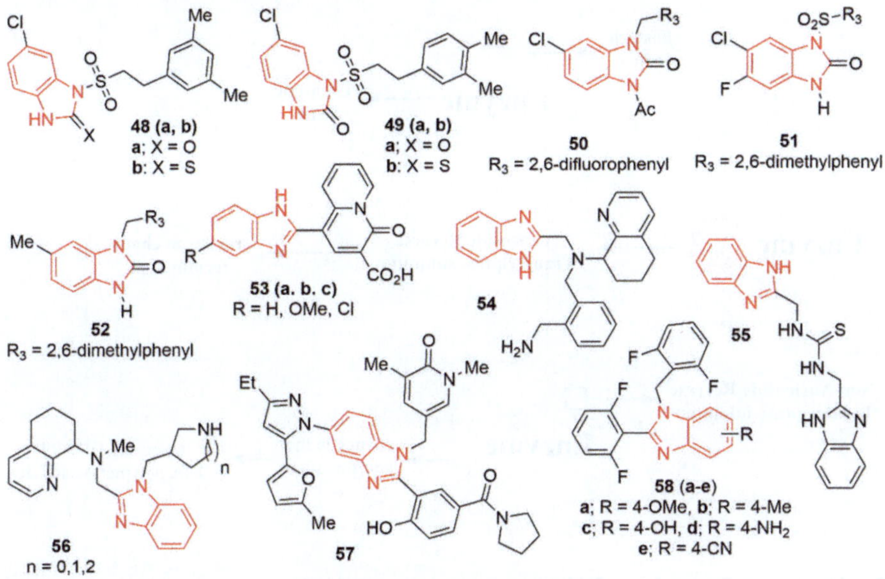

Fig. 10 Anti-HIV activity of benzimidazole derivatives [93–103]

[97]. In continuation, *N*-substituted benzimidazole (**56**) was described as a CXCR4 antagonist [98]. Furthermore, compound **57**, prevents the assembly of HIV-1 capsid, synthetic manipulations at N1, C_2, and C_{16} positions enhanced the antiviral potency [99]. In addition, nucleoside derivatives of *N*-benzylbenzimidazole (compounds **58a-e**) were investigated for inhibiting HIV-1 reverse transcriptase (RT). The best activity of those compounds was displayed with methylene azide and methoxy groups (Fig. 10) [100–103].

Anti-HIV drugs act by inhibition of reverse transcription and it is of two types. One is (nucleoside reverse transcriptase inhibitors (NRTIs) and the other is non-nucleoside reverse transcriptase inhibitors (NNRTIs). The NRTIs bind to the catalytic site and result in chain termination while NNRTIs bind to the allosteric site and cause conformational changes which result in decreasing the HIV-RT DNA polymerase activity. These NNRTIs have conformationally restrained two- and three-ring structures with different degrees of flexibility, and they are specific only for HIV-1 RT. NNRTI binding pockets have the amino acids; those have a high chance of mutation leading to drug resistance, and variants are developed. The advantage of benzimidazole derivative is utilizing linkers for attaining high flexibility (Fig. 11) [104].

Some of the benzimidazole derivatives were synthesized and screened for the anti-HIV activities against replication of HIV-1 (IIIB) and HIV-2 (ROD) in MT-4 cells. Few of the screened compounds inhibited the replication of HIV-1 and HIV-2 (EC50 = 35.40 μg/ml and CC50 > 125 μg/ml) in Mt-4 cells [105]. The anti-HIV activity of prepared *N*1-substituted 6-chloro-1,3-dihydro-2*H*-benzimidazol-2-ones

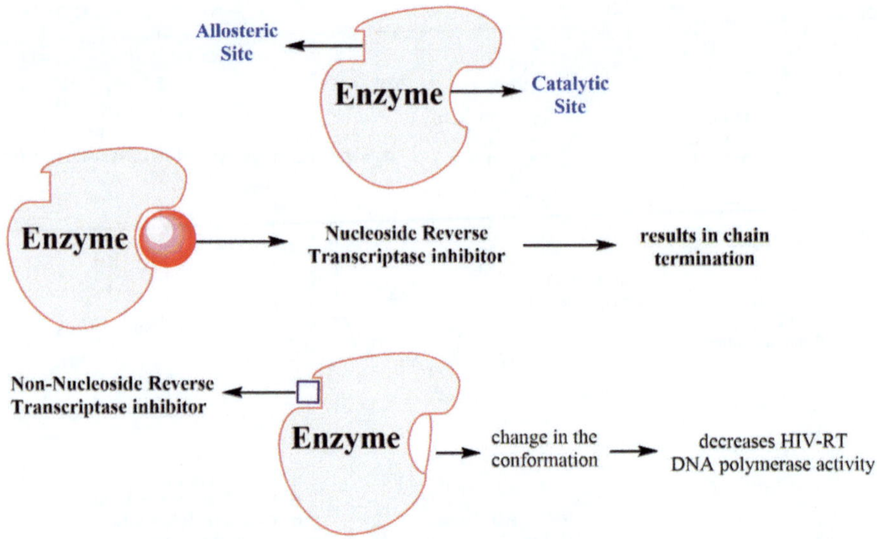

Fig. 11 Reverse Transcriptase Inhibitor acting on enzymes [104]

and their 2-thione analogues have been evaluated. Benzimidazolone system with 3,5-dimethylphenyl moiety at N-1 position has shown potent antiretroviral activity comparable with the standards against both wild and mutant strains of HIV-1 RT [106].

6 Summary/Conclusion

The benzimidazole moiety is a remarkable pharmacophore in modern synthesis and drug discovery. Attention has been progressively more specific to the preparation of benzimidazole derivatives as an origin of novel biological and medicinal agents. The recognition obtained by several researchers has proposed that functionalized benzimidazoles and related *N*-heterocycles, which are the skeletal isosteres of nucleotides, acknowledge them to relate simply with the biopolymers, acquire pharmacological and biological activity with lower toxicities. Deviations in the benzimidazole structures have afforded high biological activities that have proven effective for the evolution of new medicinal agents having revised potency and lesser toxicity. This chapter highlights the several synthesized benzimidazoles derivatives possessing diverse activities such as anti-inflammatory, antifungal, antioxidant, analgesic, anti-HIV, anticancer, antitubercular, and antiviral activity.

References

1. Guru MM, Ali MA, Punniyamurthy T (2015) Copper-mediated synthesis of substituted 2-Aryl-*N*-benzylbenzimidazoles and 2-Arylbenzoxazoles via C-H functionalization/C-N/C-O bond formation. J Org Chem 76:5295–5308
2. Mahesh D, Sadhu P, Punniyamurthy T (2015) Copper(I)-catalyzed regioselective amination of *N*-Aryl imines using $TMSN_3$ and TBHP: a route to substituted benzimidazoles. J Org Chem 80:1644–1650
3. Huang J-J, Leung M-K, Chiu T-L, Chuang Y-T, Chou P-T, Hung Y-H (2014) Novel benzimidazole derivatives as electron-transporting type host to achieve highly efficient sky-blue phosphorescent organic light-emitting diode (PHOLED) device. Org Lett 16:5398–5401
4. Karpińska MM, Matysiak J, Niewiadomy A (2011) Synthesis of Novel 4-(1H-benzimidazol-2-yl)benzene-1,3-diols and their cytotoxic activity against human cancer cell lines. Arch Pharm Res 34:1639–1647
5. Kalaitzakis E, Björnsson E (2007) A review of esomeprazole in the treatment of gastroesophageal reflux disease (GERD). Ther Clin Risk Manag 3:653–663
6. Gosse P (2006) A review of telmisartan in the treatment of hypertension: blood pressure control in the early morning hours. Vasc Health Risk Manag 2:195–201
7. Zimmermann AE, Walters JK, Katona BG, Souney PE, Levine D (2001) A review of omeprazole use in the treatment of acid-related disorders in children. Clin Ther 23:660–679
8. Ajani OO, Aderohunmu DV, Olorunshola SJ, Ikpo CO, Olanrewaju IO (2016) Facile synthesis, characterization and antimicrobial activity of 2-alkanamino benzimidazole derivatives. Orient J Chem 32:109–120
9. Horton J (2000) Albendazole: a review of anthelmintic efficacy and safety in humans. Parasitology 121:113–132
10. Brugger W, Ghielmini M (2013) Bendamustine in indolent non-hodgkin's lymphoma: a practice guide for patient management. Oncologist 18:954–964
11. Chen C, Chen C, Li B, Tao J, Peng J (2012) Aqueous synthesis of 1-H-2-substituted benzimidazoles via transition-metal-free intramolecular amination of aryl iodides. Molecules 17:12506–12520
12. Wishart DS, Feunang YD, Guo AC, Lo EJ, Marcu A, Grant JR, Sajed T, Johnson D, Li C, Sayeeda Z, Assempour N, Iynkkaran I, Liu Y, Maciejewski A, Gale N, Wilson A, Chin L, Cummings R, Le D, Pon A, Knox C, Wilson M (2018) DrugBank 5.0: a major update to the DrugBank database for 2018. Nucleic Acids Res 46:D1074-D1082
13. Pantziarka P, Bouche G, Meheus L, Sukhatme V, Sukhatme VP (2014) Repurposing drugs in oncology (ReDO)-mebendazole as an anti-cancer agent. Ecancer Med Sci 8:443–459
14. Al-Tel TH, Al-Qawasmeh RA (2010) Post groebkee-blackburn multicomponent protocol: synthesis of new polyfunctional imidazo[1,2-a]pyridine and imidazo[1,2-a] pyrimidine derivatives as potential antimicrobial agents. Eur J Med Chem 45:5848–5855
15. Fang B, Zhou C, Rao X (2010) Synthesis and biological activities of novel amine derived bis-azoles as potential antibacterial and antifungal agents. Eur J Med Chem 45:4388–4398
16. Dua R, Sonwane SK, Srivastava SK, Srivastava SD (2010) Conventional and greener approach for the synthesis of some novel substituted -4-oxothiazolidine and their 5-arylidene derivatives of 2-methyl-benzimidazole: antimicrobial activities. J Chem Pharm Res 2:415–423
17. Mungra DC, Patel MP, Patel RG (2011) Microwave-assisted synthesis of some new tetrazolo [1,5-a]quinoline-based benzimidazoles catalyzed by p-TsOH and investigation of their antimicrobial activity. Med Chem Res 20:782–789
18. Mobinikhaledi A, Foroughifar N, Kalhor M, Mirabolfathy M (2010) Synthesis and antifungal activity of novel 2-benzimidazolylimino-5-arylidene-4-thiazolidinones. J Heterocyclic Chem 47:77–80
19. Reddy VA, Reddy KR (2010) Synthesis and antimicrobial activity of some novel 4-(1H-benz[*d*]imidazol-2yl)-1,3-thiazol-2-amines. Chem Pharm Bull 58:953–956
20. Rohini R, Shanker K, Reddy PM, Ravinder V (2010) Synthesis and antimicrobial activities of a new class of 6-arylbenzimidazo[1,2-c]quinazolines. J Braz Chem Soc 21:49–57

21. Hosamani KM, Seetharamareddy HR, Keri RS, Hanamanthagouda MS, Moloney MG (2009) Microwave assisted, one-pot synthesis of 5-nitro-2-aryl substituted-1Hbenzimidazole libraries: Screening in vitro for antimicrobial activity. J Enzyme Inhibit Med Chem 24:1095–1100
22. Ranjith PK, Rajeesh P, Haridas KR, Susanta NK, Guru Rowb TN, Rishikesan R, Kumari NS (2013) Design and synthesis of positional isomers of 5 and 6-bromo-1-[(phenyl)sulfonyl]-2-[(4-nitrophenoxy)methyl]-1H-benzimidazoles as possible antimicrobial and antitubercular agents. Bioorg Med Chem Lett 23:5228–5234
23. Zhou B, Li B, Yi W, Bu X, Ma L (2013) Synthesis, antioxidant, and antimicrobial evaluation of some2-arylbenzimidazole derivatives. Bioorg Med Chem Lett 23:3759–3763
24. Soni LK, Narsinghani T, Sethi A (2012) Anti-microbial benzimidazole derivatives: synthesis and in vitro biological evaluation. Med Chem Res 21:4330–4334
25. Özkay Y, Tunalı Y, Karaca H, Işıkdağ İ (2010) Antimicrobial activity and a SAR study of some novel benzimidazole derivatives bearing hydrazone moiety. Eur J Med Chem 45:3293–3298
26. Reddy VM, Reddy KR (2010) Synthesis and antibacterial activity of some novel 6-(1H-benz[*d*]imidazol-2-yl)-8-(5-nitro-2-furyl)-3-(4-pyridyl)-7,8-dihydro[1,2,4]triazolo[3,4-b][1,3,4]thiadiazepines. Chem Pharm Bull 58:1081–1084
27. Ansari KF, Lal C (2009) Synthesis physicochemical properties and antimicrobial activity of some new benzimidazole derivatives. Eur J Med Chem 44:4028–4033
28. Jadhav GR, Shaikh MU, Kale RP, Ghawalkar AR, Gill CH (2009) Synthesis, characterization, and antimicrobial activities of clubbed [1, 2, 4]-oxadiazoles with fluorobenzimidazoles. J Heterocyclic Chem 46:980–987
29. Sharma D, Narasimhan B, Kumar P, Judge V, Narang R, Clercq ED, Balzarini J (2009) Synthesis, antimicrobial and antiviral activity of substituted benzimidazoles. J Enzyme Inhib Med Chem 24:1161–1168
30. Ansari KF, Lal C (2009) Synthesis and biological activity of some heterocyclic compounds containing benzimidazole and beta-lactam moiety. J Chem Sci 121:1017–1025
31. Kus C, Sozudonmez F, Altanlar N (2009) Synthesis and antimicrobial activity of some novel 2-[4-(substituted piperazin-/piperidin-1-ylcarbonyl)phenyl]-1H-benzimidazole derivatives. Arch Pharm Chem Life Sci 342:54–60
32. Ochal Z, Bretner M, Wolinowska R, Tyski S (2013) Synthesis and in vitro antibacterial activity of 5-halogenomethylsulfonyl-benzimidazole and benzotriazole derivatives. Med Chem 9:1129–1136
33. Kumar NS, Amandoron EA, Cherkasov A, Finlay BB, Gong H, Jackson L, Kaur S, Lian T, Moreau A, Labrière C, Reiner NE, See RH, Strynadka NC, Thorson L, Wonga EWY, Worrall L, Zoraghi R, Young RN (2012) Optimization and structure activity relationships of a series of potent inhibitors of methicillin-resistant staphylococcu saureus (MRSA) pyruvate kinase as novel antimicrobial agents. Bioorg Med Chem 20:7069–7082
34. Noolvi M, Agrawal S, Patel H, Badiger A, Gaba M, Zambre A (2014) Synthesis, antimicrobial and cytotoxic activity of novelazetidine-2-one derivatives of 1H-benzimidazole. Arabian J Chem 7:219–226
35. Garudachari B, Satyanarayana MN, Thippeswamy B, Shivakumar CK, Shivananda KN, Hegde G, Isloor AM (2012) Synthesis, characterization and antimicrobial studies of some new quinoline incorporated benzimidazole derivatives. Eur J Med Chem 54:900–906
36. Dokla EME, Abutaleb NS, Milik SN, Li D, El-Baz K, Shalaby MW, Al-Karaki R, Nasr M, Klein CD, Abouzid KAM, Seleem MN (2020) Development of benzimidazole-based derivatives as antimicrobial agents and their synergistic effect with colistin against gram-negative bacteria. Eur J Med Chem 186:111850
37. Song D, Ma S (2016) Recent development of benzimidazole-containing antibacterial agents. Chem Med Chem 11:646–659
38. Achar KCS, Hosamani KM, Seetharamareddy HR (2010) In-vivo analgesic and anti-inflammatory activities of newly synthesized benzimidazole derivatives. Eur J Med Chem 45:2048–2054

39. Sridevi C, Balaji K, Naidu A, Sudhakaran R (2009) Antioxidant and anti-inflammatory activity of some phenylpyrazolobenzimidazoloquinoxaline derivatives. J Pharma Res 2:1500–1503
40. Jesudason EP, Sridhar SK, Padma Malar EJ, Shanmugapandiyan P, Inayathullah M, Arul V, Selvaraj D, Jayakumar R (2009) Synthesis, pharmacological screening, quantum chemical and in vitro permeability studies of N-Mannich bases of benzimidazoles through bovine cornea. Eur J Med Chem 44:2307–2312
41. Shanmugapandiyan P, Denshing KS, Ilavarasan R, Anbalagan N, Nirmal R (2010) Synthesis and biological activity of 2-(thiazolidin-4-one) phenyl]-1H-phenylbenzimidazoles and 2-[4-(azetidin-2-One)-3-chloro-4-phenyl]-1H-phenyl benzimidazoles. Int J Pharm Sci Drug Res 2:115–119
42. Chhajed SS, Upasani CD (2012) Synthesis and in-silico molecular docking simulation of 3-chloro-4-substituted-1-(2-(1H-benzimidazol-2-yl)phenyl))-azetidin-2-ones as novel analgesic anti-inflammatory agent. Arabian J Chem 9:S1779–S1785
43. Chen G, Liu Z, Zhang Y, Shan X, Jiang L, Zhao Y, He W, Feng Z, Yang S, Liang G (2013) Synthesis and anti-inflammatory evaluation of novel benzimidazole and imidazopyridine derivatives. ACS Med Chem Lett 4:69–74
44. Hayashi S, Nakata E, Morita A, Mizuno K, Yamamura K, Kato A, Ohashi K (2010) Discovery of {1-[4-(2-{hexahydropyrrolo[3,4-c]pyrrol-2(1H)-yl}-1H-benzimidazolyl) piperidin-1-yl]cyclooctyl}methanol, systemically potent novel non-peptide agonist of nociceptin/orphanin FQ receptor as analgesic for the treatment of neuropathic pain: Design, synthesis, and structure-activity relationships. Bioorg Med Chem 18:7675–7699
45. Shaaban MR, Saleh TS, Mayhoub AS, Mansour A, Farag AM (2008) Synthesis and analgesic/anti-inflammatory evaluation of fused heterocyclic ring systems incorporating phenylsulfonyl moiety. Bioorg Med Chem 16:6344–6352
46. Kaplancikli ZA, Turan-Zitouni G, Özdemir A, Can OD, Chevallet P (2009) Synthesis and antinociceptive activities of some pyrazoline derivatives. Eur J Med Chem 44:2606–2610
47. Babu GS, Rajani N, Malathy PS, Srinivas B, Kulandaivelu U, Rao JV (2010) Synthesis, characterization and evaluation of novel N-(1H-benzimidazol-2-yl)-2-isatinylidene-hydrazinecarboxamide derivatives as anti-inflammatory agents. Der Pharma Chemica 2:196–204
48. Kankala S, Kankala R, Gundepaka P, Thota N, Nerella S, Gangula MR, Guguloth H, Kagga M, Vadde R, Vasam CS (2013) Regioselective synthesis of isoxazole-mercaptobenzimidazole hybrids and their in vivo analgesic and anti-inflammatory activity studies. Bioorg Med Chem Lett 23:1306–1309
49. Bamborough P, Christopher JA, Cutler GJ, Dickson MC, Mellor GW, Morey JV, Patel CB, Shewchuk LM (2006) 5-(1H-Benzimidazol-1-yl)-3-alkoxy-2 -thiophenecarbonitriles as potent, selective, inhibitors of IKK-ε kinase. Bioorg Med Chem Lett 16:6236–6240
50. Chen JJ, Thakur KD, Clark MP, Laughlin S, George KM, Bookland RG, Davis JR, Cabrera EJ, Easwaran V, De B, Zhang YG (2006) Development of pyrimidinebased inhibitors of Janus tyrosine kinase 3. Bioorg Med Chem Lett 16:5633–5638
51. Arora RK, Kaur N, Bansal Y, Bansal G (2014) Novel coumarin-benzimidazole derivatives as antioxidants and safer anti-inflammatory agents. Acta Pharm Sin B 4:368–375
52. Achar KCS, Hosamani KM, Seetharamareddy KR (2010) In-vivo analgesic and anti-inflammatory activities of newly synthesized benzimidazole derivatives. Eur J Med Chem 45:2048–2054
53. Gaba M, Gaba P, Uppal D, Dhingra N, Bahia MS, Silakari O, Mohan C (2015) Benzimidazole derivatives: search for GI-friendly anti-inflammatory analgesic agents. Acta Pharm Sin B 5:337–342
54. Cai SX (2007) Small molecule vascular disrupting agents: potential new drugs for cancer treatment. Recent Pat Anti-cancer Drug Discov 2:79–101
55. Shi LM, Fan Y, Lee JK, Waltham M, Andrews DT, Scherf U, Paull KD, Weinstein JN (2000) Mining and visualizing large anticancer drug discovery databases. J Chem Inf Comput Sci 40:367–379

56. Monks A, Scudiero D, Skehan P, Shoemaker R, Paull K, Vistica D, Hose C, Langley J, Cronise P, Wolff VA, Goodrich MG, Campbell H, Mayo J, Boyd MJ (1991) Feasibility of a high-flux anticancer drug screen using a diverse panel of cultured human tumor cell lines. Natl Cancer Inst 83:757–766
57. Waldman AD, Fritz JM, Lenardo MJ (2020) A guide to cancer immunotherapy: from T cell basic science to clinical practice. Nat Rev Immunol 20:651–668
58. Wei Q-Y, Xu Y-M, Lau ATY (2020) Recent progress of nanocarrier-based therapy for solid malignancies. Cancers 12:2783–2819
59. Demirayak S, Kayagil I, Yurttas L (2011) Microwave supported synthesis of some novel 1,3-Diarylpyrazino[1,2-a]benzimidazole derivatives and investigation of their anticancer activities. Eur J Med Chem 46:411–416
60. Refaat HM (2010) Synthesis and anticancer activity of some novel 2-substituted benzimidazole derivatives. Eur J Med Chem 45:2949–2956
61. Abdel-Mohsen HT, Ragab FAF, Ramla MM, El Diwani HI (2010) Novel benzimidazole-pyrimidine conjugates as potent antitumor agents. Eur J Med Chem 45:2336–2344
62. Moriarty E, Carr M, Bonham S, Carty MP, Aldabbagh F (2010) Synthesis and toxicity towards normal and cancer cell lines of benzimidazolequinones containing fused aromatic rings and 2-aromatic ring substituents. Eur J Med Chem 45:3762–3769
63. Bonham S, O'Donovan L, Carty MP, Aldabbagh F (2011) First synthesis of an aziridinyl fused pyrrolo[1,2-*a*]benzimidazole and toxicity evaluation towards normal and breast cancer cell lines. Org Biomol Chem 9:6700–6706
64. Hranjec M, Starčević K, Pavelić SK, Lučin P, Pavelić K, Zamola GK (2011) Synthesis, spectroscopic characterization and antiproliferative evaluation in vitro of novel Schiff bases related to benzimidazoles. Eur J Med Chem 46:2274–2279
65. Abonia R, Cortés E, Insuasty B, Quiroga J, Nogueras M, Cobo J (2011) Synthesis of novel 1,2,5-trisubstituted benzimidazoles as potentialantitumor agents. Eur J Med Chem 46:4062–4070
66. Hranjec M, Pavlović G, Marjanović M, Kralj M, Karminski-Zamola G (2011) (2010) Benzimidazole derivatives related to 2,3-acrylonitriles, benzimidazo[1,2-a] quinolines and fluorenes: Synthesis, antitumor evaluation in vitro and crystal structure determination. Eur J Med Chem 45:2405–2417
67. Sarhan AAO, Al-Dhfyan A, Al-Mozaini MA, Adra CN, Aboul-Fadl T (2010) Cell cycle disruption and apoptotic activity of 3-aminothiazolo[3,2-a]benzimidazole-2- carbonitrile and its homologues. Eur J Med Chem 45:2689–2694
68. Tong Y, Bouska JJ, Ellis PA, Johnson EF, Leverson J, Liu X, Marcotte PA, Olson AM, Osterling DJ, Przytulinska M, Rodriguez LE, Shi Y, Soni N, Stavropoulos J, Thomas S, Donawho CK, Frost DJ, Luo Y, Giranda VL, Penning TD (2009) Synthesis and evaluation of a new generation of orally efficacious benzimidazole-based poly(ADP-ribose) polymerase-1 (PARP-1) inhibitors as anticanceragents. J Med Chem 52:6803–6813
69. Chang WL, Chang CS, Chiang PC, Ho Y-F, Liu J-F, Chang K-W, Guh J-H (2010) 2-Phenyl-5-(pyrrolidin-1-yl)-1-(3,4,5-trimethoxybenzyl)-1H-benzimidazole, a benzimidazole derivative, inhibits growth of human prostate cancer cells by affecting tubulin and c-Jun N-terminal kinase. British J Pharm 160:1677–1689
70. Abdel-Aziz HA, Saleh TS, El-Zahabi HSA (2010) Facile synthesis and in-vitro antitumor activity of some pyrazolo[3,4-b]pyridines and pyrazolo[1,5-*a*]pyrimidines linked to a thiazolo[3,2-*a*]benzimidazole moiety. Arch Pharm Chem Life Sci 343:24–30
71. Wang H, Yu N, Song H, Chen D, Zou Y, Deng W, Lye PL, Chang J, Ng M, Blanchard S, Sun ET, Sangthongpitag K, Wang X, Goh KC, Wu X, Khng HH, Fang L, Goh SK, Ong WC, Bonday Z, Stunkel W, Poulsen A, Entzeroth M (2009) N-hydroxy-1,2-disubstituted-1H-benzimidazol-5-yl acrylamides as novel histone deacetylase inhibitors: Design, synthesis, SAR studies, and in vivo antitumor activity. Bioorg Med Chem Lett 19:1403–1408
72. Wang T, Sepulveda M, Gonzales P, Gately S (2013) Identification of novel HDAC inhibitors through cell based screening and their evaluation as potential anticancer agents. Bioorg Med Chem Lett 23:4790–4793

73. Rubbiani R, Kitanovic I, Alborzinia H, Can S, Kitanovic A, Onambele LA, Stefanopoulou M, Geldmacher Y, Sheldrick WS, Wolber G, Prokop A, Wölfl S, Ott I (2010) Benzimidazol-2-ylidene gold (I) complexes are thioredoxin reductase inhibitors with multiple antitumor properties. J Med Chem 53:8608–8618
74. Zheng Y, Zheng M, Ling X, Liu Y, Xue Y, An L, Gu N, Jin M (2013) Design, synthesis, quantum chemical studies and biological activity evaluation of pyrazolebenzimidazole derivatives as potent Aurora A/B kinase inhibitors. Bioorg Med Chem Lett 23:3523–3530
75. Guo C, Linton A, Jalaie M, Kephart S, Ornelas M, Pairish M, Greasley S, Richardson P, Maegley K, Hickey M, Li J, Wu X, Ji X, Xie Z (2013) Discovery of 2-((1H-benzo[d]imidazol-1-yl)methyl)-4H-pyrido[1,2-a]pyrimidin-4-ones as novel PKM2 activators. Bioorg Med Chem Lett 23:3358–3363
76. Nagarajan S, Majumder S, Sharma U, Rajendran S, Kumar N, Chatterjee S, Singh B (2013) Synthesis and anti-angiogenic activity of benzothiazole, benzimidazole containing phthalimide derivatives. Bioorg Med Chem Lett 23:287–290
77. Husain A, Rashid M, Mishra R, Parveen S, Shin DS, Kumar D (2012) Benzimidazole bearing oxadiazole and triazolo-thiadiazoles nucleus: Designand synthesis as anticancer agents. Bioorg Med Chem Lett 22:5438–5444
78. Rashid M, Husain A, Mishra R (2012) Synthesis of benzimidazoles bearing oxadiazole nucleus as anticancer agents. Eur J Med Chem 54:855–866
79. Determann R, Dreher J, Baumann K, Preu L, Jones PG, Totzke F, Schächtele C, Kubbutat MHG, Kunick C (2012) 2-Anilino-4-(benzimidazol-2-yl)pyrimidines multikinase inhibitor scaffold with antiproliferative activity toward cancer cell lines. Eur J Med Chem 53:254–263
80. El-Nassan HB (2012) Synthesis, antitumor activity and SAR study of novel [1,2,4]triazino[4,5-a]benzimidazole derivatives. Eur J Med Chem 53:22–27
81. Wang X-J, Chu N-Y, Wang Q-H, Liu C, Jiang C-G, Wang X-Y, Ikejima T, Cheng M-S (2012) Newly synthesized bis-benzimidazole derivatives exerting anti-tumor activity through induction of apoptosis and autophagy. Bioorg Med Chem Lett 22:6297–6300
82. Liu T, Sun C, Xing X, Jing L, Tan R, Luo Y, Huang W, Song H, Li Z, Zhao Y (2012) Synthesis and evaluation of 2-[2-(phenylthiomethyl)-1H-benzo[d]imidazol-1-yl)acetohydrazide derivatives as antitumor agents. Bioorg Med Chem Lett 22:3122–3125
83. Omar MA, Shaker YM, Galal SA, Ali MM, Kerwin SM, Li J, Tokuda H, Ramadan RA, El Diwani HI (2012) Synthesis and docking studies of novel antitumor benzimidazoles. Bioorg Med Chem 20:6989–7001
84. Paul K, Bindal S, Luxami V (2013) Synthesis of new conjugated coumarin benzimidazole hybrids and their anticancer activity. Bioorg Med Chem Lett 2:3667–3672
85. Husain A, Rashid M, Shaharyar M, Siddiqui AA, Mishra R (2013) Benzimidazole clubbed with triazolo-thiadiazoles andtriazolo-thiadiazines: New anticancer agents. Eur J Med Chem 62:785–798
86. Wang X-Q, Liu L-X, Li Y, Sun C-J, Chen W, Li L, Zhang H, Yang X (2013) Design, synthesis and biological evaluation of novel hybrid compounds of imidazole scaffold-based 2-benzylbenzofuranas potent anticancer agents. Eur J Med Chem 62:111–121
87. Mavrov ATs, Wesselinov D, Vassilev N, Tsenov JA (2013) Design, synthesis and antiproliferative properties of some new 5-substituted-2-iminobenzimidazole derivatives. Eur J Med Chem 63:696–701
88. Yurttas L, Demirayak S, Ciftci GA, Yıldırım SU, Kaplancık ZA (2013) Synthesis and biological evaluation of some 1,2- disubstituted benzimidazole derivatives as new potential anticancer agents. Arch Pharm Chem Life Sci 346:403–414
89. Shrivastava N, Naim MJ, Alam MJ, Nawaz F, Ahmed S, Alam O (2017) Benzimidazole scaffold as anticancer agent: synthetic approaches and structure-activity relationship. Arch Pharm Chem Life Sci 350:e1700040
90. Zeng H, Zhang H, Jang F, Zhao L, Zhang J (2011) Molecular modeling studies on benzimidazole carboxamide derivatives as PARP-1 inhibitors using 3D-QSAR and docking. Chem Biol Drug Des 78:333–352

91. Miao T-T, Tao X-B, Li D-D, Chen H, Jin X-Y, Geng Y, Wang S-F, Gu W (2018) Synthesis and biological evaluation of 2-arylbenzimidazole derivatives of dehydroabietic acid as novel tubulin polymerization inhibitors. RSC Adv 8:17511–17526
92. Brown LB, Spinelli MA, Gandhi M (2021) The interplay between HIV and COVID-19: summary of the data and responses to date. Curr Opin HIV AIDS 16:63–73
93. Gulick RM, Lalezari J, Goodrich J, Clumeck N, DeJesus E, Horban A, Nadler J, Clotet B, Karlsson A, Wohlfeiler M, Montana JB, McHale M, Sullivan J, Ridgway C, Felstead S, Dunne MW, van der Ryst E, Mayer HN (2008) Maraviroc for previously treated patients with R5 HIV-1 infection. N Engl J Med 359:1429–1441
94. Monforte A, Logoteta P, Luca LD, Iraci N, Ferro S, Maga G, Clercq ED, Pannecouque C, Chimirri A (2010) Novel 1,3-dihydro-benzimidazol-2-ones and their analogues as potent non-nucleoside HIV-1 reverse transcriptase inhibitors. Bioorg Med Chem 18:1702–1710
95. Monforte A, Rao A, Logoteta P, Ferro S, Luca LD, Barreca ML, Iraci N, Maga G, Clercq ED, Pannecouque C, Chimirri A (2008) Novel N1-substituted 1,3-dihydro-2H-benzimidazol-2-ones as potent non-nucleoside reverse transcriptase inhibitors. Bioorg Med Chem 16:7429–7435
96. Xu Y-S, Zeng C-C, Jiao Z-G, Hu L-M, Zhong R-G (2009) Design, synthesis and anti-HIV integrase evaluation of 4-oxo-4H-quinolizine-3-carboxylic acid derivatives. Molecules 14:868–883
97. Skerlj R, Bridger G, McEachern E, Harwig C, Smith C, Wilson T, Veale D, Yee H, Crawford J, Skupinska K, Wauthy R, Yang W, Zhu Y, Bogucki D, Fluri MD, Langille J, Huskens D, Clercq ED, Schols D (2011) Synthesis and SAR of novel CXCR4 antagonists that are potent inhibitors of T tropic (X4) HIV-1 replication. Bioorg Med Chem Lett 21:262–266
98. Sivan SK, Vangala R, Manga V (2013) Molecular docking guided structure based design of symmetrical N, N'-disubstituted urea/thiourea as HIV-1 gp120-CD4 binding inhibitors. Bioorg Med Chem 21:4591–4599
99. Miller JF, Turner EM, Gudmundsson KS, Jenkinson S, Spaltenstein A, Thomson M, Wheelan P (2010) Novel N-substituted benzimidazole CXCR4 antagonists as potential anti-HIV agents. Bioorg Med Chem Lett 20:2125–2128
100. Tremblay M, Bonneau P, Bousquet Y, Deroy P, Duan J, Duplessis M, Gagnon A, Garneau M, Goudreau N, Guse I, Hucke O, Kawai SH, Lemke CT, Mason SW, Simoneau B, Surprenant S, Titolo S, Yoakim C (2012) Inhibition of HIV-1 capsid assembly: Optimization of the antiviral potency by site selective modifications at N1, C2 and C16 of a 5-(5-furan-2-yl-pyrazol-1-yl)-1H-benzimidazole scaffold. Bioorg Med Chem Lett 22:7512–7517
101. Ziołkowska NE, Michejda CJ, Bujacz GD (2010) Structure of HIV-1 nonnucleoside reverse transcriptase inhibitors derivatives of N-benzyl-benzimidazole with different substituents in position 4. J Mol Structure 963:188–193
102. Gilbert BE, Knight V (1986) Biochemistry and clinical applications of ribavirin. Antimicrob Agents Chemother 30:201–205
103. Holla BS, Veerendra B, Shivananda MK, Poojary B (2003) Synthesis characterization and anticancer activity studies on some Mannich bases derived from 1,2,4-triazoles. Eur J Med Chem 38:759–767
104. Turan-Zitouni G, Sıvacı M, Kılıç S, Erol K (2001) Synthesis, spectroscopic characterization and antiproliferative evaluation in vitro of novel Schiff bases related to benzimidazoles. Eur J Med Chem 36:685–689
105. Singh A, Yadav D, Yadav M, Dhamanage A, Kulkarni S, Singh RK (2015) Molecular modelling, synthesis and biological evaluation of N-Heteroaryl compounds as reverse transcriptase inhibitors against HIV-1. Chem Biol Drug Des 85:336–347
106. Selvam P, Lakra DR, Pannecouque C (2010) Synthesis, anti-viral and cytotoxicity studies of some novel n-substituted benzimidazole derivatives. Int J Pharm Sci Res 1:105–109

Chapter 10
An Overview on Biological Activities of Oxazole, Isoxazoles and 1,2,4-Oxadiazoles Derivatives

Raghuram Gujjarappa, Sattu Sravani, Arup K. Kabi, Aakriti Garg, Nagaraju Vodnala, Ujjawal Tyagi, Dhananjaya Kaldhi, Virender Singh, Sreya Gupta, and Chandi C. Malakar

Abbreviations

SRS-A	Slow-Reacting Substance of Anaphylaxis
HIV	Human Immunodeficiency Virus
TTR	Transthyretin
DNA	Deoxyribonucleic Acid
NDM-1	New Delhi metallo-ß-lactamase-1
CYP	Cytochrome P
COX	Cyclooxygenase
CNS	Central Nervous System
ADP	Adenosine di-phosphate
OGTT	Oral Glucose Tolerance Test
PARP	Poly(ADP-ribose)polymerase
RNA	Ribonucleic Acid
SAR	Structure-Activity Relationship
DMT2	Diabetes Mellitus Type 2
PPAR	Peroxisome proliferator-activated receptor
MRSA	Methicillin-resistant Staphylococcus aureus
FATP	Fatty Acid Transport Protein

R. Gujjarappa · A. K. Kabi · N. Vodnala · U. Tyagi · D. Kaldhi · C. C. Malakar (✉)
Department of Chemistry, National Institute of Technology Manipur, Langol, Manipur, Imphal 795004, India
e-mail: cmalakar@nitmanipur.ac.in

S. Sravani · A. Garg · S. Gupta (✉)
Department of Medicinal Chemistry, National Institute of Pharmaceutical Education and Research (NIPER), Kolkata, Chunilal Bhawan, 168, Maniktala Main Road, Kolkata 700054, India

V. Singh
Department of Chemistry, Central University of Punjab, Punjab, Bathinda 151001, India

B. P. Swain (ed.), *Nanostructured Biomaterials*, Materials Horizons: From Nature to Nanomaterials, https://doi.org/10.1007/978-981-16-8399-2_10

P-GP P-glycoprotein
SAC Spindle Assembly Checkpoint

1 Introduction

Oxazole is a heterocycle of the five-membered ring which is composed of oxygen and nitrogen atoms at first and third positions, respectively, whereas for isoxazole oxygen and nitrogen atoms are at 1 and 2 positions. In the last few years, reports of biologically active compounds containing heterocyclic rings have drawn great attention from medicinal chemists. Oxazole is one of the major biologically active scaffolds found so far [1]. Surprisingly, a wide range of biological actions is associated with oxazole containing compounds, including anticancer, antibacterial, anticonvulsant, anti-allergic, anthelmintic, antiviral, antidepressant, analgesic and antioxidant properties [2]. Synthetic derivatives of oxazoles are imperative in the drug research portfolio as a result of good anti-inflammation potential [3], TRPV1 antagonist activity [4], antitubercular [5] and anti-HIV [6] activities. Additionally, oxazoles are also found to be used as fluorescent dyes, agrochemicals, corrosion inhibitors [7, 8] in polymer industries [9–11] and photography [12]. This chapter highlights the modern advancements in the progress of oxazole-based biologically active compounds.

Isoxazoles are also an essential class of heterocycles, which are generally active in the area of therapeutics and pharmaceuticals such as anticancer, insecticidal, antibacterial, antituberculosis, antifungal, antibiotic, antitumour and ulcerogenic. Moreover, marketed anti-inflammatory drugs as well as COX-2 inhibitors contain molecular scaffolds of Isoxazole. Isoxazole derivatives such as oxacillin, sulfamethoxazole, acivicin, cycloserine and sulfisoxazole have been in commercial use for the previous 40 years. In another, Cycloserine is a well-known antibiotic drug that has antibacterial, antitubercular activities, and also in medication of leprosy. Acivicin is an anti-leishmanial, antitumour drug, whilst isoxaflutole is used as an herbicidal drug [13].

Oxadiazole (known as furadiazoles) is one of the important scaffolds having a *N*-heterocyclic five-membered ring consists of two nitrogen and one oxygen atom. Oxadiazoles could arise in four distinct isomeric forms: 1,2,3-oxadiazole, 1,2,4-oxadiazole, 1,2,5-oxadiazole and 1,3,4-oxadiazole; depends on the position of nitrogen atoms. Amongst the isomers, 1,2,4-oxadiazole, one of the important five-membered *N*-heterocycles, received significant attention due to its excellent bioisosteric properties and the broad spectrum of biological and pharmaceutical applications. The fused and pendant 1,2,4-oxadiazole scaffolds have been traced in several well recognized, commercially accessible drugs. After a few decades since the chemistry of 1,2,4-oxadiazole was invented, the unique potential involved researchers around the globe, giving rise to the recognition of currently available drugs possessing 1,2,4-oxadiazole scaffolds. Nevertheless, the attention in the biological application of 1,2,4-oxadiazoles and their derivatives have been enlarged in the past twenty years [14].

1.1 Biologically Active Oxazole Related Pharmaceutical Drugs

Oxazole compounds [15–17] as the bioisostere of imidazoles [18, 19], thiazoles [18, 19], triazoles [20, 21], benzimidazoles [20, 21] as well as tetrazoles [22], have fascinated progressive consideration. Moreover, several scientists and researchers over the globe have been affording oxazole-based compounds as pharmacophore and conceivably discover novel scaffolds with excellent pharmacokinetic property, low toxicity and a wide spectrum of bioactivity (Fig. 1) [23–25].

Several oxazole-embellished natural products are known to be isolated from numerous microorganisms and marine invertebrates [26–28], which assist as core structural motifs for several pharmaceuticals that exhibit various biological activities such as antifungal, antibacterial, antiproliferative [29–33], analgesic, antileukemic, antiviral, anticancer and enzyme inhibitory activities [34–43]. The phenomenal

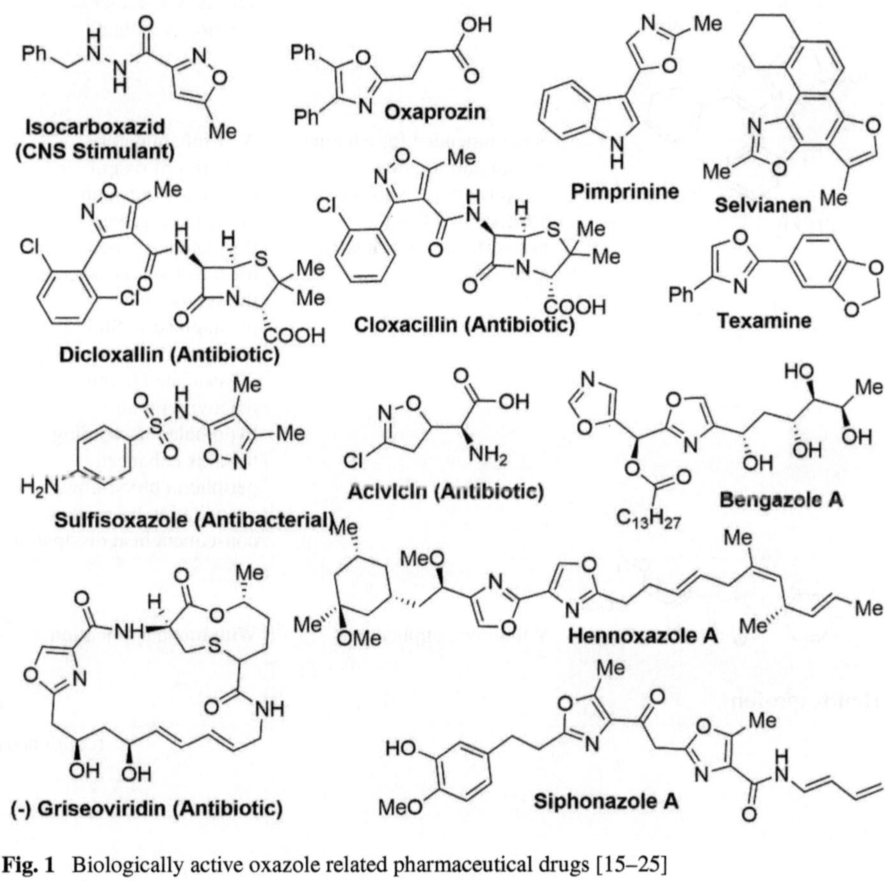

Fig. 1 Biologically active oxazole related pharmaceutical drugs [15–25]

biological functions and ubiquity of oxazoles in synthetic drugs and natural products have generated remarkable attention in the synthesis and biological evolution of functionalized cyclic oxazole scaffold.

Few of the FDA-approved oxazole containing drugs are listed in Table 1 [43].

Table 1 FDA-approved oxazole-containing drugs [43]

Structure of the drug name of the drug (Drug bank id)	Category/Indication	Mechanism of action
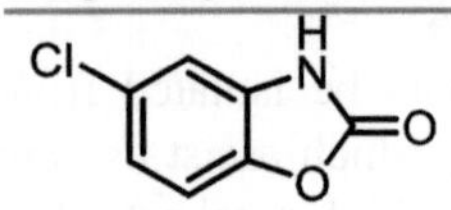 Chlorzoxazone	It is recommended for the reliever of pain related to acute painful musculoskeletal conditions	Inhibits degranulation of mast cells, eventually suppressing the liberation of histamine and anaphylaxis (SRS-A), a slow reacting substance, acts as type-I allergic reactions regulator
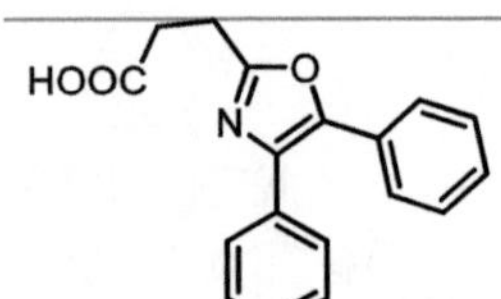 Oxaprozin	Recommended for edoema, inflammation, joint discomfort and stiffness which arises from osteoarthritis and rheumatoid arthritis	Anti-inflammatory activities of oxaprozin referred to the suppression of cyclooxygenase within the platelets which results towards the hindrance in the synthesis of prostaglandin. Shows antipyretic activities which is associated to the reactivity on the hypothalamus, leading towards enhanced peripheral blood flow, vasodilation, as well as consequent heat dissipation
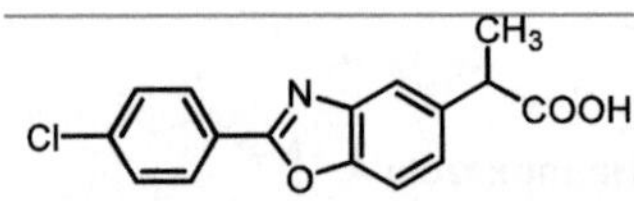 Benoxaprofen	Withdrawn application	Withdrawn application

(continued)

Table 1 (continued)

Structure of the drug name of the drug (Drug bank id)	Category/Indication	Mechanism of action
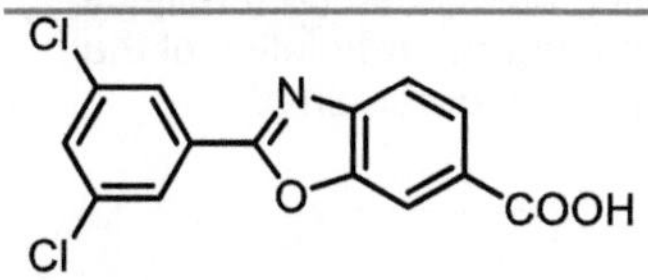 Tafamidis	STreatment for stage-1 symptomatic polyneuropathy, which may result to postpone peripheral neurologic impairment in Europe and cardiomyopathy in wild type or hereditary transthyretin-mediated amyloidosis	the specific stabilizer of TTR, and its inhibition of TTR tetramer dissociation forms the rationale for its use as a treatment to slow - but not cure - the disease progression of TTR-FAP
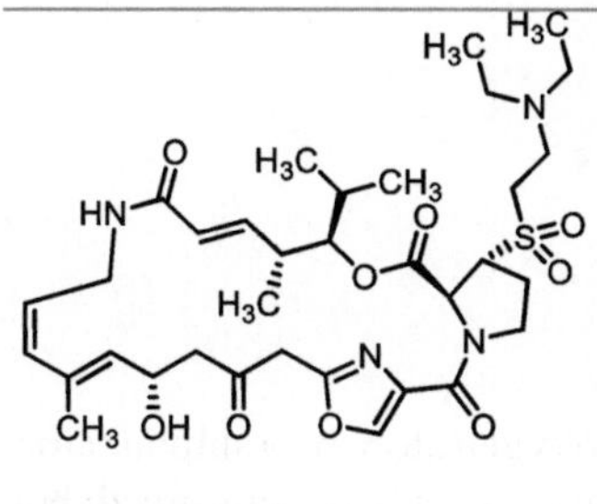 Dalfopristin	For the treatment of bacterial infections (usually in combination with quinupristin)	Inhibits the early phase of protein synthesis
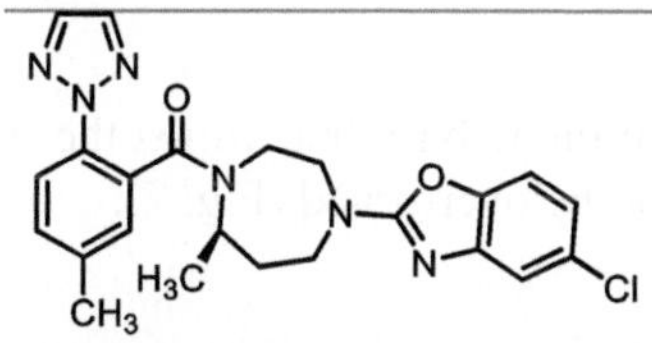 Suvorexantore	Approved for the treatment of insomnia	A dual antagonist of orexin receptors OX1R and OX2R that indorses sleep by dipping wakefulness and arousal

1.2 *SAR (Structure–Activity Relationship) of 1,2,4-Oxadiazole Compounds*

The 1,2,4-oxadiazoles showcased various essential pharmacokinetic characteristics and share decent activity in murine models of Methicillin-resistant Staphylococcus aureus (MRSA) infection (Fig. 2) [44, 45]. They are usually embrace of four cyclic ring system, assigned as A, B, C and D. The subsequent structure–activity relationship (SAR) recognized some significant explanations:

- The H-bond donor of the ring A is important, whereas for antibacterial activity, H-bond acceptors at ring A are unfavoured.
- Structural distinctions on ring C are recognized for the biological activity.

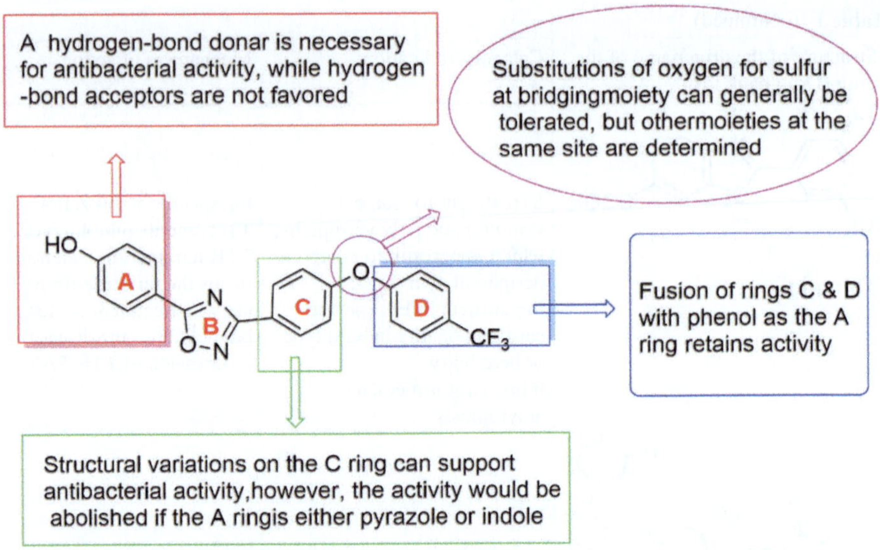

Fig. 2 SAR studies of oxadiazole compounds [44, 45]

- When other moieties are present, substitution of an oxygen atom by a sulphur atom at the connecting moiety between rings C and D can ordinarily be reinforced; but, when other moieties are present, it is unfavourable.
- Ring C and D fusions with phenol are permitted.

The structural differences on the ring C can exhibit bioactivity but, whilst the A ring is either indole or pyrazole, and the activity would be overturned (Fig. 2).

2 Antibacterial Activity of Oxazole Derivatives

The appearance of carbapenemase forming bacteria, particularly New Delhi Metallo-β-lactamase known as NDM-1 and its alternates, has elevated a major worry to the human health. NDM-1 and its variants hydrolyze a broad variety of β-lactam antibiotics, as well as carbapenemase. In this context, methyl oxazole amide, compound **1** was a comparatively potent inhibitor against NDM-1 [46–52]. The amide and thiol-groups are also essential for correlating to the active site of the NDM-1 protein [47]. In addition, Benzamide-based oxazole moieties **2** (**a–d**) displayed decent activity against *Staphylococcus aureus* ATCC (Fig. 3).

Moreover, compound **2c** was more potent than clinical Linezolid or Vancomycin. Additionally, Coumarin-based isoxazole **3**, isoxazole moiety bearing -Cl substitution at *-o* and *-p* positions of aromatic ring showed higher activity against *Pseudomonas aeruginosa* and *Bacillus cereus*. It is might specify that coumarin scaffold with isoxazoles is obviously important for the medicinal implication [48]. In

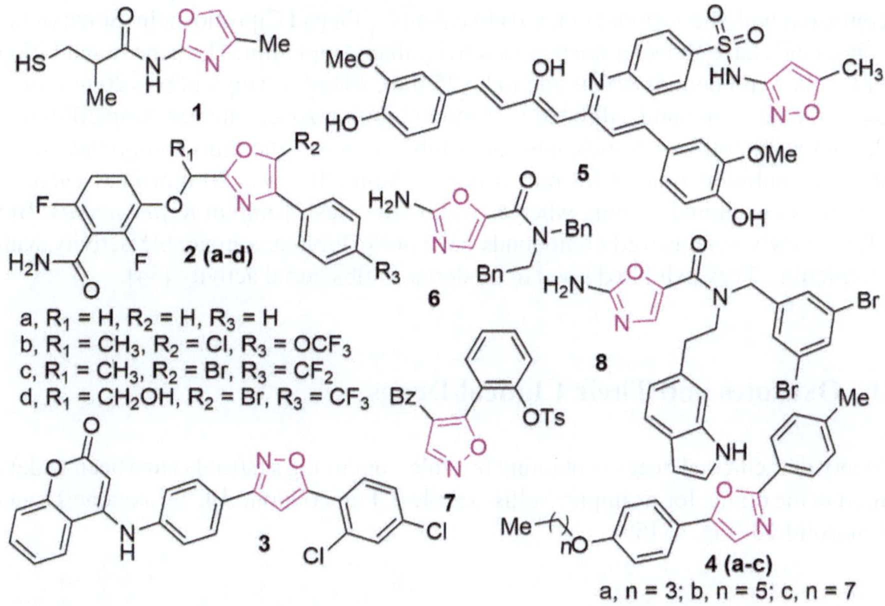

Fig. 3 Oxazole and Isoxazole as antibacterial compounds [46–52]

continuation, bisbenzyl substituted oxazoles **4** (**a-c**) were studied for antibacterial activity and compound **4c** displayed supreme antibacterial activity against *P. aeruginosa* as well as compounds **4a** and **4c** displayed comparable inhibition against *E. coli* [49]. In addition, the carbonyl group of curcumin, attached with isoxazole sulfonamide which provided compound **5** and displayed robust antibacterial activity. Again, when both carbonyl groups were involved with two isoxazole sulfonamide molecules, the activity would decline faintly [50]. In another, Enterococci are the significant pathogens for resistance, and mainly Enterococcus faecium, associated with the group of "ESKAPE" pathogens which presently originate from nosocomial infections. In continuation, D-aspartate ligase from E. faecium (Aslfm) as a prominent object for advancement of narrow-spectrum antibacterial agents which is active against multi-drug-resistant *E. faecium.* In addition, amino oxazole **6** resulting from the bacterial biotin-dependent carboxylase inhibitors exhibited little micro molar activity, and specifically, compound **8** prevents Aslfm with good activity [51]. In addition, isoxazole **7** bearing tosyloxy phenyl group was screened by the agar cup plate process for antibacterial activity, utilizing Ampicillin as a classic drug, the compound shown potential anti-*P. Aeruginosa* activity [52].

The 2-amino oxazole/4-substituted-phenyl oxazole was synthesized and assessed for antibacterial and antifungal potency. Some of the molecules have shown greater activity, i.e. zone of inhibition of 18 mm to 22 mm in Gram-positive bacteria, 16 mm to 18 mm in Gram-negative bacteria, and 16 mm to 19 mm in Fungi. It indicates that the oxazole moieties exhibit a strong antibacterial and antifungal activity [53].

Various substituted oxazoles were synthesized and assessed for antibacterial activity for various strains like *S. aureus, E. coli, P. vulgaris, K. pneumonia* and compared with the various standards like Ampicillin and Ciprofloxacin. Some of the compounds have shown more potent activity than Ampicillin. The zone of inhibition of the new compounds was found to be 25 mm, whereas Ampicillin is 20 mm in *S. aureus*. One compound exhibited a 21 mm inhibition zone, whereas Ampicillin was 22 mm in E. coli which indicates comparable activity. The third compound exhibited an inhibition zone of 22 mm, whereas Ampicillin was 20 mm in *P. vulgaris* and others exhibited 23 mm, whereas Ampicillin was 21 mm in *K. pneumonia*. But all the newly synthesized compounds have not exhibited comparable activity with Ampicillin. They exhibited good to moderate antibacterial activity [54].

3 Oxazoles and Their Clinical Drugs

Plenty of medicinal drugs containing oxazole-containing scaffolds have been widely used in the clinic, for example; Sulfisoxazole **9**, Furazolidone **10**, Toloxatone **11** and Linezolid **13** (Fig. 4) [55].

4 Antifungal Activity of Oxazole Derivatives

The installation of the oxazole motif into the indole frame, compound **15**, exhibited good activity against *Alternaria brassicicola*. Nevertheless, compound **14** comprising the oxadiazolyl group comparatively gave faint inhibitory action [56]. Streptochlorin **18** was originally synthesized from lipophilic extracts of Streptomyces sp. mycelium and a series of improved streptochlorin analogues such as **16**, **17**, **19** and **21** were tested for potency against seven phytopathogenic fungi. All of the compounds, on the other hand, demonstrated reasonable activity [57]. Additionally, 2-(4-ethyl-2-pyridyl)-1*H*-imidazole based 1,3,4-oxadiazole scaffolds (compounds **21**, **22**, **23**, and **24**) were verified against several fungal strains and compounds **22**, **23**

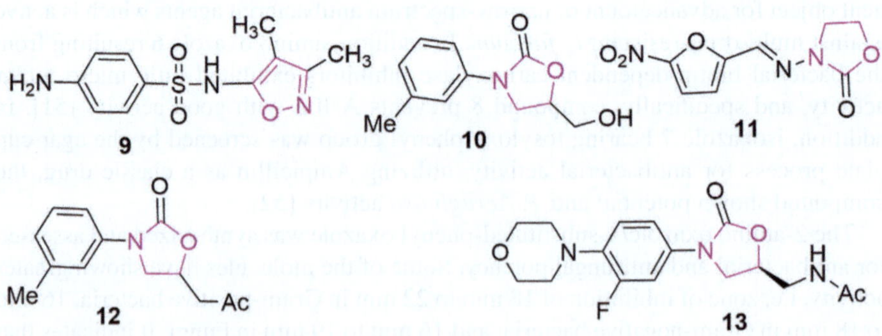

Fig. 4 Clinical oxazole drugs [55]

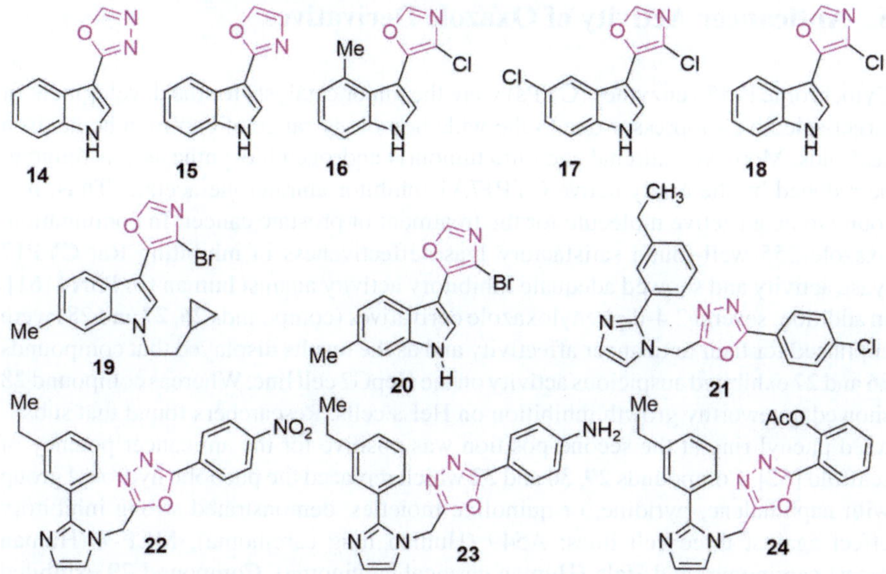

Fig. 5 Oxazole related antifungal agents [56-58]

and **24** showcased respectable antifungal activity in contrast with fluconazole (Fig. 5) [58].

The suppressors of the dual-specificity protein phosphatase CDC25C were discovered to be benzo[*d*]oxazsole-4,7-diones. Antifungal action of these moieties must be evaluated. Presence of arylamino, arylthio, or halogen groups will improve the antifungal activity. So, 5-arylamin-6-bromo-2-ethylbenzo[*d*]oxazole-4,7-diones and other compounds by several substituent were considered and evaluated for the antifungal activity. Two compounds were found to be more effective when associated with the standard drug, 5-Fluorocytosine. The MIC of two compounds was found to be 1.6 μg/ml and 0.8 μg/ml in Candida albicans Berkhout KCCM 50,235, 3.2 μg/ml and 3.2 μg/ml in Candida tropicalis Berkout KCCM 50,662, 3.2 μg/ml and 3.2 μg/ml in Candida krusei Berkhout KCCM 11,655, 1.6 μg/ml and1.6 μg/ml in Cryptococcus neoformans KCCM 50,564, 1.6 μg/ml and 0.8 μg/ml in Aspergillus niger KCTC 1231, 3.2 μg/ml and 1.6 μg/ml in Aspergillus flavus KCCM 11,899, whereas 5-Fluorocytosine was found to be 3.2 μg/ml, 3.2 μg/ml, 3.2 μg/ml, 3.2 μg/ml, 1.6 μg/ml and 1.6 μg/ml, respectively. This indicates that the Benzoxazole derivatives are highly potent antifungal agents [59].

Phenyl thiazole moiety is an insecticide. It contains phenyl thiazole in its structure. Replacement of this with its biosphere like oxazole/thiazole ring enhances the activity of the drug transforming the insecticide into fungicide. This indicates the role of oxazole in the antifungal activity exhibiting compounds [60].

5 Anticancer Activity of Oxazole Derivatives

Cytochrome P-450 enzymes (CYPs) were the major catalysts for the development of target-selective suppressor due to the wide homology range of common heme–iron scaffolds. Moreover, adrenal and intra tumoural androgen biosynthesis was found to be reduced by the orally active CYP17A1 inhibitor abiraterone acetate. Thus, it is found to be an active molecule for the treatment of prostate cancer. In continuation, oxazole 255 well-found satisfactory lyase effectiveness in inhibiting Rat CYP17 lyase activity and showed adequate inhibitory activity against human CYP3A4 [61]. In addition, several 2,4-diphenyloxazole derivatives (compounds **26**, **27** and **28**) were appraised for their anticancer affectivity and as the results displayed that compounds **26** and **27** exhibited auspicious activity on the HepG2 cell line. Whereas compound **28** showed noteworthy growth inhibition on HeLa cells. Researchers found that substituted phenyl ring at the second position was positive for the anticancer potency of scaffold [62]. Compounds **29**, **30** and **32** which replaced the phenolic hydroxyl group with naphthalene, pyridine, or quinoline moieties, demonstrated strong inhibitory effect against three cell lines: A549 (Human lung carcinoma), MCF-7 (Human breast carcinoma) and Hela (Human cervical carcinoma). Compound **29** exhibited exceptional inhibitory movement over MCF-7 (Human breast carcinoma) cell lines. Furthermore, structure–activity relationships (SAR) revealed that the C-5 position of the oxazole ring is linked to naphthalen-2-yl and quinolin-3-yl, which is important for 5-aryl-2-methyloxazole potency and selectivity [63]. In another, mono-substituted oxazole **33** afforded reduced activity in contrast with furan derivatives of transcription in transfected PC-3 cells, whereas 3,5-bis(trifluoromethyl)benzoyl aniline substituted compound **34** showed better activity (Fig. 6) [64].

In addition, MPS1 (protein kinase monopolar spindle 1) is a critical component of the spindle assembly checkpoint (SAC) signal, which is improperly expressed in a variety of human malignancies. Moreover, MPS1 is the topmost 25 genes which are over-expressed in tumours with chromosomal uncertainty. PTEN-deficient breast tumour cells are mainly reliant on MPS1 for survival, which makes it the remarkable target in oncology. In addition, the 1*H*-pyrrolo[3, 2-c]-pyridine moiety-based oxazoles (compounds **35** and **36**) verified potent as well as ligand-efficient binding to MPS1. The crystal structures of MPS1 with oxazoles **35** and **36** further revealed that the oxazole scaffold maintained connection with the active site Lys553 side chain [65]. In addition, a new 2,5-disubstituted oxazole **37** was isolated from Aspongopus chinensis (an insect from the Pentatomidae family) and exhibited noteworthy activity against various tumour cell lines. Multidrug resistance (MDR) is a complicated abnormality caused by the overexpression of transmembrane proteins from the ATP binding holder transporter family. P-glycoprotein (P-GP), one of these transporters, is frequently intertwined with MDR. In another, substituted naphthalenyl oxazole derivatives (compounds **31**, **38**, **39**, **40** and **41**) were measured as P-glycoprotein (P-GP) substrate as it encouraged ATP cell reduction. The SAR studies has shown that existence of F, H or OH substituents were encouraging for bioactivity, whereas Br was found to be contrary in the ATPase assay (Fig. 6) [66].

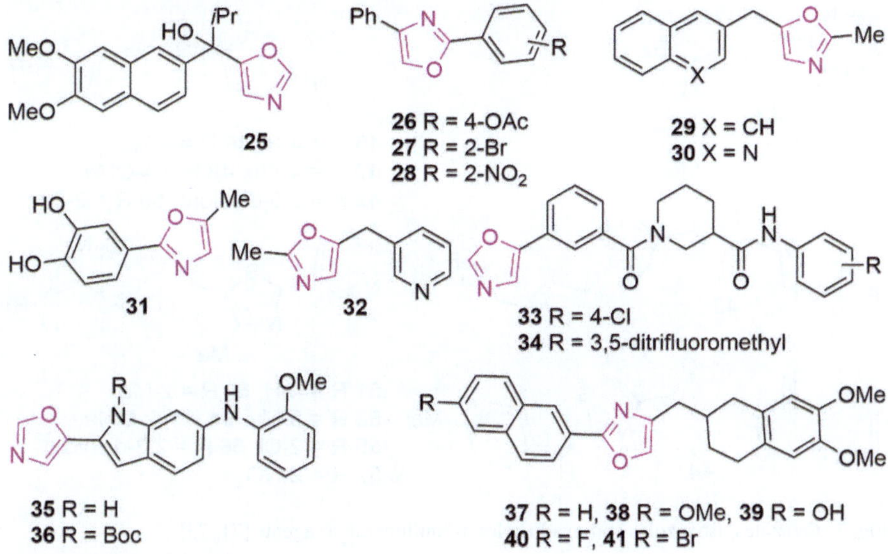

Fig. 6 Mono and bis-substituted oxazoles as anticancer agents [61–64]

Combrestatin A-4 (CA-4) exhibits an effective antitumour activity but the solubility issues limited its use in anticancer therapy. These drawbacks were resolved by the addition of imidazole and oxazole rings. In various cell lines, including human 518A2 melanoma, human HT-29 colon carcinoma and EA. HY926 endothelial hybrid cells, the substitution of a halogen atom on the oxazole ring improved anticancer activity [67]. Oxazole ring plays prominent role in anticancer activity.

Pongamol, a chalcone derivative isolated from *Derris indica,* has many pharmacological activities. One of the activities exhibited by it is the antitumour activity. There is no documented evidence of this traditional medicine. The oxazole ring addition to this natural drug enhances its antitumour activity. So pongamol derivatives of Oxazole and Pyrazole were synthesized and antitumour activity was estimated over three different human cancer cell lines, HeLa, IMR-32 and Jurkat. The oxazole derivative and the pyrazole derivative have shown increased activity compared to the natural drug Pongamol [68].

6 Antitubercular Activity of Oxazole Derivatives

Through the establishment of strains of drug-resistant of *Mycobacterium tuberculosis,* several attempts were made for the improvement of antitubercular agents possessing higher potency, less adverse effects/toxicity and less multi-drug resistances [69, 70]. Fascinated by the aminothiazole compounds showing excellent therapeutic index and dominant activity, amino oxazole compound **42** (Fig. 7) was

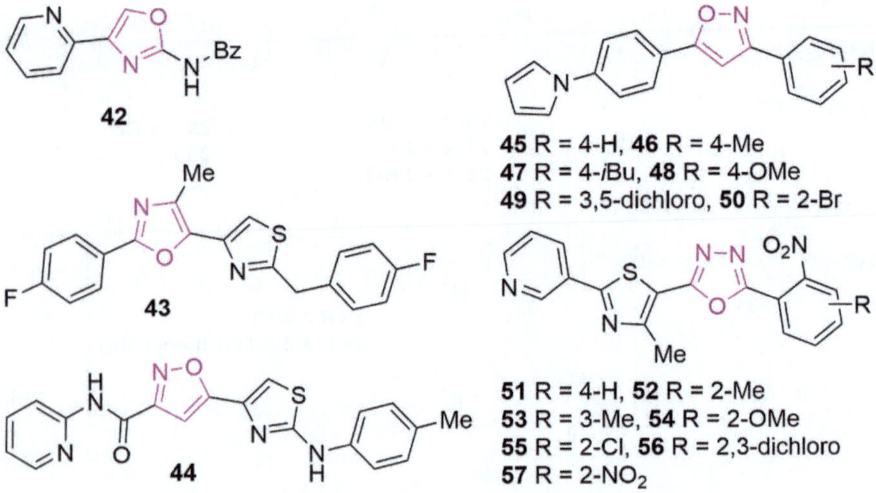

Fig. 7 Oxazoles, isoxazoles and oxadiazoles as antitubercular agents [71–73]

screened for biological evolution and found that compound **42** exhibited reasonable potency against *Mycobacterium tuberculosis* H37Rv. The replacement of oxazole scaffold through oxadiazole might lead into a 2-to-fourfold recovery in strength comparative to oxazole **42** [71]. In addition, the tri-substituted oxazole **43** with a thiazole ring was used, and equivalent action against inert Mycobacterium TB H37Ra and Mycobacterium bovis BCG strains was observed when compared to the conventional antibiotic Rifampicin. Additionally, Isoxazole **44** containing pyridine and thiazole rings was initiate to own high-inhibitory potency against vulnerable strains of *M. tuberculosis*. In another, Phenylisoxazoles (compounds **45**, **46**, **47**, **48**, **49** and **50**) bearing pyrrole rings were screened. The antitubercular action of these moieties displayed that compound **47** with electron-rich isopropyl group exhibited the notable potency for *M. tuberculosis* H37Rv strain, whereas moieties containing methyl & methoxy (compounds **46** & **48**) provided comparatively abridged activity. Nevertheless, unsubstituted compound **45** and electron-poor group substituted ones **49** and **50** displayed minimum bioactivity. Precisely, the most potent compound **47** exhibited a good safety profile against the A549 cell line [72].

2-Pyridinyl functionalized thiazolyl-5-aryl-1,3,4-oxadiazoles (compounds **51-57**) were evaluated in the search for more safe and effective anti-tubercular medicines. In addition, 2-phenyl substituted compounds **52**, **54**, **55** and **57** showed auspicious activity against Mycobacterium bovis BCG, and additionally, these compounds also exhibited little cytotoxicity over four human cancer cell lines (THP-1, HeLa, PANC-1, HCT116). Nevertheless, compound **51** with no substitution and other position substituted compounds **53** and **56** gave extremely reduced antitubercular activity [73].

Mycobacterial infections are the most common infectious disease globally. Furthermore, tuberculosis (TB) is witnessed amongst the top ten reasons of death worldwide. This is also the dominating fact of death from a single-infectious agent. The emergence of drug-resistant mycobacterial strains, which need the use of more toxic and less effective medications as well as therapy prolonging, is a source of particular concern. However, 4-methyl-2-aryl-5-(2-aryl/benzyl thiazol-4-yl) oxazole was synthesized and antitubercular activity was evaluated against various strains of M. tuberculosis HA37Ra (MTB, ATCC 25,177) and *M. bovi* s BCG (BCG, ATCC 35,743) in liquid medium using Rifampicin as a standard drug. In these series of compounds, five active compounds were found. It has been found that 3-Cl and 4-F substituted benzyl rings increased the antitubercular potency and some compounds exhibited comparable activity with Rifampicin. One of the molecules possess excellent antibacterial activity [69].

Oxazoles with carboxylic group substituents exhibit antitumour activity against *Mycobacterium tuberculosis* with low toxicity. One of the compounds exhibited anti-TB activity showing MIC of 0.07 & 0.14 mM against *Mycobacterium tuberculosis* and multi-drug-resistant *Mycobacterium tuberculosis* [74].

7 Anti-inflammatory and Analgesic Activity of Oxazole Derivatives

There is growing interest in the progress of specific inhibitors for FAAH (fatty acid amide hydrolase) that point to the cytosolic port Cys269 in medication for inflammatory, pain, or sleep disorders, due to the therapeutic potency of specific inhibitors for FAAH (fatty acid amide hydrolase) that point to the cytosolic port Cys269 in medication for inflammatory, pain, or sleep disorders. Oxazole bromide **58** (Fig. 8) showed noteworthy potency in inhibition of FAAH, whereas the compound **59** having nitrile gave slightly weak activity [75–83]. The irretrievable inhibitors of FAAH exhibited that predictable compassion to the position of the electrophile starter, but those were successfully showed amazing drifts in specious sensitivity towards Cys269 that would not be simply projected [76]. In continuation, tri-substituted oxazole compound **60** displayed excellent efficiency in frequent preclinical models together with the spinal nerve ligation (SNL) pain models and complete Freund's adjuvant (CFA). Moreover, no insightful properties were detected for this brain penetrant FAAH inhibitor, and compound **60** is found to be potent, specific reversible non-covalent modifying FAAH inhibitor [77]. In addition, Mofezolac **63** bearing isoxazole scaffold was an effective and selective COX-1 inhibitor and was broadly active in medication. Several considerations were focussed on its analogues to explore novel COX-1 inhibitors to overthrown its side effects [78]. In further, indole containing isoxazole **62** showed similar or more Secretory phospholipase A2 (sPLA2) inhibitory activity when compared to positive control Ursolic Acid. Additionally, this compound can bind the proximity of active site amino acid residues; HIS-47, TYR-21, GLY-22, PHE-5, GLY-29, CYS-44, CYS-28, PHE-98, ASP-48 and TYR-51. Furthermore, compound **62** is significant for correlations and binding capabilities with sPLA2 could be liable for its sPLA2-inhibitory effect [79]. In addition, in animal models of nonspecific inflammatory responses or Th1-type immune responses, isoxazole compound **61** proved an efficient anti-inflammatory drug. The use of compound **61** in the form of ointment is the added value in the management of skin inflammation [80]. On the other hand, compound **64**, 4,5-Diaryl-isoxazole-3-carboxylic acid, inhibits leukotriene biosynthesis and acts as an effective anti-inflammatory agent [81]. Tri-substituted indole derivative of isoxazole **65** [82], and 3,5-disubstituted isoxazole furfuryl derivative **66** [83], have also exhibited significant effect as anti-inflammatory agent.

Oxazole derivatives were synthesized, their anti-inflammatory effect was tested, and they were compared to the standard medicine nimesulide using the HRBC membrane stabilization technique. The percentage protection of the newly synthesized compound was in the values of 55.1, 50, 59 and 60%, and the nimesulide protection percentage was found to be 61% at 50 μg/ml. These compounds have exhibited comparable activity with the standard drug, nimesulide. But these compounds don't exhibit any increased anti-inflammatory or analgesic activity upon increasing the concentration to 100 μg/ml [84].

Quinolyl oxazoles with various substitutions were synthesized, and they are highly effective inhibitors of phosphodiesterase 4 (PDE4). Some of the compounds

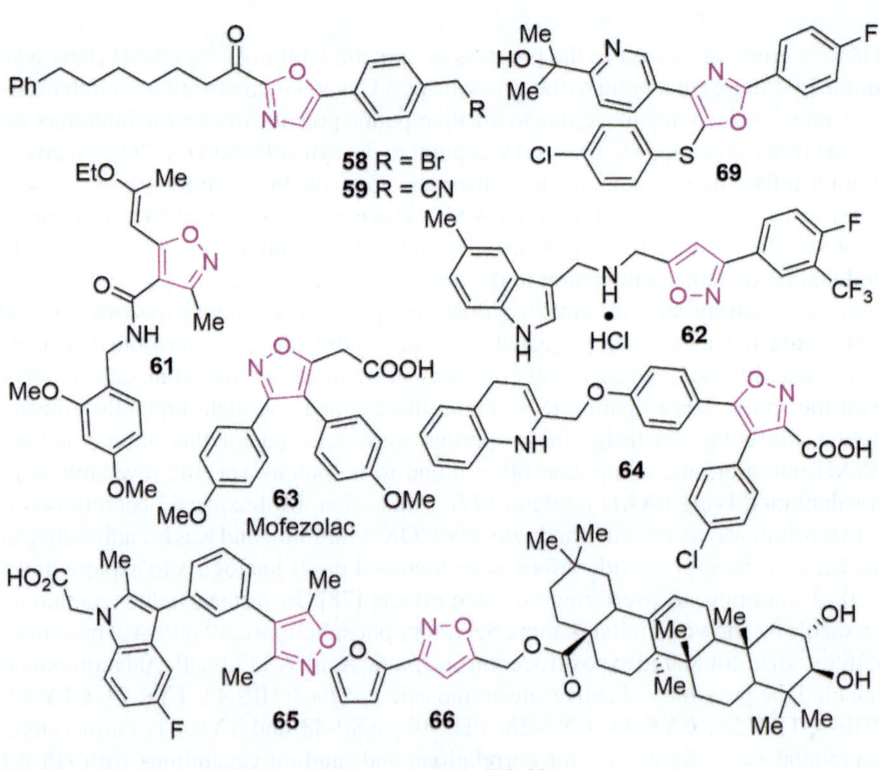

Fig. 8 Anti-inflammatory and analgesic compounds [75–83]

were considered to be more potent against PDE4 IC_{50} values of 1 to 1.4 nm. *N*-benzylcarboxamide has shown the highest selectivity against phosphodiesterase 4. Further optimization led to highly selective PDE4 inhibitors with picomolar potency with the values of 0.05, 0.03, 0.06 and 0.04 nm. This data shows that the oxazole ring containing compounds exhibit good anti-inflammatory activity [67].

8 Antidiabetic Activity of Oxazole Derivatives

The G-protein coupled receptor 40 (GPR40) is broadly populated in pancreatic β cells and recognizes endogenous fatty acids, leading in an increase in insulin output when glucose levels are high [85–87]. Furthermore, compound **67** consisting of Isoxazole (Fig. 9) showed better impact as GPR40 agonist, exceptional pharmacokinetic possessions across species, and minimum central nervous system (CNS) dispersion. OGTT study in human GPR40 knock-in mice showed this compound decreases the plasma glucose levels [88]. In continuation, bis-substituted isoxazole **68** containing thiophene moiety increases the assembly of mRNAs encrypting a select group of β cell proteins vital for glucose detecting and insulin gene transcription [89]. In addition, the tri-substituted oxazole compound **69** was displayed noteworthy activity on the GPR40 receptor. In continuation, arylsulfonyl 3-(pyridin-2-yloxy) aniline compound **70** consisting 1,2,4-oxadiazole scaffold showed exciting activity as GPR119 agonist [90].

PPARs (peroxisome proliferator-activated receptors) are also important therapeutic targets for Type 2 Diabetes Mellitus therapy (DMT2). Furthermore, most existing PPAR ligands comprise a thiazolidinedione (TZD) structure or a carboxylic acid (CA) that is crucial for activity. Furthermore, the 1,2,4-Oxadiazole compound **71** might bind to Peroxisome proliferator-activated receptors (PPARα and PPARδ) via an acetamide scaffold and an adjacent methyl group. In addition, skeletal muscle

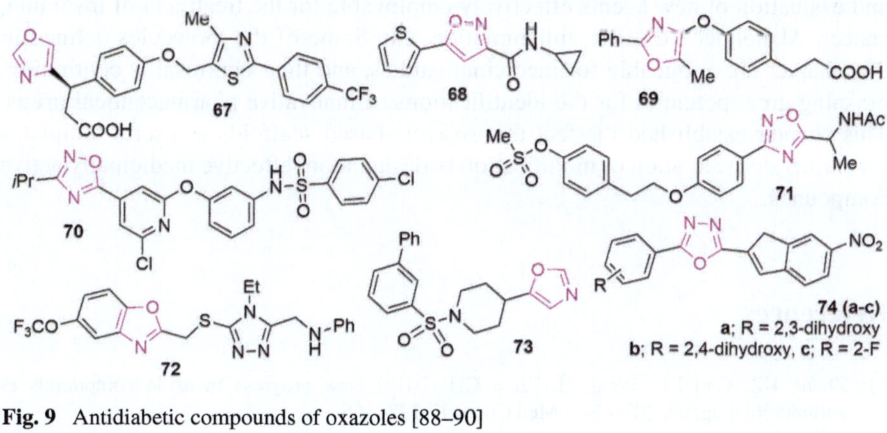

Fig. 9 Antidiabetic compounds of oxazoles [88–90]

significantly interesting target tissues are extensively working for the treatment of insulin resistance. It is worth mentioning that one of the most imperative protein targets for insulin resistance is fatty acid transport protein 1 (FATP1), a member of Acyl-CoA synthetase. FATP1 is a transmembrane protein and highly over-expressed in skeletal muscle. Thus, inhibition of FATP1 could lead to the interruption of free fatty acid transport in insulin resistant cells. Fascinatingly, compound 71 contains benzoxazole scaffold has shown significant FATP1 inhibition in mouse and human thus reflected as a potential candidate for free fatty acid regulator in insulin resistant condition [91]. In continuation, oxadiazole compounds **74a**, **74b** and **74c** were found to be more active with antidiabetic activity in contrast with the typical drug Acarbose [92]. Furthermore, suppression of the intracellular enzyme 11-beta-hydroxysteroid dehydrogenase type 1 (11βHSD1) has been proposed as a therapeutic strategy for the treatment of DMT2. The compound **73** comprising 5-oxazole with piperidyl ring at fourth position was found to express stimulating inhibitory action against the 11β-HSD1 enzyme [93].

The 1,3-dioxane carboxylic acid derivatives were synthesized and they act as dual agonists for PPAR α/ϒ. As a lipophilic heterocyclic tail, substituted oxazole must be incorporated. These compounds were produced and tested in animal models for their agonistic activity on the PPAR receptor, as well as hypoglycaemic and hypolipidemic effectiveness. One of the compounds of this series exhibited potent hypoglycaemic, hypolipidemic and insulin-sensitizing effects [94].

9 Summary/Conclusion

Oxazole-based scaffold and its analogues are well-accepted as promising moieties in the development and progress of novel drugs revealing immense biological and pharmaceutical activities. It has been acknowledged from the foregoing deliberations that various oxazole-embedded molecules can have appreciable attention in the synthesis and evaluation of new agents effectively employable for the treatment of insomnia, cancer, Alzheimer's disease, inflammation, etc. Some of the molecules defined in this chapter are applicable for medicinal studies, and their appraisal is continuing, grasping great potential for the identifications of innovative pharmaceutical drugs. This chapter established the fact that oxazole-based scaffolds as useful templates for further derivatization or modification to design more effective medicinally active compounds.

References

1. Zhang HZ, Gan LL, Wang H, Zhou CH (2017) New progress in azole compounds as antimicrobial agents. Mini-Rev Med Chem 17:122–166

2. Peng XM, Cai GX, Zhou CH (2013) Recent developments in azole compounds as antibacterial and antifungal agents. Curr Top Med Chem 13:1963–2010
3. Kean WF (2004) Oxaprozin: kinetic and dynamic profile in the treatment of pain. Curr Med Res Opin 20:1275–1277
4. Perner RJ, Koenig JR, Didomenico S, Gomtsyan A, Schmidt RG, Lee CH, Hsu M, McDonald HA, Gauvin DM, Joshi S, Turner TM, Reilly RM, Kym PR, Kort ME (2010) Synthesis and biological evaluation of 5-substituted and 4,5-disubstituted-2-arylamino oxazole TRPV1 antagonists. Bioorg Med Chem 18:4821–4829
5. Dominguez XA, de la Fuente G, Gonzalez AG, Reina M, Timon I (1988) Two new oxazoles from amyris texana p wilson. Heterocycles 27:35–38
6. Bell FW, Cantrell AS, Hoegberg M, Jaskunas SR, Johansson NG, Jordan CL, Kinnick MD, Lind P, Morin J, Noreen JR, Oberg B, Palkowitz JA, Parrish CA, Pranc P, Sahlberg C, Ternansky RJ, Vasileff RT, Vrang L, West SJ, Zhang H, Zhou XX (1995) Phenethylthiazolethiourea (PETT) compounds, a new class of HIV-1 reverse transcriptase inhibitors 1 synthesis and basic structure-activity relationship studies of PETT analogs. J Med Chem 38:4929–4936
7. Ibrahimi BE, Guo L (2020) Azole-based compounds as corrosion inhibitors for metallic materials. IntechOpen. https://doi.org/10.5772/intechopen.93040
8. Tang JS, Verkade JG (1996) Chiral auxiliary-bearing isocyanides as synthons: synthesis of strongly fluorescent (+)-5-(3,4-dimethoxyphenyl)-4-[[N- [(4S)-2-oxo-4-(phenylmethyl)-2-oxazolidinyl]]carbonyl]oxazole and its enantiomer. J Org Chem 61:8750–8754
9. Zhao Q, Liu S, Li Y, Wang QJ (2009) Design, synthesis, and biological activities of novel 2-cyanoacrylates containing oxazole, oxadiazole, or quinoline moieties. J Agric Food Chem 57:2849–2855
10. Min J, Lee JW, Ahn Y-H, Chang Y-T (2007) Combinatorial dapoxyl dye library and its application to site selective probe for human serum albumin. J Comb Chem 9:1079–1083
11. Grotkopp O, Ahmad A, Frank W, Muller TJ (2011) Blue-luminescent 5-(3-indolyl)oxazoles via microwave-assisted three-component coupling-cycloisomerization-Fischer indole synthesis. Org Biomol Chem 9:8130–8140
12. Turchi IJ, Dewar MJS (1975) Chemistry of oxazoles. Chem Rev 75:389–437
13. Zhu J, Mo J, Lin H-Z, Chen Y, Sun H-P (2018) The recent progress of isoxazole in medicinal chemistry. Bioorg Med Chem 26:3065–3075
14. Biernacki K, Daśko M, Ciupak O, Kubiński K, Rachon J, Demkowicz S (2020) Novel 1,2,4-oxadiazole derivatives in drug discovery. Pharmaceuticals 13:111–155
15. Bai Y, Chen W, Chen H, Huang H, Xiaoa F, Deng G-J (2015) Copper-catalyzed oxidative cyclization of arylamides and β-diketones: new synthesis of 2,4,5-trisubstituted oxazoles. RSC Adv 5:8002–8005
16. Glöckner S, Tran DN, Ingham RJ, Fenner S, Wilson ZE, Battilocchio C, Ley SV (2014) The rapid synthesis of oxazolines and their heterogeneous oxidation to oxazoles under flow condition. Org Biomol Chem 13:207–214
17. Ibrar A, Khan I, Abbas N, Farooq U, Khan A (2016) Transition-metal-free synthesis of oxazoles: valuable structural fragments in drug discovery. RSC Adv 6:93016–93047
18. Srivastava BK, Soni R, Joharapurkar A, Sairam KV, Patel JZ, Goswami A, Shedage SA, Kar SS, Salunke RP, Gugale SB, Dhawas A, Kadam P, Mishra B, Sadhwani N, Unadkat VB, Mitra P, Jain MR, Patel PR (2008) Bioisosteric replacement of dihydropyrazole of 4S-(-)-3-(4-chlorophenyl)-N-methyl-N'-[(4-chlorophenyl)-sulfonyl]-4-phenyl-4,5-dihydro-1H-pyrazole-1-caboxamidine (SLV-319) a potent CB1 receptor antagonist by imidazole and oxazole. Bioorg Med Chem Lett 18:963–968
19. Kaspadya M, Narayanaswamy VK, Raju M, Rao GK (2009) Synthesis, antibacterial activity of 2,4-disubstituted oxazoles and thiazoles as bioisosteres. Lett Drug Des Discovery 6:21–28
20. Zhou CH, Wang Y (2012) Recent researches in triazole compounds as medicinal drugs. Curr Med Chem 19:239–280
21. Meanwell NA (2013) The influence of bioisosteres in drug design: tactical applications to address developability problems. In: Meanwell N (eds) Tactics in contemporary drug design. Topics in Medicinal Chemistry, vol 9. Springer, Berlin, Heidelberg

22. Dai LL, Cui SF, Damu GLV, Zhou CH (2013) Recent advances in the synthesis and application of tetrazoles. Chin J Org Chem 33:224–244
23. Mayer JCP, Sauer AC, Iglesias BA, Acunha TV, Back DF, Rodrigues OED, Dornelles L (2017) Ferrocenylethenyl-substituted 1,3,4-oxadiazolyl-1,2,4-oxadiazoles: synthesis, characterization and DNA-binding assays. J Organomet Chem 841:1–11
24. Sysak A, Obmińska-Mrukowicz B (2017) Isoxazole ring as a useful scaffold in a search for new therapeutic agents. Eur J Med Chem 137:292–309
25. Demmer DS, Bunch L (2015) Benzoxazoles and oxazolopyridines in medicinal chemistry studies. Eur J Med Chem 97:778–785
26. Lindquist N, Fenical W, Van Duyne GD, Clardy J (1991) Isolation and structure determination of diazonamides A and B, unusual cytotoxic metabolites from the marine ascidian Diazona chinensis. J Am Chem Soc 113:2303–2304
27. Cruz-Monserrate HC, Vervoort RL, Bai D, Newman SB, Howell G, Los JT, Mullaney MD, Williams GR, Fenical PW, Hamel E (2003) Mol Pharmacol 63:1273–1280
28. Kobayashi J, Tsuda M, Fuse H, Sasaki T, Mikami Y (1997) Halishigamides A-D, New cytotoxic oxazole-containing metabolites from okinawan sponge *Halichondria* sp. J Nat Prod 60:150–154
29. Azman AM, Mullins RJ (2013) Heterocyclic chemistry in drug discovery. In: Li JJ (ed) New York, Wiley, 231p
30. Cocito C (1979) Antibiotics of the virginiamycin family, inhibitors which contain synergistic components. Microbiol Rev 43:145–192
31. Moura KCG, Carneiro PF, Pinto MC, da Silva JA, Malta VR, de Simone CA, Dias GG, Jardim GA, Cantos J, Coelho TS, da Silva PE, da Silva Jr EN (2012) 1,3-Azoles from ortho-naphthoquinones: synthesis of aryl substituted imidazoles and oxazoles and their potent activity against Mycobacterium tuberculosis. Bioorg Med Chem 20:6482–6488
32. Antonio J, Molinski TF (1993) Screening of marine invertebrates for the presence of ergosterol-sensitive antifungal compounds. J Nat Prod 56:54–61
33. Ryu CK, Lee RY, Kim NY, Kim YH, Song AL (2009) Synthesis and antifungal activity of benzo[d]oxazole-4,7-diones. Bioorg Med Chem Lett 19:5924–5926
34. Desroy N, Moreau F, Briet S, Le Fralliec G, Floquet S, Durant L, Vongsouthi V, Gerusz V, Denis A, Escaich S (2009) Towards Gram-negative antivirulence drugs: new inhibitors of HldE kinase. Bioorg Med Chem 17:1276–1289
35. Heng S, Gryncel KR, Kantrowitz ER (2009) A library of novel allosteric inhibitors against fructose 1,6-bisphosphatase. Bioorg Med Chem 17:3916–3922
36. Copp BR (2003) Antimycobacterial natural products. Nat Prod Rep 20:535–557
37. Lack NA, Axerio-Cilies P, Tavassoli P, Han FQ, Chan KH, Feau C, LeBlanc E, Guns ET, Guy RK, Rennie PS, Cherkasov A (2011) Targeting the binding function 3 (BF3) site of the human androgen receptor through virtual screening. J Med Chem 54:8563–8573
38. Hashimoto H, Imamura K, Haruta J, Wakitani K (2002) 4-(4-cycloalkyl/aryl-oxazol-5-yl)benzenesulfonamides as selective cyclooxygenase-2 inhibitors: enhancement of the selectivity by introduction of a fluorine atom and identification of a potent, highly selective, and orally active COX-2 inhibitor JTE-522(1). J Med Chem 45:1511–1517
39. Momose Y, Maekawa T, Yamano T, Kawada M, Odaka H, Ikeda H, Sohda T (2002) Novel 5-substituted 2,4-thiazolidinedione and 2,4-oxazolidinedione derivatives as insulin sensitizers with antidiabetic activities. J Med Chem 45:1518–1534
40. Brown P, Davies DT, O'Hanlon PJ, Wilson JM (1996) The chemistry of Pseudomonic acid. 15. Synthesis and antibacterial activity of a series of 5-alkyl, 5-alkenyl, and 5-heterosubstituted oxazoles. J Med Chem 39:446–457
41. Yang WS, Shimada K, Delva D, Patel M, Ode E, Skouta R, Stockwell BR (2012) Identification of simple compounds with microtubule-binding activity that inhibit cancer cell growth with high potency. ACS Med Chem Lett 3:35–38
42. Shaw AY, Henderson MC, Flynn G, Samulitis B, Han H, Stratton SP, Chow HH, Hurley LH, Dorr RT (2009) Characterization of novel diaryl oxazole-based compounds as potential agents to treat pancreatic cancer. J Pharmacol Exp Ther 331:636–647

43. Wishart DS, Feunang YD, Guo AC, Lo EJ, Marcu A, Grant JR, Sajed T, Johnson D, Li C, Sayeeda Z, Assempour N, Iynkkaran I, Liu Y, Maciejewski A, Gale N, Wilson A, Chin L, Cummings R, Le D, Pon A, Knox C, Wilson M (2018) DrugBank 5.0: a major update to the DrugBank database for 2018. Nucleic Acids Res 46:D1074–D1082
44. Janardhanan J, Chang M, Mobashery S (2016) The oxadiazole antibacterials. Curr Opin Microbiol 33:13–17
45. Ding D, Boudreau MA, Leemans E, Spink E, Yamaguchi T, Testero SA, O'Daniel PA, Lastochkin E, Chang M, Mobashery S (2015) Exploration of the structure–activity relationship of 1,2,4-oxadiazole antibiotics. Bioorg Med Chem Lett 25:4854–4857
46. Stokes NR, Baker N, Bennett JM, Chauhan PK, Collins I, Davies DT, Gavade M, Kumar D, Lancett P, Macdonald R, MacLeod L, Mahajan A, Mitchell JP, Nayal N, Nayal YN, Pitt GRW, Singh M, Yadav A, Srivastava A, Czaplewski LG, Haydon DJ (2014) Design, synthesis and structure–activity relationships of substituted oxazole-benzamide antibacterial inhibitors of FtsZ. Bioorg Med Chem Lett 24:353–359
47. Li N, Xu Y, Xia Q, Bai C, Wang T, Wang L, He D, Xie N, Li L, Wang J, Zhou HG, Xu F, Yang C, Zhang Q, Yin Z, Guo Y, Chen Y (2014) Simplified captopril analogues as NDM-1 inhibitors. Bioorg Med Chem Lett 24:386–389
48. Patel D, Kumari P, Patel NB (2017) Synthesis and biological evaluation of coumarin based isoxazoles, pyrimidinthiones and pyrimidin-2-ones. Arab J Chem 10:3990–4001
49. Tomi IHR, Tomma JH, Al-Daraji AHR, Al-Dujaili AH (2015) Synthesis, characterization and comparative study the microbial activity of some heterocyclic compounds containing oxazole and benzothiazole moieties. J Saudi Chem Soc 19:392–398
50. Lal J, Gupta SK, Thavaselvam D, Agarwal DD (2013) Biological activity, design, synthesis and structure activity relationship of some novel derivatives of curcumin containing sulphonamides. Eur J Med Chem 64:579–588
51. Škedelj V, Perdih A, Brvar M, Kroflič A, Dubbée V, Savage V, O'Neill AJ, Solmajer T, Bešter-Rogač M, Blanot D, Hugonnet JE, Magnet S, Arthur M, Mainardi JL, Stojan J, Zega A (2013) Discovery of the first inhibitors of bacterial enzyme D-aspartate ligase from Enterococcus faecium (Aslfm). Eur J Med Chem 67:208–220
52. Kendre BV, Landge MG, Bhusare SR (2015) Synthesis and biological evaluation of some novel pyrazole, isoxazole, benzoxazepine, benzothiazepine and benzodiazepine derivatives bearing an aryl sulfonate moiety as antimicrobial and anti-inflammatory agents. Arab J Chem 12:2091–2097
53. Rawat BS, Shukla SK (2016) Synthesis and evaluation of some new thiazole/oxazole derivatives for their biological activities. World J Pharm Pharm Sci 5:1473–1482
54. Kakkar S, Narasimhan B (2019) A comprehensive review on biological activities of oxazole derivatives. BMC Chem 13:16
55. Zhang HZ, GanL L, Wang H, Zhou CH (2017) New progress in azole compounds as antimicrobial agents. Mini-Rev Med Chem 17:122–166
56. Pedras MSC, Abdoli A (2013) Metabolism of the phytoalexins camalexins, their bioisosteres and analogues in the plant pathogenic fungus Alternaria brassicicola. Bioorg Med Chem 21:4541–4549
57. Zhang MZ, Chen Q, Xie CH, Mulholland N, Turner S, Irwin D, Gu YC, Yang GF, Clough J (2015) Synthesis and antifungal activity of novel streptochlorin analogues. Eur J Med Chem 92:776–783
58. Wani MY, Ahmad A, Shiekh RA, Al-Ghamdi KJ, Sobral AJFN (2015) Imidazole clubbed 1,3,4-oxadiazole derivatives as potential antifungal agents. Bioorg Med Chem 23:4172–4180
59. Ryu C, Lee R, Kim NY, Kim YH, Song AL (2009) Synthesis and antifungal activity of benzo[d]oxazole-4,7-diones. Bioorg Med Chem Lett 19:5924–5926
60. Yan Z, Liu A, Huang M, Liu M, Pei H, Huang L, Yi H, Liu W, Hu A (2018) Design, synthesis, DFT study and antifungal activity of the derivatives of pyrazolecarboxamide containing thiazole or oxazole ring. Eur J Med Chem 149:170–181
61. Rafferty SW, Eisner JR, Moore WR, Schotzinger RJ, Hoekstra WJ (2014) Highly-selective 4-(1,2,3-triazole)-based P450c17a 17,20-lyase inhibitors. Bioorg Med Chem Lett 24:2444–2447

62. Mathew JE, Divya G, Vachala SD, Mathew JA, Jeyaprakash RS (2013) Synthesis and characterization of novel 2,4-diphenyloxazole derivatives and evaluation of their in vitro antioxidant and anticancer activity. J Pharm Res 6:210–213
63. Wu C, Liang ZW, Xu YY, He WM, Xiang JN (2013) Gold-catalyzed oxazoles synthesis and their relevant antiproliferative activities. Chin Chem Lett 24:1064–1066
64. Bell JL, Haak AJ, Wade SM, Kirchhoff PD, Neubig RR, Larsen SD (2013) Optimization of novel nipecotic bis(amide) inhibitors of the Rho/MKL1/SRF transcriptional pathway as potential anti-metastasis agents. Bioorg Med Chem Lett 23:3826–3832
65. Naud S, Westwood IM, Faisal A, Sheldrake P, Bavetsias V, Atrash B, Cheung KMJ, Liu M, Hayes A, Schmitt J, Wood A, Choi V, Boxall K, Mak G, Gurden M, Valenti M, Brandon AH, Henley A, Baker R, McAndrew C, Matijssen B, Burke R, Hoelder S, Eccles SA, Raynaud FI, Linardopoulos S, Montfort RLM, Blagg J (2013) Structure-based design of orally bioavailable 1H-Pyrrolo[3,2-c]pyridine inhibitors of mitotic kinase monopolar spindle 1 (MPS1). J Med Chem 56:10045–10065
66. Colabufo NA, Contino M, Cantore M, Capparelli E, Perrone MG, Cassano G, Gasparre G, Leopoldo M, Berardi F, Perrone R (2013) Naphthalenyl derivatives for hitting P-gp/MRP1/BCRP transporters. Bioorg Med Chem 21:1324–1332
67. Romagnoli R, Baraldi PG, Prencipe F, Oliva P, Baraldi S, Salvador MK, Lopez-Cara LS, Brancale A, Ferla S, Hamel E, Ronca R, Bortolozzi R, Mariotto E, Porcù E, Basso G, Viola G (2017) Synthesis and biological evaluation of 2-methyl-4,5-disubstituted oxazoles as a novel class of highly potent antitubulin agents. Sci Rep 7:46356
68. Rao PR, Chaturvedi V, Babu KS, Reddy PP, Rao VRS, Sreekanth P, Sreedhar AS, Rao JM (2012) Synthesis and anticancer effects of pongamol derivatives on mitogen signaling and cell cycle kinases. Med Chem Res 21:634–641
69. Abhale YK, Sasane AV, Chavan AP, Shekh SH, Deshmukh KK, Bhansali S, Nawale L, Sarkar D, Mhaske PC (2017) Synthesis and antimycobacterial screening of new thiazolyl-oxazole derivatives. Eur J Med Chem 132:333–340
70. Azzali E, Machado D, Kaushik A, Vacondio F, Flisi S, Cabassi CS, Lamichhane G, Viveiros M, Costantino G, Pieroni M (2017) Substituted N-phenyl-5-(2-(phenylamino)thiazol-4-yl)isoxazole-3-carboxamides are valuable antitubercular candidates that evade innate efflux machinery. J Med Chem 60:7108–7122
71. Meissner A, Boshoff HI, Vasan M, Duckworth BP, Barry CE, Aldrich CC (2013) Structure-activity relationships of 2-aminothiazoles effective against Mycobacterium tuberculosis. Bioorg Med Chem 21:6385–6397
72. Joshi SD, Dixit SR, Kirankumar MN, Aminabhavi TM, Raju KVSN, Narayan R, Lherbet C, Yang KS (2013) Synthesis, antimycobacterial screening and ligand-based molecular docking studies on novel pyrrole derivatives bearing pyrazoline, isoxazole and phenyl thiourea moieties. Bioorg Med Chem 21:6385–6397
73. Dhumal ST, Deshmukh AR, Bhosle MR, Khedkar VM, Nawale LU, Sarkar D, Mane RA (2016) Synthesis and antitubercular activity of new 1,3,4-oxadiazoles bearing pyridyl and thiazolyl scaffolds. Bioorg Med Chem Lett 26:3646–3651
74. Asif M (2016) A brief overview on recent advances in the development of anti-tubercular compounds containing different heterocyclic ring systems. Appl Micro Open Access 2
75. Otrubova K, Brown M, McCormick MS, Han GW, O'Neal ST, Cravatt BF, Stevens RC, Lichtman AH, Boger DL (2013) Rational design of fatty acid amide hydrolase inhibitors that act by covalently bonding to two active site residues. J Am Chem Soc 135:6289–6299
76. Otrubova K, Cravatt BF, Boger DL (2014) Design, synthesis, and characterization of α-ketoheterocycles that additionally target the cytosolic port Cys269 of fatty acid amide hydrolase. J Med Chem 57:1079–1089
77. Chobanian HR, Guo Y, Liu P, Chioda MD, Fung S, Lanza TJ, Chang L, Bakshi RK, Dellureficio JP, Hong Q, McLaughlin M, Belyk KM, Krska SW, Makarewicz AK, Martel EJ, Leone JF, Frey L, Karanam B, Madeira M, Alvaro R, Shuman J, Salituro G, Terebetski JL, Jochnowitz N, Mistry S, McGowan E, Hajdu R, Rosenbach M, Abbadie C, Alexander JP, Shiao LL, Sullivan KM, Nargund RP, Wyvratt MJ, Lin LS, DeVita RJ (2014) Discovery of MK-4409, a novel

oxazole FAAH inhibitor for the treatment of inflammatory and neuropathic pain. ACS Med Chem Lett 5:717–721

78. Perrone M. G., Vitale P., Panella A., Fortuna C. G., and Scilimati A. (2015) General role of the amino and methylsulfamoyl groups in selective cyclooxygenase (COX)-1 inhibition by 1,4-diaryl-1,2,3-triazoles and validation of a predictive pharmacometric PLS model. Eur J Med Chem 94:252–264
79. Pedada SR, Yarla NS, Tambade PJ, Dhananjaya BL, Bishayee A, Arunasree KM, Philip GH, Dharmapuri G, Aliev G, Putta S, Rangaiah G (2016) Synthesis of new secretory phospholipase A2-inhibitory indole containing isoxazole derivatives as anti-inflammatory and anticancer agents. Eur J Med Chem 112:289–297
80. Mączyński M, Artym J, Kocięba M, Kochanowska I, Ryng S, Zimecki M (2016) Anti-inflammatory properties of an isoxazole derivative-MZO-2. Pharmacol Rep 68:894–902
81. Banoglu E, Çelikoğlu E, Völker S, Olgaç A, Gerstmeier J, Garscha U, Çalışkan B, Schubert US, Carotti A, Macchiarulo A, Werz O (2016) 4,5-Diarylisoxazol-3-carboxylic acids: A new class of leukotriene biosynthesis inhibitors potentially targeting 5-lipoxygenase-activating protein (FLAP). Eur J Med Chem 113:1–10
82. Alonso JA, Andrés M, Bravo M, Buil MA, Calbet M, Castro J, Eastwood PR, Esteve C, Ferrer M, Forns P, Gómez E, González J, Lozoya E, Mir M, Moreno I, Petit S, Roberts RS, Sevilla S, Vidal B, Vidal L, Vilaseca P (2014) Structure–activity relationships (SAR) and structure–kinetic relationships (SKR) of bicyclic heteroaromatic acetic acids as potent CRTh2 antagonists III: The role of a hydrogen-bond acceptor in long receptor residence times. Bioorg Med Chem Lett 24:5127–5133
83. Chouaïb K, Romdhane A, Delemasure S, Dutartre P, Elie N, Touboul D, Ben jannet H, Hamza MA (2016) Regiospecific synthesis, anti-inflammatory and anticancer evaluation of novel 3,5-disubstituted isoxazoles from the natural maslinic and oleanolic acids. Ind Crop Prod 85:287–299
84. Patil VC, Bhusari KP, Khedekar PB (2014) Synthesis and anti-inlfammatory activity of some 3-substituted thiazolyl and oxazolyl indole derivatives. World J Pharm Pharm Sci 3:1292–1306
85. Zahanich I, Kondratov I, Naumchyk V, Kheylik Y, Platonov M, Zozulya S, Krasavin M (2015) Phenoxymethyl 1,3-oxazoles and 1,2,4-oxadiazoles as potent and selective agonists of free fatty acid receptor 1 (GPR40). Bioorg Med Chem Lett 25:3105–3111
86. Taha M, Ismail NH, Imran S, Wadood A, Rahim F, Saad SM, Khan KM, Nasir A (2016) Synthesis, molecular docking and a-glucosidase inhibition of 5-aryl-2-(6¢-nitrobenzofuran-2¢-yl)-1,3,4-oxadiazoles. Bioorg Chem 66:117–123
87. Shioi R, Okazaki S, Noguchi-Yachide T, Ishikawa M, Makishima M, Hashimoto Y, Yamaguchi T (2017) Switching subtype-selectivity: Fragment replacement strategy affords novel class of peroxisome proliferator-activated receptor a/d (PPARa/d) dual agonists. Bioorg Med Chem Lett 27:3131 3134
88. Liu J, Wang Y, Ma Z, Schmitt M, Zhu L, Brown SP, Dransfield PJ, Sun Y, Sharma R, Guo Q, Zhuang R, Zhang J, Luo J, Tonn GR, Wong S, Swaminath G, Medina JC, Lin DCH, Houze JB (2014) Optimization of GPR40 agonists for type 2 diabetes. ACS Med Chem Lett 5:517–521
89. Kalwat MA, Huang Z, Wichaidit C, McGlynn K, Earnest S, Savoia C, Dioum EM, Schneider JW, Hutchison MR, Cobb MH (2016) Isoxazole alters metabolites and gene expression, decreasing proliferation and promoting a neuroendocrine phenotype in β-cells. ACS Chem Biol 11:1128–1136
90. Zhang J, Li AR, Yu M, Wang Y, Zhu J, Kayser F, Medina JC, Siegler K, Conn M, Shan B, Grillo MP, Eksterowicz J, Coward P, Liu J (2013) Discovery and optimization of arylsulfonyl 3-(pyridin-2-yloxy)anilines as novel GPR119 agonists. Bioorg Med Chem Lett 23:3609–3613
91. Matsufuji T, Ikeda M, Naito A, Hirouchi M, Kanda S, Izumi M, Harada J, Shinozuka T (2013) Arylpiperazines as fatty acid transport protein 1 (FATP1) inhibitors with improved potency and pharmacokinetic properties. Bioorg Med Chem Lett 23:2560–2565
92. Taha M, Rahim F, Imran S, Ismail NH, Ullah H, Selvaraj M, Javid MT, Salar U, Ali M, Khan KM (2017) Synthesis, α-glucosidase inhibitory activity and in silico study of tris-indole hybrid scaffold with oxadiazole ring: As potential leads for the management of type-II diabetes mellitus. Bioorg Chem 74:30–40

93. Yoon DS, Wu SC, Seethala R, Golla R, Nayeem A, Everlof JG, Gordon DA, Hamann LG, Robl JA (2014) Discovery of pyridyl sulfonamide 11-beta-hydroxysteroid dehydrogenase type 1 (11β-HSD1) inhibitors for the treatment of metabolic disorders. Bioorg Med Chem Lett 24:5045–5049
94. Pingali H, Jain M, Shah S, Makadia P, Zaware P, Goel A, Patel M, Giri S, Patel H, Patel P (2008) Design and synthesis of novel oxazole containing 1,3-Dioxane-2-carboxylic acid derivatives as PPAR a/c dual agonists. Bioorg Med Chem 16:7117–7127

Chapter 11
An Overview on Biological Activities of 1,2,3-Triazole Derivatives

Arup K. Kabi, Sattu Sravani, Raghuram Gujjarappa, Aakriti Garg, Nagaraju Vodnala, Ujjawal Tyagi, Dhananjaya Kaldhi, Virender Singh, Sreya Gupta, and Chandi C. Malakar

Abbreviations

NSAID	Non-steroidal Anti-inflammatory Drugs
HIV	Human Immunodeficiency Virus
RSV	Respiratory Syncytial Virus
DNA	Deoxyribonucleic Acid
EGFR	Endothelial Growth Factor Receptor
COX	Cyclooxygenase
AA	Arachidonic Acid
U.S. FDA	United State Food and Drug Administration
ADP	Adenosine di-phosphate
LOX	Lipoxygenase
TNF-α	Tumour Necrosis Factor Alpha
RNA	Ribonucleic Acid
SAR	Structure–Activity Relationship
MTB	Mycobacterium Tuberculosis
DPPH	2,2-Diphenyl 1 picrylhydrazyl
TcTS	T. cruzi trans-sialidase
WHO	World Health Organization

A. K. Kabi · R. Gujjarappa · N. Vodnala · U. Tyagi · D. Kaldhi · C. C. Malakar (✉)
Department of Chemistry, National Institute of Technology Manipur, Imphal, Langol, Manipur 795004, India
e-mail: cmalakar@nitmanipur.ac.in

S. Sravani · A. Garg · S. Gupta (✉)
Department of Medicinal Chemistry, National Institute of Pharmaceutical Education and Research (NIPER), Chunilal Bhawan, 168, Maniktala Main Road, Kolkata 700054, India

V. Singh
Department of Chemistry, Department of Chemistry, Central University of Punjab, Bathinda, Punjab 151001, India

B. P. Swain (ed.), *Nanostructured Biomaterials*, Materials Horizons: From Nature to Nanomaterials, https://doi.org/10.1007/978-981-16-8399-2_11

HDAC	Histone Deacetylases
COVID-19	Coronavirus Disease-2019
MDCK	Madin Darby Canine Kidney
EBV	Epstein-Barr Virus
UV	Ultraviolet

1 Introduction

1,2,3-Triazoles and their derivatives are among the topmost immense *N*-heterocyclic scaffolds because of the widespread spectrum of pharmacological and biological activities [1–4]. Triazoles are the five-membered heterocycles embedded with three consecutive nitrogen-atoms that can be synthesised simply utilizing 'click' chemistry through ruthenium- or copper-catalysis between azides and terminal alkynes by cycloaddition reactions [5]. However, the 'linker' feature of 1,2,3-triazoles was established and distinct 1,2,3-triazole-based conjugates and hybrids were prepared and appraised as prime compounds for various biological and pharmacological targets due to their specificity, reliability and biocompatibility [6].

Furthermore, 1,2,3-triazole-based scaffolds have several convenient attributes such as immense chemical reliability (generally inactive to basic or acidic hydrolysis and reducing and oxidizing reactions even at higher temperature), effective dipole moment (4.8–5.6 Debye), capability of hydrogen bonding and aromatic character, which enhances their aptitude and solubility to interact with biomolecular targets as well as highly stable to metabolic degradation [7]. 1,2,3-triazole and its derivatives exhibit numerous biological activities such as anti-proliferative, anti-HIV, anti-inflammation, antimicrobial, anticonvulsant, anti-leishmanial, anti-trypanosomal, etc. [8–15]. Moreover, these motifs are usually steady for hydrolysis in both acidic and basic conditions along with oxido-reductive environment, making them steady to metabolic deterioration and designates upraised aromatic compensation. Furthermore, these enormous properties reveal the functionalised 1,2,3-triazoles skeletally appearing the amide bond and capable toward mimicking an *E* or a *Z* amide bond. Several well-known pharmacophore that possesses 1,2,3-triazole motif are existing in the market as an anticancer medication carboxyamidotriazole, wide spectrum cephalosporin antibiotic cefatrizine, β-lactam antibiotic tazobactam and anticonvulsant drug Rufinamide, etc. [5]. In this chapter, modern advancements in the improvement of 1,2,3-triazole-based pharmacologically active molecules are highlighted (Fig. 1).

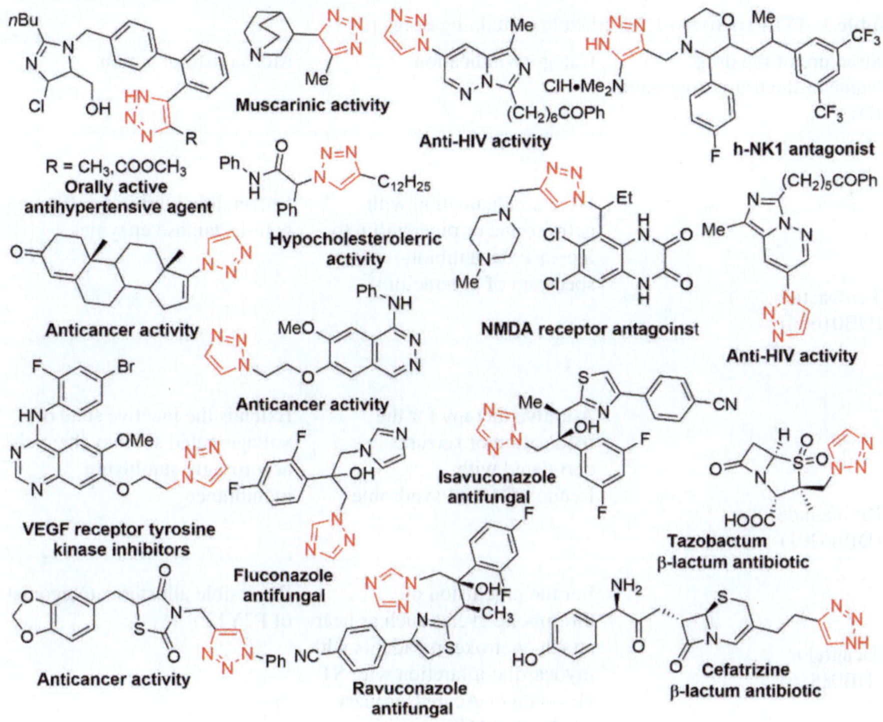

Fig. 1 Triazole embedded bioactive and drug molecules [1–6]

1,2,3-triazoles are the bioisosteres of amide bonds because of their electronic and structural resemblance and possess higher stability towards metabolic degradation. The pharmacological activities exhibited by 1,2,3-triazoles are anticancer, antimicrobial, anti-HIV, antimalarial, anti-inflammatory, antifungal, antiallergic, antiepileptic, anti-leishmanial, antituberculosis and anthelmintic activities [16]. They are proficient in forming hydrogen bonding interaction, enhancing their solubility and capacity to combine with biomolecular targets [17].

Few of the FDA approved 1,2,3-triazole containing drugs are listed in Table 1 [2].

2 Anticancer Activity of 1,2,3-Triazole Derivatives

Cancer is the principal health threat around the globe and affects a wide majority of the global community. In this contrast, several anticancer agents conveyed for medication of diverse types of cancers act through distinct mechanisms [18]. Nevertheless, the main side effect related to these anticancer agents is cytotoxicity towards the healthy cells because of inadequacy in specificity for the abnormal cells. 1,2,3-Triazoles have been investigated for the last 30 years as well as inspected mostly for

Table 1 FDA approved 1,2,3-triazole containing drugs [2]

Structure of the drug Name of the drug (drug bank ID)	Category/indication	Mechanism of action
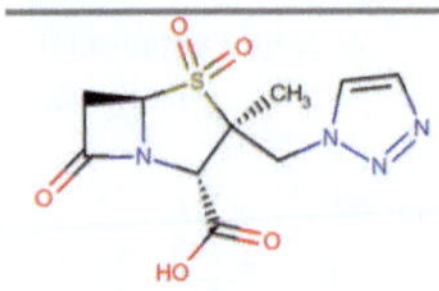 Tazobactam (DB01606)	Used amalgamation with ceftolozane or piperacillin to increase the antibiotic spectrum of piperacillin	Irreversible inhibitory effect on beta-lactamase enzymes
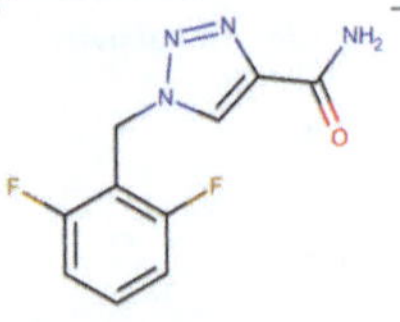 Rufinamide (DB06201)	Additive therapy for the medication of seizures correlated with Lennox-Gastaut syndrome	Extends the inactive state of voltage-gated sodium channels accordingly stabilising membranes
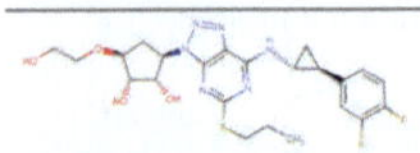 Ticagrelor (DB08816)	For the preclusion of thrombotic events such as heart attack or stroke in patients with myocardial infarction with ST elevation or Acute Coronary syndrome (ACS)	Reversible allosteric antagonist of P2Y12
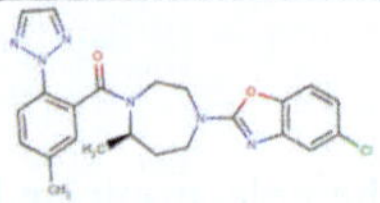 Suvorexant (DB09034)	For the treatment of insomnia	Dual antagonist of orexin receptors OX1R and OX2R
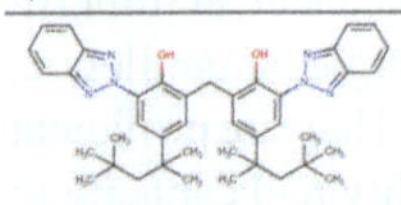 Bisoctrizole (DB11262)	Designated as a sunscreen agent in cosmetic production	An organic ultraviolet A (UVA) strainer which consumes both UVA and UVB radiation
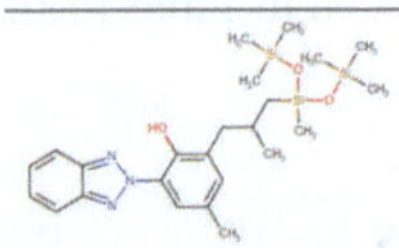 Drometrizole trisiloxane (DB11585)	An active component in several sunscreens for the indication of protection of skin	Absorbing the harmful UV radiation of sunlight

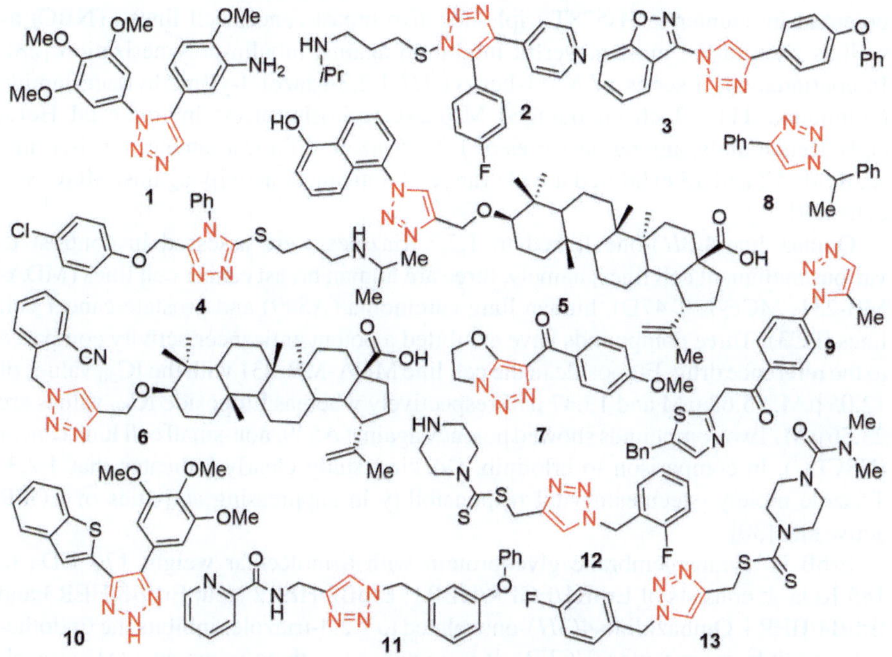

Fig. 2 Anticancer activity of 1,2,3-triazoles scaffolds [19–29]

the chemotherapeutic potential. However, numerous scientists and researchers all over the world have accomplished intensive exploration in this specific field. In this portion, we highlight modern advancements in the progress of 1,2,3-triazole-based pharmacologically active antineoplastic compounds (Fig. 2) [19–22].

Interestingly, the heterocyclic triazole such as 2-methoxy-5-(1-(3,4,5-trimethoxyphenyl)-1*H*-1,2,3- triazol-5-yl)aniline **1**, repressed tubulin polymerization. In addition, to explore the mechanism of cell death, DNA fractionation methodology was enrolled which exposed that the triazole-based compounds **2** and **4** encouraged DNA impair via apoptosis due to the biochemical ambience for these moieties and exceptional cytotoxic activity [23]. Additionally, 3-(4-(4-phenoxyphenyl)-1*H*-1,2,3-triazol-1-yl)benzo [*d*]isoxazole (PTB) (compound **3**) was establish as atop most efficient antineoplastic agent against MV4-11 cells [24]. In addition, 4-[phenyl-1-(1-phenyl–ethyl)]-1*H*-1,2,3-triazole analogue **8** exhibited decent cytotoxicity counter to HL 60 cells. In continuation, two triazole-based compound, **3**{1 *N*(2-cyanophenyl)-1*H*- 1,2,3-triazol-4yl}methyloxy betulinic acid **6** and **7**{1 *N*(5- hydroxy-naphth-1-yl)-1*H*-1,2,3-triazol-4yl}methyloxy betulinic acid **5** showed remarkable anticancer activity against leukaemia cell line HL-60 [25]. In addition, compound, **7** displayed as the most effective derivative against K562 and A431 human tumour cell lines [26]. In continuation, compound **9** was found vastly effective to the leukaemia K-562 cell line and SK-MEL-5, which is one of a series of melanoma cell lines [27]. Additionally, compound **10** showed huge potential in counter to Hs578T triple-negative breast cancer cell lines (TNBC) as well as showed the most powerful inhibition against tubulin polymerization [28]. In continuation, a series of *N*-((1-benzyl-1*H*-1,2,3-triazol-4-yl)methyl)acrylamide (compound **11**), which encouraged M-phase (mitosis) arrest in immortal HeLa cells. Remarkably, among the series of 1,2,3-triazole-dithiocarbamate-urea hybrids, scaffolds **12** and **13** exhibited a wide range of anticancer activity against MGC-803 cells [29].

Quinazoline-4(3*H*)-one linked to 1,2,3-triazoles were assessed in contrast to various malignant cell lines; namely, three are human breast cancer cell lines (MDA-MB-231, MCF-7, T-47D), human lung carcinoma (A549) and prostate cancer cell lines (PC3). Three compounds have exhibited a potent anticancer activity compared to the reference drug, Etoposide in the cell line MDA-MB-231 with the IC_{50} values of 12.05 μM, 15.62 μM and 13.47 μM respectively whereas Etoposide IC_{50} values are 23.56 μM. Two compounds showed potency against A549, non-small cell lung cancer (NSCLC), in comparison to erlotinib. Docking study clearly indicates that 1,2,3-Triazole moiety executes a vital responsibility in suppressing activities of EGFR active site [30].

ErbB is a transmembrane glycoprotein with a molecular weight 170 KDa to 185 KDa. It consists of ErbB1/EGFR/HER1; ErbB2/HER2 Neu; ErbB3/HER3 and ErbB4/HER4. Quinazoline-4(3*H*)-one related to 1,2,3-triazoles inhibits the Endothelial growth factor receptor (EGFR). It is composed of three fragments: (a) extracellular *N*-terminal, (b) transmembrane domain and (c) intracellular tyrosine kinase. The extracellular *N*-terminal comprises four domains- I, II, III and IV. It binds to several ligands but the intracellular tyrosine kinase is highly preserved [31]. with the presence of Leucine, the domain I and III of extracellular *N*-terminal are rich whereas, in the presence of Cysteine, the domains II and IV are extremely observed [32].

EGFR (Endothelial growth factor receptor) can be found in two states, i.e. the Open state which is active and closed state which is inactive. Until a ligand is bound, both these states exist in equilibrium and it stabilises the open state, once the ligand is bound. For the Closed state; The domains II and IV have interaction in between them, existing on the extracellular *N*-terminal; however, any interaction with the ligand is not shown by the domains I and III. In open state, the domains I and III interact with the ligand when the domains II and IV move away (Fig. 3) [31].

In homodimerization and heterodimerization of the receptors on the extracellular N-terminal of EGFR, domain II is involved. The formation of dimer is asymmetric as well as characteristic, the small N-lobe of the one kinase binds with the large C-lobe of the other kinase domain. The activation of CDKs (cyclin-dependent kinases) and Src family kinases leads to this dimer formation [32].

All the series of the compounds occupied the active site of EGFR. The hook shape conformation in the binding model of inhibitors is because of the existence of triazole rings. The 1,2,3-triazole ring interacts with the amino acids present in the active site of EGFR receptors involves various interactions like electrostatic interaction, Pi-anion interaction, H-bonding and Van der waals interaction [30].

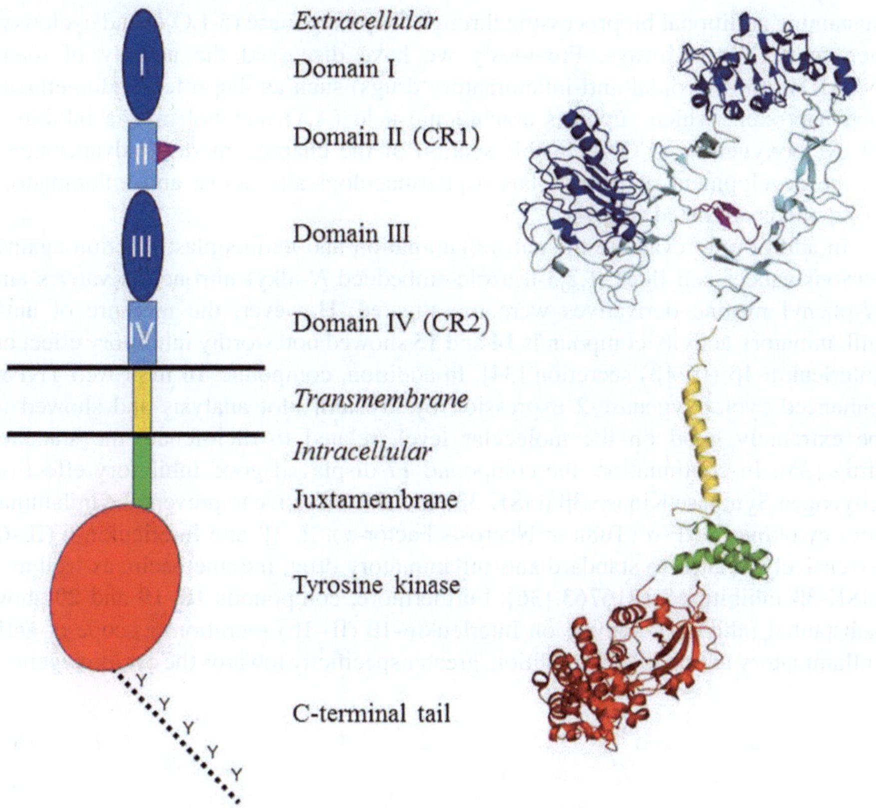

Fig. 3 Structural representation of EGFR [31]

Novel bis alkynes with di-substituted triazoles were synthesised and they were evaluated for the anticancer activities and identified that the bis alkynes themselves have some anticancer activity. In order to increase their anti-proliferative activity, many substitutions were done on the bis alkyne. The introduction of one triazole moiety in the bis alkyne has shown the deleterious effect on its anti-proliferative activity but introduction of one more triazole moiety makes it a good anticancer exhibiting compound. One of the bis triazole moiety is highly potent which exhibits the anti-proliferative activity in nanomolar range [33].

3 Anti-inflammatory Activity of 1,2,3-Triazole Derivatives

Amongst all biological processes, inflammation is the most important part of the immune system's response to infection and injury. It Comprises The release of arachidonic acid (AA) from phospholipids (are main components of the plasma membrane), sustaining additional bioprocessing through 5-lipoxygenase (5-LOX) and cyclooxygenase (COX) pathways. Previously we have discussed the activity of many NSAIDs (non-steroidal anti-inflammatory drugs) such as ibuprofen, indomethacin and naproxen, which suppress arachidonic acid (AA) metabolism via inhibition of cyclooxygenase (COX). In this section of the chapter, modern advancements in the development of triazole-based pharmacologically active anti-inflammatory compounds are listed (Fig. 4).

In addition, to evaluate the anti-inflammation and antineoplastic action against various cancer cell lines, 1,2,3-triazole-embedded *N*-alkyl nitrone derivatives and *N*-phenyl nitrone derivatives were investigated. However, the measure of anti-inflammatory activity compounds **14** and **15** showed noteworthy inhibitory effect on Interleukin-1β (IL-1β) secretion [34]. In addition, compound **16** improved TNFα-enhanced cyclooxygenase-2 expression by Western blot analysis and showed to be extremely good on the molecular level, related to diclofenac, the standard drug [35]. In continuation, the compound **17** displayed good inhibitory effect on Glycogen Synthase Kinase-3β (GSK-3β) and also effective to prevent the inflammatory cytokines TNF-α (Tumour Necrosis Factor-α), IL-1β and Interleukin-6 (IL-6) extensively, related to standard anti-inflammatory drug, indomethacin, as well as a GSK-3β inhibitor, SB216763 [36]. Furthermore, compounds **18**, **19** and **20** show substantial inhibitory activity on Interleukin-1β (IL-1β) secretion, a scope of anti-inflammatory action [34]. In addition, greater specificity towards the cyclooxygenase

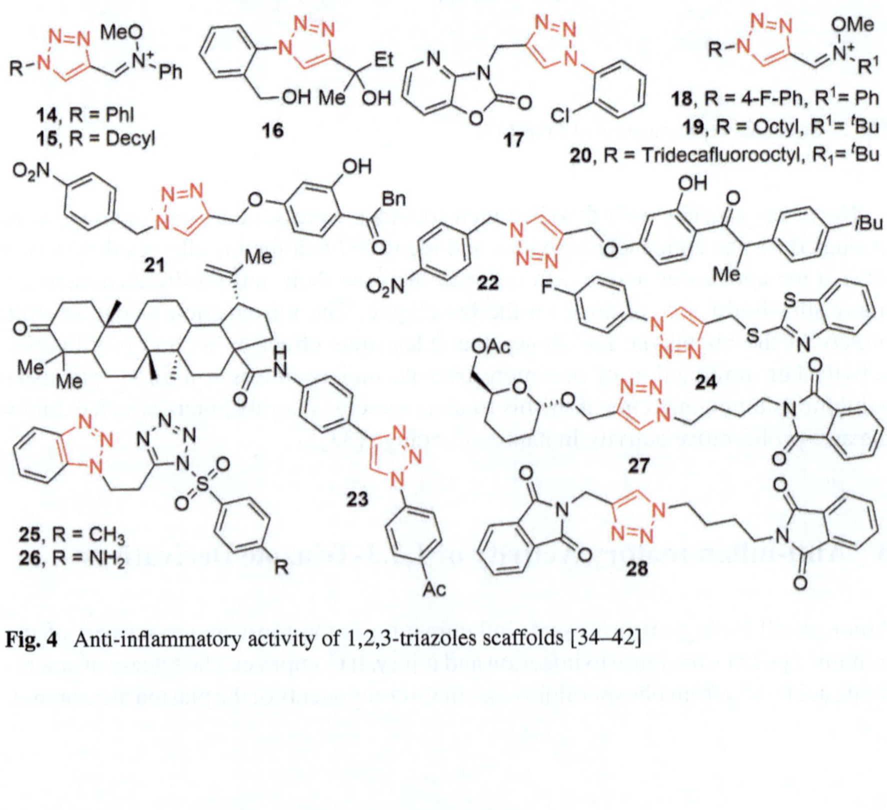

Fig. 4 Anti-inflammatory activity of 1,2,3-triazoles scaffolds [34–42]

-2 (COX-2) binding site were showed by a series of 1-{4-[(1H-1,2,3- triazol-4-yl)methoxy]-2-hydroxyphenyl}-2-phenylethanone derivatives **21**, an efficient anti-inflammatory compound [37]. In continuation, the protein–ligand relations between COX-2 (PDB code 4PH9) and desired compounds were availed for docking investigations and compound **22** anticipated the extremely good binding attitude to the active site of 4PH9 protein [38]. Additionally, a noteworthy anti-inflammatory activity showcased by betulonic conjugate **23**, suggested at the highest dose in comparison to the reference drug indomethacin. However, from the SAR study, upon altering of hexyl substitution at the triazole scaffolds onto a benzyl moiety and to a higher extent such as methoxyphenyl and acetyl phenyl substitution, revealed the possibility of enhancing the anti-inflammatory activity in triazoles [39]. In addition, compound **24** established good potential for the selective cyclooxygenase-2 (COX-2) inhibitory strain with COX-2/COX-1, however, the compound controlled major anti-inflammatory action is associated with the reference drug ibuprofen [40]. In continuation, 1-[2-(1*H*-tetrazol-5-yl)ethyl]-1*H*-benzo[*d*][1,2,3]triazoles were anticipated to the effective anti-inflammatory as well as anti-nociceptive activities. The related compounds **25** and **26** produced greater anti-inflammation compared to reference drugs [41]. However, new 1,2,3-triazole-based phthalimide analogues were investigated for anti-inflammatory activities and the compounds **27** and **28** exhibited best activity [42].

Ibuprofen, an NSAID containing free carboxylic group, is responsible for some side effects like inhibition of Prostaglandin synthesis in GIT. So, modifications were done to the carboxyl group of the Ibuprofen. One of the modifications is the introduction of the five-membered heterocycle, 1,2,3-triazole. These novel triazole-containing compounds were synthesised and their anti-inflammatory movement was evaluated.

The anti-inflammatory actions of these molecules were monitored at the dose of 10 mg/kg body weight by a suitable method; namely, carrageenan-induced rat paw edema model and their obtained outcomes are correlated with the standard drug, Ibuprofen. One of these compounds has exhibited potential actions 94.01% after 3 h, 96.35% after 4 h, 95.62% after 5 h and 94.17% after 6 h when compared with the standard drug in 3 h 93.16%, in 4 h 95.62%, in 5 h 95.62% and in 6 h 94.70%. The other molecules have exhibited comparable to moderate anti-inflammatory activity. The side effects of the Ibuprofen compounds were reduced [38].

The 4,5-Dihydro-1*H*-1,2,3-triazoles were prepared and their activity was evaluated against standard drug, Ibuprofen. The anti-inflammatory activity was observed at various hours and the percentage inhibition was calculated. Two compounds have exhibited good anti-inflammatory activity with the % inhibition of 63.28% at 1 h and 67.12% at 3 h; 63.43% at 1 h and 68.07% at 3 h respectively and the Ibuprofen activity was 64.06% at 1 h and 68.10% at 3 h [43].

4 Anti-tubercular Activity of 1,2,3-Triazole Derivatives

Over the last few decades, there has been considerable growth in occurrences of tuberculosis specifically, via drug-resistant *Mycobacterium tuberculosis* (M. tb), a species of pathogenic bacteria. Anti-tubercular medications such as rifampicin, pyrazinamide, isoniazid, streptomycin and ethambutol (known as antibiotics) are very often inadequate. Hence, there is a need for developing more advanced anti-tubercular compounds, in this context, here we are providing few active compounds against tuberculosis (Fig. 5) [44–54].

In exploration of novel potent molecules in contrast to *M. bovis BCG* and *M. tuberculosis* (MTB) H37Ra, interestingly 1,2,3-triazoles-based benzothiazinone (compounds **29** and **34**) were exhibit maximum potential against *M. bovis* BCG and MTB [44]. In addition, for the inhibition of H37Rv strain, an approach was bashed by employing 1,2,3-triazoles including fluorine-embedded benzimidazole series. Additionally, these compounds were potentially valuable for the anti-tubercular activity and mostly the compounds **30**, **31** and **32** displayed better activity associated with standard rifampicin [45]. In continuation, compound **33** exhibited two-fold greater competence than the reference drug econazole, signifying that the scaffold is expected to be improved for the anti-tubercular activity [46]. Furthermore, *N*-substituted-phenyl-1,2,3,-triazole-4-carbaldehydes were screened and the compounds **35** and **36** exhibited the best inhibition [47]. In continuation, 1,2,3-triazole conjugates of 2-mercaptobenzothiazole were considered and screened to the *M. tuberculosis* H37Rv strain for anti-tubercular activity. The related compounds **37**, **38** and **39** inhibited the growth of *M. tuberculosis* H37Rv strain [48]. In addition, compound **40** was exposed as an effective compound counter to *M. tuberculosis* (MTB) [49]. Furthermore, compound **41**, i.e. 1-dodecyl-4-phenethyl-1*H*-1,2,3-triazole exhibited inhibition against *M. tuberculosis* H37Rv [50]. Additionally, compound **42** showed a

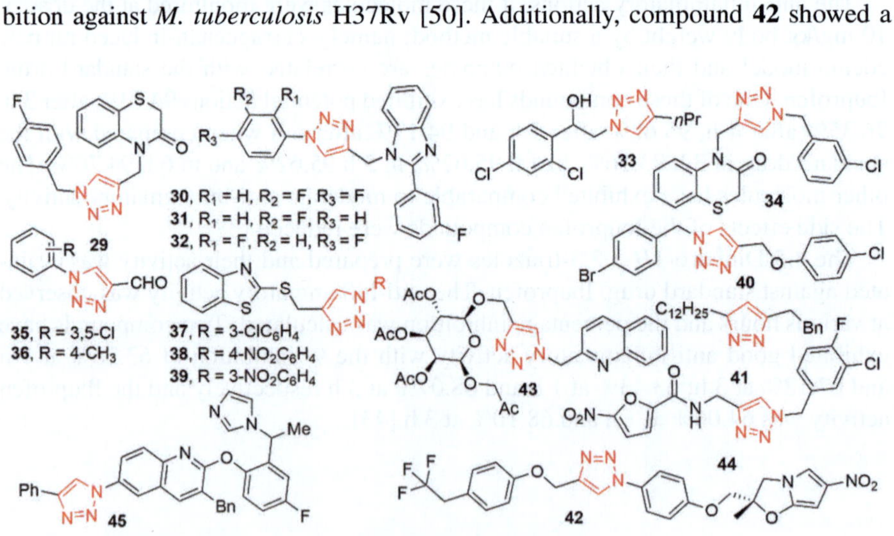

Fig. 5 1,2,3-triazoles moieties with anti-tubercular activity [44–54]

harmonious activity in combining studies with no CYP-inhibition and first line-TB drugs [51]. In continuation, quinoline-derived triazole sugar hybrid, **43** show-cased potential as an effective compound in contrast to *M. tuberculosis* H37Rv strain through LRP (Luciferase Reporter Phage) assay [52]. Additionally, 5-nitrofuran–triazole conjugates screened for antibacterial as well as anti-tubercular action. In continuation, amidst all the tested scaffolds, **44** exhibited auspicious bioactivities against tuberculosis [53]. However, compound **45** inhibited *M. tuberculosis* H37Rv. The SAR was proposed that electrostatic interactions and hydrogen bonding interaction of polar functional groups of anti-mycobacterial compounds and amino acids of ATP-synthase of bacteria, which might be the plausible incentive for the potent activity [54].

Some 1,2,3-triazoles were prepared and these molecules were evaluated for anti-tubercular action in contrast to a virulent strain of MTB (MTB H37Ra; ATCC 25,177). The parameter taken into consideration was inhibition of growth of the virulent strain. Many compounds have exhibited the activity against MTB with the MIC range of 5.8–29.9 μg mL-1. These compounds showed movement with the IC_{50} of 0.2 μg/mL to 8.3 μg/mL. But these compounds have not exhibited the comparable activity with Rifampicin. Antioxidant activity is important for anti-tubercular drugs because of oxidative stress, which can lead to chronic inflammation. So, the antioxidant properties for these compounds were evaluated by employing 2,2-diphenyl-1-picrylhydrazyl (DPPH) radical scavenging assay. Five compounds have exhibited excellent antioxidant activity when compared to BHT and vitamin C (ascorbic acid) with the IC_{50} values of 10.1 μg/mL, 10.9 μg/mL, 12.1 μg/mL, 11.0 μg/mL and 11.4 μg/mL whereas the IC_{50} values of the standard drugs were 16.5 μg/mL and 12.7 μg/mL respectively.

The reported mechanism of 1,2,3-triazole derivatives showed inhibition to the DprE1 (decaprenylphosphoryl-β-D-ribose-2′-epimerase) enzyme of MTB and it is involved in the biosynthetic pathway of an important component of mycobacterial cell wall, i.e. decaprenyl phosphoryl-D-arabinose (DPA). These compounds were docked with the enzyme DprE1. The docking studies have revealed that the triazole ring exhibits various interactions by virtue of the amino acid residues present in the active site of DprE1 through hydrophobic bonds and Vander Waals interactions and they have shown the binding energy of −77.97 kcal/mol, −73.79 kcal/mol and −66.58 kcal/mol. The compounds with the triazole ring exert a good anti-tubercular activity [55].

5 Anti-leishmanial and Anti-trypanosomal Activity of 1,2,3-Triazole Derivatives

Trypanosomatids are protist protozoan parasites that affect millions of people all over the world and cause many significant human and animal diseases. Leishmania which is a genus of trypanosomes, thus blameworthy for leishmaniasis diseases. According to WHO, more than 0.9–1.6 million new cases arise every year and 21 species are well-known sources for disease in people. In several countries, kinetoplastids, a ubiquitous group of flagellated protozoa, cause trypanosoma cruzi (chagas' disease), Leishmania species (leishmaniasis) and African trypanosomes (African sleeping sickness). These tropical diseases are specifically injurious in susceptible populations, predominantly amidst immunocompromised people or children which forcefully diminish human potential, moreover keeping people in poverty [56–68].

In briefly, 1,4-diaryl-1,2,3-triazole scaffolds were investigated in contrast to *Leishmania amazonensis promastigotes* and the analogues **46** and **51** exhibited extreme potency counter to Leishmania amazonensis (*L. amazonensis*) and Leishmania infantum (*L. infantum*) [56]. Additionally, triazolopyridopyrimidines **47** and **48** were discovered as more effective and screened against *L. infantum amastigotes*, related to the reference drug miltefosine [57]. In continuation, compounds **49** and **50**, related to the alkyltriazoles, exhibited dynamic activity counter to promastigote and amastigote forms in comparison with standard pentamidine and amphotericin B [58]. In addition, the 4-(3-nitrobenzyl)-1,2,3-triazole 5'-substituted guanosine analogue **52** displayed extreme potential on axenic amastigotes [59]. Furthermore, 1,2,3- triazole-derived *O*-benzylquercetin glycoconjugates were screened for anti-leishmanial action and among all the investigated analogues, **53** exhibited potent effectiveness related to leishmaniasis [60]. In continuation, *N*-[(1-benzyl-1*H*-1,2,3-triazole-4-yl)methyl] moiety, functionality on the C-2 amine of thiadiazole screened for anti-leishmanial action counter to the standard promastigote and 4-methylbenzyl analogue **54** was found to be the most effective compound against promastigotes [61]. Additionally, 1,2,3-triazolylsterols were screened against *L. donovani* and compound **55** was found to be potent against *Leishmania donovani* (*L. donovani*) and showed 5-times potent than the reference drugs [62]. Furthermore, it was established that diamidines **56** showed good potential against anti-trypanosomal activity than the standard melarsoprol, curing all infected mice [63]. Additionally, difluoromethylene 1,2,3-azole derivative **57** exhibited inhibition against the parasite growth expressively [64]. In addition, compound **58** contrary to the anti-parasite infections; exhibited promising effects [65]. In continuation, 5'-aryl-5'-deoxyguanosine analogues screened against *L. donovani* and compound **59** was found to be the most effective in the series without cytotoxicity [66]. In addition, 1,2,3-triazoles-linked 1,4-naphthoquinones based on nor-α-lapachone and nor-β-lapachone, moreover among the screened naphthoquinone analogues, nor-α-lapachone derivatives (compounds **60** and **61**) exhibited the highest anti-leishmanial activity [67]. However, sialic acid-6-O-galactose coupled to 1,2,3-triazole and the sialic acid galactopyranoside analogue **62** exhibited potential effect on *Trypanosoma cruzi trans-sialidase* (TcTS) inhibition (Fig. 6) [68].

However, the new analogues of long-chain alkyl [1,2,3-triazoles] and two alkylphosphocholine derivatives comprising azide scaffold and some peptide-based triazoles were screened for the inhibition of cysteine protease rCPB2.8. rCPB2.8 is one of the targets for anti-leishmanial drugs [69].

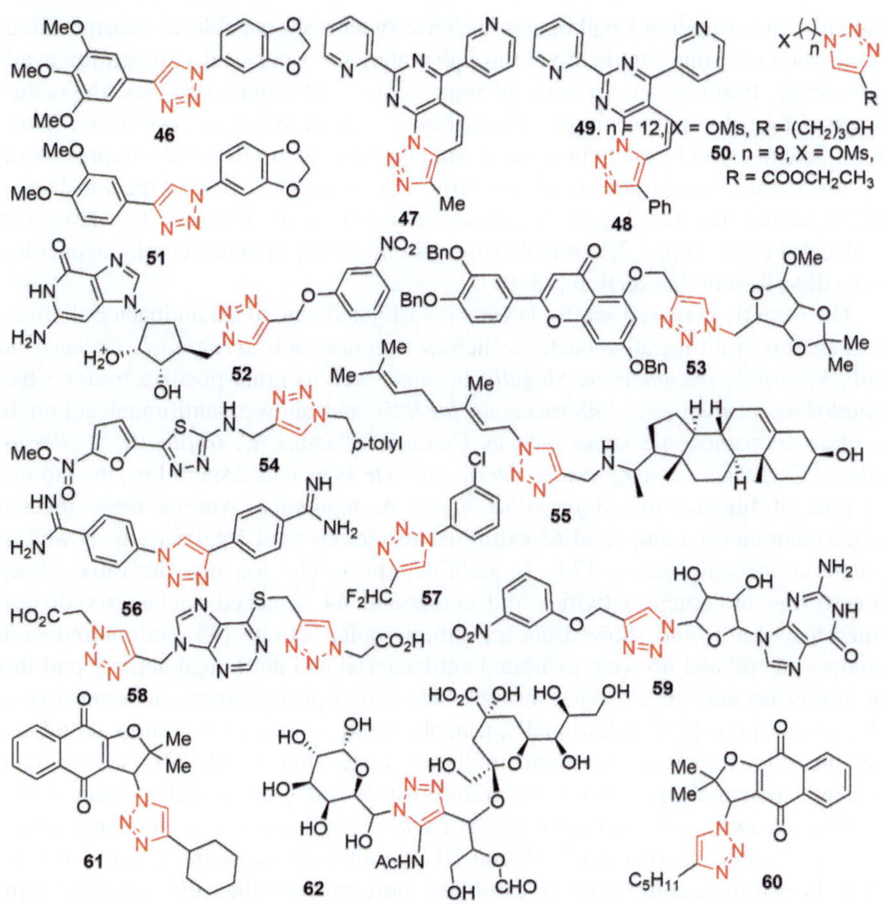

Fig. 6 1,2,3-triazoles moieties with anti-leishmanial and anti-trypanosomal activity [56–68]

Oyelere et al. reported that aryltriazolylhydroxamates with 1,2,3-triazole as a surface-recognition cap-group-linking and spacer-group chain lengths exhibited their potency as histone deacetylase (HDAC) inhibitors. Nevertheless, the addition of 1,2,3-triazole enhanced anti-leishmanial as well as antimalarial activities [70].

6 Antimicrobial Activity of 1,2,3-Triazole Derivatives

Antimicrobial resistance (AMR) is one of the top 10 worldwide public health emergencies, in agreement with the World Health Organization (WHO). It causes more than 23,000 deaths and 2 million infections per year [71]. It has been reported globally, the struggle of pathogenic bacteria (which are capable of causing infections when entering into the body through water, air or physically) regarding available drugs. Furthermore, because of imperceptive antifungal activities, the occurrence of fungal infections improved swiftly as well as increased resistance. Therefore, scientists and researchers have mainly been focused on the improvement of novel antimicrobial agents for bacterial infections [72]. In this regard, diverse 1,2,3-triazole-based analogues display auspicious antimicrobial activities. However, molecules containing 1,2,3-triazole ring systems as novel antimicrobial agents has been described in this section [73–91].

The recently prepared scaffolds were investigated toward the antibacterial efficacies against gram-negative bacteria such as *Pseudomonas aeruginosa, Escherichia coli, Klebsiella pneumoniae, Shigella boydii*as well as gram-positive bacteria like *Staphylococcus aureus, Enterococcus faecalis* and showed antifungal action in contrast to pathogenic yeast such as *Candida albicans, C. tropicalis, C. Parapsilosis, Cryptococcus neoformans, Dermatophyte* as well as Aspergillus, the moulds (a type of fungus) like *Aspergillus niger, A. fumigates.* Among these investigated compounds, compound **63** exhibited auspicious antifungal activity as well as potent antibacterial action [74]. In addition, spirocyclic triazolyl derivatives keep noteworthy biological activities and compound **64** screened against six diverse microbial strains and showed decent antimicrobial activity [75]. In continuation, compounds **65** and **66** were exhibited antibacterial and antifungal activity and this is mostly because of existence of halo-substituted phenyl rings and a piperazine ring in the frame [76]. Additionally, triazole-based moieties, i.e. compound **67** and **68** displayed important minimum inhibitory concentrations (MIC) and exhibited dynamic potential against several gram-positive and gram-negative bacteria like *Staphylococcus aureus, Bacillus cereus, Escherichia coli* and *Pseudomonas aeruginosa* [77]. Indeed, compounds **69** and **70** exhibited enhanced antibacterial activity [78]. In addition, compound **71** exhibited more potency than the standard ampicillin counter to gram-positive (*Staphylococcus aureus*) and gram-negative bacteria (*Escherichia coli*) and exhibited considerable cytotoxicity and antifungal activities [79]. In continuation, 3,4-dichlorobenzyl analogue **72** and consistent hydrochloride **74** exposed potent activity against anti-*E. coli* compared to reference drugs Norfloxacin and Chloromycin. In addition, it was proved by initial data that, compound **72** could efficiently interpolate into calf thymus DNA to form 828-DNA conjugate by blocking DNA replication and establishing antimicrobial actions [80]. Indeed, compound **73** evaluated towards anti-bio-film, antimicrobial and bactericidal actions and showcase maximum potency and as well as promising cytotoxicity (Fig. 7) [81].

In addition, compounds **75** and **76** seemed more auspicious antibacterial agents against gram-positive bacteria (*S. aureus, S. epidermidis* and *B. subtilis*) as well as gram-negative bacteria (*E. coli, P. aeruginosa, S. typhi* and *K. pneumoniae*) [82]. Further, coumarin-based triazole derivatives (compound **77**) exhibited comparable antifungal activity in comparison with reference drugs [83]. In addition, compounds **78** and **79** displayed decent potency in contrast to bacteria and fungi [84]. In continuation, compound **80** (with 4-pentylphenyl substituent), showed durable inhibitory

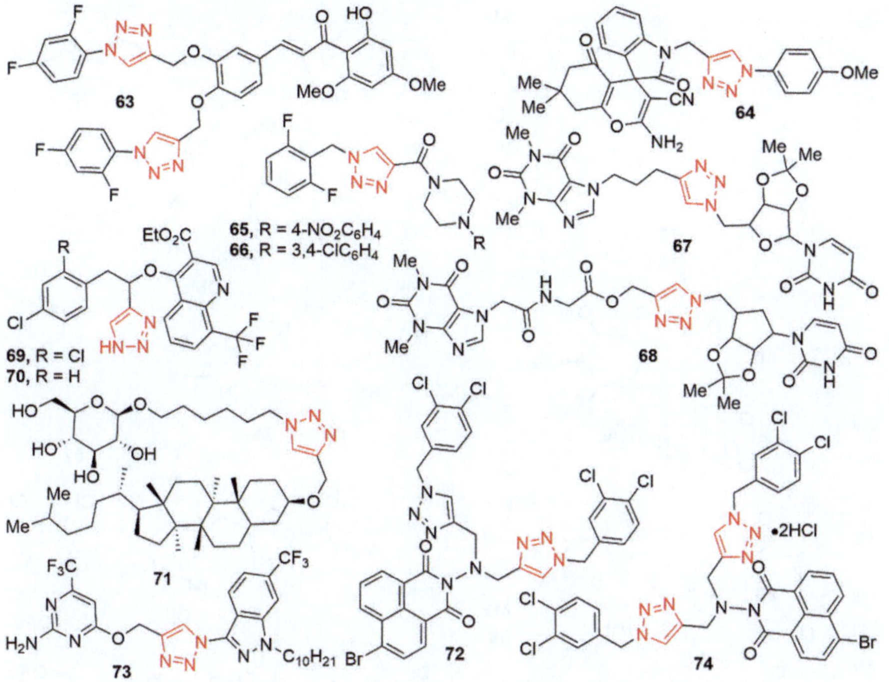

Fig. 7 Antimicrobial activity of 1,2,3-triazoles-based compounds [73–81]

action counter to bacteria and **85** (with 4- bromophenyl substituent) has a greater antifungal potency. Additionally, the above-mentioned analogues displayed a decent cytotoxic activity against a number of verified tumour cell lines, nevertheless non-toxic to the non-tumour liver cells [85]. Indeed, compound **81** exhibited outstanding antibacterial strain related to the standard drug Vancomycin, whereas **88** displayed better potency contrary to yeast [86].

In continuation, compounds **82** and **87** resulted in a stimulating antimicrobial strain in contrast to the screened bacteria and fungi omitting *E. coli*, which displayed slight counter action [87]. In addition, compounds **90** and **89** were exhibited effective antibacterial and antifungal strain respectively [88]. Indeed, compound **90** drew two-fold more antifungal activity than the standard miconazole, as well as extremely potent to diverse bacterial strains [89]. In addition, compound **86** showcased an extremely good discriminatory toxicity contrary to microorganisms [90]. However, 1,2,3-triazole attached carboxylic acid, bromoquinoline analogues were investigated and among all compounds, **89** displayed effective potential to antimicrobial and antifungal strains (Fig. 8) [91].

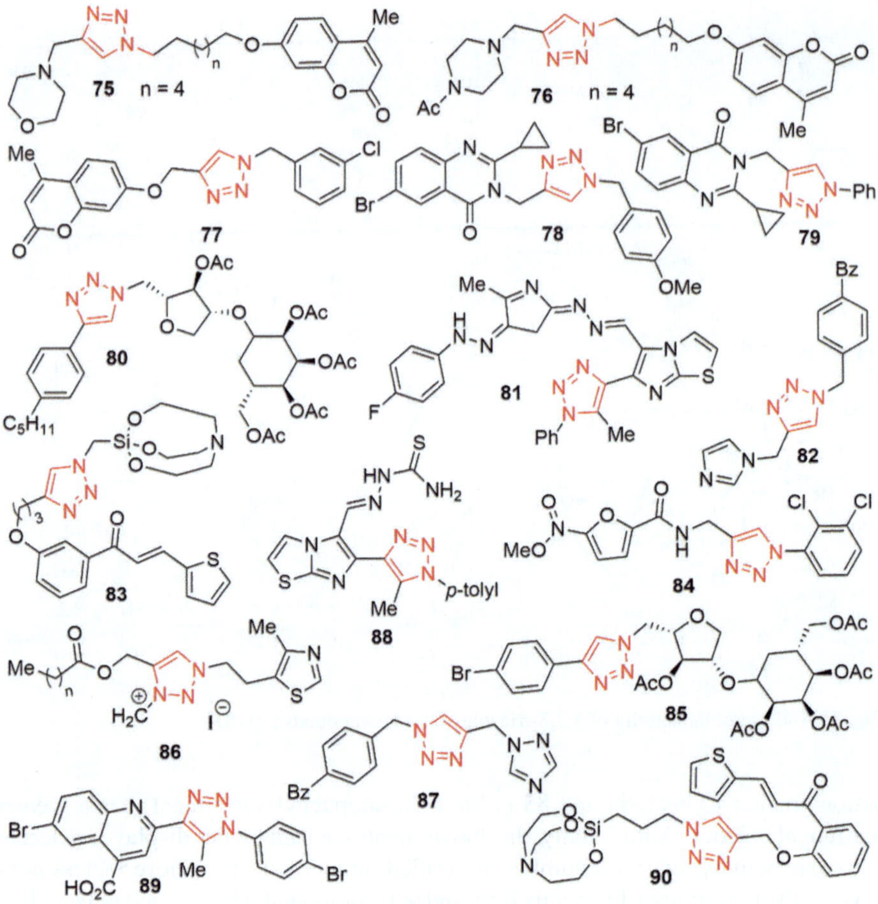

Fig. 8 Antimicrobial activity of 1,2,3-triazoles moieties [82–91]

7 Anti-viral Activity of 1,2,3-Triazole Derivatives

Viral toxicities are regarded as leading precarious infections, as they kill over million people yearly around the globe. Recently, coronavirus disease (COVID-19), an extremely infectious disease, killed more than 4 million people all over the world. In contrast, recent developments of anti-viral drugs show that there is a public health crisis to discover the extremely potent and ingenious agents. With this intent, several 1,2,3-triazole-based compounds were produced and evaluated for anti-viral action [92–98].

The compound **91** exhibited potency against influenza A replication which was higher to the standard drug ribavirin in terms of anti-viral activity as well as displayed human immunodeficiency virus type-1 reverse transcriptase (HIV-1 RT) activity

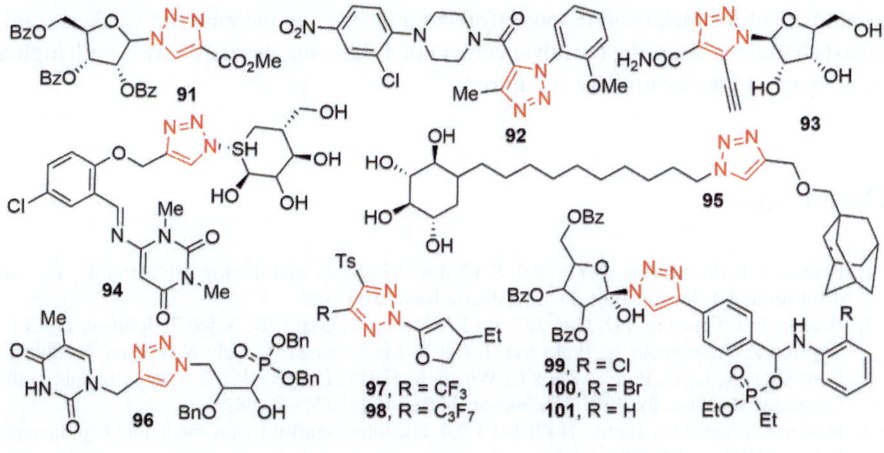

Fig. 9 1,2,3-triazoles moieties with anti-viral activity [92–98]

[92]. Indeed, analogue **92** showed powerful inhibitory action against the replication of H5N1 (RG14) (influenza a virus), amantadine-resistant A/WSN/33 (H1N1) and oseltamivir-resistant A/WSN/1933 (H1N1, 274Y) virus strains. In addition, 5-ethynyl nucleoside **93** showed anti-viral activities counter to most of the viruses [93]. In continuation, the interactivity of circulating tumour DNA (CT-DNA) with the sugar-triazoles was screened, revealed that the compound **94** can interrelate via groove binding with circulating tumour DNA (CT-DNA) [94]. In addition, longer alkyl chain compound **95** potentially abridged RNA copies multiplication, with 10 μM treatment diminish viral RNA [95]. Furthermore, phosphonate **96**, exhibited anti-viral action in MDCK (Madin Darby canine kidney cell cultures) counter to Influenza A H3N2 subtype [96]. Additionally, compounds **97** and **98** exhibited high selectivity indices (polymerase chain reaction (PCR) method) for the cytotoxic activity and anti-EBV activity [97]. However, compounds **99** and **100** exhibited diffident inhibition in contrast to respiratory syncytial virus (RSV) and compound **101** showed uncertain inhibition counter to Coxsackievirus B4 (Fig. 9) [98].

8 Summary/Conclusion

1,2,3-triazole and its derivatives are recognised as prosperous motifs in medicinal chemistry and biosciences. Moreover, excellent features of 1,2,3-triazole accelerate its broad spectrum of implementation from bioconjugation to material science. However, among the synthesised triazoles they resemble properties in which these scaffolds act as a linker, displaying CH-π interaction with enzymes. Nevertheless, the unique features of these moieties in the hybrid molecules make them a remarkable motif, a significant aspect in drug delivery and design. Scientists and researchers around the globe make continuous efforts to improve and increase the 1,2,3-triazole-based chemistry, enabling the advancement of 1,2,3-triazoles derivatives with higher selectivity and lesser toxicity in future.

References

1. Praveena KSS, Murthy NYS, Pal S (2015) Synthesis and biological activities of 1,4-disubstituted-1,2,3-triazoles. J Chem Pharm Res 7:506–522
2. Wishart DS, Feunang YD, Guo AC, Lo EJ, Marcu A, Grant JR, Sajed T, Johnson D, Li C, Sayeeda Z, Assempour N, Iynkkaran I, Liu Y, Maciejewski A, Gale N, Wilson A, Chin L, Cummings R, Le D, Pon A, Knox C, Wilson M (2018) DrugBank 5.0: a major update to the DrugBank database for 2018. Nucleic Acids Res 46:D1074–D1082
3. Haider S, Alam MS, Hamid H (2014) 1,2,3-Triazoles: scaffold with medicinal significance. Inflam Cell Signal 1:e95
4. Kappe CO, Eycken EVD (2010) Click chemistry under non-classical reaction conditions. Chem Soc Rev 39:1280–1290
5. Agalave SG, Maujan SR, Pore VS (2011) Click chemistry: 1,2,3-Triazoles as Pharmacophores 6:2696–2718
6. Khurshed B, Jiangyu Z, Aisaa HA (2019) 1,2,3-Triazole-containing hybrids as leads in medicinal chemistry: a recent overview. Bioorg Med Chem 27:3511–3531
7. Vatmurge NS, Hazra BG, Pore VS, Shirazi F, Chavan PS, Deshpande MV (2008) Synthesis and antimicrobial activity of beta-lactam-bile acid conjugates linked via triazole. Bioorg Med Chem Lett 18:2043–2047
8. Da Silva FDC, De Souza MCBV, Frugulhetti IIP, Castro HC, Souza SLDO, De Souza TML, Rodrigues DQ, Souza AMT, Abreu PA, Passamani F, Rodrigues CR, Ferreira VF (2009) Synthesis, HIV-RT inhibitory activity and SAR of 1-benzyl-1H-1,2,3-triazole derivatives of carbohydrates. Eur J Med Chem 44:373–383
9. Cocklin S, Gopi H, Querido B, Nimmagadda M, Kuriakose S, Cicala C, Ajith S, Baxter S, Arthos J, Martin-Garcia J, Chaiken IM (2007) Broad-spectrum antihuman immunodeficiency virus (HIV) potential of a peptide HIV type 1 entry inhibitor. J Virol 81:3645–3648
10. Yu WJ, Rao Q, Wang M, Tian Z, Lin D, Liu XR, Wang JX (2006) The Hsp90 inhibitor 17-allylamide-17- demethoxygeldanamycin induces apoptosis and differentiation of Kasumi-1 harboring the Asn822Lys KIT mutation and down-regulates KIT protein level. Leukemia Res 30:575–582
11. Nahrwold M, Bogner T, Eissler S, Verma S, Sewald N (2010) "Clicktophycin- 52": a bioactive cryptophycin-52 triazole analogue. Org Lett 12:1064–1067
12. Fichtali I, Chraibi M, Aroussi FE, Ben-Tama A, Hadrami EME, Benbrahim KF, Stiriba SE (2016) Synthesis of some 1,2,3-triazoles derivatives and evaluation of their antimicrobial activity. Der Pharma Chem 8:236–242
13. Abdel-Wahab BF, Mohamed HA, Awad GEA (2015) Synthesis and biological activity of some new 1,2,3-triazole hydrazone derivatives. Eur Chem Bull 4:106–109
14. Kharb R, Sharma PC, Yar MS (2011) Pharmacological significance of triazole scaffold. J Enzyme Inhib Med Chem 26:1–21
15. Ferreira VF, da Rocha DR, da Silva FC, Ferreira PG, Boechat NA, Magalhães JL (2013) Novel 1H–1,2,3-, 2H–1,2,3-, 1H–1,2,4- and 4H–1,2,4-triazole derivatives: a patent review (2008–2011). Expert Opin Ther Pat 23:319–331
16. Sekhar DC, Rao DVV, Uppalla LK, Jha A (2017) Synthesis, molecular modeling and anticancer activity of novel 1,2,3-triazole hybrids-a click chemistry approach. J Chem Pharm Res 9:1–10
17. Prachayasittikul V, Pingaew R, Anuwongcharoen N, Worachartcheewan A, Nantasenamat C, Prachayasittikul S, Ruchirawat S, Prachayasittikul V (2015) Discovery of novel 1,2,3-triazole

derivatives as anticancer agents using QSAR and in silico structural modification. Springerplus 4:571

18. Zhang Z, Zhou L, Xie N et al (2020) Overcoming cancer therapeutic bottleneck by drug repurposing. Sig Transduct Target Ther 5:113
19. Odlo K, Hentzen J, Chabert JFD, Ducki S, Gani OABSM, Sylte I, Skrede M, Flørenes VA, Hansen TV (2008) 1,5-Disubstituted 1,2,3-triazoles as cis-restricted analogues of combretastatin A-4: synthesis, molecular modeling and evaluation as cytotoxic agents and inhibitors of tubulin. Bioorg Med Chem 16:4829–4838
20. Bathula SNP, Vadla R (2011) Bioactivity of 1,4-disubstituted 1,2,3-triazoles as cytotoxic agents against the various human cell lines. Asian J Pharm Clin Res 4:66–67
21. Salameh BA, Leffler H, Nilsson UJ (2005) 3-(1,2,3-Triazol-1-yl)-1-thiogalactosides as small, efficient, and hydrolytically stable inhibitors of galectin-3. Bioorg Med Chem Lett 15:3344–3346
22. Stefely JA, Palchaudhuri R, Miller PA, Peterson RJ, Moraski GC, Hergenrother PJ, Miller MJ (2010) N-((1-Benzyl-1H-1,2,3-triazol-4-yl)methyl) arylamide as a new scaffold that provides rapid access to antimicrotubule agents: synthesis and evaluation of antiproliferative activity against select cancer cell lines. J Med Chem 53:3389–3395
23. Philip S, Purohit MN, La KK, Eswar MS, Raizaday T, Prudhvi S, Pujar GV (2014) Design, synthesis and in vitro anti-cancer activity of novel 1,2,4-triazole derivatives. Int J Pharm Sci 6:185–189
24. Ashwini N, Garg M, Mohan CD, Fuchs JE, Rangappa S, Anusha S, Swaroop TR, Rakesh KS, Kanojia D., Madan V, Bender A, Koeffler HP, Basappa, Rangappa BKS (2015) Synthesis of 1,2-benzisoxazole tethered 1,2,3-triazoles that exhibit anticancer activity in acute myeloid leukemia cell lines by inhibiting histone deacetylases, and inducing p21 and tubulin acetylation. Bioorg Med Chem 23:6157–6165
25. Majeed R, Sangwan PL, Chinthakindi PK, Khan I, Dangroo NA, Thota N, Hamid A, Sharma PR, Saxena AK, Koul S (2013) Synthesis of 3-O-propargylated betulinic acid and its 1,2,3-triazoles as potential apoptotic agents. Eur J Med Chem 63:782–792
26. Yan S-J, Liu Y-J, Chen Y-L, Liu L, Lin J (2010) An efficient one-pot synthesis of heterocycle-fused 1,2,3-triazole derivatives as anti-cancer agents. Bioorg Med Chem Lett 20:5225–5228
27. Pokhodylo N, Shyyka O, Matiychuk V (2013) Synthesis of 1,2,3-triazole derivatives and evaluation of their anticancer activity. Sci Pharm 81:663–676
28. Penthala NR, Madhukuri L, Thakkar S, Madadi NR, Lamture G, Eoff RL, Crooks PA (2015) Synthesis and anti-cancer screening of novel heterocyclic-(2H)- 1,2,3-triazoles as potential anti-cancer agents. Med Chem Commun 6:1535–1543
29. Duan Y-C, Zheng Y-C, Li X-C, Wang M-M, Ye X-W, Guan Y-Y, Liu G-Z, Zheng J-X, Liu H-M (2013) Design, synthesis and antiproliferative activity studies of novel 1,2,3-triazole-dithiocarbamate-urea hybrids. Eur J Med Chem 64:99–110
30. Safavi M, Ashtari A, Khalili F, Mirfazli SS, Saeedi M, Ardestani SK, Ranjbar PR, Tehrani MD, Larijani B, Mahdavi M (2018) Novel quinazolin-4(3*H*)-one linked to 1,2,3-triazoles: synthesis and anticancer activity. Chem Biol Drug Des 92:1373–1381
31. Seshacharyulu P, Ponnusamy MP, Haridas D, Jain M, Ganti A, Batra SK (2012) Targeting the EGFR signaling pathway in cancer therapy. Expert Opin Ther Targets 16:15–31
32. Collin AA, Hubert P, Crémel G, Bennasroune A. Role of ErbB receptors in cancer cell migration and invasion. Front Pharmacol 6:283
33. Elamari H, Slimi R, Chabot GG, Quentin L, Scherman D, Girard C (2013) Synthesis and *in vitro* evaluation of potential anticancer activity of mono- and bis-1,2,3-triazole derivatives of bis-alkynes. Eur J Med Chem 60:360–364
34. Jiang B, Huang X, Yao H, Jiang J, Wu X, Jiang S, Wang Q, Lu T, Xu J (2014) Discovery of potential anti-inflammatory drugs: diaryl-1,2,4-triazoles bearing N-hydroxyurea moiety as dual inhibitors of cyclooxygenase-2 and 5-lipoxygenase. Org Biomol Chem 12:2114–2127
35. Kim TW, Yong Y, Shin SY, Jung H, Park KH, Lee YH, Lim Y, Jung K-Y (2015) Synthesis and biological evaluation of phenyl-1H-1,2,3-triazole derivatives as anti-inflammatory agents. Bioorg Chem 59:1–11

36. Sarigol D, Uzgoren-Baran A, Tel BC, Somuncuoglu EI, Ozadali-Sari IKK, Unsal-Tan O, Okay G, Ertan M, Tozkoparan B (2015) Novel thiazolo[3,2-b]-1,2,4- triazoles derived from naproxen with analgesic/anti-inflammatory properties: synthesis, biological evaluation and molecular modeling studies. Bioorg Med Chem 23:2518–2528
37. Abdel-Megeed AM, Abdel-Rahman HM, Alkaramany GES, El-Gendy MA (2009) Design, synthesis and molecular modeling study of acylated 1,2,4-triazole-3- acetates with potential anti-inflammatory activity. Eur J Med Chem 44:117–123
38. Angajala KK, Vianala S, Macha R, Raghavender M, Thupurani MK, Pathi PJ (2016) Synthesis, anti-inflammatory, bactericidal activities and docking studies of novel 1,2,3-triazoles derived from ibuprofen using click chemistry. Springerplus 5:423–438
39. Paprocka R, Wiese M, Eljaszewicz A, Helmin-Basa A, Gzella A, Modzelewska-Banachiewicz B, Michalkiewicz J (2015) Synthesis and anti-inflammatory activity of new 1,2,4-triazole derivatives. Bioorg Med Chem Lett 25:2664–2667
40. Shafi S, Alam MM, Mulakayala N, Mulakayala C, Vanaja G, Kalle AM, Pallu R, Alam MS (2012) Synthesis of novel 2-mercapto benzothiazole and 1,2,3- triazole based bis-heterocycles: their anti-inflammatory and anti-nociceptive activities. Eur J Med Chem 49:324–333
41. Ahmadi F, Ghayahbashi MR, Sharifzadeh M, Alipoiur E, Ostad SN, Vosooghi M, Khademi HR, Amini M (2014) Synthesis and evaluation of anti-inflammatory and analgesic activities of new 1,2,4-triazole derivatives. Med Chem 11:69–76
42. Assis SP, da Silva MT, de Oliveira RN, Lima VL (2012) Synthesis and antiinflammatory activity of new alkyl-substituted phthalimide 1H-1,2,3-triazole derivatives. Sci World J 2012:925925
43. Singh RJ (2009) 1, 2, 3-Triazole derivatives as possible anti-inflammatory agents. Rasayan J Chem 2:706–708
44. Shaikh MH, Subhedar DD, Arkile M, Khedkar VM, Jadhav N, Sarkar D, Shingate BB (2016) Synthesis and bioactivity of novel triazole incorporated benzothiazinone derivatives as antitubercular and antioxidant agent. Bioorg Med Chem Lett 26:561–569
45. Gill C, Jadhav G, Shaikh M, Kale R, Ghawalkar A, Nagargoje D, Shiradkar M (2008) Clubbed [1,2,3] triazoles by fluorine benzimidazole: a novel approach to H37Rv inhibitors as a potential treatment for tuberculosis. Bioorg Med Chem Lett 18:6244–6247
46. Flieger J, Tatarczak-Michalewska M, Wujec M, Pitucha M, Swieboda R (2015) RPHPLC analysis and in vitro identification of antimycobacterial activity of novel thiosemicarbazides and 1,2,4-triazole derivatives. J Pharm Biomed Anal 107:501–511
47. Costa MS, Boechat N, Rangel EA, Silva FC, de Souza AMT, Rodrigues CR, Castro HC, Junior IN, Lourenco MCS, Wardell SMSV, Ferreira VF (2006) Synthesis, tuberculosis inhibitory activity, and SAR study of N-substitutedphenyl-1,2,3-triazole derivatives. Bioorg Med Chem 14:8644–8653
48. Kucukguzel I, Tatar E, Kucukguzel SG, Rollas S, Clercq ED (2008) Synthesis of some novel thiourea derivatives obtained from 5-[(4-aminophenoxy) methyl]-4-alkyl/aryl-2,4-dihydro-3H-1,2,4-triazole-3-thiones and evaluation as antiviral/anti-HIV and anti-tuberculosis agents. Eur J Med Chem 43:381–392
49. Shaikh MH, Subhedar DD, Nawale L, Sarkar D, Khan FAK, Sangshetti JA, Shingate BB (2008) (2015) 1,2,3-Triazole derivatives as antitubercular agents: synthesis, biological evaluation and molecular docking study. Med Chem Commun 6:1104–1116
50. Menendeza C, Gaua S, Lherbet C, Rodriguez F, Inard C, Pasca MR, Baltas M (2011) Synthesis and biological activities of triazole derivatives as inhibitors of InhA and antituberculosis agents. Eur J Med Chem 46:5524–5531
51. De P, Yoya GK, Constant P, Bedos-Belval F, Duran H, Saffon N, Daff M, Baltas M (2011) Design, synthesis, and biological evaluation of new cinnamic derivatives as antituberculosis agents. J Med Chem 54:1449–1461
52. Kumar KK, Seenivasan SP, Kumar V, Das TM (2011) Synthesis of quinoline coupled [1,2,3]-triazoles as a promising class of anti-tuberculosis agents. Carbohyd Res 346:2084–2090
53. Tehrani KH, Mashayekhi V, Azerang P, Minaei S, Sardari S, Kobarfard F (2015) Synthesis and antimycobacterial activity of some triazole derivatives–new route to functionalized triazolopyridazines. Iran J Pharm Res 14:59–68

54. Upadhayaya RS, Kulkarni GM, Vasireddy NR, Vandavasi JK, Dixit SS, Sharma V, Chattopadhyaya J (2009) Design, synthesis and biological evaluation of novel triazole, urea and thiourea derivatives of quinoline against Mycobacterium tuberculosis. Bioorg Med Chem 17:4681–4692
55. Shaikh MH, Subhedar DD, Nawale L, Sarkar D, Khan FAK, Sangshetti JN, Shingate BB (2015) 1,2,3-Triazole derivatives as antitubercular agents: synthesis, biological evaluation and molecular docking study. Med Chem Commun 6:1104–1116
56. Cassamale TB, Costa EC, Carvalho DB, Cassemiro NS, Tomazela CC, Marques MCS, Ojeda M, Matos MFC, Albuquerque S, Arruda CCP, Baroni ACM (2016) Synthesis and antitrypanosomastid activity of 1,4-diaryl-1,2,3- triazole analogues of neolignans veraguensin, grandisin and machilin G. J Braz Chem Soc 27:1217–1228
57. Adam R, Bilbao-Ramos P, Abarca B, Ballesteros R, González-Rosende ME, Dea-Ayuela MA, Estevan F, Alzuet-Piña G (2015) Triazolopyridopyrimidines: an emerging family of effective DNA photocleavers. DNA binding. Antileishmanial activity. Org Biomol Chem 13:4903–4917
58. Gontijo VS, Espuri PF, Alves RB, de Camargos LF, dos Santos FV, de Souza JWA, Marques MJ, Freitas RP (2015) Leishmanicidal, antiproteolytic, and mutagenic evaluation of alkyltriazoles and alkylphosphocholines. Eur J Med Chem 101:24–33
59. Daligaux P, Pomel S, Leblanc K, Loiseau PM, Cave C (2016) Simple and efficient synthesis of 5'-aryl-5'deoxyguanosine analogs by azide-alkyne click reaction and their antileishmanial activities. Mol Divers 20:507–519
60. Dwivedi P, Mishra KB, Mishra BB, Singh N, Singh RK, Tiwari VK (2015) Click inspired synthesis of antileishmanial triazolyl O-benzylquercetin glycoconjugates. Glycoconj J 32:127–140
61. Tahghighi A, Razmi S, Mahdavi M, Foroumadi P, Ardestani SK, Emami S, Kobarfard F, Dastmalchi S, Shafiee A, Foroumadi A (2012) Synthesis and anti-leishmanial activity of 5-(5-nitrofuran-2-yl)-1,3,4-thiadiazol-2-amines containing N-[(1-benzyl-1H-1,2,3-triazol-4-yl)methyl] moieties. Eur J Med Chem 50:124–128
62. Porta EOJ, Carvalho PB, Avery MA, Tekwani BL, Labadie GR (2014) Click chemistry decoration of amino sterols as promising strategy to developed new leishmanicidal drugs. Steroids 79:28–36
63. Bakunov SA, Bakunova SM, Wenzler T, Ghebru M, Werbovetz KA, Brun R, Tidwell RR (2010) Synthesis and antiprotozoal activity of cationic 1,4-diphenyl-1H-1,2,3-triazoles. J Med Chem 53:254–272
64. Ferreira SB, Costa MS, Boechat N, Bezerra RJS, Genestra MS, Canto-Cavalheiro MM, Kover WB, Ferreira VF (2007) Synthesis and evaluation of new difluoromethyl azoles as antileishmanial agents. Eur J Med Chem 42:1388–1395
65. Corrales RCNR, de Souza NB, Pinheiro LS, Abramo C, Coimbra ES, da Silva AD (2011) Thiopurine derivatives containing triazole and steroid: synthesis, antimalarial and antileishmanial activities. Biomed Pharmacother 65:198–203
66. Daligaux P, Pome S, Leblanc K, Loiseau PM, Cavé C (2016) Simple and efficient synthesis of 5'-aryl-5'-deoxyguanosine analogs by azide-alkyne click reaction and their antileishmanial activities. Mol Divers 20:507–519
67. Guimaraes TT, Pinto MC, Lanza JS, Melo MN, Monte-Neto RL, Diogo EB, Ferreira VF, Camara CA, Valença WO, de Oliveira RN, Frézard F, da Silva Júnior EN (2013) Potent naphthoquinones against antimony-sensitive and resistant Leishmania parasites: synthesis of novel α- and nor-α-lapachone-based 1,2,3-triazoles by copper catalyzed azide-alkyne cycloaddition. Eur J Med Chem 63:523–530
68. Campo VL, Sesti-Costa R, Carneiro ZA, Silva JS, Schenkman S, Carvalho I (2012) Design, synthesis and the effect of 1,2,3-triazole sialylmimetic neoglycoconjugates on Trypanosoma cruzi and its cell surface transsialidase. Bioorg Med Chem 20:145–156
69. Dheer D, Singh V, Shankar R (2017) Medicinal attributes of 1,2,3-triazoles: current developments. Bioorg Chem 71:30–54
70. Gontijo VS, Espuri PF, Alves RB, Camargos LF, Santos FVD, Judice WAS, Marques MJ, Freitas RP (2015) Leishmanicidal, antiproteolytic, and mutagenic evaluation of alkyltriazoles and alkylphosphocholines. Eur J Med Chem 101:24–33

71. Holpuch A (2016) UN meeting tackles the "fundamental threat" of antibiotic-resistant superbugs. Guardian
72. Grange JM, Zumla A (2002) The global emergency of tuberculosis: what is the cause? J R Soc Health 122:78–81
73. Mady MF, Awad GAE, Jørgensen KB (2014) Ultrasound-assisted synthesis of novel 1,2,3-triazoles coupled diaryl sulfone moieties by the CuAAC reaction, and biological evaluation of them as antioxidant and antimicrobial agents. Eur J Med Chem 84:433–443
74. Kant R, Kumar D, Agarwal D, Gupta RD, Tilak R, Awasthi SK, Agarwal A (2016) Synthesis of newer 1,2,3-triazole linked chalcone and flavone hybrid compounds and evaluation of their antimicrobial and cytotoxic activities. Eur J Med Chem 113:34–49
75. Singh H, Sindhu J, Khurana JM, Sharma C, Aneja KR (2014) Ultrasound promoted one pot synthesis of novel fluorescent triazolyl spirocyclic oxindoles using DBU based task specific ionic liquids and their antimicrobial activity. Eur J Med Chem 77:145–154
76. Jadhav RP, Raundal HN, Patil AA, Bobade VD (2015) Synthesis and biological evaluation of a series of 1,4-disubstituted 1,2,3-triazole derivatives as possible antimicrobial agents. J Saudi Chem Soc 21. https://doi.org/10.1016/j.jscs.2015.03.003
77. Ruddarraju RR, Murugulla AC, Kotla R, Tirumalasetty MCB, Wudayagiri R, Donthabakthuni S, Maroju R, Baburao K, Parasa LS (2016) Design, synthesis, anticancer, antimicrobial activities and molecular docking studies of theophylline containing acetylenes and theophylline containing 1,2,3- triazoles with variant nucleoside derivatives. Eur J Med Chem 123:379–396
78. Garudachari B, Isloor AM, Satyanarayana MN, Fun HK, Hegde G (2014) Click chemistry approach: regioselective one-pot synthesis of some new 8- trifluoromethylquinoline based 1,2,3-triazoles as potent antimicrobial agents. Eur J Med Chem 74:324–332
79. Aly MRES, Saad HA, Mohamed MA (2015) Click reaction based synthesis, antimicrobial, and cytotoxic activities of new 1,2,3-triazoles. Bioorg Med Chem Lett 25:2824–2830
80. Lv JS, Peng XM, Kishore B, Zhou CH (2014) 1,2,3-Triazole-derived naphthalimides as a novel type of potential antimicrobial agents: synthesis, antimicrobial activity, interaction with calf thymus DNA and human serum albumin. Bioorg Med Chem Lett 24:308–313
81. Nagender P, Reddy GM, Kumar RN, Poornachandra Y, Kumar CG, Narsaiah B (2014) Synthesis, cytotoxicity, antimicrobial and anti-biofilm activities of novel pyrazolo[3,4-*b*]pyridine and pyrimidine functionalized 1,2,3-triazole derivatives. Bioorg Med Chem Lett 24:2905–2908
82. Kushwaha K, Kaushik N, Lata Jain SC (2014) Design and synthesis of novel 2H-chromen-2-one derivatives bearing 1,2,3-triazole moiety as lead antimicrobials. Bioorg Med Chem Lett 24:1795–1801
83. Shaikh MH, Subhedar DD, Shingate BB, Khan FAK, Sangshetti JN, Khedkar VM, Nawale L, Sarkar D, Navale GR, Shinde SS (2016) Synthesis, biological evaluation and molecular docking of novel coumarin incorporated triazoles as antitubercular, antioxidant and antimicrobial agents. Med Chem Res 25:790–804
84. Vani KV, Ramesh G, Rao CV (2016) Synthesis of new triazole and oxadiazole derivatives of quinazolin-4(3H)-one and their antimicrobial activity. J Heterocycl Chem 53:719–726
85. Petrova KT, Potewar TM, Correia-da-Silva P, Barros MT, Calhelha RC, Ciric A, Sokovic M, Ferreira ICFR (2015) Antimicrobial and cytotoxic activities of 1,2,3-triazole-sucrose derivatives. Carbohydr Res 417:66–71
86. Abdel-Wahab BF, Khidre RE, Awad GEA (2016) Design and synthesis of novel 6- (5-methyl-1H-1,2,3-triazol-4-yl)-5-[(2-(thiazol-2-yl)hydrazono)methyl]imidazo[2,1-*b*]thiazoles as antimicrobial agents. J Heterocycl Chem 54:489–494
87. Fichtali I, Laaboudi W, Hadrami EME, Aroussi FE, Ben-Tama A, Benlemlih M, Stiriba SE (2016) Synthesis, characterization and antimicrobial activity of novel benzophenone derived 1,2,3-triazoles. J Mater Environ Sci 7:1633–1641
88. Singha G, Arora A, Rani S, Maurya IK, Aulakh D, Wriedt M (2016) Heteroaryl chalcone allied triazole conjugated organosilatranes: synthesis, spectral analysis, antimicrobial screening, photophysical and theoretical investigations. RSC Adv 6:82057–82081

89. Kamal A, Hussaini SMA, Sucharitha ML, Poornachandra Y, Sultana F, Kumar CG (2015) Synthesis and antimicrobial potential of nitrofuran-triazole congeners. Org Biomol Chem 13:9388–9397
90. Tejero R, López D, López-Fabal F, Gómez-Garcés JL, Fernández-García M (2015) (2015) Antimicrobial polymethacrylates based on quaternized 1,3-thiazole and 1,2,3-triazole side-chain groups. Polym Chem 6:3449–3459
91. Senthil S, Gopi R (2015) N-Substituted-1,2,3-triazoles: synthesis, characterization and antimicrobial activity studies. Der Pharma Chem 7:15–23
92. de Ferreira MLG, Pinheiro LCDS, Santos-Filho O, Pecanha MDS, Sacramento CQ, Machado V, Ferreira V, Souza TML, Boechat N (2014) Design, synthesis, and antiviral activity of new 1H–1,2,3-triazole nucleoside ribavirin analogs. Med Chem Res 23:1501–1511
93. Krajczyk A, Kulinska K, Kulinski T, Hurst BL, Day CW, Smee DF, Ostrowski T, Januszczyk P, Zeidler J (2014) Antivirally active ribavirin analogues—4,5- disubstituted 1,2,3-triazole nucleosides: biological evaluation against certain respiratory viruses and computational modelling. Antivir Chem Chemother 23:161–171
94. Hemamalini A, Singam ERA, Mudedla SK, Subramanian V, Das TM (2015) Design and synthesis of sugar-triazole based uracil appended sugar-imine derivatives—an application in DNA binding studies. New J Chem 39:4575–4582
95. Howe JD, Smith N, Lee MJR, Ardes-Guisot N, Vauzeilles B, Désiré J, Baron A, Blériot Y, Sollogoub M, Alonzi DS, Butters TD (2013) Novel imino sugar aglucosidase inhibitors as antiviral compounds. Bioorg Med Chem 21:4831–4838
96. Głowacka IE, Balzarini J, Andrei G, Snoeck R, Schols D, Piotrowska DG (2014) Design, synthesis, antiviral and cytostatic activity of ω-(1H–1,2,3-triazol-1-yl)(polyhydroxy)alkylphosphonates as acyclic nucleotide analogues. Bioorg Med Chem 22:3629–3641
97. Kanishchev OS, Gudz GP, Shermolovich YG, Nesterova NV, Zagorodnya SD, Golovan AV (2011) Synthesis and biological activity of the nucleoside analogs based on polyfluoroalkyl-substituted 1,2,3-triazoles. Nucleos Nucleot Nucleic Acids 30:768–783
98. Ouahrouch A, Taourirte M, Schols D, Snoeck R, Andrei G (2016) Design, synthesis, and antiviral activity of novel ribonucleosides of 1,2,3-triazolylbenzylaminophosphonates. Arch Pharm Chem Life Sci 349:30–41

Chapter 12
Graphene-Derived Nanomaterials and Their Application in COVID-19 Related Prevention, Treatment, and Diagnosis

Naorem Aruna Devi and Bibhu Prasad Swain

1 Introduction

Currently, the whole world has been suffering from a difficult health situation because of a pandemic due to COVID-19, which is caused by a virus called SARS-COV-2. COVID-19 was first reported from Wuhan, China, on December 31, 2019 [1, 2] and World Health Organization (WHO) declared the COVID-19 as a pandemic on March 11, 2020 [3]. However, researchers found out that there is more than one strain, and that mutations have led to changes in how infectious and deadly it is. WHO has introduced a new naming convention for the identification of the various variants by using the Greek alphabet. The new naming was given as Delta, Gamma, Beta, and Alpha for B.1.617.2, P.1, B.1.351, and B.1.1.7 variants which were found in India, Brazil, South Africa, and the United Kingdom, respectively, for the first time. It is terrifying because this variant is rapidly emerging and becoming more infectious and explosively spreading around worldwide so quickly. Moreover, detection of Delta plus (Delta-AY.1) variant, a sub-lineage of Delta variant, which is essentially a mutated version of the B.1.617.2 variant or strain and was first found in Europe as a "variant of concern" has been being reported in countries across the world like China, Russia, Nepal, Poland, Japan, Switzerland, Portugal, UK and USA [4]. In April 2021, the Delta Plus case was detected for the first time in India, in the state of Maharashtra [5]. This strain is characterized by the K417N mutation in the spike protein of the SARS-CoV2 virus that causes the Covid-19 disease. Recently, it has been found that Delta plus is more contagious, deadlier, and is potentially resistant to monoclonal antibody therapy, a potent intravenous infusion of antibodies to neutralize the virus [4]. Till now, 177,559,790 coronavirus positive cases were reported with 3,840,527 deaths and 162,032,091 recoveries as of June 16, 2021 worldwide affecting 220 countries and territories [6, 7] and the spread of coronavirus is increasing so on from

N. A. Devi · B. P. Swain (✉)
Department of Physics, National Institute of Technology Manipur, Langol, Imphal-West, Imphal, Manipur 795004, India

B. P. Swain (ed.), *Nanostructured Biomaterials*, Materials Horizons: From Nature to Nanomaterials, https://doi.org/10.1007/978-981-16-8399-2_12

one person to another person day by day. In India, total 29,739,631 covid-19 positive cases were reported with 379,601 deaths and 28,462,351 total recoveries, i.e., on June 16, 2021 [6]. To fight against the spread of this virus, measures have been taken up all over the world which include lockdown, social and physical distancing, etc., causing social and economic hardship. Countries have been consistently working hard in search of a solution for this pandemic, which includes detection, diagnosis, treatment, and prevention of coronavirus. These can be done by developing improved and fast biosensors, diagnostic devices, vaccines, and antiviral drugs [8–11]. Besides this, to prevent transmission of COVID-19, face masks and other physical protective appliances have been developed. Symptoms of COVID-19 include tiredness, headache, sore throat fever, dry cough, etc. [12]. This virus can be transmitted from one individual to another through respiratory droplets which can be due to coughing and sneezing or touching the infected surface area. Previous studies have shown that there exist various sizes of respiratory droplets; the most commonly found droplets were in sub-5 μm size. The droplets whose size is >5 μm does not travel very far, it get settled down due to gravitational force within 1–2 m [13]. On the other hand, the droplets which are lightweight and small in size, remain in the air for a long period, due to which causes the virus to spread rapidly. Therefore, wearing mask helps in preventing exposure to respiratory droplets. Further, it is very important to maintain cleanliness in human and public places such as shopping complexes, parks, offices, airports, hospitals using disinfectants and various decontamination and sanitization methods to prevent COVID-19. Many antiviral and antibacterial materials including graphene, metal/metal-oxide nanoparticles (NPs) [14–16] have been developed broadly to battle against bacterial diseases. Among them, graphene/GDNMs-based materials exhibit amazing features for fighting against bacterial/viral diseases.

In these few years, technology based on graphene and GDNMs including graphene oxide (GO), reduced graphene oxide (rGO), nanoporous-graphene, etc., has captivated tremendous consideration because of its exclusive chemical, mechanical, electrical, and physical characteristics that make it a superior material for energy storage, reinforcement for constructions and aeronautics, chemical sensor, and optoelectronics applications, etc. [17–21]. Further, as a result of their high biocompatibility, tunable size, and high-surface area, GDNMs-based materials have been utilized in many applications which include ionic sieving [22], molecular separation [23], desalination [24], gas-phase separation [25, 26], water sterilization [27–29], biosensors [30], and other biomedical appliance [13, 31, 32]. Graphene is a 2-D hexagonal honeycomb lattice structure of sp^2-hybridized carbon atoms exhibiting hydrophobic nature [32, 33]. While GO is a functional derivative of graphene that contains a large number of oxygen groups and possesses hydrophilic characteristics [34]. Whereas oxygen content in rGO is less and it possesses high-hydrophilic characteristics [35]. In addition, in graphene, the layers stick together on top of each other due to the interaction of π-π-stacking [36]. Conversely, this spacing rises considerably in the GO's and rGO's interlayers as a result of oxygen functional groups exist into it [37]. Many previous researchers have found the effectiveness of GDNMs-based materials in both antimicrobial and antiviral activities [31, 32, 38–40]. Hu et al. [41] reported the first study of antibacterial action using GO and rGO materials toward E. coli bacteria, where GDNMs materials inhibit the growth of E. coli bacteria infection. GDNMs exhibit outstanding performances which include a high-surface region contributing excellent association against viral protein, well-established chemistry of surface allowing the development of multirole forums, and an effective photothermal process under near-infrared (NIR) laser irradiation, which lead to enhance the local temperature. Its electronic movements associated with viruses or bacteria or germs are responsible for the development of antiviral properties in graphene and GDNMs materials [41–43]. Further, the electronic movement produces reactive oxygen species (ROS), affects lipid membrane, decreases metabolism, causes cytoplasmic efflux, loss of glutathione, induces oxidative stress, and at the end, it kills the bacteria [44]. Song et al. have explained that the hydrogen bonding and electrostatic interactions are responsible for the adsorption of the lipid bilayer of the feline virus over the rGO's and GO's surface [45]. Therefore, the attachment of GDNMs damaged the membrane of the virus [46, 47] which proved its effectiveness in challenging against viruses [48]. Despite that the modification of the surface of graphene could be done through binding with those antivirals with a negative charge which includes heparan sulfate and heparin drugs [49, 50] which will cause to increase the attraction of graphene-modified with antiviral toward the viruses with positive charge residues, which are found to be useful in the development of therapeutic or diagnostic types of equipment [51]. Likewise, modified-rGO accompanying sulfate derivatives successfully eradicate orthopoxvirus, swine fever, and herpesvirus strains [52].

Recent advancements related to GDMNs have shown that it can be considered as a potential candidate to fulfill the requirements which include the advancement of personal protective equipment (PPE) with a higher level, which will increase its protection against virus and transmission due to COVID-19. GDNMs have the capacity of interacting and binding with microorganisms, DNA, and RNA which contributes to developing engineered textiles for the utilization in PPE. Face masks that are commonly used by health workers and people in risky areas give only physical barriers which reduce the risk of getting infected but cannot disable the virus. Due to the COVID-19 pandemic, masks are being worn in every country to prevent the spreading of the virus in crowded places, offices, hospitals, etc. [53–56]. Further, cloth masks have been used globally due to the scarcity of N95 masks to prevent transmission through the air [56]. Although such masks provide certain protection against virus spread, given that high demand for PPE is likely to continue especially because of the unableness to avoid or complication in isolation of infected individuals that are contagious amidst early days of infection when symptoms are mildest or absent, PPE and masks with improved protective properties are required [57, 58]. To further enhance the protective properties of the mask and PPE, scientists of NMs have introduced the incorporation of virus-disabling characteristics in addition to cloth-filtering and standard propylene- characteristics to develop more advanced PPEs [31, 59]. Being a single layer of carbon atoms, graphene has a surprisingly high-surface area as well as it interacts with the bacterium whose size is 100 nm in a very unique way [31]. Moreover, the interaction of an organism with the graphene surface has been found to cause the organism to lose its integrity [60]. Moreover, due to the high-viral inhibition ability of graphene, graphene associates directly toward viruses, generally through redox reactions, electrostatic interactions, and hydrogen bonding [45], as well as GDMNS, has intrinsic properties that allow it to adsorb and break down membranes and large lipids which includes blanketed viruses, of which the coronavirus is also one of them [47, 48]. GDNMs show a significant gateway in enhancing the effectiveness of face masks to bring the world closer to the objective to obstruct the transmission of COVID-19 infection. For example, recently, graphene-based prototype filters have been created as a disposable part of reusable respiratory masks [61–63], where the multilayers electrospun cellulose acetate and poly(lactic acid) were employed to construct the filter. Bonbouton Company introduced an affordable and recyclable graphene-based mask in which to enhance the antibacterial/antiviral properties of the mask, its filter was integrated with graphene. Due to the hydrophobic nature of graphene as well as allowing excellent breathability, the layer of graphene operates as a natural filter [64]. Further, a composite of Ag NPs and GO's ink was introduced by Zen Graphene Solutions Ltd. to efficiently eliminate the other strains of SARs-CoV-2 [65]. British planar TECH and Thai startup IDEATI deal for the development of inexpensive shielded masks by employing graphene, in which cotton will be used and the filter of the mask will be made from graphene and other carbon materials. The inventors also mentioned that graphene has antibacterial and antistatic characteristics due to which washing of such masks can be done as well as till 10 times, and it can be reusable. The idea of designing improved masks, gloves, and other protective accessories by using graphene materials especially considering for the health care workers to fight against the current COVID-19 pandemic has been brought in by the Italian company Directa Plus also. In addition, Graphene, the European company launches the production of alcohol-based graphene hydrogel disinfectants which were later distributed among people with old age, military, police, doctors, etc., without taking any costs [63].

In addition, SAR-CoV-2 can be transmitted to a person by touching their eyes, mouth, or nose immediately after their exposure to the infected surfaces or kinds of stuff. Apart from direct contact, the main plausible medium for transmitting this virus including other bacteria and germs is through fomites. Recently, Cambridge-based graphene specialist Roark Industries proclaimed that there was evidence of graphene-based coatings to have 100% efficacy of possibility to eradicate the membrane of viruses immediately thereafter the exposure. This antiviral coating has been designed to be utilized in crowded areas which include public transit, doorknobs, offices, hospitals, and shops regions [66]. Therefore, GDMNs have the huge potentiality for developing antiviral surface and layers-coating to prevent from getting the infection and to restrict the mass-spreading through deadly and infectious viruses like coronavirus. Furthermore, by far the most crucial challenge is to develop effective, fast, ultraprecise, portable, and scalable testing techniques/devices so that they can help in the detection and diagnosis of COVID-19 patients at the earliest. Currently, the testing depends on the reverse transcription-polymerase chain reaction (RT-PCR) method which needs costly appliances, also delay in the procedure, as well as it also needs the involvement of operative staff in each activity regarding testing which includes collecting of nasopharyngeal swab sample, data analysis plus sample-treatment [67]. Hence, the development of rapid, cost-effective as well as trustworthy techniques is required which will not need complicated devices nor staff operators for the utilization in the testing of coronavirus disease during this present epidemic. Regardless of recent progress of point-of-care (POC) rapid RT-PCR test [68–70], nucleic tests have also been found to give wrong results, due to which it may resist the strategy of containment and from getting treatment [71]. It is necessary to develop ultrasensitive, fast, and affordable testing techniques as soon as possible so that the past and present infection status of the patient can be detected [3, 38]. Recently, scientists have focused on the development and contribution of GDNMs-based diagnostic sensors to fight against COVID-19 [3, 13, 31, 38, 72]. Researchers suggested that antibody-conjugated GDMNs can quickly identify desired virus proteins and perhaps bound with electronic characteristics of nanomaterial for signal amplification [31, 72, 73]. GDNMs will not only be beneficial for screening of large population, cost-effective, and point-of-care but they will also be beneficial in developing eco-friendly sensors. Thus, GDNMs-based materials could be designed especially aiming at a specific viral antigen as well as antiviral surface coatings which will be effective in the war against COVID-19 infection and other contagious viruses. Moreover, the development of an air-purifier based on graphene is undergoing which can eradicate the SARS-CoV-2 virus [74, 75]. Although the promising outcomes obtained regarding other biomedical applications which include antibacterial, antifungal, and activation of the immune system make graphene and GDNMs a potential candidate in combating virus-related diseases, the recent advancement of GDNMs toward antiviral applications is still in the primitive phase. The present chapter has reviewed and discussed the recent advancement as well as importance related to GDNMs which can help in detecting, decontaminating, preventing, and protecting from SARS-CoV-2 infection. The structure of coronaviruses as well as the items based on GDNMs which can be used in combating COVID-19 has been presented in Fig. 1.

2 Role of Graphene in Prevention and Protection Against COVID-19

COVID-19 is commonly transmitted via very small size respiratory droplets [77]. The survivability of SAR-CoV-2 over different surfaces varies. It is found to be higher in the surfaces like stainless steel: 48 h and plastic: 72 h in comparison with the cardboard: 24 h and copper: 4 h surfaces. [78]. Further, the stability of the virus is higher in smooth surfaces than in rough surfaces like cloths: 2 h, wood: 2 h, tissue papers: 3 h, etc. It was found that the virus can survive on the outer surface of the surgical masks even after 7 days [79]. Therefore, the chance of virus spread is high if we touch the surface which may be contaminated. With the rapid increase in the number of coronavirus positive case reports worldwide, it is the need of the hour to develop effective equipment to prevent COVID-19 infection which can control the spread of the virus through physical contact. In general, GDNMs possess characteristics that are antiviral in nature like against DNA and RNA viruses, negative

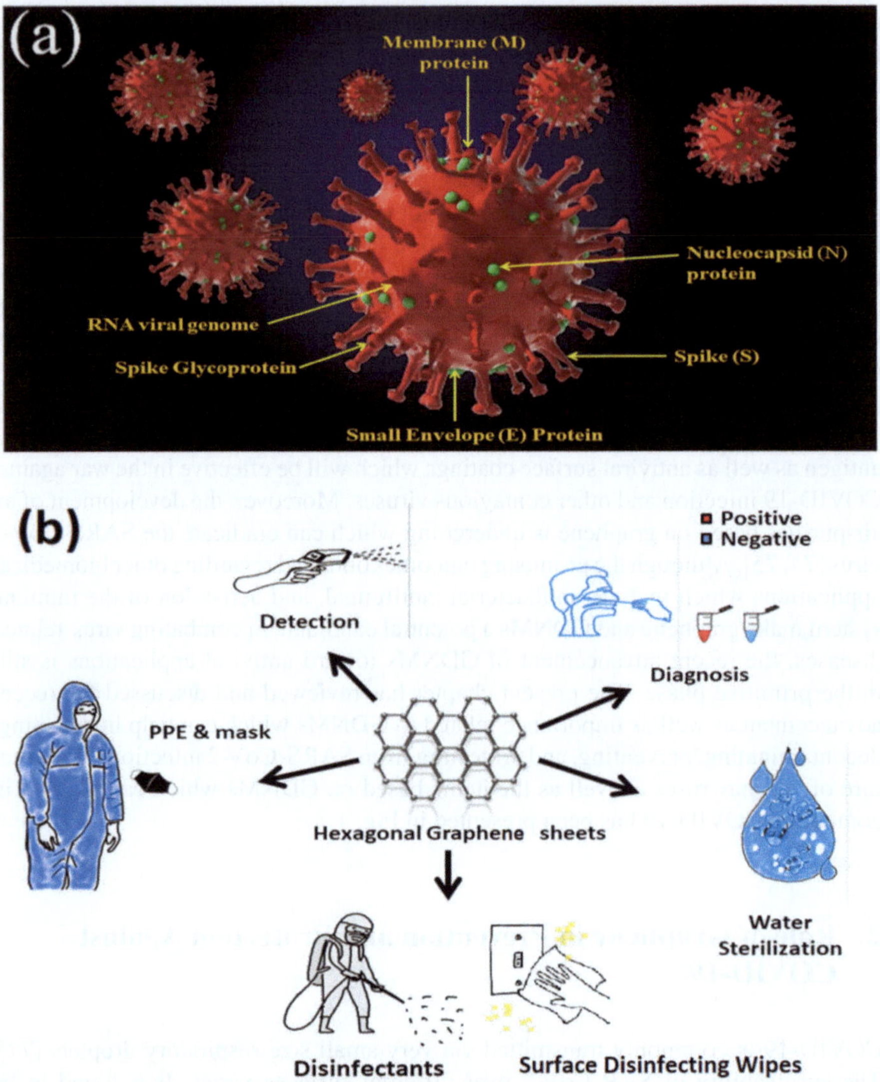

Fig. 1 **a** Morphology of coronaviruses. Reprinted with permission from [76]. Copyright (2020) American Chemical Society. **b** Schematic diagram displaying the materials made of GDNMs and their application for prevention and protection against COVID-19

and positive sense viruses, non-enveloped and viruses enveloped, etc. [38]. Graphene can help in three ways to prevent the spread of viruses because of its characteristic which is thin but conductive and strong:

- The sharp-edged of graphene NPs damages the cells and viruses mechanically as they pass through them.

- The negative charge of graphene along with its high-mobile electrons traps as well electrostatically deactivates the viruses and cells.
- Graphene stimulates cells to produce oxygen free radicals, thereby destroying the virus and impairing its cellular metabolic pathways. [80]
- In the year 2012, the first proof of GDNMs-based antiviral efficacy has been reported, where rGO/tungsten oxide has been employed for photoinactivation of bacteriophages under visible light irradiation which confirmed the inhibition activity of GDNMs against bacteria, viruses, fungi, etc. [81]. It was observed that graphene possesses antiviral properties to inhibit pathogens. Negatively charged sulfates are adsorbed over the graphene with the maximum ligand contact region as a result of a large surface area. These sulfates then come in contact with the positively charged virus particles leading to blockage of microorganisms [51]. Recently, Meredith et al. studied the antivirus action of the composites GO/Ag NPs toward two enveloped RNA viruses which are OC43 coronavirus and influenza A virus. It was discovered that the GO/Ag NPs composite produced through adding Ag nanospheres has eradicated each virus in one minute of exposure and further explained that GO/AgNPs composite materials can defeat RNA viruses very fast [81]. Therefore, more study of GDNMs in the biomedical area is required to deal with the necessity regarding numerous demands and challenges. Herein, a review on the advantage and development of GDMNs-coating to prevent coronavirus reported by some previous researchers was highlighted below.

2.1 Graphene-Coated Face Masks and PPE Kits

Usage of PPE has been expanded tremendously to curb the current COVID-19 pandemic all over the world. Because of this, nowadays the whole world is suffering lack of PPEs which include a respirator, face shields, face masks, wearable protective clothing, etc. Wearing a face mask is one of the best protective routes to avoid COVID-19 infection. It is necessary to cover the faces of both non-infected and infected persons in public, mostly in locations where keeping a physical distance is hardly possible or unattainable. Individuals who are infected can prevent the virus from spreading which is developed, while sneezing or coughing, and the individuals who are not infected can prevent themselves by wearing a face mask against the viruses that exist in the air like aerosols or droplets, materials, or unintended touching of mucus membranes. Wearing a mask is a must to prevent airborne and spreading of virus through direct contact [82]. Figure 2 shows the illustration diagram of virus spread from one individual to another individual through the inhalation of aerosol/droplets that came out of the infected people caused by not wearing of protective face mask. Recently, respirators N95 and surgical masks can contribute various heights of safety as well as was commonly utilized by health workers, patients, and people in high-threat zone [82, 83]. Although wearing any type of face mask can efficiently defend any person from getting infected, however, this face mask itself

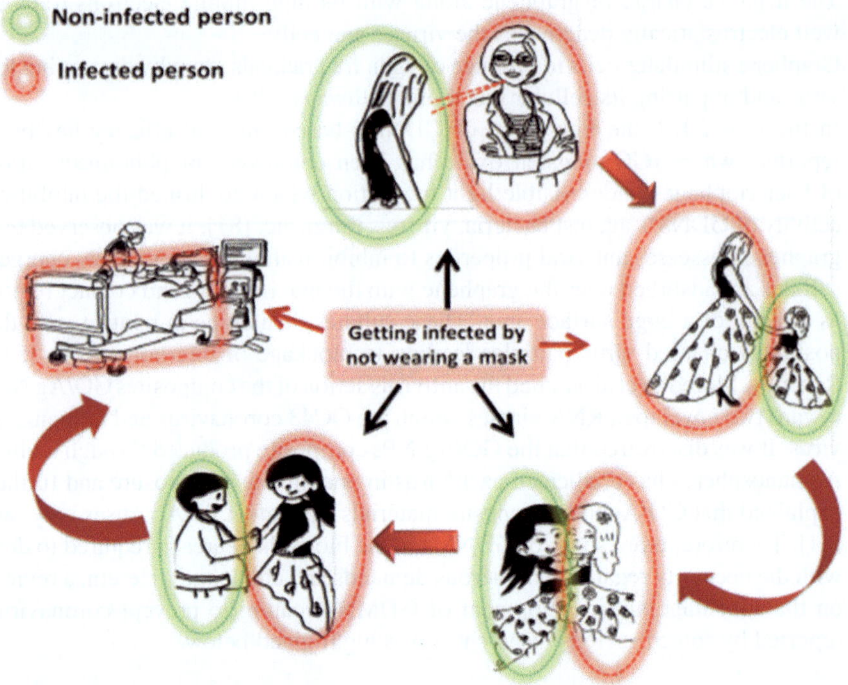

Fig. 2 Illustration diagram of transmission of coronavirus among one person to another person caused by not wearing a mask

emerges as fomite since the exterior layer of such face masks has hydrophobic characteristics due to which this virus can retain upon the exterior layer of the fabric. From this, it indicates that either these masks should be thrown away immediately after use or they have to go through a sterilization process before using it again and as for surgical mask it has a larger pore size which may not be able to prevent aerosolized viruses including coronavirus. None of the options are tempting because sterilizing the face masks derived from the polymer is difficult even by steaming, at the same time disposal of these face masks causes environmental damage. Moreover, due to the low-melting point viz. < 130 °C, these face masks cannot be sterilized and can't be reutilized [84]. The N95 respirator masks are found to be more efficient in protecting toward coronavirus in comparison to other surgical or cloth face-masks, as well as it is widely utilized by doctors and other medical staffs [82, 85]. The N95 certification showed that under standard conditions this mask can protect against the total bacteria in a salt's aerosol by 95%, and further this material has also been found to block bacteria whose diameter size is 300 nm (nm). However, the N95 mask is expensive as well as there is a shortage of supplies; moreover, its filtration effectiveness is reduced by up to 85% against the pathogen of less than 300 nm size [86, 87]. Due to the size of coronaviruses being 60–125 nm, the N95 mask does not fully provide the necessary protection against the current COVID-19 virus [88]. Consequently, there is a paramount necessity to design more effective filtration masks as well as PPE with high grade [84]. Over the years, various respirators filters and face masks conjugated with GDNMs have been produced since graphene NMs exhibit good electrical, anti-static, and antimicrobial characteristics to manage the COVID-19 pandemic. These filters and masks possess properties of self-sterilization and self-cleaning. Zhong et al. [76] designed a face mask based on graphene that can be recycled and reused by depositing laser-induced graphene (LIG) over the surgical masks which are available commercially using dual laser-induced forward transfer method for enhancing its and photothermal and superhydrophobic characteristics and the authors further demonstrated that the superhydrophobic nature of graphene makes it similar to the leaf of a lotus, in which the water droplets do not stick easily over the mask's surface. Moreover, these masks coated with graphene could be easily sterilized using sunlight (40–100 s). This happens as a result of absorption of light by graphene even more than 95% within the spectrum 300 nm–2500 nm, thereby increasing the temperature of the graphene-coated masks up to 70 °C in 40 s and > 80 °C in 100 s. This is sufficient enough to eradicate almost all kinds of pathogens, making it reusable or recyclable even after damage. In addition, masks coated with graphene helps in improving their self-cleaning capacity with 140° as a static contact angle [76]. Conversely, this photothermal efficiency cannot be seen in the other bare masks, because the absorption of sunlight by them is very weak. The authors further explained that the solar illumination of these masks is very slowly such that the temperature is unable to reach 50 °C even after 5 min. An image of a face mask-coated with graphene NMs has been depicted in Fig. 3a. It can be seen that the bare white mask is coated with the black color indicating the transfer of graphene upon the layer of the mask. Figure 3b exhibits the SEM image of the materials with graphene coating within the surgical mask. The Raman spectra of the mask-coated with graphene were shown in Fig. 3c, in which the chemical elements and phonon vibration modes that exist in the graphene-coated mask were identified and investigated. The peaks were obtained at 2665 cm^{-1} (2D-band), 1570 cm^{-1} (G-band), and 1330 cm^{-1} (D-Band), respectively, which confirms the incorporation of graphene. Figure 3d displays the water contact angle of the mask-coated with graphene which obtained 141°. In Fig. 3e, the comparison of self-cleaning properties between the pristine blue mask, and the black graphene-coated mask was shown. Further, the study has also been done to find out the health protection efficiency against Escherichia coli bacteria at an inactivation temperature over 60 °C, where LIG masks were compared with commercially activated carbon fiber (ACF) masks and the melt-blown fabrics (MBF) [89]. Amazingly, the antibacterial activity came out to be best for LIG (81.57%) as compared to ACF (2%) and MBF (9.13%). After 10 min of solar illumination, LIG material attained 62 °C, which enhances antibacterial efficiency (99.998%), whereas ACF attained only 52 °C with less antibacterial efficacy (67.24%). However, MBF acquired only 36 °C, but the antibacterial efficiency (85.3%) enhances, suggesting that temperature isn't the only key factor [89]. Further, Bonbouton Company had introduced masks decorated with graphene that can be reusable and effective in blocking the droplets which contain viruses [64]. ZEN Graphene Solution Ltd. in collaboration

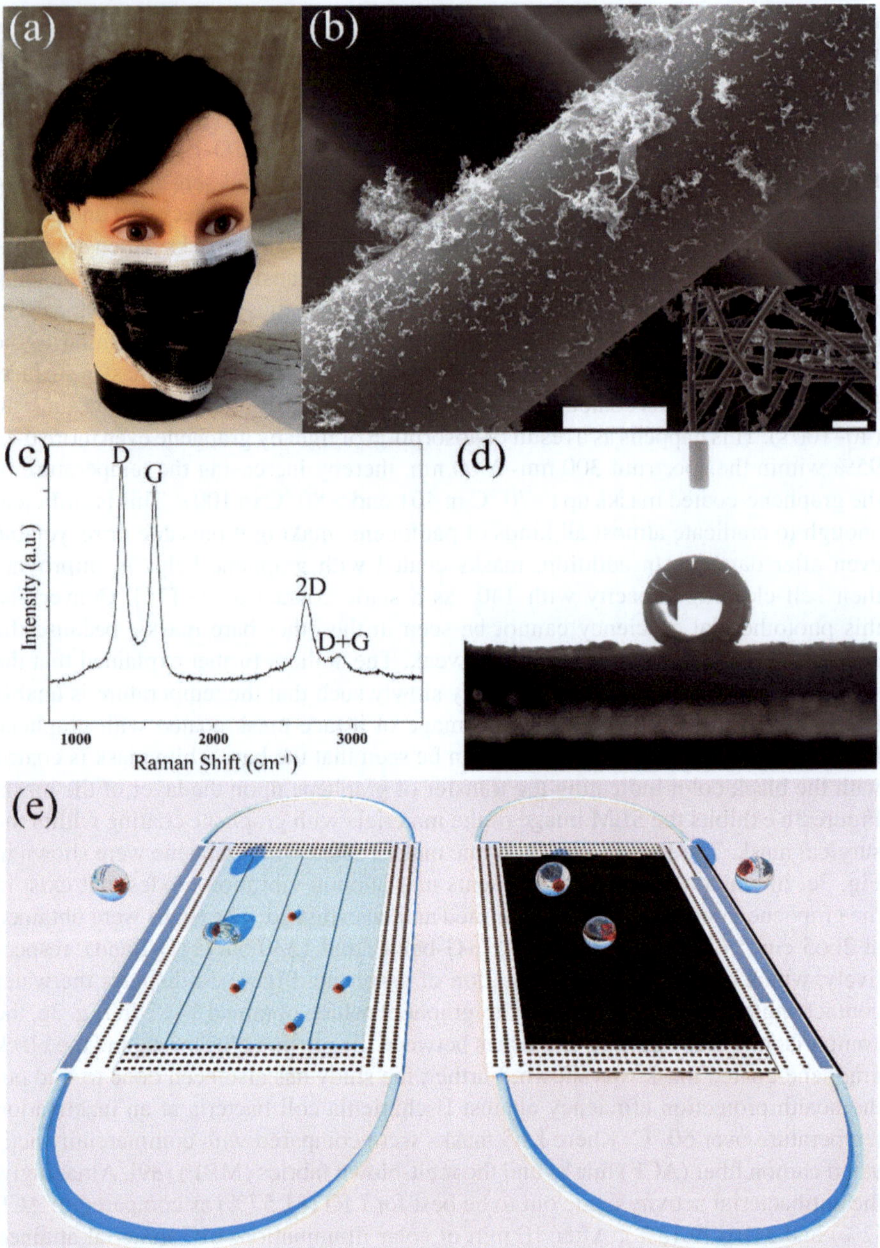

Fig. 3 **a** Illustrated image of the GDNMs-coated face mask through laser fabrication. **b** SEM images of the GDNMs-coated face mask at 10 μm scale bar and 100 μm scale bar (inset). **c** Raman spectroscopy for GDNMs-coated-mask. **d** Water wettability measurement over the surface of GDNMs-coated-mask. **e** Image representation showing the comparison between the bare mask (blue-left) and GDNMs-coated-mask (black-right) based on its self-sterilizing characteristics. Reprinted with permission from [76]. Copyright (2020) American Chemical Society

with Graphene Composites Ltd (GC) created a potential virucidal graphene-based composite ink that is to be spread over the textile's surface which includes N95 face masks and personal protective equipment to effectively increase the protection against COVID-19 [65, 90, 91]. Under the collaboration, ZEN synthesized AgNPs/GO ink which had successfully killed the earlier strains of COVID-19 also including the influenza A and B viruses. It was reported that after the completion of the test clinically, the ink based on graphene is likely to be integrated with the textiles which will further be contained within filters and mask introduced by GC. Further, planar TECH and IDEATI's 2 AM have also designed graphene/carbon NMs-based cotton material face mask which provides the advantage of exhibiting both antibacterial and antistatic properties through graphene [92]. The mask-coated with graphene has been proven to protect (99.95%) against the staphylococcus aureus bacteria in 24 h, simultaneously maintaining the mask to remain fresh and clean. In addition, this graphene-coated mask also resists dust and found useful to fight against particles of size less than 2.5 microns in diameter which are remain floating in the air. This mask can be washed and reused up to 10 times without losing its antibacterial or antistatic properties [92]. At recent, LIGC Applications have designed respirator mask based on graphene and named it "Guardian G-Volt" which are sterilizable as well recyclable, these masks found their capability on competing against ideal N95-respirator masks [93]. These masks have superior characteristics which include anti-bacterial, dust-repellent; in addition, they are effective to block airborne particles of PM2.5. Moreover, the efficacy of the Guardian G-Volt mask was found out to be 99% and 80% against pathogens with sizes >300 nm and sizes <300 nm, respectively. In addition to this, an electrical filter based on graphene was also installed into this mask which was developed by laser-induced graphene (LIG) process to trap germs, bacteria, and viruses like the SARS-CoV-2 virus. The Guardian G-Volt mask has the potential to conduct the electricity for self-sterilization of its surface, through plugging the power into a portable battery pack via a USB port after which a low level of electrical charge will pass through the mask which will repel each dust or pathogens captured into it. It also has an LED light system that warns the user when the mask needs to be changed. It can then be heated and sterilized in the at-home dock [93]. Some research teams have deposited a few layers of carbon sheets over the commercial unwoven masks by using a laser manufacturing technique by which the masks turned out to be superhydrophobic, thereby reducing the risks of contagious droplets to adhere into the mask, whereas due to high light-absorption characteristics of graphene it has the ability to self-sterilized by exposing to sunlight [94]. Maio et al. observed that the incubation of COVID-19 virus in GO suspension drastically degraded viral infectivity even at the minimum concentration of GO (0.06 mg/mL), as reflected by the decrease in the cytopathic effect of the virus and increased cell survival [95]. GO also remarkably reduces a load of viral particles as assessed by immunofluorescent labeling with an anti-SARS-CoV-2 spike protein antibody and reduced cellular cytotoxicity as measured by both Crystal violet staining and lactate dehydrogenase (LDH) release in the supernatant, further demonstrating that suspension of GO interacts with SARS-CoV-2 viral particles and lessens viral infectivity in the in vitro live virus model of SARS-CoV-2 infection of VERO cells [95]. So from the results observed by the previous researchers, it can be seen that graphene-coated face masks can be used for protection against coronavirus as well as can be reused for the long term. However, the national public health agency of Canada notified that inhalation of graphene may cause damage to the lung. In 2016, some researchers also reported that a little amount of graphene NPs aerosols might pass down through a mouth and nose, thereby penetrating within the lungs. However, in the year 2018, researchers observed that brief exposure to a little portion of aerosolized graphene doesn't particularly harmful to lung cells in a model and some other researchers also suggested that a tiny amount of graphene into lungs will be fine, though a huge amount will be hazardous [80].

Another protective measure used by healthcare frontline workers to prevent coronavirus is wearable PPE. Personal protective clothing has been designed for the utilization in the protection against several hazardous as well as in threatening circumstances which may affect or even end life. Generally, PPE kits aren't antibacterial or antiviral. Traditional wearable protective equipment has some drawbacks which include heaviness, diminishing scope of vision, heat stress, difficulty to move freely, low-heat dissipation, high-physical stress, shortage in breathing, bulky nature as well as low protection from microorganisms, viruses, etc. Utilization of NMs-based and antiviral coating NMs in the fabrication of PPE enables to enhance its protectiveness [82]. Graphene, owing to its superior qualities, modified graphene-based fibers can play a vital role to prevail the impediments mentioned above for traditional protective equipment along with improving its antibacterial efficiency, UV resistance, mechanical strength, and electrical characteristics as well as preventing its heat diffusion [96]. A review on the uses of graphene as modified personal protective equipment clothing was first reported by Bhattacharjee et al. [96]. Zhao et al. fabricated the modified antibacterial GO cotton fabric, where GO coating was done over cotton via various techniques such as radiation-induced crosslinking using (APS as agent), chemical crosslinking, and adsorption process which exhibited outstanding antibacterial properties through all the composites [43]. It was further observed that in protection against both gram-positive and gram-negative bacteria, all the GO-based cotton fabric composites offered higher antibacterial efficiency than the bare cotton fabric [43]. Ye et al. [97] investigated GO/poly(diallyl dimethylammonium chloride) (PDDA) composite's, GO/polyvinylpyrrolidone (PVP) composite's, rGO's, and GO's antiviral activities with graphite and graphite oxide precursors. Further, the study observed a wide range in antiviral efficiency of GO toward an RNA virus, PEDV, and a DNA virus, PRV. The authors also demonstrated that the sharp-edged structure of GO with a negative charge is responsible for the antiviral characteristics exhibited by GO. Researchers also found that 3D printing based on GDNMs could contribute to the development of PPE manufacturing [38]. It has also been observed that the combination of GO and Ag NPs exhibiting antivirals properties can potentially trap and kill viruses [81]. Moreover, the mono or multi-layered GDNMs could be utilized as a mist spray to prevent the viruses to access as well as they can be coated over textiles to enhance protection. So, multilayers of GDNMs should be applied to PPE for better protection and to keep it dry which will help in preventing the aerosol spreading of coronavirus in medical healthcare workers. Thus, a previous study on

GDNMs displayed inherent antiviral properties suggesting that materials based on GDNMs will not only save the individuals against the virus, but also eradicate the virus. Hence, GDNMs can be considered as a potential candidate for the development of several products which include PPEs, face masks, including various other diagnosis-related instruments or materials, etc., to prevent aerosol transmission into hospitals and other infected areas.

2.2 Graphene-Based Surface Coating, Disinfectants, and Sterilization

COVID-19 virus stays alive over surfaces including plastic, glass, metal, etc., for about 9 days after exposure. Several types of research have been done toward analyzing the blocking capacity of viruses in various environmental circumstances. For example, exhalations of viral particles stay infectious even after 3 h of exhalation [62]. In addition to direct contact and droplets, fomites are also one of the possible factors for transmitting the COVID-19 infection including other viral diseases. Fomites are stationery products that could transmit the virus or bacteria to another individual via unexpected transmit through exposure to virus present over lifeless materials. ATMs, lift buttons, doorknobs, touchscreen, tables, and other workplace-related materials are some common examples of fomites. Though, the survivability relies upon the surface's nature, relative humidity, condition of the environment, and a specific strain of the virus. Recently, Riddell et al. investigated the survivability rate of the coronavirus upon the stuff present which has been used in our day-to-day life which includes vinyl, cotton, stainless steel, paper money, etc. [98]. Surprisingly, coronavirus is active for about 28 days on non-porous substrates when incubated in the absence of light, at 20 °C, at 50% relative humidity. The stability in cotton is lesser which underlined the importance of frequent hand-wash and sanitization, demanding for the study of new functional coating which would help in shortening the stability of microorganisms and viruses. Currently, most of the decontamination methods widely used include cleaning, degradation, sanitization, sterilization, chemical disinfection measures, solvent/detergent, or halogenated disinfection strategies but these methods suffers from various limitations including production of toxic byproducts along with lower disinfect effectiveness to suppress the virus [32, 45, 99]. In addition, various techniques like HEPA filters, ion generation, battle decontamination system, ozone, ultraviolet germicidal irradiation, hydrogen peroxide decontamination (STERRAD Sterilization system), and photocatalytic oxidation have been utilized in decontaminating as well as disinfecting the public areas such as shops, health care center, offices, airports, etc. [100]. Such techniques, mechanisms and products are effective to certain points, yet most of them aren't appropriate for sensitive tools particularly in a biomedical application. Hence, there is an urgent need to introduce a new system with more smart materials and technologies to protect against the coronavirus. In this regard, GDNMs have the potential to decontaminate and disinfect the COVID-19 virus. Graphene is indissoluble in anything except in some solvents. Based on these properties, Atkinson explained that the deposition of GDNMs coating over the substrate like doorknobs can lead it to remain effective and present unless a solvent is used to remove forcibly which will provide prolonged protection up to 60 days. Further, this coating can be delivered through a direct spray bottle [66]. Dawid et al. [101] reported porous foams based on GO/polyethyleneimine for effectual elimination of hazardous cations from water, which can also work as an antiviral coating of surfaces. A crucial feature for the understanding of graphene-based smart coatings concerns their incorporation with the actual surfaces. A tremendous research attempt was devoted to developing GDNMs-based inks or dispersions along large processibility, where such inks can be applied over the surfaces via various processes including roll-to-roll processing, coating, spraying, dip-coating, screen printing, (ink-jet) printing, etc. [102]. The ink that is going to be used for deposition and the features of the surface decide which method will be more suitable. For example, if transparency is needed then spray-coating could be a favorable choice [103], whereas if there is the requirement of industrial-scale coating of textiles then the pad-dry-cure method might be the preferable choice as described in the previous studies related to e-textiles [104]. Andreia and their co-workers reported electrospun poly(lactide-co-glycolide) PLGA–chitosan mats fabricated with GO/Ag which shows excellent antimicrobial properties and the authors further suggested that combination of GO/Ag nanocomposites over PLGA–chitosan mats surface gives possibilities in developing scalable, low cost, and antiviral coating materials onto the surface of solid [105]. To boost the antimicrobial properties, GDNMs find their application in developing antiviral coatings with ad hoc functional groups. Keeping in mind the huge amount of concentration of –COOH functional groups present in the structure of the virus and the low survival rate of this virus over the surface of copper (Cu) surfaces, Srivastava et al. [38] suggested that GDNMs-SO_3 coatings integrated with copper NPs can be considered as a favorable material for the advancement of the antiviral surface coating. The authors further demonstrated that such materials can be efficiently utilized in trapping and damaging the structure of the virus as well as in minimizing its survivability over diversely coated surfaces [38]. Thus, the coatings of GDNMs-based composites can play a vital role in diminishing the survivability of viruses over various high-touch areas. Furthermore, the mist spray-based on graphene can be introduced to help in cleaning and sanitizing the purpose of various surfaces and bodies of humans. In addition, it can find its application in preventing against S-protein of COVID-19 infection significantly through developing mouth spray or nasal. Moreover, graphene and GDNMs-coated disinfecting surface wipes are also one of the preferable measures in disinfecting various contaminated surfaces. Amidst the COVID-19 pandemic situation, human wastewater has become a new problem for the world as it was reported that the coronavirus in fecal samples was detected confirming the presence of coronavirus in human wastewater. At recent, coronavirus is found to present in numerous sewage in several countries including France, Italy, Australia, The Netherlands [109], and the U.S [106–111]. Recently in Jaipur, Arora et al. [112] investigated the untreated wastewater specimens, collected from the municipal wastewater treatment plants (WWTPs) and hospital wastewater, where they confirmed the first case of coronavirus being found in untreated wastewater in India [112]. So, it is in urgent need to develop an efficient wastewater treatment, moreover, it's very crucial to prevent the spreading of the virus via contagious wastewater. Graphene-based membranes have shown their application in removing viruses and bacteria from water. The previous studies described that graphene or GDNMs-based membranes can be effectively employed for the removal of a toxic substance, microorganism, molecules, ions, viruses, etc., out of the water, through tuning their interlayer spacing [27, 45, 113]. So graphene can be a potential material for the separation of the SARS-CoV-2 virus from water. Thus, separation of coronavirus from water could be feasible employing graphene-based filter membranes by controlling their microstructural properties and interlayer spacing. However, so far very few people have worked in GDNMs-based wastewater treatment applications, so the appropriate way is to be determined to develop graphene-based membrane and filter for the removal of virus out of water henceforth. Moreover, photocatalyst based on GDNMs can help in deteriorating and deactivating the SARS-CoV-2 virus in water. Numerous research related to photocatalytic inactivation of pathogens has shown great decontaminant properties against various microorganisms, which not only destroys them but also sterilizes [114, 115]. In addition, an effective air-conditioner and air-purifier appliance are necessary to develop to filter and clean the germs and viruses including coronavirus that exists in air through designing modified multi-layered graphene or GDNMs-based positive charge filters. Recently, Pang et al. reported the incorporation of GDNMs in air purification products to mitigate the threat of virus infection [116]. Recently, Stanford et al. developed an air filter that can be self-sterilized using graphene NMs [74]. The air filter is derived from LIGs based on LIG; synthesis of graphene foam was done utilizing the heating surface of a common carbon material (polyimide-sheets) with industrial CO-based LASER cutting which. Such filter has shown excellent antimicrobial properties which can eradicate various microorganisms, bacteria, and viruses through a joule heating process, in addition, the filter was found to achieved 300 °C or >300 °C. Moreover, LIGs-based air filter can eradicate other particles and pathogens which include mycotoxins, fungi, spores, allergens, prions, exotoxins, endotoxins, etc. Furthermore, gel or lotion based on GDNMs could be introduced as disinfecting agents to prevent as well as to eliminate the COVID-19 virus. Several new products based on GDNMs can be developed to fight against COVID-19. Thus, further effort is required to acknowledge such specific requirements regarding numerous challenging and exciting opportunities related to bio-medical. Figure 4 illustrated the GDNMs coating over the material's surfaces which will help to prevent from widespread of coronavirus and from getting infected.

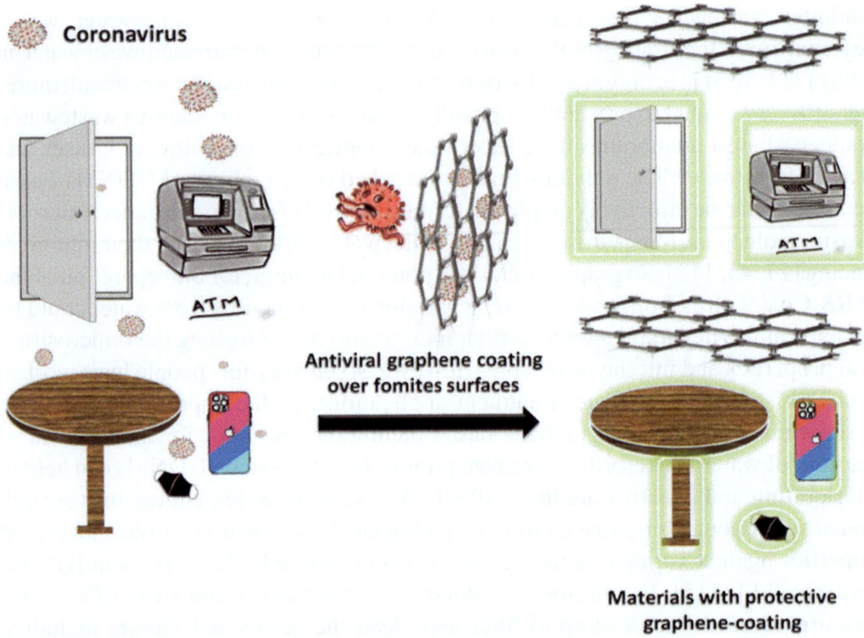

Fig. 4 Diagrammatic representation of GDNMs coated over the surface of fomites to control the spread of coronavirus

3 Diagnosis

For now, the most important thing is to develop an efficient test system for the detection of the COVID-19 virus at the earliest. Various researchers have explained the advantages of proper tracing, testing, and isolation and quarantine of the community to manage and prevent the transmission of the virus from one individual to another. Currently, the RT-PCR test is a widely used testing technique for the detection of current COVID-19 infection, in which the genetic material of the virus is examined. This testing method has some shortcomings such as taking a long time (2–7 days) to provide results due to different factors [3]. Graphene exhibits excellent electrical, electrochemical, piezoelectric properties due to which it can be used in various applications such as piezoelectric biosensors, field-effect transistor (FET)-based biosensors, and electrochemical biosensors for fast, low cost, accurate, and primitive stage identification of viruses. Materials based on GDNMs can be employed as Surface Plasmon Resonance (SPR) substrate for designing ultra-sensitive diagnostic tools of viral infections [38]. It was reported that if bacteria come in contact with graphene's surface then they lose their integrity [31, 60], although effectiveness on viruses has been less well studied. Recently, numerous researchers have focus on the development of smart graphene-based sensors to diagnose viral diseases [3, 31, 38, 45]. Recently, some scientists from Korea have designed a graphene-based biosensor that showed effectiveness in identifying the current COVID-19 infection in just 1 min. This biosensor is composed of FET- developed on high-standard graphene NMs which exhibits high electronic properties. Antibodies specific to the spike SARS-CoV-2 proteins, the spiky outgrowths on the surface of the virus, are attached to the surface of graphene. The synergy of SARS-CoV- 2 with antibodies causes a change in the current strength via the transistor, which can be easily determined through a measuring instrument [63]. Similarly, the company, Grolltex created an efficient and cheap biosensor based on graphene using plastic as a substrate that has the ability for detection against 12 distinct viruses simultaneously, which includes the COVID-19 virus [63].

3.1 Testing

Controlling transmission is challenging as it is difficult to identify an infectious person. The community transmission of the SARS-CoV-2 virus is most likely due to the absence of symptoms. Identifying asymptomatic cases is challenging as individuals are not aware that they are infected until they get tested during their infection period. Asymptomatic persons seem to account for approximately 40–45% of COVID infections, in addition, they can spread the virus to another individual for longer than 14 days. Those who have not shown symptoms also need to be included in the testing program due to the threat for silent transmission via asymptomatic individuals. So, it is of utmost necessity to focus on the increase in the testing programs of COVID-19 all over the world including those people without any symptoms of SARS-CoV-2 infection. Further, the huge number of tests required to be done poses difficulties to the healthcare systems in reporting the results of the RT-PCR test to the patient, where there have been delays in many cases up to around 7–10 days in reporting positive results [3] which further leads to delay of taking required steps to begin quarantine and observation of patients as well as requires the involvement of trained-staffs in each analyzing processes. Moreover, the RT-PCR methods only determine active carriers of the virus. Detecting recovering individuals based upon coronavirus disease antibody demonstration is just as significant because it is capable of providing health officials with important details concerning the possible consequences of reopening measures [117]. Serologic assays identify circulating antibodies peculiar to SARS-CoV-2 antigens, such as the outer spike protein and the nucleocapsid protein [67]. Though, it is impossible in distinguishing between asymptomatic carriers and immune persons by employing antibody detection. Hence, a system that can determine both the viral and antibody status of a person is needed to efficiently reduce the chance of community spread via the SARS-CoV-2 virus. So, it is imperative to develop ultra-sensitive, fast, cheap, telemedicine COVID-19 testing kits which will not involve sophisticated equipment or trained staff to detect the present and past contagiousness status of a patient [118]. The POC COVID-19 testing has been developed, yet all the testing equipment that is available commercially can only give qualitative outcomes. Quantitative analysis of COVID-19 biomarkers utilizing a telemedicine device could present seroconversion information related to the time course of a disease and give predictive information of disease severity. Regarding this, electrochemical biosensors become beneficial because they have the efficiency to detect rapidly and easily use in POC applications [11]. Recently, Rebeca et al. [3] designed a portable, wireless electrochemical-multiplexed device to perform ultra-rapid detection of SARS-CoV-2 RapidPlex. This method can quantitatively detect biomarkers specific to SARS-CoV-2 in both saliva and blood which include specific immunoglobulins (Igs), SARS-CoV-2 nucleocapsid protein (NP) against CRP, and SARS-CoV-2 spike protein (S1) (S1-IgM and S1-IgG), between biologically relevant scales. It also employs capture antigens and antibodies immobilized on mass-production, economic laser-engraved graphene (LEG) electrodes [3]. Such a device can monitor the disease further via identifying the disease stage, which permits to identify the person's infectiousness, vulnerable, either-or immune [3]. Moreover, Saliva-compatible POC attempts can be beneficial because saliva has rich information and the patients themselves can collect it for telemedicine essays in an easy and non-invasive manner [119]. Moreover, sensors and detectors based on graphene have received significant attention amid the COVID-19 pandemic, in which the viral variant of SARS-CoV-2 can be rapidly detected. These sensors have immense potential toward replacing the RT-PCR kit, because of their highly sensitive and fast detection in just a few minutes. These appliances can also be set up near the entrance area which includes railway stations, airports, bus stands, workplaces, and borders region. Due to graphene exhibiting, a large surface area to volume ratio, excellent electronic conductivity, as well as ultra-high sensitivity toward environmental change makes it potential material in developing novel and extremely sensitive sensors related to COVID-19 infection and other pathogens. Zuo et al. [120] detected SARS-CoV in sputum by fabricating piezoelectric immunosensor successfully, in which a horse polyclonal antibody against SARS-CoV has bounded upon the surface of the piezoelectric crystal in an organized arrangement via protein. The authors further explained that when the antigen sample was atomized into an aerosol, the antibody on the crystal adsorbed SARS antigen and the change in the mass of piezoelectric crystal results in a change in frequency. The previous study on piezoelectric biosensor devices showed good reproducibility and it revealed that it can be reusable (approx. 100 times) without losing its detectable action [121].

3.2 Detection (Biosensor)

Recently, sensing devices based on GDNMs have contributed very encouraging outcomes. Graphene NMs have been considered for the development of high-efficiency biosensors because of their optical nature and electrochemical stability, excellent electrocatalytic behavior, good thermal, and mechanical properties [122]. Graphene-based mediums have been utilized for designing biosensors to inactivate microorganisms or viruses. Janire et al. investigated the immobilization of biomolecules onto graphene surfaces using the surface chemical engineering method [123]. Afsahi et al. detected the Zika virus using a portable biosensor based on graphene [124]. Similarly, the influenza virus was detected employing a GO-based biosensor by Joshi et al. [125], in which the detection limits for the target virus were obtained as 33 and 26 PFU/mL in the samples of saliva and PBS, respectively. Pant et al. [126] have designed FET biosensor based on rGO to detect rotavirus; thereby suggesting that the biosensors based on GDNMs-based FET could be utilized for the detection of the SARS-CoV-2 virus rapidly and sensitively. To the best of our knowledge, recently, Seo et al. [127] reported the fabrication of graphene-coated FET device for detection of COVID-19 virus, which was composed of graphene-coating sheets of the FET transistor with a specific antibody using a bifunctional linker containing pyrene which ties with graphene through π–π stacking interaction, and an activated carboxylic acid to build a stable amide bond with the antibody that allowed the ad hoc detection against the SARS-CoV-2 spike-protein. Further, the features and effectiveness of this device have been evaluated against nasopharyngeal swab samples, cultured-virus, and antigen-protein from COVID-19-infected patients. This designed device has shown inspiring results and is capable to identify the spike-protein of COVID-19 virus in phosphate-buffered up to the 1 and 100 fg mL^{-1} in clinical samples within a minute. Moreover, graphene-based FET devices possess the ability to detect SARS-CoV-2 in the cultured form up to $\sim$1.6 $\times$ 10 pfu mL^{-1}, and 2.42 $\times$ 10 pfu mL^{-1} in clinical samples. Such advanced technology derived from graphene coating brings lots of advantages which include fast response, not require preparing the sample, high sensitivity as well as can be utilized in line with many nasal swab tests over conventional techniques. Moreover, such FET-based biosensors have the capability of distinguishing between the COVID-19 virus and other members of the coronavirus family. In addition, the authors also claim to detect the presence of viruses in the air [127]. The widespread production of a FET sensor derived from the incorporation of graphene with the SARS–CoV-2 spike antibody encourages the crucial role of graphene NMs for diagnostic scope [128]. Alternatively, electrochemical biosensors based on GDNMs have gained great attention for the detection of viruses including the current COVID-19 virus. The integration of the biosensors into electrochemical devices has been developed quickly in the last few years which allow the development of POC test of specific microorganisms. Huang et al. [73] designed an electrochemical immunosensor based on AgNPs/graphene composite materials which showed highly specific and sensitive to AIV H7. Recently, Qiu et al. have introduced a smart graphene-based multiplexed device for the detection of the spike protein of COVID-19 virus, the C-reactive protein, and the associated immunoglobulins in either saliva or blood [129]. Layqah et al. fabricated the electrochemical immunosensor by [130] employing gold NPs and modified-carbon electrodes to monitor the proteins of MERS-CoV and human coronavirus (HCoV) in spiked nasal specimens. These graphene-based sensors hold massive potential in the detection of COVID-19 infection [31, 38, 82], though much-advanced research is still needed to conduct for the development of high-quality diagnostic tools. The illustrated mechanism of GDNMs-based electrochemical device for the detection of COVID-19 virus was presented in Fig. 5a. In addition, the converse GDNMs-based

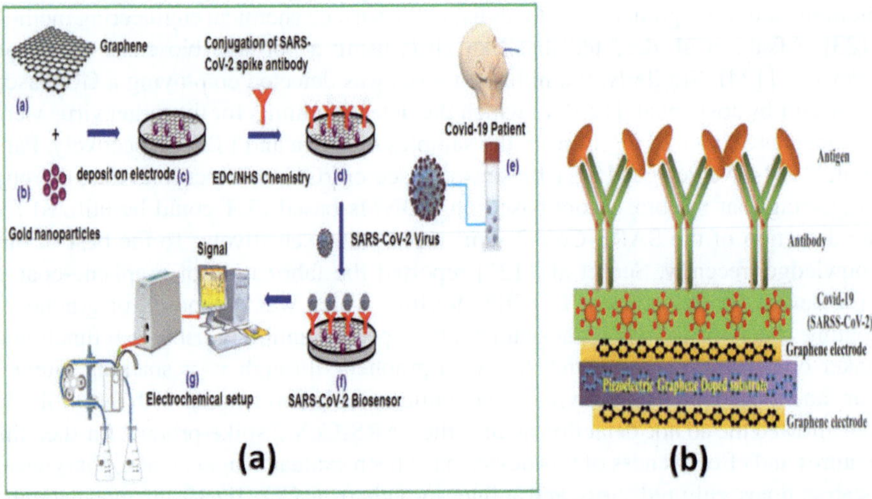

Fig. 5 **a** Speculative representation of electrochemical biosensors based on GDNMs for detecting the COVID-19 infection. **b** Diagrammatic representation of COVID-19 absorption through piezoelectric GDNMs. (Reprinted from reference [38], Copyright (2020), with permission from Elsevier)

piezoelectric sensors can be another option to design the coronavirus-based biosensors. Lian et al. fabricated a piezoelectric sensor employing aptamer/graphene integrated gold electrode for fast and accurate detection by employing Staphylococcus aureus as a microbiological bacterium. [131]. Therefore, the generation of piezoelectricity in graphene through numerous strategies can result in the advancement of high-efficiency graphene-based piezoelectric biosensors which can rapidly and sensitively detect various microorganisms and viruses which include the COVID-19 virus also. The previous study also suggested that the bare and doped graphene could be utilized as an electrode and quartz crystal in the piezoelectric crystal microbalance, respectively, to increase the sensitivity of the virus by the rise in the binding of antibody/antigen into the surface of electrically conductive graphene as well as improved alterations in the mass of the piezoelectric graphene crystal [38]. Figure 5b shows the hypothetical diagram of showing how piezoelectric crystal microbalance sensor-based on GDNMs could detect various viruses including the COVID-19 virus.

3.3 Treatment

The previous study claim that graphene-based nano-drugs integrated with antivirals could be an effective and beneficial composition to combat COVID-19. GDNMs have broad applications; drug delivery application is one of them for antiviral drugs such as reverse transcriptase inhibitors associated with graphene quantum dots for

HIV treatment and GO/hypericin against reovirus [132, 133]. It was reported and explained that the interplay between the viruses and rGO sulfated derivatives are polymer density-dependent and sulfation degree [51]. The smaller the size and the higher the degree of sulfation, the greater is the effect on the herpes virus. This was demonstrated in the role of a collective conclusion of the easiest bending and cooperative encapsulation by two or more small GO sheets [51]. GO flakes can confine and wrap microorganisms by enclosing them in an insulating carbon blanket [134]. Previous studies based upon GO explained in what manner such NMs can be convenient in examining coronavirus helicase inhibitors being drug nominees for antivirus remedy [135]. Helicases are enzymes that disjoined double-stranded nucleic acids into single-stranded at the time of virus proliferation and replication. Therefore, helicases have allowed for the use in antiviral remedies. Nevertheless, the traditional method for helicase activity is prolonged and inefficacious due to lengthy preparation time and procedural problems. GDNMs have a superior binding of single-stranded DNA over their surface, which can help in obtaining an economic monitor for helicase activity. More notably, it was found that GDNMs testing is high throughput and provides parallel assays for screening of helicase inhibition by drugs. Recently researchers explained in what manner the SARS-CoV-2 Spike S1 protein receptor-binding domain could associate with heparin and transform conformation. This has significance in developing a favorable therapeutic through reprocessing heparin and glycosamminoglycans-based antivirals [136], containing GO's sulfated derivatives. The graphene's light absorbance demolishes the virus particles after capturing them. Deokar et al. prepared sulfonated magnetic NPs with rGO, in which it captured and destroyed herpes simplex virus type 1 (HSV-1) in a photothermal manner utilizing NIR light [137], thereby explained how GDNMs capture can be linked with NIR treatments of lungs. In fact absorbers like GDNMs are in the NIR tissue transparency window and enable deep penetration in the body of incident light, as demonstrated for lung metastases treatment. In 2014, researchers demonstrated there is no dissimilarity between GO and its sulfated rGO derivative regarding the antiviral action against HSV-1. The GDNMs have the same negative charge density, which is possibly the main reason inhibiting the virus [97, 138]. GDNMs are intrinsically able to adsorb charged lipids thereby demolishing membranes after interaction with their aromatic plane [46, 47]. Chen et al. have used this property to describe GO action against feline coronavirus compared to its ineffectiveness against non-enveloped infectious bursal disease virus [39]. Moreover, Du et al. studied the antiviral action of Ag NPs/GO nanocomposites against porcine reproductive and respiratory syndrome virus and where GDNMs-based composite was shown to exhibit a wide antiviral action. In addition, it is reported that AgNPs/GO nanocomposite treatment increases the generation of interferon-α (IFN-α) and IFN-stimulating genes (ISGs), leading to the prevention of rapid growth of the virus. [139]. GDNMs have previously displayed certain exciting outcomes in biomedical applications and have been broadly used for biological activities. Thus, from previous findings, GDNMs can be considered as an effective material in fighting against COVID-19 by developing graphene-based diagnostic devices, antiviral face-mask, PPE, surface coating, and many other components which will help in prevention and protection from this virus and other emerging and remerging diseases.

4 Challenges and Limitations

- Typical challenges such as graphene instability or aggregation need to be addressed before its usage in drug delivery or vaccination. In addition, in vivo tests on dosage, surface chemistry and exposure route of the drug-graphene composite are necessary for full commercialization.
- Much effort and research are still needed to establish mature handling, processing, production, and scale-up route mapping of GDNM-based products.
- There has not been any clinical trial yet for graphene in spite of its effectiveness as an antiviral component against COVID-19.
- There is a theoretical possibility of lungs damage from using a graphene-coated mask. This happens when the graphene particles seep through the filter layers of the mask and goes into the lungs. And the body is unable to remove such particles swiftly. Even so, the risk of lungs damage from contracting coronavirus is higher than the theoretical risk of lungs damage from using a graphene-coated mask. Therefore, it would be wise to wear the graphene-coated mask rather than facing the complication of getting a COVID-19 infection.

5 Conclusion

The COVID-19 pandemic has caused a lot of unparalleled loss of lives and economy all over the world. Till now, there is a tremendous rise in the spreading of SARS-CoV-2 infection in the whole world non-stop which leads to an increase in the number of SARS-CoV-2 positive patients and mortality rate every day. So, the development of smart technology and materials are very much necessary in this current pandemic for fighting against the worldwide troublemaker known as the SARS-CoV-2 virus. GDNMs have shown good performance to fight against COVID-19 where GDNMs could be used for the establishment of (i) face masks with small size pores, ii) various PPE and various medical-related protective appliances, iii) Coating with antiviral and antibacterial to protect from getting infected and widespread through high-touch surfaces, iv) wastewater treatment, v) electrochemical, FET, piezoelectric-based biosensors for rapid, early and accurate detection and diagnosis, and vi) SPR substrates for sensitive diagnosis of viruses and biomolecules which will help in prevention, detection, sterilization, diagnosis, treatment and restriction of COVID-19 virus-spread. Despite the advancement of GDNMs-based sensors successfully for diagnostics and drug screening, as of now the medical components and devices based on GDNMs have not progressed to the clinical trials stage indicating that the track of GDNMs toward biomedical applications is still in an early stage. A very crucial topic of research of GDNMs would be progressing the effort to the stage of clinical trials and manufacture of antiviral surface coatings protective components, diagnostic devices, etc., for sale in the market which will help in combating the current SARS-CoV-2 epidemic. Amidst the COVID-19 pandemic situation, we have discussed the antiviral properties of graphene and reviewed the recent advancement related to GDNMs in fighting against the COVID-19 pandemic. Further research and development of GDNMs are needed as they possess high potential for breakthrough innovations for future advancement of equipment for diagnosing, treatment, and prevention of various emerging and remerging diseases.

References

1. Russell SM, Alba-Patiño A, Barón E, Borges M, Gonzalez-Freire M, de la Rica R (2020) Biosensors for managing the COVID-19 cytokine storm: challenges ahead. ACS Sens 5:1506–1513
2. Chan WCW (2020) Nano research for COVID-19. ACS Nano 14:3719–3720
3. Torrente-Rodríguez RM, Lukas H, Tu J, Min J, Yang Y, Xu C, Rossiter HB, Gao W (2020) SARS-CoV-2 RapidPlex: a graphene-based multiplexed telemedicine platform for rapid and low-cost COVID-19 diagnosis and monitoring. Matter 3:1981–1998
4. Delta plus India (2021) Scientists say too early to tell risk of Covid-19 variant. https://www.bbc.com/news/world-asia-india-57564560
5. Delta Plus (2021) List of states where this new Covid variant has been reported so far. https://www.hindustantimes.com/india-news/delta-plus-list-of-states-where-this-new-covid-variant-has-been-reported-so-far-101624546024157.html
6. https://www.worldometers.info/coronavirus/countries-where-coronavirus-has-spread/
7. World Health Organization: Coronavirus Disease (COVID-19) Pandemic. https://covid19.who.int/
8. Gao Q, Bao L, Mao H, Wang L, Xu K, Yang M, Li Y, Zhu L, Wang N, Lv Z, Gao H, Ge X, Kan B, Hu Y, Liu J, Cai F, Jiang D, Yin Y, Qin C, Li J, Gong X, Lou X, Shi W, Wu D, Zhang H, Zhu L, Deng W, Li Y, Lu J, Li C, Wang X, Yin W, Zhang Y, Qin C (2020) Development of an inactivated vaccine candidate for SARS-CoV-2. Science 369:77–81
9. Gil C, Ginex T, Maestro I, Nozal V, Barrado-Gil L, Cuesta-Geijo MÁ, Urquiza J, Ramírez D, Alonso C, Campillo NE, Martinez A (2020) COVID-19: drug targets and potential treatments. J Med Chem 63:12359–12386
10. Kaur M, Tiwari S, Jain R (2020) Protein-based biomarkers for non-invasive Covid-19 detection. Sens Bio-Sens Res 29:100362
11. Jalandra R, Yadav AK, Verma D, Dalal N, Sharma M, Singh R, Kumar A, Solanki PR (2020) Strategies and perspectives to develop SARS-CoV-2 detection methods and diagnostics. Biomed Pharmacother 129:110446
12. Singal CMS, Jaiswal P, Seth P (2020) SARS-CoV-2 more than a respiratory virus: its potential role in neuropathogenesis. ACS Chem Neurosci 11:1887–1899
13. Maqbool I, Rehman F, Soomro F, Bhatti Z, Ali U, Jatoi AH, Lal B, Iqbal M, Phulpoto S, Ali A, Thebo KH (2021) Graphene-based materials for fighting coronavirus disease 2019: challenges and opportunities. ChemBioEng Rev 8:67–77
14. Kumar A, Vemula PK, Ajayan PM, John G (2008) Silver-nanoparticle-embedded antimicrobial paints based on vegetable oil. Nat Mater 7:236–241
15. Wei C, Lin WY, Zainal Z, Williams NE, Zhu K, Kruzic AP, Smith RL, Rajeshwar K (1994) Bactericidal activity of TiO_2 photocatalyst in aqueous media: toward a solar-assisted water disinfection system. Environ Sci Technol 28:934–938

16. Shen H, Zhang L, Liu M, Zhang Z (2012) Biomedical applications of graphene. Theranostics 2:283–294
17. Wang J, Mu X, Sun M, Mu T (2019) Optoelectronic properties and applications of graphene-based hybrid nanomaterials and van der Waals heterostructures. Appl Mater Today 16:1–20
18. Devi NA, Nongthombam S, Sinha S, Bhujel R, Rai S, Singh WI, Dasgupta P, Swain BP (2020) Investigation of chemical bonding and supercapacitivity properties of Fe_3O_4-rGO nanocomposites for supercapacitor applications. Diam Relat Mater 104:107756
19. Nongthombam S, Devi NA, Sinha S, Bhujel R, Rai S, Ishwarchand W, Laha S, Swain BP (2020) Reduced graphene oxide/gallium nitride nanocomposites for supercapacitor applications. J Phys Chem Solids 109406
20. Sinha S, Devi NA, Nongthombam S, Bhujel R, Rai S, Sarkar G, Swain BP (2020) Investigation of optical, electrical, and electrochemical properties of polyaniline/rGO/Ag_2O nanocomposite. Diam Relat Mater 107:107885
21. Devi NA, Swain BP (2021) Investigation of metal-oxide/reduced graphene-oxide nanocomposites for gas sensor applications. In: Swain BP (eds) Nanostructured materials and their applications. materials horizons: from nature to nanomaterials. Springer, Singapore
22. Ali A, Aamir M, Thebo KH, Akhtar J (2020) Laminar graphene oxide membranes towards selective ionic and molecular separations: challenges and progress. Chem Rec 20:344–354
23. Yang Q, Su Y, Chi C, Cherian CT, Huang K, Kravets VG, Wang FC, Zhang JC, Pratt A, Grigorenko AN, Guinea F, Geim AK, Nair RR (2017) Ultrathin graphene-based membrane with precise molecular sieving and ultrafast solvent permeation. Nat Mater 16:1198
24. Sun M, Li M, Zhang X, Wu C, Wu Y (2020) Graphene oxide modified porous P84 co-polyimide membranes for boron recovery by bipolar membrane electrodialysis process. Sep Purif Technol 232:115963
25. Kim HW, Yoon HW, Yoon S-M, Yoo BM, Ahn BK, Cho YH, Shin HJ, Yang H, Paik U, Kwon S, Choi J-Y, Park HB (2013) Selective gas transport through few-layered graphene and graphene oxide membranes. Science 342:91–95
26. Ali A, Pothu R, Siyal SH, Phulpoto S, Sajjad M, Thebo KH (2019) Graphene-based membranes for CO_2 separation. Mater Sci Energy Technol 2:83–88
27. Musico YLF, Santos CM, Dalida MLP, Rodrigues DF (2014) Surface modification of membrane filters using graphene and graphene oxide-based nanomaterials for bacterial inactivation and removal. ACS Sustainable Chem Eng 2:1559–1565
28. Nahyoon NA, Liu L, Rabe K, Thebo KH, Yuan L, Sun J, Yang F (2018) Significant photocatalytic degradation and electricity generation in the photocatalytic fuel cell (PFC) using novel anodic nanocomposite of Fe, graphene oxide, and titanium phosphate. Electrochim Acta 271:41–48
29. Janjhi FA, Chandio I, Memon AA, Ahmed Z, Thebo KH, Pirzado AAA, Hakro AA, Iqbal M (2021) Functionalized graphene oxide based membranes for ultrafast molecular separation. Sep Purif Technol 274:117969
30. Lee YM, Jung B, Kim YH, Park AR, Han S, Choe W-S, Yoo PJ (2014) Nanomesh-structured ultrathin membranes harnessing the unidirectional alignment of viruses on a graphene-oxide film. Adv Mater 26:3899–3904
31. Palmieri V, Papi M (2020) Can graphene take part in the fight against COVID-19? Nano Today 33:100883
32. Raghav PK, Mohanty S (2020) Are graphene and graphene-derived products capable of preventing COVID- 19 infection? Med Hypotheses 144:110031
33. Novoselov KS, Geim AK, Morozov SV, Jiang D, Zhang Y, Dubonos SV, Grigorieva IV, Firsov AA (2004) Electric field effect in atomically thin carbon films. Science 306:666–669
34. Banerjee AN (2018) Graphene and its derivatives as biomedical materials: future prospects and challenges. Interface Focus 8:20170056
35. Pei S, Cheng HM (2012) The reduction of graphene oxide. Carbon 50:3210–3228
36. Eigler S, Hirsch A (2014) Chemistry with graphene and graphene oxide – challenges for synthetic chemists. Angew Chem Int Ed 53:7720–77381

37. Sheng Y, Tang X, Peng E, Xue J (2013) Graphene oxide based fluorescent nanocomposites for cellular imaging. J Mater Chem B 1:512–521
38. Srivastava AK, Dwivedi N, Dhand C, Khan R, Sathish N, Gupta MK, Kumar R, Kumar S (2020) Potential of graphene-based materials to combat COVID-19: properties, perspectives and prospects. Mater Today Chem 18:100385
39. Chen Y-N, Hsueh Y-H, Hsieh C-T, Tzou D-Y, Chang P-L (2016) Antiviral activity of graphene-silver nanocomposites against non-enveloped and enveloped viruses. Int J Environ Res Public Health 13:430
40. Kumar P, Huo P, Zhang R, Liu B (2019) Antibacterial properties of graphene-based nanomaterials. Nanomaterials 9:737
41. Hu W, Peng C, Luo W, Lv M, Li X, Li D, Huang Q, Fan C (2010) Graphene-based antibacterial paper. ACS Nano 4:4317–4323
42. Han W, Wu Z, Li Y, Wang Y (2019) Graphene family nanomaterials (GFNs)—promising materials for antimicrobial coating and film: a review. Chem Eng J 358:1022–1037
43. Zhao J, Deng B, Lv M, Li J, Zhang Y, Jiang H, Peng C, Li J, Shi J, Huang Q, Fan C (2013) Graphene oxide-based antibacterial cotton fabrics. Adv Healthc Mater 2:1259–1266
44. Ji H, Sun H, Qu X (2016) Antibacterial applications of graphene-based nanomaterials: recent achievements and challenges. Adv Drug Delivery Rev 105:176–189
45. Song Z, Wang X, Zhu G, Nian Q, Zhou H, Yang D, Qin C, Tang R (2015) Virus capture and destruction by label-free graphene oxide for detection and disinfection applications. Small 11:1171–1176
46. Rui L, Liu J, Li J, Weng Y, Dou Y, Yuan B, Yang K, Ma Y (2015) Reduced graphene oxide directed self-assembly of phospholipid monolayers in liquid and gel phases. Biochim Biophys Acta Biomembr 1848:1203–1211
47. Frost R, Jönsson GE, Chakarov D, Svedhem S, Kasemo B (2012) Graphene oxide and lipid membranes: interactions and nanocomposite structures. Nano Lett 12:3356–3362
48. Chen YN, Hsueh YH, Te HC, Tzou DY, Chang PL (2016) Antiviral activity of graphene-silver nanocomposites against non-enveloped and enveloped viruses. Int J Environ Res Public Health 13:430
49. Naskalska A, Dabrowska A, Szczepanski A, Milewska A, Jasik KP, Pyrc K (2019) Membrane protein of Human coronavirus NL63 Is responsible for interaction with the adhesion receptor. J Virol 93:e00355-e419
50. Jones ST, Cagno V, Janeček M, Ortiz D, Gasilova N, Piret J, Gasbarri M, Constant DA, Han Y, Vuković L, Král P, Kaiser L, Huang S, Constant S, Kirkegaard K, Boivin G, Stellacci F, Tapparel C (2020) Modified cyclodextrins as broad-spectrum antivirals. Sci Adv 6:eaax9318
51. Ziem B, Azab W, Gholami MF, Rabe JP, Osterrieder N, Haag R (2017) Size-dependent inhibition of herpesvirus cellular entry by polyvalent nanoarchitectures. Nanoscale 9:3774–3783
52. Ziem B, Thien H, Achazi K, Yue C, Stern D, Silberreis K, Gholami MF, Beckert F, Gröger D, Mülhaupt R, Rabe JP, Nitsche A, Haag R (2016) Highly efficient multivalent 2D nanosystems for inhibition of orthopoxvirus particles. Adv Healthc Mater 5:2922–2930
53. Feng S, Shen C, Xia N, Song W, Fan M, Cowling BJ (2020) Rational use of face masks in the COVID-19 pandemic. Lancet Respir Med 8:434–436
54. Sunjaya AP, Jenkins C (2020) Rationale for universal face masks in public against COVID-19. Respirology 25:678–679
55. Sommerstein R, Fux CA, Vuichard-Gysin D, Abbas M, Marschall J, Balmelli C, Troillet N, Harbarth S, Schlegel M, Widmer A (2020) Swissnoso: risk of SARS-CoV-2 transmission by aerosols, the rational use of masks, and protection of healthcare workers from COVID-19. Antimicrob Resist Infect 9:1–8
56. Clase CM, Fu EL, Joseph M, Beale RCL, Dolovich MB, Jardine M, Mann JFE, Pecoits-Filho R, Winkelmayer WC, Carrero JJ (2020) Cloth masks may prevent transmission of COVID-19: an evidence-based, risk-based approach. Ann Intern Med 173:489–491
57. Arons MM, Hatfield KM, Reddy SC, Kimball A, James A, Jacobs JR, Taylor J, Spicer K, Bardossy AC, Oakley LP, Tanwar S, Dyal JW, Harney J, Chisty Z, Bell JM, Methner M, Paul

P, Carlson CM, McLaughlin HP, Thornburg N, Tong S, Tamin A, Tao Y, Uehara A, Harcourt J, Clark S, Brostrom-Smith C, Page LC, Kay M, Lewis J, Montgomery P, Stone ND, Clark TA, Honein MA, Duchin JS, Jernigan JA (2020) Public health–Seattle and King County and CDC COVID-19 investigation team. Presymptomatic SARS-CoV-2 infections and transmission in a skilled nursing facility. N Engl J Med 382:2081–2090
58. Liu Z, Chu R, Gong L, Su B, Wu J (2020) The assessment of transmission efficiency and latent infection period in asymptomatic carriers of SARS-CoV-2 infection. Int J Infect Dis 99:325–327
59. O'Dowd K, Nair KM, Forouzandeh P, Mathew S, Grant J, Moran R, Bartlett J, Bird J, Pillai SC (2020) Face masks and respirators in the fight against the COVID-19 pandemic: a review of current materials. Adv Fut Perspect. Mater (Basel) 13:3363–3390
60. Palmieri V, Bugli F, Lauriola MC, Cacaci M, Torelli R, Ciasca G, Conti C, Sanguinetti M, Papi M, Spirito MD (2017) Bacteria meet graphene: modulation of graphene oxide nanosheet interaction with human pathogens for effective antimicrobial therapy. ACS Biomater Sci.Eng 3:619–627
61. Ahmed MK, Afifi M, Uskoković V (2020) Protecting healthcare workers during COVID-19 pandemic with nanotechnology: a protocol for a new device from Egypt. J Infect Public Health 13:1243–1246
62. Reina G, Iglesias D, Samorì P, Bianco A (2021) Graphene: a disruptive opportunity for COVID-19 and future pandemics? Adv Mater 33:2007847
63. Graphene protection against coronavirus: graphene-producing companies embarked on a warpath with COVID-19 (2020). https://rusgraphene.com/graphene-protection-against-coronavirus
64. Bonbouton's Graphene Masks (2021). http://go.affoa.org/bonbouton-graphene-mask/
65. Graphene ink may be used to fight coronavirus (2020). https://www.mining.com/graphene-ink-may-be-used-to-fight-novel-coronavirus/
66. https://www.pcimag.com/articles/107939-graphene-based-coating-proves-effective-in-destroying-covid-19-membranes, (2021)
67. Li Z, Yi Y, Luo X, Xiong N, Liu Y, Li S, Sun R, Wang Y, Hu B, Chen W, Zhang Y, Wang J, Huang B, Lin Y, Yang J, Cai W, Wang X, Cheng J, Chen Z, Sun K, Pan W, Zhan Z, Chen L, Ye F (2020) Development and clinical application of a rapid IgM-IgG combined antibody test for SARS-CoV-2 infection diagnosis. J Med Virol 92:1518–1524
68. Cheng MP, Papenburg J, Desjardins M, Kanjilal S, Quach C, Libman M, Dittrich S, Yansouni CP (2020) Diagnostic testing for severe acute respiratory syndrome-related coronavirus 2: a narrative review. Ann Intern Med 172:726–734
69. Huang Z, Tian D, Liu Y, Lin Z, Lyon CJ, Lai W, Fusco D, Drouin A, Yin X, Hu T, Ning B (2020) Ultra-sensitive and high-throughput CRISPR-powered COVID-19 diagnosis. Biosens Bioelectron 164:112316
70. van Dongen JE, Berendsen J, Steenbergen R, Wolthuis R, Eijkel J, Segerink LI (2020) Point-of-care CRISPR/Cas nucleic acid detection: recent advances, challenges and opportunities. Biosens Bioelectron 166:112445
71. Yang Y, Yang M, Shen C, Wang F, Yuan J, Li J, Zhang M, Wang Z, Xing L, Wei J, Peng L, Wong G, Zheng H, Wu W, Liao M, Feng K, Li J, Yang Q, Zhao J, Zhang Z, Liu L, Liu Y (2020) Evaluating the accuracy of different respiratory specimens in the laboratory diagnosis and monitoring the viral shedding of 2019-nCoV infections. medRxiv
72. Cordaro A, Neri G, Sciortino MT, Scala A, Piperno A (2020) Graphene-based strategies in liquid biopsy and in viral diseases diagnosis. Nanomaterials (Basel) 10:1014
73. Huang J, Xie Z, Xie Z, Luo S, Xie L, Huang L, Fan Q, Zhang Y, Wang S, Zeng T (2016) Silver nanoparticles coated graphene electrochemical sensor for the ultrasensitive analysis of avian influenza virus H7. Anal Chim Acta 913:121–127
74. Stanford MG, Li JT, Chen Y, McHugh EA, Liopo A, Xiao H, Tour JM (2019) Self-sterilizing laser-induced graphene bacterial air filter. ACS Nano 13:11912–11920
75. Graphene-Info: the graphene experts. https://www.graphene-info.com/

76. Zhong H, Zhu Z, Lin J, Cheung CF, Lu VL, Yan F, Chan C-Y, Li G (2020) Reusable and recyclable graphene masks with outstanding superhydrophobic and photothermal performances. ACS Nano 14:6213–6221
77. Stadnytskyi V, Bax CE, Bax A, Anfinrud P (2020) The airborne lifetime of small speech droplets and their potential importance in SARS-CoV-2 transmission. PNAS 117:11875–11877
78. van Doremalen N, Bushmaker T, Morris DH, Holbrook MG, Gamble A, Williamson BN, Tamin A, Harcourt JL, Thornburg NJ, Gerber SI, Lloyd-Smith JO, de Wit E, Munster VJ (2020) Aerosol and surface stability of SARS-CoV-2 as compared with SARS-CoV-1. New Eng J Med 382:1564–1567
79. Chin A, Chu J, Perera M, Hui K, Yen HL, Chan M, Peiris M, Poon L (2020) Stability of SARS-CoV-2 in different environmental conditions. Lancet Microbe 1:e10
80. Are graphene-coated face masks a COVID-19 miracle– or another health risk? (2021). https://theconversation.com/are-graphene-coated-face-masks-a-covid-19-miracle-or-another-health-risk-159422
81. Crane MJ, Devine S, Jamieson AM (2021) Graphene oxide/silver nanoparticle ink formulations rapidly inhibit influenza A virus and OC43 coronavirus infection in vitro. bioRxiv, pp 432893
82. Chaudhary V, Royal A, Chavali M, Yadav SK (2021) Advancements in research and development to combat COVID-19 using nanotechnology. Nanotechnol Environ Eng 6:8
83. Tebyetekerwa M, Xu Z, Yang S, Ramakrishna S (2020) Electrospun nanofibers-based face masks. Adv Fiber Mater 2:161–166
84. Lustig SR, Biswakarma JJH, Rana D, Tilford SH, Hu W, Su M, Rosenblatt MS (2020) Effectiveness of common fabrics to block aqueous aerosols of virus-like nanoparticles. ACS Nano 14:7651–7658
85. Sportelli MC, Izzi M, Kukushkina EA, Hossain SI, Picca RA, Ditaranto N, Cioffi N (2020) Can nanotechnology and materials science help the fight against SARS-CoV-2? Nanomaterials (Basel) 10:802
86. El-Atab N, Qaiser N, Badghaish H, Shaikh SF, Hussain MM (2020) A flexible nanoporous template for the design and development of reusable anti-COVID-19 hydrophobic face masks. ACS Nano 14:7659–7665
87. Liao L, Xiao W, Zhao M, Yu X, Wang H, Wang Q, Chu S, Cui Y (2020) Can N95 respirators be reused after disinfection? How many times? ACS Nano 14:6348–6356
88. Kumar S, Nyodu R, Maurya VK, Saxena SK (2020) Morphology, genome organization, replication, and pathogenesis of severe acute respiratory syndrome coronavirus 2 (SARS-CoV-2). In: Saxena S (eds) Coronavirus disease 2019 (COVID-19). Medical virology: from pathogenesis to disease control. Springer, Singapore
89. Huang L, Xu S, Wang Z, Xue K, Su J, Song Y, Chen S, Zhu C, Tang BZ, Ye R (2020) Self-reporting and photothermally enhanced rapid bacterial killing on a laser-induced graphene mask. ACS Nano 14:12045–12053
90. A COVID-19 Virucidal Graphene-based Composite Ink for More Effective PPE (2020). https://statnano.com/news/67657/A-COVID-19-Virucidal-Graphene-based-Composite-Ink-for-More-Effective-PPE
91. Technology against COVID-19: Nano Insights into Prevention, Diagnosis, and Treatment. https://statnano.com/technology-against-covid-19-nano-insights
92. Graphene-enhanced Face Mask (2020). www.planartech.com/store/p1/graphene-enhanced-face-mask-pack-of-3.html
93. Guardian G-Volt masks would use graphene and electrical charge to repel viruses and bacteria (2020). https://www.dezeen.com/2020/03/06/guardian-g-volt-face-mask-graphene-coronavirus-bacteria/
94. Reusable Graphene-coated Surgical Mask Sterilizes Itself with Sunlight (2020). https://statnano.com/news/67681/Reusable-Graphene-coated-Surgical-MaskSterilizes-Itself-with-Sunlight

95. Maio FD, Palmieri V, Babini G, Augello A, Palucci I, Perini G, Salustri A, Spirito MD, Sanguinetti M, Delogu G, Rizzi LG, Cesareo G, Soon-Shiong P, Sali M, Papi M (2020) Graphene nanoplatelet and graphene oxide functionalization of face mask materials inhibits infectivity of trapped SARS-CoV-2. medRxiv
96. Bhattacharjee S, Joshi R, Chughtai AA, Macintyre CR (2019) Graphene modified multifunctional personal protective clothing. Adv Mater Interfaces 6:1900622
97. Ye S, Shao K, Li Z, Guo N, Zuo Y, Li Q, Lu Z, Chen L, He Q, Han H (2015) Antiviral activity of graphene oxide: how sharp edged structure and charge matter. ACS Appl Mater Interfaces 7:21571–21579
98. Riddell S, Goldie S, Hill A, Eagles D, Drew TW, Riddell S, Goldie S, Hill A, Eagles D, Drew TW (2020) The effect of temperature on persistence of SARS-CoV-2 on common surfaces. Virol. J. 17:145
99. Weiss C, Carriere M, Fusco L, Capua I, Regla-Nava JA, Pasquali M, Scott JA, Vitale F, Unal MA, Mattevi C, Bedognetti D, Merkoçi A, Tasciotti E, Yilmazer A, Gogotsi Y, Stellacci F, Delogu LG (2020) Toward nanotechnology-enabled approaches against the COVID-19 pandemic. ACS Nano 14:6383–6406
100. Torres AE, Lyons AB, Narla S, Kohli I, Parks-Miller A, Ozog D, Hamzavi IH, Lim HW (2020) Ultraviolet-C and other methods of decontamination of filtering facepiece N-95 respirators during the COVID-19 pandemic. Photochem Photobiol Sci 19:746–751
101. Pakulski D, Czepa W, Witomska S, Aliprandi A, Patroniak V, Ciesielski A, Samorì P (2018) Graphene oxide-branched polyethylenimine foams for efficient removal of toxic cations from water. J Mater Chem A 6:9384
102. Witomska S, Leydecker T, Ciesielski A, Samorì P (2019) Production and patterning of liquid phase-exfoliated 2D sheets for applications in optoelectronics. Adv Funct Mater 29:1901126
103. Aliprandi A, Moreira T, Anichini C, Stoeckel MA, Eredia M, Sassi U, Bruna M, Pinheiro C, Laia CAT, Bonacchi S, Samorì P (2017) Hybrid copper-nanowire–reduced-graphene-oxide coatings: a "Green Solution" toward highly transparent, highly conductive, and flexible electrodes for (Opto)electronics. Adv Mater 29:1703225
104. Karim N, Afroj S, Tan S, He P, Fernando A, Carr C, Novoselov KS (2017) Scalable production of graphene-based wearable E-Textiles. ACS Nano 11:12266
105. De Faria AF, Shaulsky E, Chavez LHA, Elimelech M (2015) Antimicrobial electrospun biopolymer nanofiber mats functionalized with graphene oxide-silver nanocomposites. ACS Appl Mater Interfaces 7:12751
106. Ahmed W, Angel N, Edson J, Bibby K, Bivins A, O'Brien JW, Choi PM, Kitajima M, Simpson SL, Li J, Tscharke B, Verhagen R, Smith WJM, Zaugg J, Dierens L, Hugenholtz P, Thomas KV, Mueller JF (2020) First confirmed detection of SARS-CoV-2 in untreated wastewater in Australia: a proof of concept for the wastewater surveillance of COVID-19 in the community. Sci Total Environ 728:138764
107. Rosa GL, Iaconelli M, Mancini P, Ferraro GB, Veneri C, Bonadonna L, Lucentini L, Suffredini E (2020) First detection of SARS-CoV-2 in untreated wastewaters in Italy. Sci Total Environ 736:139652
108. Wurtzer S, Marechal V, Mouchel J -M, Moulin L (2020) Time course quantitative detection of SARS-CoV-2 in Parisian wastewaters correlates with COVID-19 confirmed cases. medRxiv
109. Medema G, Heijnen L, Elsinga G, Italiaander R, Brouwer A (2020) Presence of SARS-Coronavirus-2 RNA in sewage and correlation with reported COVID-19 prevalence in the early stage of the epidemic in the Netherlands. Environ Sci Technol Lett 7:511–516
110. Wu F, Zhang J, Xiao A, Gu X, Lee WL, Armas F, Kauffman K, Hanage W, Matus M, Ghaeli N, Endo N, Duvallet C, Poyet M, Moniz K, Washburne AD, Erickson TB, Chai PR, Thompson J, Alm EJ (2020) SARS-CoV-2 titers in wastewater are higher than expected from clinically confirmed cases. mSystems 5:4
111. Usman M, Farooq M, Hanna K (2020) Existence of SARS-CoV-2 in wastewater: implications for its environmental transmission in developing communities. Environ Sci Technol 54:7758–7759

112. Arora S, Nag A, Sethi J, Rajvanshi J, Saxena S, Shrivastava SK, Gupta AB (2020) Sewage surveillance for the presence of SARS-CoV-2 genome as a useful wastewater based epidemiology (WBE) tracking tool in India. Water Sci Technol 82:2823–2836
113. Mahmoud KA, Mansoor B, Mansour A, Khraisheh M (2015) Functional graphene nanosheets: the next generation membranes for water desalination. Desalination 356:208–225
114. Deng Y, Li Z, Tang R, Ouyang K, Liao C, Fang Y, Ding C, Yang L, Su L, Gong D (2020) What will happen when microorganisms "Meet" photocatalysts and photocatalysis? Environ Sci Nano 7:702–723
115. Ahmed Z, Rehman F, Ali U, Ali A, Iqbal M, Thebo KH (2021) Recent advances in MXene-based separation membranes. ChemBioEng Rev 8:110–120
116. Pang YX, Hong Q, Lu C, Lin B, Wen W, Wu X, Yan Y, Zhao H, Sharmin N, Lester E, Wu T, Pang CH (2021) The impact of COVID-19 on the global graphene industry. J Biomed Eng & Med Dev 6:153
117. Sethuraman N, Jeremiah SS, Ryo A (2020) Interpreting diagnostic tests for SARSCoV-2. JAMA 323:2249–2251
118. Morales-Narváez E, Dincer C (2020) The impact of biosensing in a pandemic outbreak: COVID-19. Biosens Bioelectron 163:112274
119. Pasomsub E, Watcharananan SP, Boonyawat K, Janchompoo P, Wongtabtim G, Suksuwan W, Sungkanuparph S, Phuphuakrat A (2021) Saliva sample as a non-invasive specimen for the diagnosis of coronavirus disease 2019: a cross-sectional study. Clin Microbiol Infect 27:285.e1-285.e4
120. Zuo B, Li S, Guo Z, Zhang J, Chen C (2004) Piezoelectric immunosensor for SARS-associated coronavirus in sputum. Anal Chem 76:3536–3540
121. Pohanka M (2018) Piezoelectric biosensor for the determination of tumor necrosis factor alpha. Talanta 178:970–973
122. Pumera M (2011) Graphene in biosensing. Mater Today 14:308–315
123. Peña-Bahamonde J, Nguyen HN, Fanourakis SK, Rodrigues DF (2018) Recent advances in graphene-based biosensor technology with applications in life sciences. J Nanobiotechnology 16:75
124. Afsahi S, Lerner MB, Goldstein JM, Lee J, Tang X, Bagarozzi DA Jr, Pan D, Locascio L, Walker A, Barron F (2018) Novel graphene-based biosensor for early detection of Zika virus infection. Biosens Bioelectron 100:85–88
125. Joshi SR, Sharma A, Kim G-H, Jang J (2020) Low cost synthesis of reduced graphene oxide using biopolymer for influenza virus sensor. Mater Sci Eng C 108:110465
126. Pant M, Kharkwal H, Singh K, Joshi H (2017) Detection of rota virus with the help of nanomaterial based field effect transistor (BIO-FET). Biosens J 6:2
127. Seo G, Lee G, Kim MJ, Baek S-H, Choi M, Ku KB, Lee C-S, Jun S, Park D, Kim HG, Kim S-J, Lee J-O, Kim BT, Park EC, Kim SI (2020) Rapid detection of COVID-19 causative virus (SARS-CoV-2) in human nasopharyngeal swab specimens using field-effect transistor-based biosensor. ACS Nano 14:5135–5142
128. Bellassai N, Spoto G (2016) Biosensors for liquid biopsy: circulating nucleic acids to diagnose and treat cancer. Anal Bioanal Chem 408:7255–7264
129. Qiu G, Gai Z, Tao Y, Schmitt J, Kullak-Ublick GA, Wang J (2020) Dual-functional plasmonic photothermal biosensors for highly accurate severe acute respiratory syndrome coronavirus 2 detection. ACS Nano 14:5268–5277
130. Layqah LA, Eissa S (2019) An electrochemical immunosensor for the corona virus associated with the Middle East respiratory syndrome using an array of gold nanoparticle-modified carbon electrodes. Mikrochim Acta 186:224
131. Lian Y, He F, Wang H, Tong F (2015) A new aptamer/graphene interdigitated gold electrode piezoelectric sensor for rapid and specific detection of Staphylococcus aureus. Biosens Bioelectron 65:314–319
132. Iannazzo D, Pistone A, Ferro S, Luca LD, Monforte AM, Romeo R, Buemi MR, Pannecouque C (2018) Graphene quantum dots based systems as HIV inhibitors. Bioconjug Chem 29:3084–3093

133. Du X, Xiao R, Fu H, Yuan Z, Zhang W, Yin L, He C, Li C, Zhou J, Liu G (2019) Hypericin-loaded graphene oxide protects ducks against a novel duck reovirus. Mater Sci Eng C 105:110052
134. Palmieri V, Papi M, Conti C, Ciasca G, Maulucci G, Spirito MD (2016) The future development of bacteria fighting medical devices: the role of graphene oxide. Expert Rev Med Devices 13:1013–1019
135. Jang H, Kim Y, Kwon H, Yeo W, Kim D, Min D (2010) A graphene-based platform for the assay of duplex-DNA unwinding by helicase. Angew Chemie Int Ed 49:5703–5707
136. Mycroft-West C, Su D, Elli S, Li Y, Guimond S, Miller G, Turnbull J, Yates E, Guerrini M, Fernig D, Lima M, Skidmore M (2020) The 2019 coronavirus (SARS-CoV-2) surface protein (Spike) S1 receptor binding domain undergoes conformational change upon heparin binding. bioRxiv
137. Deokar AR, Nagvenkar AP, Kalt I, Shani L, Yeshurun Y, Gedanken A, Sarid R (2017) Graphene-based "Hotplate" for the capturing and destruction of the herpes simplex virus type. Bioconjug Chem 28:1115–1122
138. Sametband M, Kalt I, Gedanken A, Sarid R (2014) Herpes simplex virus type-1 attachment inhibition by functionalized graphene oxide. ACS Appl Mater Interfaces 6:1228–1235
139. Du T, Lu J, Liu L, Dong N, Fang L, Xiao S, Han H (2018) Antiviral activity of graphene oxide-silver nanocomposites by preventing viral entry and activation of the antiviral innate immune response. ACS Appl Bio Mater 1:1286–1293

MIX
Papier aus verantwortungsvollen Quellen
Paper from responsible sources
FSC® C105338

If you have any concerns about our products,
you can contact us on
ProductSafety@springernature.com

In case Publisher is established outside the EU,
the EU authorized representative is:
Springer Nature Customer Service Center GmbH
Europaplatz 3, 69115 Heidelberg, Germany

Printed by Libri Plureos GmbH
in Hamburg, Germany